T0210972

Lecture Notes in Computer Science 13153

More information about this subseries at https://link.springer.com/bookseries/7407

Weili Wu · Hongwei Du (Eds.)

Algorithmic Aspects in Information and Management

15th International Conference, AAIM 2021
Virtual Event, December 20–22, 2021
Proceedings

 Springer

Editors
Weili Wu (iD)
The University of Texas at Dallas
Richardson, TX, USA

Hongwei Du (iD)
Harbin Institute of Technology
Shenzhen, China

ISSN 0302-9743   ISSN 1611-3349   (electronic)
Lecture Notes in Computer Science
ISBN 978-3-030-93175-9   ISBN 978-3-030-93176-6   (eBook)
https://doi.org/10.1007/978-3-030-93176-6

LNCS Sublibrary: SL1 – Theoretical Computer Science and General Issues

This Springer imprint is published by the registered company Springer Nature Switzerland AG
The registered company address is: Gewerbestrasse 11, 6330 Cham, Switzerland

Preface

The 15th International Conference on Algorithmic Aspects in Information and Management (AAIM 2021), was held online during December 20–22, 2021. The conference was held virtually due to the COVID-19 pandemic.

The AAIM conference series, which started in 2005 in Xi'an, China, aims to stimulate various fields for which algorithmics has become a crucial enabler, and to strengthen the ties of various research communities of algorithmics and applications. AAIM 2021 seeks to address emerging and important algorithmic problems by focusing on the fundamental background, theoretical technological development, and real-world applications associated with information and management analysis, modeling, and data mining. Special considerations are given to algorithmic research that was motivated by real-world applications.

We would like to thank the two keynote speakers, Ovidiu Daescu from the University of Texas at Dallas, Texas, USA and Amo Tong from the University of Delaware, Delaware, USA, for their contributions to the conference.

We would like to express our appreciation to all members of the Program Committee and the external referees whose efforts enabled us to achieve a high scientific standard for the proceedings. We would also like to thank all members of the Organizing Committee for their assistance and contribution which was essential to the success of the conference. In particular, we would like to thank Anna Kramer and her colleagues at Springer for meticulously supporting us in the timely production of this volume. Last but not least, our special thanks go to all the authors and participants for their contributions to the success of this event.

November 2021

Weili Wu
Hongwei Du

Organization

General Chairs

Ding-Zhu Du   University of Texas at Dallas, USA
Zhenhua Duan   Xidian University, China

Program Chairs

Weili Wu   University of Texas at Dallas, USA
Hongwei Du   Harbin Institute of Technology, China

Program Committee Members

Wolfgang Bein   University of Nevada, USA
Annalisa De Bonis   Università degli Studi di Salerno, Italy
Sergiy Butenko   Texas A&M University, USA
Gruia Calinescu   Illinois Institute of Technology, USA
Xujin Chen   University of Chinese Academy of Science, China
Zhi-Xiang Chen   University of Texas Rio Grande Valley, USA
Zhi-Zhong Chen   Tokyo Denki University, Japan
Maggie Cheng   Illinois Institute of Technology, USA
Yongxi Cheng   Xi'an Jiaotong University, China
Chuangyin Dang   City University of Hong Kong, China
Bhaskar Dasgupta   University of Illinois at Chicago, USA
Rudolf Fleischer   Heinrich Heine University Düsseldorf, Germany
Bin Fu   University of Texas-Rio Grande Valley, USA
Hong Gao   Harbin Institute of Technology, China
Smita Ghosh   Santa Clara University, USA
Shuyang Gu   Texas A&M University, USA
Qianping Gu   Simon Fraser University, Canada
Jianxiong Guo   Beijing Normal University Zhuhai, China
Sun-Yuan Hsieh   National Cheng Kung University, China
Kazuo Iwama   Kyoto University, Japan
Liying Kang   Shanghai University, China
Michael Khachay   Krasovsky Institute of Mathematics and Mechanics, Russia
Joong-Lyul Lee   University of North Carolina at Pembroke, USA
Chia-Wei Lee   National Taitung University, China
Xiang Li   Santa Clara University, USA
Xianyue Li   Lanzhou University, China
Yi Li   University of Texas at Tyler, USA
Chuanwen Luo   Beijing Forestry University, China

Viet Hung Nguyen	University of Clermont Auvergne, France
Yan Qiang	Taiyuan Institute of Technology, China
Yan Shi	University of Wisconsin-Platteville, USA
Pavel Skums	Georgia State University, USA
Jack Snoeyink	University of North Carolina at Chapel Hill, USA
Zhiyi Tan	Zhejiang University, China
Weitian Tong	Georgia Southern University, USA
Guangmo Tong	University of Delaware, USA
Lidong Wu	University of Texas at Tyler, USA
Wen Xu	Texas Women's University, USA
Wenguo Yang	University of Chinese Academy of Science, China
Boting Yang	University of Regina, Canada
Chihao Zhang	Chinese University of Hong Kong, China
Nan Zhang	Xidian University, China
Zhao Zhang	Zhejiang Normal University, China
Huaming Zhang	University of Alabama in Huntsville, USA
Fay Zhong	California State University, East Bay, USA
Yuqing Zhu	California State University, Los Angeles, USA

Contents

Approximation Algorithms

Constant-Approximation for Prize-Collecting Min-Sensor Sweep Coverage
with Base Stations. 3
   Wei Liang and Zhao Zhang

Approximation Algorithm for the Capacitated Correlation Clustering
Problem with Penalties 15
   Sai Ji, Gaidi Li, Dongmei Zhang, and Xianzhao Zhang

Approximation Algorithms for the Maximum Bounded Connected
Bipartition Problem. 27
   Yajie Li, Weidong Li, Xiaofei Liu, and Jinhua Yang

An Approximation Algorithm for Solving the Heterogeneous Chinese
Postman Problem 38
   Jianping Li, Lijian Cai, Junran Lichen, Pengxiang Pan,
   Wencheng Wang, and Suding Liu

On Stochastic k-Facility Location 47
   Yicheng Xu, Chunlin Hao, Chenchen Wu, and Yong Zhang

The Complexity of Finding a Broadcast Center. 57
   Hovhannes A. Harutyunyan and Zhiyuan Li

An Online Algorithm for Data Caching Problem in Edge Computing 71
   Xinxin Han, Guichen Gao, Yang Wang, and Yong Zhang

Scheduling

Scheduling on Multiple Two-Stage Flowshops with a Deadline. 83
   Jianer Chen, Minjie Huang, and Yin Guo

Single Machine Scheduling with Rejection to Minimize
the Weighted Makespan. 96
   Lingfa Lu, Liqi Zhang, and Jinwen Ou

Maximizing Energy Efficiency for Charger Scheduling of WRSNs 111
   Yi Hong, Chuanwen Luo, Zhibo Chen, Xiyun Wang, and Xiao Li

A New Branch-and-Price Algorithm for Daily Aircraft Routing
and Scheduling Problem 123
   Yu Si, Suixiang Gao, and Wenguo Yang

Optimizing Mobile Charger Scheduling for Task-Based Sensor Networks . . .   134
   Xiangguang Meng, Jianxiong Guo, Xingjian Ding, and Xiujuan Zhang

Semi-online Early Work Maximization Problem on Two Hierarchical
Machines with Partial Information of Processing Time.   146
   Man Xiao, Xiaoqiao Liu, and Weidong Li

Nonlinear Combinatorial Optimization

Streaming Algorithms for Maximizing DR-Submodular Functions
with d-Knapsack Constraints   159
   Bin Liu, Zihan Chen, and Hongmin W. Du

Stochastic Submodular Probing with State-Dependent Costs.   170
   Shaojie Tang

Bi-criteria Adaptive Algorithms for Minimizing Supermodular Functions
with Cardinality Constraint.   179
   Xiaojuan Zhang, Qian Liu, Min Li, and Yang Zhou

Improved Algorithms for Non-submodular Function
Maximization Problem.   190
   Zhicheng Liu, Hong Chang, Donglei Du, and Xiaoyan Zhang

Fixed Observation Time-Step: Adaptive Influence Maximization.   200
   Yapu Zhang, Shengminjie Chen, Wenqing Xu, and Zhenning Zhang

Measured Continuous Greedy with Differential Privacy   212
   Xin Sun, Gaidi Li, Yapu Zhang, and Zhenning Zhang

Network Problems

Robust t-Path Topology Control Algorithm in Wireless Ad Hoc Networks. . .   229
   Xiujuan Zhang, Yongcai Wang, Deying Li, Wenping Chen,
   and Xingjian Ding

Multi-attribute Based Influence Maximization in Social Networks   240
   Qiufen Ni, Jianxiong Guo, and Hongmin W. Du

A Parallel Algorithm for Constructing Multiple Independent Spanning
Trees in Bubble-Sort Networks.   252
   Shih-Shun Kao, Ralf Klasing, Ling-Ju Hung, and Sun-Yuan Hsieh

A Fast FPTAS for Two Dimensional Barrier Coverage Using Sink-Based
Mobile Sensors with MinSum Movement.   265
   Wenjie Zou, Longkun Guo, Chunlin Hao, and Lei Liu

Time Sensitive Sweep Coverage with Multiple UAVs 277
   Huizhen Wang

Recursive Merged Community Detection Algorithm Based
on Node Cluster 289
   Ailian Wang, Liang Meng, and Lu Cui

Purchase Preferences - Based Air Passenger Choice Behavior Analysis
from Sales Transaction Data 303
   Xinghua Li, Suixiang Gao, Wenguo Yang, Yu Si, and Zhen Liu

Blockchain, Logic, Complexity and Reliability

A Multi-window Bitcoin Price Prediction Framework on Blockchain
Transaction Graph 317
   Xiao Li and Linda Du

Sensitivity-Based Optimization for Blockchain Selfish Mining 329
   Jing-Yu Ma and Quan-Lin Li

Design and Implementation of List and Dictionary in XD-M Language 344
   Yajie Wang, Nan Zhang, and Zhenhua Duan

Reliable Edge Intelligence Using JPEG Progressive 356
   Haobin Luo, Xiangang Du, Luobing Dong, Guowei Su, and Ruijie Chen

A Game-Theoretic Analysis of Deep Neural Networks 369
   Chunying Ren, Zijun Wu, Dachuan Xu, and Wenqing Xu

Energy Complexity of Satisfying Assignments in Monotone Circuits:
On the Complexity of Computing the Best Case 380
   Janio Carlos Nascimento Silva, Uéverton S. Souza,
   and Luiz Satoru Ochi

Miscellaneous

The Independence Numbers of Weighted Graphs with Forbidden Cycles 395
   Ye Wang and Yan Li

Wegner's Conjecture on 2-Distance Coloring 400
   Junlei Zhu, Yuehua Bu, and Hongguo Zhu

An Efficient Oracle for Counting Shortest Paths in Planar Graphs 406
   Ye Gong and Qian-Ping Gu

Restrained and Total Restrained Domination in Cographs 418
   Xue-gang Chen and Moo Young Sohn

An Order Approach for the Core Maintenance Problem
on Edge-Weighted Graphs   426
  Bin Liu, Zhenming Liu, and Feiteng Zhang

Fixed-Parameter Tractability for Book Drawing with Bounded Number
of Crossings per Edge   438
  Yunlong Liu, Yixuan Li, and Jingui Huang

Author Index   451

Approximation Algorithms

Constant-Approximation
for Prize-Collecting Min-Sensor Sweep
Coverage with Base Stations

Wei Liang and Zhao Zhang[✉]

College of Mathematics and Computer Science, Zhejiang Normal University,
Jinhua 321004, Zhejiang, China
hxhzz@sina.com

Abstract. A sweep cover is a set of routes for mobile sensors such that
each target point v is visited by mobile sensors at least once in every time
period t_v. In this paper, we propose a new sweep coverage paradigm,
prize-collecting min-sensor sweep coverage problem (PCMSSC). Instead
of requiring each target point to be periodically visited, PCMSSC selec-
tively discards some target points and bears the loss arising therefrom.
To be precise, a target point v which is not covered incurs a penalty $\pi(v)$.
The objective is to minimize the sum of the cost of mobile sensors and
the total penalty on uncovered target points. For PCMSSC with constant number of stations, we present a polynomial time algorithm with
approximation ratio at most 5. For this purpose, we propose the *prize-
collecting forest with k components problem* and design a 2-approximation
algorithm for it, which might be interesting in its own sense.

Keywords: Sweep-coverage · Prize-collecting · Approximation
algorithm

1 Introduction

Coverage problems are among the most fundamental issues in wireless sensors
networks (WSN). Due to the fast development of WSN during the past two
decades, coverage problem is extensively studied under various models [1,2,7,
14,15,17,19]. Many concerns are put on coverage quality, energy consumption,
node mobility, data collection, coverage scheduling, etc.

The first paper studying sweep cover problem is [5], in which Cheng *et al.* considered the problem with the objective to minimize the number of mobile sensors
(call the *min-sensor sweep cover problem* (MSSC)) and proved a lower bound
of approximation ratio 2. Gorain and Mandal [10] proposed a 3-approximation
algorithm for MSSC when all target points have the same sweep-period and all
mobile sensors move at the same velocity.

Since a mobile sensor is often powered by a battery with limited energy, some
researchers considered the MSSC problem with base stations for replenishment
[4,8,11,14].

© Springer Nature Switzerland AG 2021
W. Wu and H. Du (Eds.): AAIM 2021, LNCS 13153, pp. 3–14, 2021.
https://doi.org/10.1007/978-3-030-93176-6_1

In real applications, satisfying the covering requirement of all target points might be too expensive. It might be more economic to pay for some uncovered points. In this paper, we propose a new sweep coverage problem, *prize-collecting min-sensor sweep coverage with base stations problem* (PCMSSC$_{BS}$) and design an approximation algorithm for PCMSSC$_{BS}$ with a theoretically guaranteed approximation ratio.

1.1 Related Work

The sweep coverage problem was first proposed by Cheng et al. [5] with the aim of minimizing the number of mobile sensors.

By reducing the traveling salesman problem (TSP) to MSSC, it was shown that MSSC cannot be approximated within factor 2. Early works focused on the case when mobile sensors work independently in the sense that every sensor is responsible for a distinct set of target points [6,12]. Later, Gorain and Mandal [10] found that it would be more effective if mobile sensors are grouped to cooperatively accomplish the sweep coverage task. They proposed a 3-approximation algorithm for MSSC on a graph whose edge lengths are metric. So far, this is still the best known approximation ratio for MSSC. There are also a lot of heuristic algorithms for MSSC, such as Du's MinExpand algorithm [6], Wang's MinMobileGrowth algorithm [18], and Huang's ant colony optimization-based algorithm [3], etc.

Liang *et al.* [13] considered a budgeted version of the sweep coverage problem, the goal of which is to find the routes for a given set of mobile sensors such that the total weight of target points whose sweep-cover requirements are satisfied is maximized. They studied the case when all target points are on a line. When sensors have the same velocity, an optimal solution can be computed in polynomial time. When the sensors have different velocities, the problem was proven to be NP-hard, and they proposed a $\frac{1}{2}$-approximation algorithm when sensors have a constant number of velocities, and a $(\frac{1}{2} - \frac{1}{2e})$-approximation algorithm for the general case.

There are also some other variants of sweep cover problems. For example, due to energy limitation of mobile sensors, Nie *et al.* [16] considered the *general energy restricted min-sensor sweep coverage problem* by assuming that energy consumption of mobile sensors may vary in different road sections. They designed a constant-factor approximation algorithm.

1.2 Our Contribution

Assume that all target points have the same sweep-period t, all mobile sensors are homogeneous and move at the same velocity a. Our contributions are summarized as follows:

- We propose a new sweep coverage problem: *prize-collecting min-sensor sweep coverage with base stations* (PCMSSC$_{BS}$). This problem is more practical from the economic point of view: paying penalties for those distant points might save cost. Having base stations is also common in real practice, and the number of base stations will not be too much.

- We design a 5-approximation algorithm for PCMSSC$_{BS}$. In fact, the result is stronger, it is a 5-LMP algorithm. And r-LMP will be formally defined in Sect. 2.
- As a step stone for the 5-LMP for PCMSSC$_{BS}$, we propose the *prize-collecting forest with k components problem* (PCF$_k$), which might be interesting in its own sense, and design a 2-LMP for PCF$_k$.

The rest of this paper is organized as follows: The problem is defined in Sect. 2, together with some preliminary results. A 2-LMP algorithm for PCF$_k$ is presented in Sect. 3, based on which a 5-LMP approximation algorithm for PCMSSC$_{BS}$ is presented in Sect. 4. Section 5 concludes the paper.

2 Problem Formulation and Preliminaries

In this paper, we focus on the sweep coverage problem on a graph with metric edge lengths. The related terminologies are defined as follows.

Definition 1 (sweep-cover). Suppose G is a graph on vertex set V and edge set E, each vertex $v \in V$ is associated with a sweep period t_v. For a set of routes scheduled for a set of mobile sensors moving with speed a, a vertex v is said to be *sweep-covered* if v is visited at least once in every time period t_v, where v gets visited if and only if a mobile sensor goes through the location of vertex v.

MSSC requires *all* vertices to be sweep covered [10]. But in many applications, visiting distant vertices might be too costly. This consideration leads to the prize-collecting min-sensor coverage problem (PCMSSC). Note that in real applications, it is common that mobile sensors are dispatched from base stations. In this paper, we consider the PCMSSC problem with constant number of base stations.

Definition 2 (prize-collecting min-sensor sweep coverage with base stations (PCMSSC$_{BS}$)). Given a graph $G = (V, E)$ with a metric edge weight function $w : E \mapsto \mathbb{R}^+$, a set of base stations B located at some vertices of V, each vertex v is associated with a sweep period t_v and a penalty $\pi(v)$, determine a set of mobile sensors $\mathcal{S}$ dispatched from B and design their trajectories to minimize the cost of mobile sensors plus the total penalty on vertices which are not sweep-covered, that is, $c \cdot |\mathcal{S}| + \sum_{v \notin \mathcal{C}(\mathcal{S})} \pi(v)$, where $\mathcal{C}(\mathcal{S})$ is the set of vertices sweep-covered by $\mathcal{S}$, and c is the cost of a mobile sensor.

Note that not all base stations are needed to dispatch mobile sensors, therefore, the problem also requires us to determine those base stations to be used. Throughout this paper, we assume that all vertices have the same sweep periods $t_v \equiv t \ (\forall v \in V)$.

For Problem 2, we shall present a *Lagrangian multiplier preserving algorithm with factor* 5, which is stronger than a 5-approximation.

Definition 3 (Lagrangian multiplier preserving algorithm with factor r (r-LMP)). An algorithm for Problem 2 is said to be an r-LMP, if for any instance I of Problem 2, the algorithm can find in polynomial-time a set of sensors $\mathcal{S}'$ with

$$c \cdot |\mathcal{S}'| + r \cdot \sum_{v \notin \mathcal{C}(\mathcal{S}')} \pi(v) \leq r \cdot opt(I),$$

where $opt(I)$ is the optimal value for Problem 2 on instance I.

To solve Problem 2, we need a solution to the following problem.

Definition 4 (prize-collecting forest with k components ($\mathbf{PCF}_k$)). Given a graph $G = (V, E)$, a positive integer k, a set $R = \{r_1, r_2, \ldots, r_k\} \subseteq V$ of k roots, a weight function w on E, and a penalty function π on V, the goal is to find a forest F with k components, each component contains exactly one root, such that $w(F) + \pi(V \backslash V(F))$ is minimized, where $V(F)$ is the set of vertices in F, $\pi(V \backslash V(F)) = \sum_{v \in V \backslash V(F)} \pi(v)$ is the penalty on those vertices not covered by F, and $w(F) = \sum_{e \in F} w(e)$.

3 A Primal-Dual Algorithm for $\mathbf{PCF}_k$

In this section, we present a 2-LMP algorithm for Problem 4 using primal-dual method. Our algorithm is inspired by [9] for PCF_1, with more cares devoted to how to deal with the challenge brought by the required number of components.

We first write out an integer program for PCF_k. Given a graph $G = (V, E)$, define an indicator variables x_e for whether $e \in E$ is selected or not. Use $\delta(S)$ to denote the set of edges having exactly one end vertex in S. For an edge set F, denote $x(F) = \sum_{e \in F} x_e$. And define a variable p_T to indicate whether T is the vertex set that is punished. Then PCF_k can be formulated as the following integer program:

$$\min \sum_{e \in E} w_e x_e + \sum_{T \subseteq V \backslash R} \left(\sum_{v \in T} \pi_v \right) p_T \tag{1}$$

$$\text{s.t. } x(\delta(S)) + \sum_{T \subseteq V \backslash R : S \subseteq T} p_T \geq 1, \quad \forall S \subseteq V \backslash R,$$

$$x_e \in \{0, 1\}, \quad \forall e \in E,$$

$$p_T \in \{0, 1\}, \quad \forall T \subseteq V \backslash R,$$

where the second term of the objective says that if T is the vertex set to be punished, then all vertices in T are punished; the first constraint says that any vertex set S that should be punished is contained in the biggest vertex set T that should be punished.

Relax 1 into a liner program (LP) by replacing the variable constraints by $x_e \geq 0$ and $p_T \geq 0$, and write out its dual LP:

$$\max \sum_{S \subseteq V \backslash R} y_S \tag{2}$$

$$\text{s.t.} \sum_{S \subseteq V \backslash R : e \in \delta(S)} y_S \leq w_e, \quad \forall e \in E$$

$$\sum_{S \subseteq T} y_S \leq \sum_{v \in T} \pi_v, \quad \forall T \subset V \backslash R$$

$$y_S \geq 0, \quad \forall S \subseteq V \backslash R$$

The details of the primal-dual algorithm for PCF_k are described in Algorithm 1. It maintains a forest F and use $\mathcal{C}$ to denote the collection of connected components of $G[F]$. Some components in $\mathcal{C}$ are *active*, meaning that their dual variables are allowed to be increased (the dual variable corresponding to $C \in \mathcal{C}$ is $y_{V(C)}$, or abbreviated as y_C), and those *non-active* components in $\mathcal{C}$ have their dual variables frozen. We use label $\lambda(C) = 1$ to indicate that component $C \in \mathcal{C}$ is active, and $\lambda(C) = 0$ to indicate a non-active component C. Initially, $F = \emptyset$, and every root forms a non-active component of $\mathcal{C}$, and every vertex in $V \backslash R$ forms an active component of $\mathcal{C}$. Stating from dual feasible solution $\{y_S \equiv 0\}$, the algorithm simultaneously increases dual variables, keeping dual feasibility all the time, and takes primal variables corresponding to the first tightened dual constraints to construct primal solutions. Note that there are two types of constraints in (2). Depending on which constraint becomes tight first (a constraint is *tight* if equality holds), there are two types of operations. If a first-type constraint becomes tight at edge e (see line 6 of the algorithm), then e is added into F. Since only active components are allowed to increase their dual variables,

such edge e has its two ends in different connected components of $G[F]$, (3)

so adding e will merge two components into one. If the merged component contains a root, then it is labeled as non-active, otherwise, it is active. Because only active components can have their dual variables increased, so e cannot have its two ends in two non-active components, and thus if the merged component is non-active, then it contains exactly one root. If a second-type constraint becomes tight at an active component $C \in \mathcal{C}$ (see line 7 of the algorithm), then C is *deactivated*. To simplify the calculation, $d(v)$ is used to record the accumulated amount of increase on vertex v, that is, $\sum_{S:v \in \delta(S)} y_S$, and $h(C)$ is used to record the accumulated amount of increase on component C, that is $\sum_{S:S \subseteq V(C)} y_S$. The above process (in the while loop) terminates when all connected components of $G[F]$ are non-active. By Property 3, the resulting F is a forest. Denote by F_R the sub-forest of F consisting of all those connected components of $G[F]$ having nonempty intersection with R. The final output is obtained from F_R by a reverse deletion of edges.

In fact, in each iteration, either an edge is added to decrease the number of connected components, or an active component is deactivated and thus the

---

Algorithm 1. $PCF_k(G, R)$

Input: A graph $G = (V, E)$ with edge length function w on E and penalty function π on V, and a set of k roots $R = \{r_1, \ldots, r_k\} \subseteq V$.

Output: A forest F_R with k components, each component contains one root.

1: $l \leftarrow 0$; $F \leftarrow \emptyset$; $\mathcal{C} \leftarrow \{\{v\} : v \in V\}$; $y_C \leftarrow 0$ and $h(C) \leftarrow 0$ for each $C \in \mathcal{C}$;

2: **for** each $C = \{v\} \in \mathcal{C}$, set $\lambda(C) \leftarrow 1$ if $v \in V \backslash R$ and $\lambda(C) \leftarrow 0$ if $v \in R$;

3: set $d(v) \leftarrow 0$ for ever $v \in V$;

4: **while** $\exists C \in \mathcal{C}$ with $\lambda(C) = 1$ **do**

5:     $l \leftarrow l + 1$;

6:     $\varepsilon_1 \leftarrow \min_{e=uv, u \in C \in \mathcal{C}, v \in C' \in \mathcal{C}, C \neq C'} \left\{ \frac{w_e - d(u) - d(v)}{\lambda(C) + \lambda(C')} \right\}$, $\hat{e} \leftarrow \arg \varepsilon_1$;

7:     $\varepsilon_2 \leftarrow \min_{C \in \mathcal{C} : \lambda(C) = 1} \left\{ \sum_{v \in C} \pi_v - h(C) \right\}$, $\hat{C} \leftarrow \arg \varepsilon_2$;

8:     $\varepsilon \leftarrow \min\{\varepsilon_1, \varepsilon_2\}$;

9:     **for** every $C \in \mathcal{C}$ with $\lambda(C) = 1$ **do**

10:        $h(C) \leftarrow h(C) + \varepsilon$; $y_C \leftarrow y_C + \varepsilon$; for every $v \in C$, $d(v) \leftarrow d(v) + \varepsilon$;

11:    **end for**

12:    **if** $\varepsilon = \varepsilon_2$ **then**

13:        $\lambda(\hat{C}) \leftarrow 0$

14:    **else**

15:        $F \leftarrow F \cup \{\hat{e}\}$

16:        $\mathcal{C} \leftarrow \mathcal{C} \cup \{C \cup C'\} \backslash \{C, C'\}$ where C, C' are the components containing the two ends of $\hat{e}$;

17:        $h(C \cup C') \leftarrow h(C) + h(C')$

18:        If $C \cup C'$ contains some root, then $\lambda(C \cup C') \leftarrow 0$, otherwise $\lambda(C \cup C') \leftarrow 1$;

19:    **end if**

20: **end while**

21: $F_R \leftarrow$ the set of connected components of $G[F]$ containing roots.

22: **for** edges e in F_R in reverse order of their addition into F **do**

23:    **if** e is an edge incident with a leaf non-active component C_e of the contract forest where components refer to the iteration when e is added **then**

24:        $F_R \leftarrow F_R - e - C$;

25:    **end if**

26: **end for**

27: **return** $F_R \leftarrow$ the set of connected components of $G[F]$ containing roots.

---

number of active components is reduced, it can be easily checked that the Algorithm 1 runs in time $O(nm)$.

The following lemma shows that the output F_R of Algorithm 1 is a 2-LMP.

Lemma 1. *For the forest F_R output by Algorithm 1,*

$$\sum_{e \in F_R} w_e + 2 \sum_{v \in V \backslash V(F_R)} \pi_v \leq 2 \cdot opt,$$

where opt is the optimal value for the PCF_k instance.

Proof. Since the set of variables $\{y_S\}_{S \subseteq V \setminus R}$ is a feasible solution to the dual LP 2 throughout the algorithm, by Duality Theory, we have

$$\sum_{S \subseteq V \setminus R} y_S \leq opt. \tag{4}$$

Suppose at the end of the algorithm, the sets of connected components of $G[F]$ not containing roots are $C_1, \ldots, C_q$. Then $V \setminus V(F_R)$ is the disjoint union of $V(C_1) \cup \cdots \cup V(C_q)$. For each $i = 1, \ldots, q$, the reason for the deactivation of C_i is because the second-type constraint corresponding to C_i is tight. So

$$\sum_{v \in V \setminus V(F_R)} \pi_v = \sum_{j=1}^{q} \sum_{v \in V(C_j)} \pi_v = \sum_{j=1}^{q} \sum_{S: S \subseteq V(C_j)} y_S. \tag{5}$$

For an edge $e \in F_R$, it is added into F because the first-type constraint corresponding to e is tight. So,

$$\sum_{e \in F_R} w_e = \sum_{e \in F_R} \sum_{S: e \in \delta(S)} y_S = \sum_{S: S \subseteq V} y_S \cdot |F_R \cap \delta(S)|. \tag{6}$$

Combining expressions (4), (5) and (6), to prove the lemma, it suffices to prove

$$\sum_{S: S \subseteq V} y_S \cdot |F_R \cap \delta(S)| + 2 \sum_{j=1}^{q} \sum_{S: S \subseteq V(C_j)} y_S \leq 2 \sum_{S: S \subseteq V \setminus R} y_S. \tag{7}$$

In the following, we prove that inequality (7) holds for $\{y_S\}$ in every iteration of the while loop. Initially, all $y_S \equiv 0$, and thus inequality (7) trivially holds. Suppose (7) is true at the beginning of the ith iteration, we show that in the ith iteration,

the left-side increase of (7) is no more than the right-side increase of (7). (8)

Let $\mathcal{C}^{active}$ and $\mathcal{C}^{non}$ be the set of active components and the set of non-active components at the beginning of the ith iteration, and let $\mathcal{C}^{(1)}, \mathcal{C}^{(2)}$ be the subsets of $\mathcal{C}^{active}$ which are contained in $V(F_R)$ and $V \setminus V(F_R)$, respectively. Note no component $C \in \mathcal{C}^{(1)}$ can be contained in a C_j, and any component $C \in \mathcal{C}^{(2)}$ has $F_R \cap \delta(V(C)) = \emptyset$. So the left-hand increase of (7) is $\sum_{C \in \mathcal{C}^{(1)}} \varepsilon \cdot |F_R \cap \delta(V(C))| + 2 \sum_{C \in \mathcal{C}^{(2)}} \varepsilon$. While the the right-hand increase of 7 is $2\varepsilon |\mathcal{C}^{active}|$. So, to prove (8), it suffices to prove

$$\sum_{C \in \mathcal{C}^{(1)}} |F_R \cap \delta(V(C))| \leq 2 |\mathcal{C}^{(1)}|. \tag{9}$$

Consider one component of F_R, call it T_R. Let $\mathcal{C}_{T_R}$ be the set of connected components at the beginning of the ith iteration which are contained in $V(T_R)$, and denote $\mathcal{C}_{T_R}^{active}$ and $\mathcal{C}_{T_R}^{non}$ the subsets of active and non-active components in $\mathcal{C}_{T_R}$, respectively. Contracting every component of $\mathcal{C}_{T_R}$ into a super node, the subgraph spanned by T_R is a tree. By the reverse deletion step, all but at most one leaf node (namely the component of $\mathcal{C}_{T_R}^{non}$ containing the root) are active. So all but at most one non-active node has degree at least two in the contracted tree. It follows that

$$\sum_{C \in \mathcal{C}_{T_R}} |T_R \cap \delta(V(C))| = \sum_{C \in \mathcal{C}_{T_R}^{active}} |T_R \cap \delta(V(C))| + \sum_{C \in \mathcal{C}_{T_R}^{non}} |T_R \cap \delta(V(C))|$$

$$\geq \sum_{C \in \mathcal{C}_{T_R}^{active}} |T_R \cap \delta(V(C))| + 1 + 2(|\mathcal{C}_{T_R}^{non}| - 1). \quad (10)$$

On the other hand, using Shaking Hands Lemma on the contracted tree,

$$\sum_{C \in \mathcal{C}_{T_R}} |T_R \cap \delta(V(C))| = 2|E(T_R)| = 2(|\mathcal{C}_{T_R}| - 1), \quad (11)$$

where the second equality comes from the fact that the contracted subgraph is a tree. Combining inequalities (10) and (11), we have

$$\sum_{C \in \mathcal{C}_{T_R}^{active}} |T_R \cap \delta(V(C))| < 2|\mathcal{C}_{T_R}^{active}| \quad (12)$$

Then inequality (9) follows from summing (12) over all components of F_R. As analyzed before, this implies the validity of the lemma.

4   5-LMP for PCMSSC

The algorithm for PCMSSC is described in Algorithm 2. It takes the 2-LMP algorithm $PCF_k(G, R)$ in Sect. 3 as a subroutine and consists of two stages. In the first stage, for each guessed positive integer k and each guessed set of roots R with $R \subseteq B$ and $|R| = k$, it calls $PCF_k(G, R)$ to calculate a forest F_R with k components. The vertex set of each component of F_R form one group with exactly one base station. In the second stage, for each group, find a Hamiltonian cycle using Remark 1.

Suppose the Hamiltonian cycle has length L, then $\lceil L/at \rceil$ mobile sensors are uniformly deployed along the cycle and move in the same direction with speed a, forming a sweep coverage for this group. If a group has only one vertex, namely a base station, then a mobile sensor is stationed at this vertex, providing a continuous monitoring of this base station.

Remark 1 (finding Hamiltonian cycle). Given a minimum spanning tree T on a graph G with a metric edge length w, a Hamiltonian cycle can be constructed in the following way: double every edge of T to form an Eulerian graph F; follow an Eulerian tour of F, and short-cut repeated vertices; the resulting Hamiltonian cycle H has $w(H) \leq w(F) = 2w(T)$.

Algorithm 2. 5-LMP for PCMSSC$_{BS}$

---

input: A graph $G = (V, E)$ with metric edge length w, and a set of base stations B.
output: A schedule of routes for a set of mobile sensors.

1: **for** $k = 0, 1, 2, \ldots, |B|$ **do**
2:     **for** each $R \subseteq B$ with $|R| = k$ **do**
3:         $F_R \leftarrow PCF_R(G, R)$ with components $T_1^{(R)}, T_2^{(R)}, \ldots, T_k^{(R)}$;
4:         $n_R \leftarrow 0$;
5:         **for** $i = 1, 2, \ldots, k$ **do**
6:             **if** $|V(T_i^{(R)})| \geq 2$ **then**
7:                 $n_R \leftarrow n_R + \left\lceil \frac{2w(T_i^{(R)})}{at} \right\rceil$;
8:             **else**
9:                 $n_R \leftarrow n_R + 1$;
10:             **end if**
11:         **end for**
12:     **end for**
13: **end for**
14: $(\hat{k}, \hat{R}) \leftarrow \arg\min_{k \in \{0,1,\ldots,|B|\}, R \subseteq B, |R| = k} \{c \cdot n_R + 5 \sum_{v \in V \setminus V(F_R)} \pi(v)\}$;
15: **for** each subtree $T^{(\hat{R})}$ of $F_{\hat{R}}$ **do**
16:     **if** $|V(T^{(\hat{R})})| \geq 2$ **then**
17:         Construct a Hamiltonian cycle $C(\hat{R})$ on $V(T^{(\hat{R})})$ using Remark 1;
18:         Uniformly deploy $\lceil w(C^{(\hat{R})})/at \rceil$ mobile sensors along $C^{(\hat{R})}$ and let them
           move along $C^{(\hat{R})}$ in the same direction at speed a;
19:     **else**
20:         Deploy one mobile sensor at the unique vertex of $T^{(\hat{R})}$;
21:     **end if**
22: **end for**
23: **return** the above deployment.

To analyze the performance of Algorithm 2, we assume, w.l.o.g, that

$$\frac{at}{c} = \frac{4}{5}. \tag{13}$$

This can be done by scaling both c and π without changing the objective $c \cdot |\mathcal{S}| + r \cdot \sum_{v \notin \mathcal{C}(\mathcal{S})} \pi(v)$ to be minimized.

Theorem 1. *Algorithm 2 is a 5-LMP for PCMSSC and the running time is* $O(|B|^{|B|+1} nm)$.

Proof. Suppose an optimal solution uses k^* mobile sensors and X^* is the set of non-covered vertices. During time interval $[0, t]$, the trajectories of these mobile sensors form a subgraph of G spanning vertex set $V \setminus X^*$. Suppose this subgraph contains b^* base stations from B. Let $m^* = \min\{k^*, b^*\}$. Deleting some edges if necessary, we can modify this subgraph into a forest containing m^* components, still spanning $V \setminus X^*$, and each component contains at least one base station.

Let the set of these base stations be R^* and denote the modified forest as $F_{R^*}^*$. Note that

$$w(F_{R^*}^*) \leq k^* a t. \tag{14}$$

When R^* is guessed by Algorithm 2, a forest F_{R^*} is obtained. Since $\text{PCF}_{|R^*|}(G, R^*)$ is a 2-LMP algorithm, and by (14), we have

$$w(F_{R^*}) + 2\pi(V\backslash V(F_{R^*})) \leq 2(k^* a t + \pi(V\backslash V(F_{R^*}^*))) = 2(k^* a t + \pi(X^*)), \tag{15}$$

For each component $T_i^{(R^*)}$ of F_{R^*}, the number of mobile sensors deployed to cover group $V(T_i^{(R^*)})$ is at most

$$\frac{w(C_i^{(R^*)})}{at} + 1 \leq \frac{2w(T_i^{(R^*)})}{at} + 1.$$

Denote by $\mathcal{S}^*$ the set of mobile sensors deployed when R^* is guessed. Then

$$c|\mathcal{S}^*| + 5\pi(V\backslash\mathcal{C}(\mathcal{S}^*)) \leq c \cdot \sum_{i=1}^{m^*} \left(\frac{2w(T_i^{(R^*)})}{at} + 1 \right) + 5\pi(V\backslash V(F_{\mathcal{R}^*}))$$

$$= \frac{2cw(F_{R^*})}{at} + cm^* + 5\pi(V\backslash V(F_{\mathcal{R}^*}))$$

$$\leq \frac{5w(F_{R^*})}{2} + ck^* + 5\pi(V\backslash V(F_{\mathcal{R}^*}))$$

$$= \frac{5}{2}(w(F_{R^*}) + 2\pi(V\backslash V(F_{\mathcal{R}^*}))) + ck^*,$$

where the second inequality uses (13) and the fact $m^* \leq k^*$. Making use of 15,

$$c|\mathcal{S}^*| + 5\pi(V\backslash\mathcal{C}(\mathcal{S}^*)) \leq 5(k^* a t + \pi(X^*)) + ck^*,$$

Again using (13), $5k^* a t + ck^* = 5ck^*$, and thus

$$c|\mathcal{S}^*| + 5\pi(V\backslash\mathcal{C}(\mathcal{S}^*)) \leq 5(ck^* + \pi(X^*)) = 5opt.$$

For the time complexity, there are $\sum_{k=0}^{|B|} \binom{|B|}{k} = O(|B|^{|B|+1})$ guesses for R. For each guessed R, computing F_R requires time $O(mn)$,

finding a Hamiltonian cycle using Remark 1 needs time $O(n)$, and all the other steps can be easily done in time $O(n)$. Hence the time complexity is $O(|B|^{|B|+1}nm)$.

5   Conclusion and Future Work

In this paper, we proposed the prize-collecting min-sensor sweep coverage problem with base stations, and designed a 5-LMP algorithm. As a step stone, we presented a 2-LMP for the prize-collecting forest problem with exactly k components.

Note that the time complexity of the algorithm depends exponentially on the number of base stations. Since we have assumed a constant number of base-stations, the algorithm runs in polynomial time. If there is no base station, then guessing the number of connected components of the forest might be time consuming. Whether there is a constant approximation algorithm for the prize-collecting sweep cover problem without base stations is a topic which might be theoretically challenging.

Acknowledgment. This research work is supported in part by NSFC (U20A2068, 11771013), and ZJNSFC (LD19A010001).

References

1. Gorain, B., Mandal, P.S.: Approximation algorithms for sweep coverage in wireless sensor networks. J. Parallel Distrib. Comput. **74**(08), 2699–2707 (2014)
2. Bhowmick, S., Inamdar, T., Varadarajan, K.: Fault-tolerant covering problems in metric spaces. Algorithmica **83**, 413–446 (2021)
3. Cao, J.-N., Huang, P., Lin, F., Liu, C., Gao, J., Zhou, J.: ACO-based sweep coverage scheme in wireless sensor networks. J. Sens. **2015**, 484–902 (2015)
4. Chen, Q., Huang, X., Ran, Y.: Approximation algorithm for distance constraint sweep coverage without predetermined base stations. Discrete Math. Algorithms Appl. **10**(05), 1850064 (2018)
5. Cheng, W., Li, M., Liu, K., Liu, Y., Li, X., Liao, X.: Sweep coverage with mobile sensors. In: 2008 IEEE International Symposium on Parallel and Distributed Processing, pp. 1–9 (2008)
6. Du, J., Li, Y., Liu, H., Sha, K.: On sweep coverage with minimum mobile sensors. In: IEEE 16th International Conference on Parallel and Distributed Systems, pp. 283–290 (2010)
7. Fan, H., Li, M., Sun, X., Wan, P.-J., Zhao, Y.: Barrier coverage by sensors with adjustable ranges. ACM Trans. Sens. Netw. **11**, 1–20 (2014)
8. Gao, X., Fan, J., Wu, F., Chen, G.: Cooperative sweep coverage problem with mobile sensors. IEEE Trans. Mob. Comput. 1 (2020). https://doi.org/10.1109/TMC.2020.3008348
9. Goemans, M.X., Williamson, D.P.: A general approximation technique for constrained forest problems. SIAM J. Comput. **24**(2), 296–317 (1995)
10. Gorain, B., Mandal, P.S.: Approximation algorithm for sweep coverage on graph. Inf. Process. Lett. **115**(9), 712–718 (2015)
11. Gorain, B., Mandal, P.S.: Solving energy issues for sweep coverage in wireless sensor networks. Discret. Appl. Math. **228**, 130–139 (2017)
12. Li, M., Cheng, W., Liu, K., He, Y., Li, X., Liao, X.: Sweep coverage with mobile sensors. IEEE Trans. Mob. Comput. **10**(11), 1534–1545 (2011)
13. Liang, D., Shen, H.: Efficient algorithms for max-weighted point sweep coverage on lines. Sensors (Basel, Switzerland) **21**(4), 1457 (2021)
14. Liang, J., Huang, X., Zhang, Z.: Approximation algorithms for distance constraint sweep coverage with base stations. J. Comb. Optim. **37**(4), 1111–1125 (2018). https://doi.org/10.1007/s10878-018-0341-3
15. Liang, W., Zhang, Z., Huang, X.H.: Minimum power partial multi-cover on a line. Theoret. Comput. Sci. **864**, 118–128 (2021)

16. Nie, Z., Hongwei, D.: An approximation algorithm for general energy restricted sweep coverage problem. Theoret. Comput. Sci. **864**, 70–79 (2021)

17. Ran, Y., Huang, X., Zhang, Z., Du, D.-Z.: Approximation algorithm for minimum power partial multi-coverage in wireless sensor networks. J. Glob. Optim. **80**(3), 661–677 (2021). https://doi.org/10.1007/s10898-021-01033-y

18. Wang, C., Ma, H.: Data collection with multiple controlled mobile nodes in wireless sensor networks. In: IEEE 17th International Conference on Parallel and Distributed Systems, pp. 489–496 (2011)

19. Wu, W., Zhang, Z., Lee, W., Du, D.-Z.: Optimal Coverage in Wireless Sensor Networks. Springer, Heidelberg (2020). https://doi.org/10.1007/978-3-030-52824-9

Approximation Algorithm for the Capacitated Correlation Clustering Problem with Penalties

Sai Ji[1], Gaidi Li[2(✉)], Dongmei Zhang[3], and Xianzhao Zhang[4]

[1] Academy of Mathematics and Systems Science, Chinese Academy of Sciences,
Beijing 100190, People's Republic of China
jisai@amss.ac.cn

[2] Department of Operations Research and Information Engineering,
Beijing University of Technology, Beijing 100124, People's Republic of China
ligd@bjut.edu.cn

[3] School of Computer Science and Technology, Shandong Jianzhu University,
Jinan 250101, People's Republic of China
zhangdongmei@sdjzu.edu.cn

[4] School of Mathematics and Statistics, Linyi University,
Linyi 276005, People's Republic of China
zhangxianzhao@lyu.edu.cn

Abstract. Correlation clustering problem is an elegant clustering problem and has many applications in protein interaction networks, cross-lingual link detection, etc. In this paper, we introduce the capacitated correlation clustering problem with penalties by combining the capacitated correlation clustering problem and the correlation problem with penalties. There are two main contributions in this paper. The first one is that we give an integer programming for the capacitated correlation clustering problem with penalties. The second one is that we provide an LP-based $(4/(4-5\alpha), 8/\alpha)$-bi-criteria approximation algorithm for this problem, where parameter $\alpha \in (0, 4/9]$.

Keywords: Correlation clustering · Capacitated · Penalties · Approximation algorithm · LP-rounding

1 Introduction

Correlation clustering is a classical clustering problem, which has applications in protein interaction networks, cross-lingual link detection, and communication networks, etc. The correlation clustering problem was first introduced by Bansal et al. [5], which was motivated from a document clustering problem in which one has a pairwise similarity function f learned from past data, and the goal is to partition the current set of documents in a way that correlates with f as much as possible. In the correlation clustering problem, we are given a complete graph $G = (V, E)$. Each edge $(u, v) \in E$ is labeled by $+$ or $-$ based on the similarity of vertex u and vertex v. If two vertices are similar, the label is positive and vice versa. Compared with other clustering problems, we do not have to restrict the number of clusters in the correlation clustering problem. The goal of the problem is to partition set V into several clusters such that the vertices in the same

© Springer Nature Switzerland AG 2021
W. Wu and H. Du (Eds.): AAIM 2021, LNCS 13153, pp. 15–26, 2021.
https://doi.org/10.1007/978-3-030-93176-6_2

cluster are similar and the vertices between two clusters are dissimilar. Let each positive edge whose endpoints lie in the same cluster and each negative edge whose endpoints lie in two clusters be an agreement. On the contrary, let each positive edge whose endpoints lie in two different clusters and each negative edge whose endpoints lie in one cluster be a disagreement. Form the purpose of the correlation clustering problem, there are two different versions of this problem: minimizing disagreements and maximizing agreements. The goal of the former one is to minimize the number of disagreements while the latter one is to maximize the number of agreements. In this paper, we only focus on the minimizing disagreements.

The correlation clustering problem is NP-hard [5], and many approximation algorithms have been proposed for this problem [4, 6, 15, 18]. The first constant approximation algorithm for correlation clustering problem is provided by Bansal et al. [5] with an approximation ratio of 17433. Charikar et al. [8] gave a nature integer programming for the correlation clustering and proved that the integrality gap of the LP formulation is 2. Then, they provided a 4-approximation algorithm for this problem based on LP-rounding technique. They also presented an $O(\log n)$-approximation algorithm for the correlation clustering on general graphs. Chawla et al. [9] provided an LP-rounding 2.06-approximation algorithm based on the LP formulation given by Charikar et al. [8]. At present, the 2.06-approximation algorithm is the best deterministic approximation algorithm for the correlation algorithm, which achieving an approximation ratio almost matching the integrality gap 2.

Besides the correlation clustering, several meaningful variants of the correlation clustering problem have also been studied extensively [1–3, 13, 16, 17]. In the above problems, we are particularly interested in the capacitated correlation clustering problem and the correlation clustering problem with penalties.

Capacity constraint is a natural constraint in combinatorial optimization problems [7, 11, 12, 19]. In capacitated correlation clustering problem, we are given a complete graph $G = (V, E)$ and an upper bound U on capacity. The goal of this problem is to partition the vertices into several clusters, subject to a capacitated constraint, so as to minimize the number of disagreements. Puleo and Milenkovic [17] first introduced the capacitated correlation clustering problem and provided a 6-approximation algorithm for this problem based on the 4-approximation algorithm for this problem [8].

Penalty constraint is also an important constraint in combinatorial optimization problems [10, 14, 20]. In the correlation clustering problem with penalties, there is a penalty cost for each vertex. Each vertex needs to be clustered or penalized. The goal of the correlation clustering problem with penalties is to select a penalized set and then partition the remain vertices into several clusters so as to minimize the sum of the total number of disagreements and the penalty cost. Aboud and Rabani [1] first introduced the correlation clustering problem with penalties and provided a 9-approximation algorithm based on primal-dual schema for the problem.

In this paper, we introduce the capacitated correlation clustering problem with penalties. In this problem, we are given a labeled completed graph, there is a penalty cost for each vertex. Each vertex needs to be clustered or penalized. Moreover, there is an upper bound U on capacity. The goal of this problem is to select a penalized set and then partition the remain vertices into several clusters, subject to a capacity constraint, so as to minimize the sum of the total number of disagreements and the penalty cost. Here the number of disagreements is the sum of the number of positive edges whose two endpoints are not penalized and lie in different clusters and the number of negative edges whose two endpoints are not penalized and lie in a same clusters. There are two main contributions in this paper.

(1) We provide an integer programming for the capacitated correlation clustering problem with penalties. The difficulty in the construction process is that once a vertex is penalized, it does not affect the number of vertices in any cluster, so we need to carefully identify the number of vertices in each cluster.
(2) Given a parameter $\alpha \in (0, 4/9]$, we provide an LP-based $(4/(4 - 5\alpha), 8/\alpha)$-bi-criteria approximation algorithm and its theoretical analysis for the capacitated correlation clustering problem with penalties. We skillfully select the vertices to be penalized and cluster the remaining vertices, so as to analyze the upper bound of the penalty cost, the upper bound on the number of disagreements and the upper bound on the number of vertices in each cluster.

The rest of this paper is organized as follows. In Sect. 2, we first provide the definition of the capacitated correlation clustering problem with penalties, then we give an integer programming as well as its LP-relaxation for the problem. In Sect. 3, we provide our bi-criteria approximation algorithm and the theoretical analysis. Some conclusions are given in Sect. 4.

2 Preliminaries

In this section, we first give a detailed definition of the capacitated correlation clustering problem with penalties. Then, we provide an integer programming and its LP relaxation for the capacitated correlation clustering problem with penalties. For each integer t, denote set $[t] := \{1, 2, \ldots, t\}$, then the capacitated correlation clustering problem and the capacitated correlation clustering problem can be defined as follows.

Definition 1 (Capacitated Correlation clustering problem). *Given a labeled complete graph $G = (V, E)$, an upper bound U. The correlation clustering problem is to find a partition $V = \{V_1, V_2, \ldots, V_t\}$ of V with $|V_i| \leq U, i \in [t]$ such that*

$$\frac{1}{2} \sum_{v \in V_i, i \in [t]} (|\{(u, v) \in E^+, u \in V \backslash V_i\}| + |\{(u, v) \in E^-, u \in V_i\}|)$$

is minimized, where E^+ is the set of positive edges and E^- is the set of negative edges.

Definition 2 (Capacitated correlation clustering problem with penalties). *Given a labeled complete graph $G = (V, E)$, an upper bound U, and a penalty cost p_v for each vertex $v \in V$. The capacitated correlation clustering problem with penalties is to find a subset $P \subseteq V$ as well as a partition $V = \{V_1, V_2, \ldots, V_t\}$ of $V \backslash P$ with $|V_i| \leq U, i \in [t]$ such that*

$$\frac{1}{2} \sum_{v \in V_i, i \in [t]} (|\{(u, v) \in E^+, u \in V \backslash (V_i \cup P)\}| + |\{(u, v) \in E^-, u \in V_i\}|) + \sum_{v \in P} p_v$$

is minimized, where E^+ is the set of positive edges and E^- is the set of negative edges.

First, we introduce the following four types of binary variables:

- For each edge $(u, v) \in E$, let x_{uv} indicates whether vertex u and vertex v lie in the same cluster. If vertex u and vertex v lie in the same cluster, then we let $x_{uv} = 0$. Otherwise, let $x_{uv} = 1$.
- For each edge $(u, v) \in E$, let w_{uv} indicates whether both vertex u and vertex v are both clustered. If both vertex u and vertex v are clustered, then we let $w_{uv} = 0$. Otherwise, let $w_{uv} = 1$.
- For each vertex $v \in V$, let y_v indicate whether vertex v is penalized. If vertex v is penalized, then we let $y_v = 1$. Otherwise, let $y_v = 0$.
- For each edge $(u, v) \in E$, let z_{uv} indicate whether edge (u, v) is a disagreement. If edge (u, v) is a disagreement, then we let $z_{uv} = 1$. Otherwise, let $z_{uv} = 0$.

Based on above variables, we can give an integer programming for the capacitated correlation clustering problem with penalties.

$$
\begin{aligned}
\min \quad & \sum_{(u,v)\in E} z_{uv} + \sum_{v\in V} p_v y_v \\
\text{s. t.} \quad & x_{uv} + x_{vw} \geq x_{uw}, & \forall u, v, w \in V, \\
& w_{uv} \leq y_u + y_v \leq 2w_{uv}, & \forall u, v \in V, \\
& w_{uv} + z_{uv} \geq 1 - x_{uv}, & \forall (u, v) \in E^-, \\
& w_{uv} + z_{uv} \geq x_{uv}, & \forall (u, v) \in E^+, \\
& x_{uv} \geq z_{uv}, & \forall (u, v) \in E^+, \quad (1) \\
& \sum_{(u,v)\in E^+, u\in V} (1 - w_{uv} - z_{uv}) + \sum_{(u,v)\in E^-, u\in V} z_{uv} \leq U, & \forall v \in V, \\
& x_{vv}, z_{vv}, w_{vv} = 0, & \forall v \in V, \\
& x_{uv}, z_{uv}, y_u, w_{uv} \in \{0, 1\}, & \forall u, v \in V.
\end{aligned}
$$

The value of the objective function is the number of disagreements and the penalty cost of the penalized vertices. There are seven types of constraints in Programming (1). The first one is a triangle inequality, which guarantees that the solution returned by Programming (1) is a feasible solution of the correlation clustering problem. The second to the fifth constraints give the condition for a edge to become a disagreement. To be specific, for each positive edge (u, v), variable $z_{uv} = 1$ iff vertex u and vertex v are both clustered and lie in different clusters. For each negative edge (u, v), variable $z_{uv} = 1$ iff vertex u and vertex v are both clustered and lie in the same cluster. The sixth and the seventh one ensure that there are at most U vertices in each cluster. By relaxing the variables, we obtain the following LP relaxation of (1):

$$
\begin{aligned}
\min \quad & \sum_{(u,v)\in E} z_{uv} + \sum_{v\in V} p_v y_v \\
\text{s. t.} \quad & x_{uv} + x_{vw} \geq x_{uw}, & \forall u, v, w \in V, \\
& w_{uv} \leq y_u + y_v \leq 2w_{uv}, & \forall u, v \in V,
\end{aligned}
$$

$$w_{uv} + z_{uv} \geq 1 - x_{uv}, \qquad\qquad\qquad \forall (u, v) \in E^-,$$
$$w_{uv} + z_{uv} \geq x_{uv}, \qquad\qquad\qquad\quad \forall (u, v) \in E^+,$$
$$x_{uv} \geq z_{uv}, \qquad\qquad\qquad\qquad\quad \forall (u, v) \in E^+, \qquad (2)$$
$$\sum_{(u,v) \in E^+, u \in V} (1 - w_{uv} - z_{uv}) + \sum_{(u,v) \in E^-, u \in V} z_{uv} \leq U, \qquad \forall v \in V,$$
$$x_{vv}, z_{vv}, w_{vv} = 0, \qquad\qquad\qquad\quad\; \forall v \in V,$$
$$x_{uv}, z_{uv}, y_u, w_{uv} \in [0, 1], \qquad\qquad\quad \forall u, v \in V.$$

3   Bi-criteria Approximation Algorithm and Analysis

This section is the core section of this paper. In Subsect. 3.1, we provide our bi-criteria approximation algorithm for the capacitated correlation clustering problem with penalties. The theoretical analysis of our algorithm are provided in Subsect. 3.2.

3.1   Bi-criteria Approximation Algorithm

Before giving Algorithm 1, we first provide a high level description for it. There are three main phases in this algorithm. The first phase is a computational process. In this phase, we solve Programming (2) to obtain the optimal fractional solution (x^*, y^*, z^*). For each value x_{uv}^*, we can be regard it as the distance between vertex u and vertex v. The second phase is a penalized process. In this phase, we select a set P of vertices to be penalized based on the value of $y*$ and parameter $\alpha/8$. For each vertex $v \in V$, if $y_v^* \geq \alpha/8$, then we make vertex v as a penalized vertex. Otherwise, we cluster it. The last phase is an iterative clustering process, which is based on the 4-approximation for the correlation clustering problem [8]. In each iteration, we first a vertex v is selected randomly from the un-clustered vertices as a center. Then we select a set T_v from the un-clustered vertices based on the value of x^* and parameter α. At last, we decide whether to set vertex v and set T_v as a cluster according to parameter $\alpha/2$ as well as the average distance between the vertices in T_v and the center v. We repeat the clustering process until all the vertices are clustered.

3.2   Theoretical Analysis

In this section, we mainly analyze three parts. The first part is the upper bound on the penalty cost of penalized vertices returned by Algorithm 1. The second part is the upper bound on the number of disagreements returned by Algorithm 1. The final part is the upper bound on the number of vertices of each cluster returned by Algorithm 1.

Penalty Cost. Without loss of generality, we assume that $C := \{v_1, v_2, \ldots, v_k\}$. The penalized set is P and the partition of $V \backslash P$ is $\mathcal{C} := \{C_{v_1}, C_{v_2}, \ldots, C_{v_k}\}$. The penalty cost is

$$\sum_{v \in P} p_v,$$

and its upper bound is shown in Lemma 1.

Algorithm 1.

Input: A labeled complete graph $G = (V, E)$, parameter $\alpha \in (0, 4/9]$
Output: A partition of vertices
 1: Solve (2) to obtain the optimal fractional solution (x^*, y^*, z^*)
 2: Initialize $S := V$, $P := \emptyset$ and $C := \emptyset$
 3: Update $P := \{v \in V : y_v^* \geq \alpha/8\}$, $S := V \backslash P$
 4: **while** $S \neq \emptyset$ **do**
 5:    Select a vertex v from S randomly and update $C := C \cup \{v\}$
 6:    Let $T_v := \{u \in S : x_{uv}^* \leq \alpha\}$
 7:    **if** $\dfrac{\sum_{u \in T_v} x_{uv}^*}{|T_v|} \geq \dfrac{\alpha}{2}$, **then**
 8:       Let $C_v = \{v\}$
 9:    **else**
10:       Let $C_v = T_v$
11:    **end if**
12:    Update $S := S - C_v$
13: **end while**
14: **return** set P and the partition $\mathcal{C} := \{C_v : v \in C\}$ of $V \backslash P$

Lemma 1. *The penalty cost can be bounded by*

$$\frac{8}{\alpha} \sum_{v \in V} p_v y_v^*.$$

Proof. From the construction of P, for each vertex $v \in P$, we have $y_v^* \geq \alpha/8$, which indicates that

$$\sum_{v \in P} p_v \leq \frac{8}{\alpha} \sum_{v \in P} p_v y_v^* \leq \frac{8}{\alpha} \sum_{v \in V} p_v y_v^*.$$

The lemma is concluded. □

Disagreements. The number of disagreements generated by positive edges is

$$\sum_{i \in [k-1]} \left| (u, v) \in E^+ : v \in C_{v_i}, u \in \cup_{t \in [k] \backslash [i]} C_{v_t} \right|,$$

and the number of disagreements generated by negative edges is

$$\sum_{i \in [k]} \left| (u, v) \in E^-, u, v \in C_{v_i} \right|.$$

Since each cluster $C_{v_i}, i \in [k]$ must be one of the following two types:

- Type 1: $C_{v_i} := \{v_i\}$;
- Type 2: $C_{v_i} := T_{v_i}$.

Then, we analyze the upper bound of disagreements by above two types.

Type 1 of Cluster: Because cluster C_{v_i} belongs to Type 1, the number of new disagreements generated by C_{v_i} equals $|(v_i, v) \in E^+, v \in \cup_{t \in [k] \setminus [i]} C_{v_t}|$, the upper bound of on number of disagreements is analyzed by Lemma 2.

Lemma 2. *If C_{v_i} is of Type 2, then the upper bound on the number of disagreements generated by the positive edges satisfies:*

(i) *The number of disagreements from positive edges $(v_i, v) \in E^+, v \in T_{v_i}$ can be bounded by*

$$\frac{4}{\alpha} \sum_{v \in T_{v_i}} z_{vv_i}^*.$$

(ii) *The number of disagreement from positive edge $(v_i, v) \in E^+, v \in \cup_{t \in [k] \setminus [i]} C_{v_t} \setminus T_{v_i}$ can be bounded by*

$$\frac{4}{3\alpha} z_{vv_i}^*.$$

Proof. We prove (i) and (ii) of Lemma 2, respectively.

(i) From Steps 6-8 of Algorithm 1, we have

$$\frac{\sum_{v \in T_{v_i}} x_{vv_i}^*}{|T_{v_i}|} \geq \frac{\alpha}{2}.$$

From the definition of T_{v_i}, for each $v \in T_{v_i}$ we have $1 - x_{vv_i}^* \geq x_{vv_i}^*$, which indicates that

$$\frac{\sum_{(v_i,v) \in E^+, v \in T_{v_i}} x_{vv_i}^* + \sum_{(v_i,v) \in E^-, v \in T_{v_i}} (1 - x_{vv_i}^*)}{|T_{v_i}|} \geq \frac{\alpha}{2}. \tag{3}$$

Combining Step 3 of Algorithm 1, inequality (3), the second constraint, the third constraint and the fourth constraint of Programming (2), we can obtain

$$\frac{\sum_{v \in T_{v_i}} z_{vv_i}^*}{|T_{v_i}|} \geq \frac{\alpha}{2} - 2 \cdot \frac{\alpha}{8} = \frac{\alpha}{4}.$$

Then, the number of disagreements generated by positive edges $(v_i, v), v \in T_{v_i}$ is no more than $|T_{v_i}|$, and it can be bounded by

$$\frac{4}{\alpha} \sum_{v \in T_{v_i}} z_{vv_i}^*.$$

(ii) For each positive edges $(v_i, v) \in E^+, v \in \cup_{t \in [k] \setminus [i]} C_{v_t} \setminus T_{v_i}$, from the construction of T_{v_i} and the constraints of Programming (2), we have

$$z_{vv_i}^* \geq x_{vv_i}^* - 2 \cdot \frac{\alpha}{8} \geq \frac{3\alpha}{4}.$$

The lemma is concluded.   □

Type 2 of Cluster: As C_{v_i} is of Type 2, There are two kinds of disagreements generated by C_{v_i}. One is the disagreements generated by positive edges whose two endpoints lie in different clusters. The other one is the disagreements generated by negative edges whose two endpoints lie in the same cluster. The corresponding upper bound is shown in Lemmas 3 and 4.

Lemma 3. *If C_{v_i} is of Type 2, then the upper bound on the number of disagreements generated by the positive edges satisfies:*

(i) *The number of disagreement from positive edge $(v,q) \in E^+$, $v \in C_{v_i}$, $q \in \cup_{t \in [k] \setminus [i]} C_{v_t}$ with $x_{qv_i}^* \geq 3\alpha/2$ can be bounded by*

$$\frac{4}{\alpha} z_{vq}^*.$$

(ii) *The number of disagreements from positive edges $(v,q) \in E^+$, $v \in C_{v_i}$, $q \in \cup_{t \in [k] \setminus [i]} C_{v_t}$ with $\alpha \leq x_{qv_i}^* < 3\alpha/2$ can be bounded by*

$$\frac{4}{\alpha} \sum_{v \in C_{v_i}} z_{vq}^*.$$

Proof. We prove (i) and (ii) of Lemma 3, respectively.

(i) For each positive edge $(v,q) \in E^+$, $v \in C_{v_i}$, $q \in \cup_{t \in [k] \setminus [i]} C_{v_t}$ with $x_{qv_i}^* \geq 3\alpha/2$, we have

$$
\begin{aligned}
z_{vq}^* &\geq x_{vq}^* - 2 \cdot \frac{\alpha}{8} \\
&\geq x_{qv_i}^* - x_{vv_i}^* - \frac{\alpha}{4} \\
&\geq \frac{\alpha}{2} - \frac{\alpha}{4} = \frac{\alpha}{4}.
\end{aligned}
$$

Therefore, the disagreement generated by positive edge (v,q) can be bounded by

$$\frac{4}{\alpha} z_{vq}^*.$$

(ii) For each $q \in \cup_{t \in [k] \setminus [i]} C_{v_t}$ with $\alpha \leq x_{qv_i}^* < 3\alpha/2$. Denote by P_q the number of positive edges $(v,q) \in E^+$ with $v \in C_{v_i}$, and N_q the number of negative edges $(v,q) \in E^-$ with $v \in C_{v_i}$. Then, we have

$$
\begin{aligned}
&\sum_{v \in C_{v_i}} z_{vq}^* \\
&\geq \sum_{(v,q) \in E^+, v \in C_{v_i}} \left(x_{qv_i}^* - x_{vv_i}^* - 2 \cdot \frac{\alpha}{8} \right) + \sum_{(v,q) \in E^+, v \in C_{v_i}} \left(1 - x_{qv_i}^* - x_{vv_i}^* - 2 \cdot \frac{\alpha}{8} \right) \\
&\geq P_q \left(x_{qv_i}^* - \frac{\alpha}{4} \right) + N_q \left(1 - x_{qv_i}^* - \frac{\alpha}{4} \right) - \sum_{v \in C_{v_i}} x_{vv_i}^*
\end{aligned}
$$

$$\geq P_q \left(x^*_{qv_i} - \frac{\alpha}{4} \right) + N_q \left(1 - x^*_{qv_i} - \frac{\alpha}{4} \right) - \frac{\alpha}{2}(P_q + N_q)$$

$$> \left(\alpha - \frac{\alpha}{4} \right) P_q + \left(1 - \frac{3}{2}\alpha - \frac{\alpha}{4} \right) N_q - \frac{\alpha}{2}(P_q + N_q)$$

$$= \frac{\alpha}{4} P_q.$$

Therefore, the number of disagreements generated by positive edges $(v, q) \in E^+$, $v \in C_{v_i}$, $q \in \bigcup_{t \in [k] \setminus [i]} C_{v_t}$ with $\alpha \leq x^*_{qv_i} < 3\alpha/2$ can be bounded by

$$\frac{4}{\alpha} \sum_{v \in C_{v_i}} z^*_{vq}$$

The lemma is concluded.   □

Lemma 4. *If C_{v_i} is of Type 2, then the upper bound on the number of disagreements generated by the negative edges satisfies:*

(i) *The number of disagreement from negative edge $(v, q) \in E^-$, $v, q \in C_{v_i}$ with $x^*_{vv_i}, x^*_{qv_i} \leq 3\alpha/4$ can be bounded by*

$$\frac{4}{4 - 7\alpha} z^*_{vq}.$$

(ii) *The number of disagreements from negative edges $(v, q) \in E^-, v, q \in C_{v_i}$ with $x^*_{vv_i} \leq x^*_{qv_i}$ and $x^*_{qv_i} > 3\alpha/4$ can be bounded by*

$$\frac{4}{4 - 7\alpha} \sum_{v \in C_{v_i}, x^*_{vv_i} \leq x^*_{qv_i}} z^*_{vq}.$$

Proof. We prove (i) and (ii) of Lemma 4, respectively.

(i) For each negative edge $(v, q) \in E^-$, $v, q \in C_{v_i}$ with $x^*_{vv_i}, x^*_{qv_i} \leq 3\alpha/4$, we have

$$z^*_{vq} \geq 1 - x^*_{vq} - 2 \cdot \frac{\alpha}{8} \geq 1 - x^*_{vv_i} - x^*_{qv_i} - \frac{\alpha}{4} \geq 1 - \frac{3}{2}\alpha - \frac{\alpha}{4} = 1 - \frac{7\alpha}{4},$$

which indicates that the number of disagreement generated by edge (v, q) can be bounded by

$$\frac{4}{4 - 7\alpha} z^*_{vq}.$$

(ii) For each $q \in C_{v_i}$ with $x^*_{qv_i} > 3\alpha/4$. Denote by P'_q the number of positive edges $(v, q) \in E^+, v \in C_{v_i}$ with $x^*_{vv_i} \leq x^*_{qv_i}$, and N'_q the number of negative edges $(v, q) \in E^-, v \in C_{v_i}$ with $x^*_{vv_i} \leq x^*_{qv_i}$. Then, we have

$$\sum_{v \in C_{v_i}, x^*_{vv_i} \leq x^*_{qv_i}} z^*_{vq}$$

$$\geq \sum_{(v,q) \in E^+, v \in C_{v_i}, x^*_{vv_i} \leq x^*_{qv_i}} \left(x^*_{qv_i} - x^*_{vv_i} - 2 \cdot \frac{\alpha}{8} \right)$$

$$+ \sum_{(v,q)\in E^+, v\in C_{v_i}, x^*_{vv_i}\le x^*_{qv_i}} \left(1 - x^*_{qv_i} - x^*_{vv_i} - 2\cdot\frac{\alpha}{8}\right)$$

$$\ge P'_q\left(x^*_{qv_i} - \frac{\alpha}{4}\right) + N'_q\left(1 - x^*_{qv_i} - \frac{\alpha}{4}\right) - \sum_{v\in C_{v_i}} x^*_{vv_i}$$

$$> \frac{\alpha}{2}P'_q + \left(1 - \alpha - \frac{\alpha}{4}\right)N'_q - \frac{\alpha}{2}(P'_q + N'_q)$$

$$\ge \left(1 - \frac{7}{4}\alpha\right)N'_q.$$

The number of disagreements from negative edges $(v,q)\in E^-, v<q\in C_{v_i}$ with $x^*_{qv_i} > 3\alpha/4$ can be bounded by

$$\frac{4}{4-7\alpha}\sum_{v\in C_{v_i}, x^*_{vv_i}\le x^*_{qv_i}} z^*_{vq}.$$

The lemma is concluded. $\qquad\square$

Combining Lemma 1–Lemma 4, we can obtain the following Lemma.

Lemma 5. *The sum of disagreements and penalty cost can be bounded by*

$$\frac{8}{\alpha}\left[\sum_{(u,v)\in E} z^*_{uv} + \sum_{v\in V} p_v y^*_v\right].$$

The Upper Bound on the Vertices in Each Cluster of Type 2. In each cluster $C_{v_i}, i\in [k]$ of Type 2, the upper bound on the number of vertices in C_{v_i} can be analyzed by the following Lemma.

Lemma 6. *The number of vertices in C_{v_i} can be bounded by*

$$\frac{4}{4-5\alpha}U.$$

Proof. From the third, the fifth and the sixth constraints of Programming (2), we can obtain for each positive edge (u,v) we have $1 - w^*_{wu} - z^*_{uv} \ge 1 - 5\alpha/4$. Moreover, for each negative edge (u,v) we have $z^*_{uv} \ge 1 - 5\alpha/4$. Therefore, we can obtain

$$|C_{v_i}| \le \frac{4}{4-5\alpha}\left[\sum_{(u,v)\in E^+, u\in V}(1 - w^*_{uv} - z^*_{uv}) + \sum_{(u,v)\in E^-, u\in V} z^*_{uv}\right]$$

$$\le \frac{4}{4-5\alpha}U.$$

The lemma is concluded. $\qquad\square$

Combining Lemma 5 and Lemma 6, we can obtain our main result.

Theorem 1. *Algorithm 1 is a $\left(\dfrac{4}{4-5\alpha}, \dfrac{8}{\alpha}\right)$-bi-criteria approximation algorithm for the capacitated correlation clustering problem with penalties, where $\alpha \in (0, 4/9]$.*

4   Conclusions

In this paper, we introduce the capacitated correlation clustering problem with penalties. We provide an integer programming and an LP-based $(4/(4 - 5\alpha), 8/\alpha)$-bi-criteria approximation algorithm for this problem, where parameter $\alpha \in (0, 4/9]$. There are two interesting future work for the capacitated correlation clustering problem with penalties. One is to design a constant approximation algorithm for this problem. Another one is to study other interesting variants of the capacitated correlation clustering problem with penalties, such as the capacitated correlation clustering problem with penalties on general graphs and the min-max capacitated correlation clustering problem with penalties.

Acknowledgements. The first author is supported by National Natural Science Foundation of China (No. 12101594) and the Project funded by China Postdoctoral Science Foundation (No. 2021M693337). The third author is supported by National Natural Science Foundation of China (No. 11871081). The fourth author is supported by National Natural Science Foundation of China (No. 11801310).

References

1. Aboud, A., Rabani, Y.: Correlation clustering with penalties and approximating the reordering buffer management problem. Doctoral dissertation, Computer Science Department, Technion (2008)
2. Ahmadi, S., Khuller, S., Saha, B.: Min-max correlation clustering via multicut. In: Lodi, A., Nagarajan, V. (eds.) IPCO 2019. LNCS, vol. 11480, pp. 13–26. Springer, Cham (2019). https://doi.org/10.1007/978-3-030-17953-3_2
3. Ahn, K.J., Cormode, G., Guha, S., Mcgregor, A., Wirth, A.: Correlation clustering in data streams. In: Proceedings of the 32nd International Conference on Machine Learning, pp. 2237–2246 (2015)
4. Ailon, N., Avigdor-Elgrabli, N., Liberty, E., Zuylen, A.V.: Improved approximation algorithms for bipartite correlation clustering. SIAM J. Comput. **41**(5), 1110–1121 (2012)
5. Bansal, N., Blum, A., Chawla, S.: Correlation clustering. Mach. Learn. **56**(1–3), 89–113 (2004)
6. Bressan, M., Cesa-Bianchi, N., Paudice, A., Vitale, F.: Correlation clustering with adaptive similarity queries. In: Proceedings of the 32nd Annual Conference on Neural Information Processing Systems, pp. 12510–12519 (2019)
7. Castro, J., Nasini, S., Saldanha-Da-Gama, F.: A cutting-plane approach for large-scale capacitated multi-period facility location using a specialized interior-point method. Math. Program. **163**(1–2), 411–444 (2017)
8. Charikar, M., Guruswami, V., Wirth, A.: Clustering with qualitative information. J. Comput. Syst. Sci. **3**(71), 360–383 (2005)
9. Chawla, S., Makarychev, K., Schramm, T., Yaroslavtsev, G.: Near optimal LP rounding algorithm for correlation clustering on complete and complete k-partite graphs. In: Proceedings of the 47th ACM Symposium on Theory of Computing, pp. 219–228 (2015)
10. Chen, X., Hu, X., Jia, X., Li, M., Tang, Z., Wang, C.: Mechanism design for two-opposite-facility location games with penalties on distance. In: Proceedings of the 10th International Symposium on Algorithmic Game Theory, pp. 256–260 (2018)

11. Cohen-Addad, V.: Approximation schemes for capacitated clustering in doubling metrics. In: Proceedings of the 30th Annual ACM-SIAM Symposium on Discrete Algorithms, pp. 2241–2259 (2020)
12. Filippi, C., Guastaroba, G., Speranza, M.G.: On single-source capacitated facility location with cost and fairness objectives. Eur. J. Oper. Res. **289**(3), 959–974 (2021)
13. Jafarov, J., Kalhan, S., Makarychev, K., Makarychev, Y.: Correlation clustering with asymmetric classification errors. In: Proceedings of the 37th International Conference on Machine Learning, pp. 4641–4650 (2020)
14. Ji, S., Xu, D., Du, D., Wu, C.: Approximation algorithms for the fault-tolerant facility location problem with penalties. Discret. Appl. Math. **264**, 62–75 (2019)
15. Lange, J.H., Karrenbauer, A., Andres, B.: Partial optimality and fast lower bounds for weighted correlation clustering. In: Proceedings of the 35th International Conference on International Conference on Machine Learning, pp. 2892–2901 (2018)
16. Li, P., Puleo, G.J., Milenkovic, O.: Motif and hypergraph correlation clustering. IEEE Trans. Inf. Theory **66**(5), 3065–3078 (2019)
17. Puleo, G.J., Milenkovic, O.: Correlation clustering with constrained cluster sizes and extended weights bounds. SIAM J. Optim. **25**(3), 1857–1872 (2015)
18. Thiel, E., Chehreghani, M.H., Dubhashi, D.: A non-convex optimization approach to correlation clustering. In: Proceedings of the 33rd AAAI Conference on Artificial Intelligence, pp. 5159–5166 (2019)
19. Xu, Y., Möhring, R.H., Xu, D., Zhang, Y., Zou, Y.: A constant FPT approximation algorithm for hard-capacitated k-means. Optim. Eng. **21**(2), 709–722 (2020)
20. Zhang, D., Hao, C., Wu, C., Xu, D., Zhang, Z.: Local search approximation algorithms for the k-means problem with penalties. J. Comb. Optim. **37**(2), 439–453 (2019)

Approximation Algorithms for the Maximum Bounded Connected Bipartition Problem

Yajie Li[1], Weidong Li[1], Xiaofei Liu[2(✉)], and Jinhua Yang[3]

[1] School of Mathematics and Statistics, Yunnan University, Kunming, China
[2] School of Information Science and Engineering, Yunnan University, Kunming, China
[3] Dianchi College, Kunming, China

Abstract. In this paper, we study the maximum bounded connected bipartition problem (2-BCBP): given a vertex-weighted connected graph $G = (V, E; w)$ and an upper bound B, the vertex set V is partitioned into two subsets denoted as (V_1, V_2) such that both subgraphs induced by V_1 and V_2 are connected and the total weight of these two subgraphs is maximized, where the weight of the subgraph is the minimum of the sum of the weight of all vertices and B. The 2-BCBP is a hybrid variant of the maximum balanced connected partition problem on connected graphs and the maximum total early work problem in scheduling theory. In this paper, we consider the 2-BCBP and present an $\frac{8}{7}$-approximation algorithm. In particular, we consider the 2-BCBP on interval graphs and present a fully polynomial-time approximation scheme.

Keywords: Connected bipartition problem · Interval graph · Approximation algorithm · FPTAS

1 Introduction

Balanced connected graph k-partitioning (BCP$_k$) is one of the most important research fields in operations research and combinatorial optimization. Given a vertex-weighted connected graph $G = (V, E, w)$, the balanced connected graph k-partition is to partition vertex set V into k subsets $V_1, V_2, \ldots, V_k$ such that the subgraph induced by each part is connected and the weights of these k parts are as balanced as possible. In the max-min BCP$_k$ model, the objective is to maximize the minimum weight of the k subsets. In the min-max BCP$_k$ model, the objective is to minimize the maximum weight of the k subsets. Many applications have been found in many areas such as image processing, databases, operating systems, and cluster analysis [3,4,21–23].

For max-min BCP$_k$, Dyer and Frieze [11] proved that it is NP-hard on either bipartite graphs or planar graphs for the uniformly vertex-weighted graph and $k \geq 2$. Chlebíková [9] proved that for any $\epsilon > 0$ it is NP-hard to approximate

W. Wu and H. Du (Eds.): AAIM 2021, LNCS 13153, pp. 27–37, 2021.
https://doi.org/10.1007/978-3-030-93176-6_3

max-min BCP_2 for the uniformly vertex-weighted bipartite graph $G = (V, E)$ with an absolute error guarantee of $|V|^{1-\epsilon}$. Chataigner et al. [4] further proved that max-min BCP_k on k-connected graphs for any fixed $k \geq 2$ is NP-hard and that there is no $(1 + \epsilon)$-approximation algorithm for max-min BCP_2 unless $P = NP$, where $\epsilon \leq \frac{1}{|V|^2}$. In particular, when k is an input, Chataigner et al. [4] showed that max-min BCP_k does not admit an approximation algorithm with a ratio smaller than $\frac{6}{5}$, unless $P = NP$. Chlebíková [9] presented a $\frac{4}{3}$-approximation algorithms for max-min BCP_2. Chataigner et al. [4] presented two 2-approximation algorithms for max-min BCP_3 and max-min BCP_4 on 3-connected graphs and on 4-connected graphs, respectively. Chen et al. [5] presented a $\frac{5}{3}$-approximation for max-min BCP_3 on connected graphs. For max-min BCP_k on grid graphs, Becker et al. [1] proved that it is NP-hard if the graph has at least three rows (the number of columns is arbitrary). Then, Becker et al. [2] presented a polynomial-time algorithm to solve max-min BCP_k on ladders (grid graphs with only two rows). Wu [30] presented a $\frac{5}{4}$-approximation algorithm for the max-min BCP_2 problem on grid graphs and presented a fully polynomial-time approximation scheme (FPTAS) for a fixed number of rows. Wu [28] improved a polynomial-time $\frac{7}{6}$-approximation algorithm for max-min BCP_2 on grid graphs. For max-min BCP_k on interval graphs, Wu [29] proved that it is NP-hard and presented an FPTAS for this problem.

The min-max BCP_k problem is also called the minimum spanning k-forest problem. Dyer and Frieze [11] proved that it is NP-hard on either bipartite graphs or planar graphs for uniformly vertex-weighted images and $k \geq 2$. Chen et al. [5] presented a $\frac{3}{2}$-approximation for the min-max BCP_3 problem, and they also showed that the algorithm by Chlebíková [9] is also a $\frac{5}{4}$-approximation algorithm for the min-max BCP_2 problem. Chen et al. [8] presented another $\frac{3}{2}$-approximation algorithm for the min-max BCP_3 problem and then extended it to become a $\frac{k}{2}$-approximation for the min-max BCP_k problems with any constant $k \geq 3$. Furthermore, they proposed an improved $\frac{24}{13}$-approximation algorithm for the min-max BCP_4 problem. If graph G is a tree, Becker and Perl [3] gave an $O(k^3 rd(T) + k|V|)$-time exact algorithm, where $rd(T)$ is the radius of T. Then, Frederickson [13] and Frederickson and Samson [14] improved the time complexity to $O(|V|)$. More related results can be found in [16, 18, 24, 25].

If graph G is a completed graph, the min-max BCP_k problem is the parallel schedule problem introduced by Graham [15], which is another important research field in operations research and combinatorial optimization. Given a set M of m machines and a set J of n jobs such that each job has to be processed on one of the machines in nonoverlapping and nonpreemptive way, where p_j denotes the time for a machine to process job $J_j \in J$, the parallel schedule problem aims to minimize the maximum load over all machines, where the load of a machine is the total processing time of the jobs assigned to it. Early work scheduling is a new field in scheduling theory, and early work denotes a part of a job executed before its due date [7, 17, 20, 26]. Early work scheduling on parallel machines with a common due date $(P|d_j = d|X)$ schedules n jobs to m identical parallel machines such that the total early work of jobs is maximized, where early work denotes a part of a job executed before the common due date. $P|d_j = d|X$ has many practical applications in control systems, agriculture, manufacturing systems, and software engineering [6,27]. For $P_2|d_j = d|X$, i.e., $m = 2$, Sterna and Czerniachowska [27] proposed a polynomial-time approximation scheme (PTAS) based on structuring problem input. Chen et al. [7] proved that the classical largest processing time first heuristic is a $\frac{10}{9}$-approximation algorithm. Based on a dynamic programming approach, Chen et al. [6] proposed an FPTAS for $P_m|d_j = d|X$ with fixed m. For $P|d_j = d|X$, i.e., m is an input, Györgyi and Kis [17] proposed a PTAS. Recently, Li [20] improved the results of PTAS and FPTAS by Györgyi and Kis [17] and Chen et al. [6], respectively. Choi [10] presented a pseudopolynomial-time algorithm to solve the weighted $P_m|d_j = d|X$ problem with fixed m.

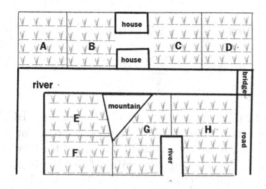

Fig. 1. An example of farmlands with mountains, rivers and houses

In the real world, due to restrictions on mountains, rivers, houses and so on, a harvester may not be able to directly reach adjacent farmlands (see Fig. 1). Therefore, when harvesters harvest crops, we not only need to consider how to harvest the most crops in a given time but also need to consider the route of the harvesters in the farmlands. Motivated by BCP_k and $P|d_j = d|X$, we propose a new problem, called the maximum bounded connected bipartition problem (2-BCBP), which is defined as follows. Given a vertex-weighted connected graph $G = (V, E; w)$ and an upper bound B, the vertex set V is partitioned into two subsets denoted as (V_1, V_2) such that both subgraphs induced by V_1 and V_2 are connected and the total weight of these two subgraphs is maximized, where the weight of the subgraph is the minimum of the sum of the weight of all vertices and B. In this paper, we present an $\frac{8}{7}$-approximation algorithm for this problem. In particular, when G is an interval graph, we present an FPTAS.

The remainder of this paper is structured as follows. In Sect. 2, we provide a formal problem statement. In Sect. 3, for the 2-BCBP, inspired by [9] and [30], we present an $\frac{8}{7}$-approximation algorithm. In Sect. 4, we consider a special case on interval graphs and present an FPTAS, based on a pseudopolynomial-time algorithm, which is a modification of the algorithm in [29].

2 Preliminaries

Let $G = (V, E)$ be a graph and $V' \subseteq V$ be a subset of V. The subgraph induced by V' is denoted by $G[V'] = (V', E')$, where $E' = \{(v, v') \in E | v \in V'$ and $v' \in V'\}$. Graph G is connected if there is a path from v to v' on G for any $v, v' \in V$, and we define that $G[\{v\}]$ is connected for any $v \in V$. In particular, graph G is called biconnected if $G[V \setminus \{v\}]$ is connected for any $v \in V$.

Given a connected graph $G = (V, E; w; B)$, where $w : V \to \mathbb{R}_{\geq 0}$ is a nonnegative vertex weight function, and B is an upper bound. The 2-BCBP partitions the vertex set V into two subsets denoted as (V_1, V_2) such that subgraphs $G[V_1]$ and $G[V_2]$ are connected. The objective is to maximize the weight $w_B(V_1, V_2)$, where

$$w_B(V_1, V_2) = \min\{ \sum_{v:v \in V_1} w(v), B\} + \min\{ \sum_{v:v \in V_2} w(v), B\}.$$

To simplify the notation, we use $w(V')$ to represent the sum cost of the vertices in set V'; i.e.,

$$w(V') = \sum_{v:v \in V'} w(v), \ \forall V' \subseteq V,$$

and let $w(\emptyset) = 0$.

Related to the 2-BCBP, there are two problems max-min BCP_2 and min-max BCP_2: one objective is to maximize the minimum weight of the 2-subset; the other objective is to minimize the maximum weight of the 2-subset. As shown in Fig. 2, for max-min BCP_2, the objective value of (V_1, V_2) is $w(V_2)$; for min-max BCP_2, the objective value of (V_1, V_2) is $w(V_1)$. For the 2-BCBP, the objective value of (V_1, V_2) is $w(V_2) + B$ by $w(V_1) > B$.

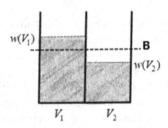

Fig. 2. A bipartition (V_1, V_2) such that both subgraphs $G[V_1]$ and $G[V_2]$ are connected.

For any instance $G = (V, E; w; B)$, let OPT be the optimal value of the 2-BCBP.

Lemma 1. *For any instance* $G = (V, E; w; B)$ *of the 2-BCBP on connected graphs, we have*

$$OPT \leq \min\{w(V), 2B\}.$$

Proof. Let (V_1^*, V_2^*) be an optimal bipartition for the 2-BCBP, so the objective value is

$$OPT = \begin{cases} \min\{w(V_1^*), B\} + \min\{w(V_2^*), B\} \le w(V_1^*) + w(V_2^*) = w(V), \\ \min\{w(V_1^*), B\} + \min\{w(V_2^*), B\} \le B + B = 2B. \end{cases}$$

Thus, the lemma holds.

3   2-BCBP

In this section, based on st-numbering, we first present a linear-time $\frac{8}{7}$-approximation algorithm for the 2-BCBP on biconnected graphs by modifying the algorithms in [30] and [9]. Using the ideas in [9], based on this algorithm, it is easy to obtain a linear-time $\frac{8}{7}$-approximation algorithm for the connected graphs.

The st-numbering method proposed by Lempel et al. [19] provides a definition for connected graphs. Given a biconnected graph $G = (V, E)$ and any edge $(s, t) \in E$, the vertices of G can be labeled from 1 to $|V|$ so that vertex s is labeled as vertex 1 and vertex t is labeled as vertex $|V|$; for any vertex $v \in V \setminus \{s, t\}$, it is adjacent both to a lower-numbered and a higher-numbered vertex. This numbering is called st-numbering for G. Even and Tarjan [12] presented an EJ algorithm to solve the st-numbering problem.

Lemma 2 [12]. *For any $(s, t) \in E$, the EJ algorithm can output st-numbering in linear-time.*

Inspired by [9] and [30], we begin our algorithm by finding two vertices s and t with the largest and the second-largest weights. If $w(s) \ge \frac{1}{2} w(V)$, let $V_1 = \{s\}$ and $V_2 = V \setminus \{s\}$. (V_1, V_2) are output and the algorithm stops. In Lemma 3, we prove that (V_1, V_2) is an optimal bipartition. Otherwise, using the EJ algorithm in [12], an st-numbering scheme is found for G; in particular, edge (s, t) is added to E if $(s, t) \notin E$. For convenience, we rearrange the indices of the vertices in V according to the st-numbering scheme, which satisfies

$$\lambda(v_i) = i, \ \forall v_i \in V.$$

Then, we find an integer k that satisfies

$$w(V(k)) \le \frac{1}{2} w(V) \text{ and } w(V(k+1)) > \frac{1}{2} w(V),$$

where $V(k) = \{v_i | 1 \le i \le k\}$ is the first k vertex in V.

If $w(V(k)) \ge w(V) - w(V(k+1))$, let $V_1 = V(k)$ and $V_2 = V \setminus V(k)$; otherwise, let $V_1 = V(k + 1)$ and $V_2 = V \setminus V(k + 1)$. (V_1, V_2) are output and the algorithm stops. We propose the detailed algorithm in Algorithm 1.

Algorithm 1: [9,30]

Input: A biconnected graph $G = (V, E; w)$ and an upper bound B.
Output: A feasible bipartition (V_1, V_2).

1 Let $s := \arg\max_{v:v\in V} w(v)$ and $t := \arg\max_{v:v\in V\setminus\{s\}} w(v)$. Set $V_1 := \{s\}$.
2 **if** $w(s) \geq \frac{1}{2}w(V)$ **then**
3   Let $V_2 := V\setminus V_1$ and output (V_1, V_2).

4 **if** $(s, t) \notin E$ **then**
5      $E := E \cup \{(s, t)\}$.

6 Using the *EJ* algorithm in [13], find an st-numbering scheme for G, then rearrange the indices of vertices in V according to the st-numbering scheme as above. Set $k := 1$.
7 **while** $w(V_1) \leq \frac{1}{2}w(V)$ **do**
8      $k := k + 1$ and $V_1 := V_1 \cup \{v_k\}$.

9 **if** $w(V_1\setminus\{v_k\}) \geq w(V) - w(V_1)$ **then**
10      $V_1 := V_1\setminus\{v_k\}$.

11 Let $V_2 := V\setminus V_1$ and output (V_1, V_2).

Lemma 3. *For any vertex v in a biconnected graph G, if $w(v) \geq \frac{1}{2}w(V)$, then $(\{v\}, V - \{v\})$ is an optimal bipartition for the 2-BCBP.*

Proof. Since graph G is biconnected, for any vertex $v \in V$, $G[V\setminus\{v\}]$ is connected for any $v \in V$, which means that $(\{v\}, V - \{v\})$ is a feasible bipartition for the 2-BCBP.

Let v^* be the vertex with $w(v^*) \geq \frac{1}{2}w(V)$. If $w(v^*) \leq B$, then $w(V\setminus\{v^*\}) \leq \frac{1}{2}w(V) \leq B$ and

$$w_B(\{v^*\}, V\setminus\{v^*\}) = \min\{w(v^*), B\} + \min\{w(V\setminus\{v^*\}), B\} = w(V).$$

This statement and Lemma 1 imply that $(\{v^*\}, V\setminus\{v^*\})$ is an optimal bipartition for the 2-BCBP. Otherwise, for $w(v^*) > B$, let (V_1, V_2) be a feasible bipartition for the 2-BCBP. Without loss of generality, we assume that $v^* \in V_1$. Then, the objective value of (V_1, V_2) is

$$w_B(V_1, V_2) = \min\{ \sum_{v:v\in V_1} w(v), B\} + \min\{ \sum_{v:v\in V_2} w(v), B\}$$

$$\leq B + \min\{ \sum_{v:v\in V_2} w(v) + \sum_{v:v\in V_1\setminus\{v^*\}} w(v), B\}$$

$$= w_B(\{v^*\}, V\setminus\{v^*\})$$

This means that $(\{v^*\}, V\setminus\{v^*\})$ is an optimal bipartition for the 2-BCBP. Therefore, the lemma holds.

Formally, we have the following theorem, and the proof is placed in Appendix, due to the limitation of space.

Theorem 1. *For the 2-BCBP on a biconnected graph, Algorithm 1 is a linear-time $\frac{8}{7}$-approximation algorithm.*

4   2-BCBP on an Interval Graph

In this section, we consider the 2-BCBP on interval graphs and present an FPTAS based on a pseudopolynomial-time algorithm, which is a modification of the algorithm in [29].

An interval is usually defined by $[l, r]$, where $l < r$ and l(or r) is the left (or right) endpoint of the interval. A graph $G = (V, E)$ is an interval graph if each of the vertices can be represented by an interval such that there is an edge between the two vertices if and only if the two intervals intersect. We assume that the input interval graph is given by its interval representation and $V = \{v_i \in [l_i, r_i] | 1 \le i \le n\}$ such that $l_i \le l_{i+1}$ for $1 \le i < n$, see Fig. 3.

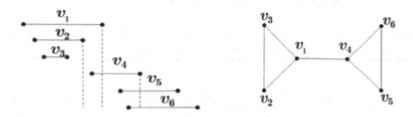

Fig. 3. An example of interval graph

Our algorithm is basically the same as the algorithm in [29], except for the way in which the output solution is determined. Consistent with the symbols in [29], let $l(v_i)$ and $r(v_i)$ be the left and right endpoints represented by the interval of vertex v_i, respectively. For any $V' \subseteq V$, let

$$r(V') = \max\{r(v_i) | v_i \in V'\}.$$

Let $V(k) = \{v_1, v_2, \dots v_k\}$ be the first k vertices in V. For any k, let $(V_1(k), V_2(k))$ be a feasible bipartition of $G[V(k)]$ induced by $V(k)$. Assuming $r(V_1(k)) \le r(V_2(k))$, we can use the pair (x, y) to record this feasible bipartition, where $x = r(V_1(k))$ and $y = w(V_1(k))$. Then, we present a pseudopolynomial-time algorithm to generate the set $\mathcal{P}_n$ of all possible pairs for feasible bipartition of $G[V(n)]$, where $w(v_i)$ is an integer for any $v_i \in V$.

Since $w(v_i)$ is an integer for any $v_i \in V$, set $\mathcal{P}_i$ contains at most $nw(V)$ pairs for any $i \in \{1, 2, \dots, n\}$. Each pair can be found in $O(1)$, so the running time of Algorithm 2 is $O(n^2 w(V))$. Thus, the following lemma is obvious.

Lemma 4. *When $w(v_i)$ is an integer for any $v_i \in V$, The 2-BCBP on an interval graph can be solved in time $O(n^2 w(V))$.*

Theorem 2. *For the 2-BCBP on an interval graph, there is an FPTAS with running time $O((1/\varepsilon)n^3)$.*

Algorithm 2: [29]

Input: An interval graph $G = (V, E; w)$ and an upper bound B, where $w(v_i)$ is an integer for any $v_i \in V$.

Output: A feasible bipartition (V_1, V_2).

1 Initially, set $\mathcal{P}_1 = \{(0,0)\}$ and $\mathcal{P}_i = \emptyset, \forall i = 2, 3, \ldots, n$.

2 **for** $i = 2$ *to* n **do**

3   **for** $(x, y) \in \mathcal{P}_{i-1}$ **do**

4   add pair (x, y) to $\mathcal{P}_i$

5   **if** $l(v_i) \leq x$ *or* $x = 0$ **then**

6   **if** $r(v_i) < r(V(i-1))$ **then**

7   add pair $(\max\{x, r(v_i)\}, y + w(v_i))$ to $\mathcal{P}_i$

8   **else**

9   add pair $(r(V(i-1)), w(V(i-1)) - y)$ to $\mathcal{P}_i$

10 Let $(x', y') := \arg\max_{(x,y):(x,y)\in\mathcal{P}_n} \min\{y, B\} + \min\{w(V) - y, B\}$ and output the bipartition (V_1, V_2) corresponding to the pair (x', y').

Proof. Given a reduction factor $f = \frac{\rho B}{2n}$. For any instance $G = (V, E, w; B)$, construct an auxiliary interval graph $G' = (V, E, w'; B)$, where $w'(v) = \lfloor \frac{w(v)}{f} \rfloor$. Using Algorithm 2, we can obtain an optimal bipartition (V_1, V_2) of G'. Then, we have

$$\min\{w'(V_1), B\} + \min\{w'(V_2), B\} \geq \min\{w'(V_1^*), B\} + \min\{w'(V_2^*), B\}, \quad (1)$$

where (V_1^*, V_2^*) is the optimal bipartition of $G = (V, E, w; B)$.

Since $\frac{w(v)}{f} \geq w'(v) = \lfloor \frac{w(v)}{f} \rfloor > \frac{w(v)}{f} - 1$, we have $w(v) \geq fw'(v) \geq w(v) - f$, $\forall v \in V$ and

$$
\begin{aligned}
OUT &= \min\{w(V_1), B\} + \min\{w(V_2), B\} \\
&\geq \min\{fw'(V_1^*), B\} + \min\{fw'(V_2^*), B\} \\
&\geq \min\{w(V_1^*), B\} + \min\{w(V_2^*), B\} - 2fn \\
&= OPT - 2fn,
\end{aligned}
$$

where the first inequality follows from inequality (1), and OPT is the objective value of the optimal bipartition (V_1^*, V_2^*).

If $w(V) \leq B$, for any feasible bipartition (V_1', V_2') of G, we have $w(V_1) \leq B$ and $w(V_2) \leq B$. Thus, we have

$$\min\{w(V_1'), B\} + \min\{w(V_2'), B\} = w(V_1) + w(V_2) = w(V),$$

and (V_1', V_2') is an optimal bipartition of G by Lemma 1. Otherwise, $w(V) > B$. Since $(\emptyset, V)$ is a feasible bipartition of G, we have $OPT \geq \min\{0, B\} + \min\{w(V), B\} = B$ and

$$f = \frac{\rho B}{2n} \leq \frac{\rho OPT}{2n}.$$

Thus,

$$OUT \geq OPT - 2fn \geq OPT - \rho B \geq OPT - \rho OPT = (1 - \rho)OPT.$$

For any $\varepsilon > 0$, we set $\rho = \frac{\varepsilon}{1+\varepsilon}$; then,

$$\frac{OPT}{OUT} \leq \frac{1}{1 - \rho} = 1 + \varepsilon.$$

By Lemma 4, the time complexity is

$$O(n^2 w'(V)) = O(n^2 \frac{w(V)}{f}) = O(n^2 \frac{w(V)}{\rho B/(2n)}) = O\left((1/\varepsilon)n^3\right).$$

5   Conclusion

In this paper, we consider the 2-BCBP, which is a hybrid variant of the maximum balanced connected partition problem on connected graphs and the maximum total early work problem in scheduling theory. We present an $\frac{8}{7}$-approximation algorithm. In particular, we consider the 2-BCBP on interval graphs and present an FPTAS.

The topic could be further studied in the following ways. It is also challenging to design a better algorithm for the 2-BCBP. The grid graph and bipartite graph for this problem are worth considering, since graphs may have different properties in the real world. Moreover, for the maximum bounded connected multiple partitions problem, algorithms under this setting could be further developed.

Acknowledgements. The work is supported in part by the National Natural Science Foundation of China [No. 12071417], and Project for Innovation Team (Cultivation) of Yunnan Province [No. 202005AE160006].

References

1. Becker, R., Lari, I., Lucertini, M., Simeone, B.: Max-min partitioning of grid graphs into connected components. Networks **32**, 115–125 (1998)
2. Becker, R., Lari, I., Lucertini, M., Simeone, B.: A polynomial-time algorithm for max-min partitioning of ladders. Theory Comput. Syst. **34**, 353–374 (2001)
3. Becker, R., Perl, Y.: Shifting algorithms for tree partitioning with general weighting functions. J. Algorithms 4(2), 101–120 (1983)
4. Chataigner, F., Salgado, L., Wakabayashi, Y.: Approximation and inapproximability results on balanced connected partitions of graphs. Discrete Math. Theor. Comput. Sci. **9**, 177–192 (2007)
5. Chen, G., Chen, Y., Chen, Z.-Z., Lin, G., Liu, T., Zhang, A.: Approximation algorithms for the maximally balanced connected graph tripartition problem. J. Comb. Optim., 1–21 (2020). https://doi.org/10.1007/s10878-020-00544-w
6. Chen, X., Liang, Y., Sterna, M., Wang, W., Blazewicz, J.: Fully polynomial time approximation scheme to maximize early work on parallel machines with common due date. Eur. J. Oper. Res. **284**, 67–74 (2020)

7. Chen, X., Wang, W., Xie, P., Zhang, X., Sterna, M., Blazewicz, J.: Exact and heuristic algorithms for scheduling on two identical machines with early work maximization. Comput. Ind. Eng. **144**, Article No. 106449 (2020)
8. Chen, Y., Chen, Z.-Z., Lin, G., Xu, Y., Zhang, A.: Approximation algorithms for maximally balanced connected graph partition. In: Li, Y., Cardei, M., Huang, Y. (eds.) COCOA 2019. LNCS, vol. 11949, pp. 130–141. Springer, Cham (2019). https://doi.org/10.1007/978-3-030-36412-0_11
9. Chlebíková, J.: Approximating the maximally balanced connected partition problem in graphs. Inf. Process. Lett. **60**, 225–230 (1996)
10. Choi, B., Park, M., Kim, K., Min, Y.: A parallel machine scheduling problem maximizing total weighted early work. Asia-Pac. J. Oper. Res. Article No. 2150007 (2021)
11. Dyer, M., Frieze, A.: On the complexity of partitioning graphs into connected subgraphs. Discret. Appl. Math. **10**, 139–153 (1985)
12. Even, S., Tarjan, R.: Computing an ST-numbering. Theoret. Comput. Sci. **2**, 339–344 (1976)
13. Frederickson, G.: Optimal algorithms for tree partitioning. In: Symposium on Discrete Algorithms, pp. 168–177 (1991)
14. Frederickson, G., Samson, Z.: Optimal parametric search for path and tree partitioning. arXiv:abs/1711.00599 (2017)
15. Graham, R.L.: Bounds for certain multiprocessing anomalies. Bell Syst. Tech. J. **45**(9), 1563–1581 (1966)
16. Guan, L., Li, W., Xiao, M.: Online algorithms for the mixed ring loading problem with two nodes. Optim. Lett. **15**(4), 1229–1239 (2020). https://doi.org/10.1007/s11590-020-01632-w
17. Györgyi, P., Kis, T.: A common approximation framework for early work, late work, and resource leveling problems. Eur. J. Oper. Res. **286**(1), 129–137 (2020)
18. Jana, S., Pandit, S., Roy, S.: Balanced connected graph partition. In: Mudgal, A., Subramanian, C.R. (eds.) CALDAM 2021. LNCS, vol. 12601, pp. 487–499. Springer, Cham (2021). https://doi.org/10.1007/978-3-030-67899-9_38
19. Lempel, A., Even, S., Cederbaum, I.: An algorithm for planarity testing of graphs. In: Rosenstiehl, P. (ed.) International Symposium 1966, Theory of Graphs, pp. 215–232. Gordon and Breach, New York; Dunod, Paris (1966)
20. Li, W.: Improved approximation schemes for early work scheduling on identical parallel machines with common due date. arXiv:abs/2007.12388 (2020)
21. Lucertini, M., Perl, Y., Simeone, B.: Image enhancement by path partitioning. In: Cantoni, V., Creutzburg, R., Levialdi, S., Wolf, G. (eds.) PAR 1988. LNCS, vol. 399, pp. 12–22. Springer, Heidelberg (1989). https://doi.org/10.1007/3-540-51815-0_37
22. Lucertini, M., Perl, Y., Simeone, B.: Most uniform path partitioning and its use in image processing. Discret. Appl. Math. **42**(2–3), 227–256 (1993)
23. Maravalle, M., Simeone, B., Naldini, R.: Clustering on trees. Comput. Stat. Data Anal. **24**(2), 217–234 (1997)
24. Miyazawa, F., Moura, P., Ota, M., Wakabayashi, Y.: Partitioning a graph into balanced connected classes: formulations, separation and experiments. Eur. J. Oper. Res. **293**, 826–836 (2021)
25. Soltan, S., Yannakakis, M., Zussman, G.: Doubly balanced connected graph partitioning. ACM Trans. Algorithms **16**, 1–24 (2020)
26. Sterna, M.: Late and early work scheduling: a survey. Omega-Int. J. Manag. Sci. **104**(10), Artical No. 102453 (2021)

27. Sterna, M., Czerniachowska, K.: Polynomial time approximation scheme for two parallel machines scheduling with a common due date to maximize early work. J. Optim. Theory Appl. **174**(3), 927–944 (2017)
28. Wu, B.Y.: A 7/6-approximation algorithm for the max-min connected bipartition problem on grid graphs. In: Akiyama, J., Bo, J., Kano, M., Tan, X. (eds.) CGGA 2010. LNCS, vol. 7033, pp. 188–194. Springer, Heidelberg (2011). https://doi.org/10.1007/978-3-642-24983-9_19
29. Wu, B.: Fully polynomial time approximation schemes for the max-min connected partition problem on interval graphs. Discret Math. Algorithm Appl. **4**, Artical No. 1250005 (2012)
30. Wu, B.: Algorithms for the minimum non-separating path and the balanced connected bipartition problems on grid graphs. J. Comb. Optim. **26**, 592–607 (2013)

An Approximation Algorithm for Solving the Heterogeneous Chinese Postman Problem

Jianping Li[1]([✉]), Lijian Cai[1], Junran Lichen[2], Pengxiang Pan[1], Wencheng Wang[1], and Suding Liu[1]

[1] Department of Mathematics, Yunnan University, East Outer Ring South Road, University Town, Kunming 650504, People's Republic of China
jianping@ynu.edu.cn
[2] Institute of Applied Mathematics, Academy of Mathematics and Systems Science, No. 55, Zhongguancun East Road, Beijing 100190, People's Republic of China

Abstract. We consider the heterogeneous Chinese postman problem (the HCPP), which is modelled as follows. Given a weighted graph $G = (V, E; w; r)$ with a length function $w : E \to R^+$ satisfying the triangle inequality, a fixed depot $r \in V$, and k vehicles having k nonuniform speeds $\lambda_1, \lambda_2, \ldots, \lambda_k$, respectively, we are asked to find k tours for these k vehicles, each starting and ending at the same depot r, such that every edge in G is traversed at least once by one of these k vehicles. The objective is to minimize the maximum completion time, where the completion time of each vehicle is the total traveling length divided by its speed.

In this paper, we design a $20.8765(1 + \delta)$-approximation algorithm to solve the HCPP for any small number $\delta > 0$.

Keywords: Combinatorial optimization · Chinese postman tours · Nonuniform speeds · Approximation algorithm

1 Introduction

The capacitated vehicle routing problem, especially including the traveling salesman problem (the TSP) and the Chinese postman problem (the CPP), form a much-studied family of combinatorial optimization problems, due to their many important practical applications [5, 9, 10], such as routing of street sweepers, snow

This paper is supported by the National Natural Science Foundation of China [Nos. 11861075, 12101593], Project for Innovation Team (Cultivation) of Yunnan Province [No. 202005AE160006], Jointed Key Project of Yunnan Provincial Science and Technology Department and Yunnan University [No. 2018FY001014] and Program for Innovative Research Team (in Science and Technology) in Universities of Yunnan Province [No. C176240111009]. Jianping Li is also supported by Project of Yunling Scholars Training of Yunnan Province.

W. Wu and H. Du (Eds.): AAIM 2021, LNCS 13153, pp. 38–46, 2021.
https://doi.org/10.1007/978-3-030-93176-6_4

plows, household refuse collection vehicles, the spraying of roads with salt-grit to prevent ice formation, the inspection of electric power lines, and gas or oil pipelines for faults.

The heterogeneous traveling salesman problem (the HTSP) addressed in 2016 by Gørtz et al. [6] is one of the important capacitated vehicle routing problems, which is modelled as follows. Given a complete weighted graph $G = (V, E)$ equipped with a length function $w : E \to R^+$ that satisfies the triangle inequality, a fixed depot $r \in V$ and k vehicles that have nonuniform speeds $\lambda_1, \lambda_2, \ldots, \lambda_k$, respectively, it is asked to find k cycles for these k vehicles, each starting and ending at the same depot r, such that every vertex in G is traveled at least once by one of these k vehicles. The objective is to minimize the maximum completion time, where the completion time of a vehicle is the total traveling length divided by its speed. Using a new approximate minimum spanning tree construction, Gørtz et al. [6] presented a constant factor approximation algorithm to solve the HTSP.

The metric traveling salesman problem ($\triangle$-TSP or metric-TSP) [11] is a special case of the HTSP, where $k = 1$, $i.e.$, only using one vehicle. Using techniques to solve the minimum spanning tree problem [8] and the minimum perfect matching problem [3] specialized to weighted graphs that have the triangle inequality, Christofides [1] in 1976 designed a 1.5-approximation algorithm to solve the $\triangle$-TSP. The k-traveling salesman problem (the k-TSP) originally considered in 1978 by Frederickson et al. [4] is also a special case of the HTSP, where $\lambda_i = 1$ for each $i \in \{1, 2, \ldots, k\}$, $i.e.$, all vehicles having a uniform speed. Using a splitting technique, Frederickson et al. [4] designed a $(\frac{5}{2} - \frac{1}{k})$-approximation algorithm to solve the k-TSP.

The k-Chinese postman problem (the k-CPP) originally addressed in 1978 by Frederickson et al. [4] is modelled as follows. Given a weighted graph $G = (V, E; w)$ equipped with a length function $w : E \to R^+$ that satisfies the triangle inequality, a fixed depot $r \in V$ and k vehicles, where we may assume that these k vehicles have uniform speeds, it is asked to find a tour for each vehicle, starting and ending at the depot r, such that every edge in G is traveled at least once by one of these k vehicles. The objective is to minimize the maximum total length among these k tours, where the total length of a vehicle is the sum of edge lengths that vehicle travels in that tour. Equivalently, the objective is to minimize the maximum completion time, where the completion time of a vehicle is the total traveled length divided by its speed for the case of these k vehicles having uniform speeds. The 1-CPP is exactly the Chinese postman problem (the CPP) addressed in 1960 by Guan [7]. Using an efficient algorithm for solving the maximum weighted matching problem [2], Edmonds and Johnson [3] in 1973 designed an optimal algorithm to solve the CPP in polynomial time. Using a splitting technique, Frederickson et al. [4] presented a $(2 - \frac{1}{k})$-approximation algorithm to solve the general k-CPP.

Motivated by some applications in reality practices of the HTSP, it is natural for us to consider one problem similar to a generalization of the k-CPP and the HTSP, where these k vehicles have nonuniform speeds. We address the heterogeneous Chinese postman problem (the HCPP), which is modelled as follows. Given a weighted graph $G = (V, E; w)$ equipped with a length function $w : E \to R^+$ that satisfies the triangle inequality, a fixed depot $r \in V$ and k vehicles that have k nonuniform speeds $\lambda_1, \lambda_2, \ldots, \lambda_k$, respectively, it is asked to find k tours $\mathcal{C} = \{C_i \mid i = 1, 2, \ldots, k\}$ for these k vehicles, each starting and ending at the same depot r, such that every edge in G is traversed at least once by one of these k vehicles. The objective is to minimize the maximum completion time, where the completion time of each vehicle is the total traveling length divided by its speed. In a mathematical way, the makespan objective is to minimize the value of $\max\{\frac{w(C_i)}{\lambda_i} \mid i = 1, 2, \ldots, k\}$.

As far as what we have known, the HCPP has not been considered in the literature. Given any small number $\delta > 0$, we shall design a $20.8765(1 + \delta)$-approximation algorithm to solve the HCPP.

This paper is organized as follows. In Sect. 2, we present some terminologies and fundamental lemmas to ensure the correctness of our algorithm; In Sect. 3, we design an approximation algorithm with constant factor to solve the HCPP; In Sect. 4, we provide our conclusion and further research.

2   Terminologies and Fundamental Lemmas

Given a graph $G = (V, E)$, a new graph $G' = (V', E')$ is called a subgraph of G if $V' \subseteq V$ and $E' \subseteq E$. If $E' = \{uv \in E \mid u, v \in V'\}$, the subgraph $G' = (V', E')$ is called the subgraph induced by V', and denoted by $G[V']$. Similarly, if $V' = \{u, v \in V \mid$ there exists an edge $uv \in E'\}$, then the subgraph $G' = (V', E')$ is called the subgraph induced by E', and denoted by $G[E']$.

Given a graph $G = (V, E)$, a walk P connecting a vertex v_{i_1} and a vertex $v_{i_{k+1}}$ is an alternating sequence $\pi := v_{i_1} e_{i_1} v_{i_2} e_{i_2} v_{i_3} \cdots v_{i_k} e_{i_k} v_{i_{k+1}}$ such that, for each integer $1 \leq j \leq k$, two end-vertices of edge e_{i_j} are v_{i_j} and $v_{i_{j+1}}$, and we may simply write this walk as $P_{v_{i_1}, v_{i_{k+1}}} = v_{i_1} v_{i_2} v_{i_3} \cdots v_{i_k} v_{i_{k+1}}$. A walk P is called a tour with k edges if $v_{i_1} = v_{i_{k+1}}$. In addition, a walk P is called a path if the vertices in P are all distinct, and we may simply denote this path P as a v_{i_1}-$v_{i_{k+1}}$ path in G.

A graph $G = (V, E)$ is called connected if, for any two vertices x and y in V, there exists a path P_{xy} connecting x and y, where each edge $e = uv$ on such a path P_{xy} may be traversed either from u to v or from v to u on that edge e. The maximal connected subgraphs of a graph G are called its connected components. When G is a weighed graph, for an x-y path P_{xy}, we denote by $w(P_{xy})$ the total length of all edges in P_{xy}, and denote by $d(x, y)$ the length of a shortest path from x to y, i.e., $d(x, y) = \min\{w(P_{xy}) \mid P_{xy}$ is a path connecting x and $y\}$.

When we plan to design some approximation algorithm to solve the HCPP, we shall use the following definition in [6], which is called the assignable subtours.

Definition 1 [6]. *Given a weighted graph $G = (V, E; w)$ with two constants $M > 0$ and $\varepsilon > 0$, where $w : E \to R^+$, let T_i be a set of tours in G, each starting and ending at the same vertex r, for each integer $i \geq 0$. Then the collection*

$\bigcup_{i \geq 0} T_i$ is said to be $(\alpha, \beta)_{M,\varepsilon}$-assignable if this collection has the following two properties

(1) $w(C) \leq \alpha \cdot (1+\varepsilon)^i M$ for each $i \geq 0$ and every $C \in T_i$;
(2) $\sum_{j \geq i} w(T_i) \leq \beta M \cdot \Lambda((1+\varepsilon)^{i-1})$ for each $i \geq 0$, where $\Lambda((1+\varepsilon)^{i-1})$ is the sum of speeds that are at least $(1+\varepsilon)^{i-1}$ and $w(T_i) = \sum_{C \in T_i} w(C) = \sum_{C \in T_i} \sum_{e \in C} w(e)$.

For convenience, we denote $\{T_i\}_{i \geq 0}$ instead of $\bigcup_{i \geq 0} T_i$.
Using Definition 1 and the ASSIGN algorithm [6], we can obtain the following

Lemma 1 [6]. *Given an $(\alpha, \beta)_{M,\varepsilon}$-assignable collection $\{T_i\}_{i \geq 0}$ of tours, each starting and ending at the same depot r, the ASSIGN algorithm can assign $\{T_i\}_{i \geq 0}$ in polynomial time to k vehicles, each of which starts and ends at the same depot r, such that the completion time of any vehicle is at most $((1+\varepsilon)\alpha + \beta)M$.*

3   The Heterogeneous Chinese Postman Problem

In this section, we consider the heterogeneous Chinese postman problem (the HCPP). Without loss of generality, we may assume that graphs considered are all connected, otherwise there is no feasible solution to the HCPP.

By Lemma 1, we want to use the following strategies to design an algorithm to solve the HCPP, *i.e.*,

(1) Find an $(\alpha, \beta)_{M,\varepsilon}$-assignable collection of subtours that cover all edges;
(2) Assign subtours in the collection to k vehicles such that the completion time on any vehicle is minimized as soon as possible.

Given a weighted graph $G = (V, E; w; k)$ with two current values M and ε (the precise value to be fixed later), we may partition the set V into subsets $V_1, V_2, \ldots$, where $V_0 = \{v \in V \mid d(r, v) \leq M\}$, and $V_i = \{v \in V \mid (1+\varepsilon)^{i-1}M < d(r, v) \leq (1+\varepsilon)^i M\}$ for each $i \geq 1$. Similarly, we may partition the set E into subsets $E_0, E_1, E_2, \ldots$, where $E_0 = \{rv \in E \mid d(r, v) \leq M\}$, and $E_i = \{vv' \in E \mid d(r, v) \leq (1+\varepsilon)^i M$ and $(1+\varepsilon)^{i-1}M < d(r, v') \leq (1+\varepsilon)^i M\}$ for each $i \geq 1$. For each $i \geq 1$, we denote $V_{\leq i} := \bigcup_{j \leq i} V_j$, $V_{\geq i} := \bigcup_{j \geq i} V_j$, $E_{\leq i} := \bigcup_{j \leq i} E_j$ and $E_{\geq i} := \bigcup_{j \geq i} E_j$.

Our approximation algorithm, denoted by the HCP algorithm, to solve the HCPP is described in details as follows.

Algorithm: HCP
INPUT: A weighted graph $G = (V, E; w; k)$ with a depot $r \in V$, a length function
    $w : E \to R^+$, k vehicles having nonuniform speeds $\lambda_1, \lambda_2, \ldots, \lambda_k$ and $\varepsilon > 0$;
OUTPUT: A set $\mathcal{C} = \{C_i \mid i = 1, 2, \ldots, k\}$ of k tours in G.
Begin
Step 1 Set $M = \max_{vv' \in E}\{\frac{d(r,v)+w(vv')+d(v',r)}{\lambda_{\max}}\}$, where $\lambda_{\max} = \max\{\lambda_i \mid i = 1, 2, \ldots, k\}$;

Step 2 Using two current values M and ε (the precise value to be fixed later), we may partition the edge-set E into subsets $E_0, E_1, E_2, \ldots$ as mentioned above; For convenience, we may assume that the number of subsets partitioned is t, i.e., $(1+\varepsilon)^{t-1}M < \max\{d(r,v) \mid v \in V\} \leq (1+\varepsilon)^t M$;

Step 3 If $(w(E_{\geq l}) > M \cdot \Lambda((1+\varepsilon)^{l-1}))$ holds for some $l \in \{1, 2, \ldots, t\}$) then
      Set $M := (1+\delta)M$, where $\delta > 0$ is a small constant, and go to Step 2;

Step 4 Set $\mathcal{S}_0^m \ (= \mathcal{S}_0) = \{r\} \cup G[E_0]$, $\gamma = \frac{\varepsilon}{(2+\varepsilon)(1+\varepsilon)}$;

Step 5 For all $i \in \{1, 2, \ldots, t\}$, determine all connected components, denoted by $\mathcal{S}_i$, in $G[E_i]$, and set $\mathcal{S}_i^m = \emptyset$ and $\mathcal{S}_i^u = \emptyset$;

Step 6 For all $i \in \{1, 2, \ldots, t\}, \eta \in \mathcal{S}_i$ do:
    If $(w(\eta) \geq \gamma \cdot (1+\varepsilon)^i M)$ then
      Set $\mathcal{S}_i^m := \mathcal{S}_i^m \cup \{\eta\}$;
    Else
      Set $\mathcal{S}_i^u := \mathcal{S}_i^u \cup \{\eta\}$;

Step 7 For all $i \in \{1, 2, \ldots, t\}, \sigma \in \mathcal{S}_i^u$ do:
    Determine a connected component $\pi(\sigma)$ in $G[E_{i-1}]$, having $\pi(\sigma) \cap \sigma \neq \emptyset$;

Step 8 For all $i \in \{0, 1, 2, \ldots, t-1\}, \tau \in \mathcal{S}_i^m$ do:
    (8.1) Set $\text{Dangle}(\tau) = \{\sigma \in \mathcal{S}_{i+1}^u \mid \pi(\sigma) = \tau\}$;
    (8.2) Find an Eulerian tour of $(\tau \cup \text{Dangle}(\tau))$ by "doubling" its edges, and split the resulting Eulerian tour into maximal paths of total length at most $2(1+\varepsilon)^{i+1}M$ each, denote them in turn by $P_1, P_2, \ldots, P_q$;
    (8.3) For each $j \in \{1, 2, \ldots, q\}$, augment P_j by adding two shortest paths from r to the end vertices of P_j (when $q = 1$, then both end-vertices are defined as the vertex closest to r in P_j), to obtain a set of subtours, each starting and ending at the same depot r, denote it by $\mathcal{T}_i(\tau)$;

Step 9 For each $i \in \{0, 1, \ldots, t\}$, set $\mathcal{T}_i = \bigcup_{\tau \in \mathcal{S}_i^m} \mathcal{T}_i(\tau)$; Using the ASSIGN algorithm [6], we may combine $\{\mathcal{T}_i\}_{i \geq 0} = \bigcup_{i \geq 0} \mathcal{T}_i$ into k tours $\mathcal{C} = \{C_i \mid i = 1, 2, \ldots, k\}$, corresponding to k vehicles;

Step 10 Output "k tours $\mathcal{C} = \{C_i \mid i = 1, 2, \ldots, k\}$ corresponding to k vehicles".
End

By the HCP algorithm, we can obtain the following

Lemma 2. *Given a weighted graph $G = (V, E; w; k)$ with a depot $r \in V$ as an instance of the HCPP, if $OPT \leq M$, then $w(E_{\geq l}) \leq M \cdot \Lambda((1+\varepsilon)^{l-1})$ holds for each integer $l \geq 0$, where OPT is an optimal value of this weighted graph $G = (V, E; w; k)$, and $\Lambda((1+\varepsilon)^{l-1})$ is the sum of speeds, each of which is at least $(1+\varepsilon)^{l-1}$.*

Proof. Consider an edge $e = vv' \in E_{\geq l}$, we assume that e is traveled by a vehicle of speed λ' in an optimal solution for the HCPP. Since $\max\{d(r,v), d(r,v')\} \geq (1+\varepsilon)^{l-1}M$, we have $\lambda' \cdot OPT \geq \max\{d(r,v), d(r,v')\} \geq (1+\varepsilon)^{l-1}M$. Since $M \geq OPT$, we deduce that a vehicle of speed λ' travels distance at most $\lambda' \cdot OPT \leq \lambda' \cdot M$. Thus, we obtain $\lambda' \cdot M \geq \lambda' \cdot OPT \geq (1+\varepsilon)^{l-1}M$, implying $\lambda' \geq (1+\varepsilon)^{l-1}$.

In an optimal solution, it is clear that every edge in $E_{\geq l}$ must be traveled by some vehicle, and then we have $w(E_{\geq l}) \leq OPT \cdot \Lambda((1+\varepsilon)^{l-1})$. Since $OPT \leq M$, we obtain $OPT \cdot \Lambda((1+\varepsilon)^{l-1}) \leq M \cdot \Lambda((1+\varepsilon)^{l-1})$, then we have $w(E_{\geq l}) \leq OPT \cdot \Lambda((1+\varepsilon)^{l-1}) \leq M \cdot \Lambda((1+\varepsilon)^{l-1})$, implying $w(E_{\geq l}) \leq M \cdot \Lambda((1+\varepsilon)^{l-1})$.

This completes the proof of the lemma. $\qquad\square$

By executing Steps 6–7 in the HCP algorithm, we can obtain the following

Lemma 3. *For any $i \geq 1$ and $\sigma \in \mathcal{S}_i^u$, there exists $\pi(\sigma) \in \mathcal{S}_{i-1}$. Moreover, $\pi(\sigma) \in \mathcal{S}_{i-1}^m$.*

Proof. For an element $\sigma \in \mathcal{S}_i^u$, it is clear that $w(\sigma) < \gamma \cdot (1+\varepsilon)^i M$. By the definition of $\mathcal{S}_i^u$, we have $v \in V_{\leq i}$ for every $v \in V(\sigma)$, implying $v \in V_i \cup V_{i-1}$. Otherwise, we may assume that $vv' \in E(\sigma)$, where $v \in V_i$ and $v' \in V_{<i-1}$, it follows that $w(\sigma) \geq w(vv') \geq d(r, v) - d(r, v') > (1+\varepsilon)^{i-1}M - (1+\varepsilon)^{i-2}M > \gamma \cdot (1+\varepsilon)^i M$, which contradicts the fact $\sigma \in \mathcal{S}_i^u$. This shows that $v \in V_i \cup V_{i-1}$ for every $v \in V(\sigma)$.

Since the given graph G is connected, we conclude that there is some edge $yx \in E(\sigma)$ satisfying $y \in V_{\leq i-1}$ and $x \in V_i$, which implies $y \in V_{i-1}$ (due to $v \in V_i \cup V_{i-1}$ for every $v \in V(\sigma)$). Hence, there exists an element $\pi(\sigma) \in \mathcal{S}_{i-1}$ to satisfy $\pi(\sigma) \cap \sigma \neq \emptyset$.

Secondly, we shall prove $\pi(\sigma) \in \mathcal{S}_{i-1}^m$. When $i = 1$, we have $\pi(\sigma) = \mathcal{S}_0$, and it is obvious to obtain that $\pi(\sigma) \in \mathcal{S}_0^m$. When $i \geq 2$, from the arguments above, we obtain $\pi(\sigma) \in \mathcal{S}_{i-1}$, thus there is $z \in \pi(\sigma)$ that satisfies $z \in V_{<i-1}$. Using the triangle inequality twice, we obtain the following

$$d(z, y) + d(y, x) \geq d(z, x) \geq d(r, x) - d(r, z).$$

Since $x \in V_i$ and $z \in V_{<i-1}$, we have $d(r, x) - d(r, z) > (1+\varepsilon)^{i-1}M - (1+\varepsilon)^{i-2}M = \varepsilon(1+\varepsilon)^{i-2}M$, implying $d(z, y) + d(y, x) > \varepsilon(1+\varepsilon)^{i-2}M$. Since $\sigma \in \mathcal{S}_i^u$, we have $d(y, x) = w(yx) \leq w(\sigma) < \gamma \cdot (1+\varepsilon)^i M$. Thus, we obtain the following

$$w(\pi(\sigma)) \geq d_{\pi(\sigma)}(z, y) \geq d(z, y) > \varepsilon(1+\varepsilon)^{i-2}M - \gamma \cdot (1+\varepsilon)^i M = \gamma \cdot (1+\varepsilon)^{i-1}M,$$

implying $\pi(\sigma) \in \mathcal{S}_{i-1}^m$.

This completes the proof of the lemma. $\qquad\square$

By executing Step 8 in the HCP algorithm and using the similar argument as in [6], we determine the two following lemmas.

Lemma 4. *For any $C \in \mathcal{T}_i(\tau)$, we have $w(C) \leq (4+4\varepsilon)(1+\varepsilon)^i M$.*

Proof. Note that every tour $C \in \mathcal{T}_i(\tau)$ consists of a path P_j (for some $1 \leq j \leq q$) and shortest paths from r to both end-vertices of P_j. Based on the construction of P_j, we have $w(P_j) \leq 2(1+\varepsilon)^{i+1}M$. Since P_j only includes edges in $E_{\leq i+1}$, it follows that the total length of two additional shortest paths is at most $2(1+\varepsilon)^{i+1}M$. Thus, we obtain $w(C) \leq 4(1+\varepsilon)^{i+1}M = (4+4\varepsilon)(1+\varepsilon)^i M$. $\qquad\square$

Lemma 5. $\sum_{C \in \mathcal{T}_i(\tau)} w(C) \leq \max\{4 + \frac{4}{\varepsilon}, 6\} \cdot w(\tau \cup \mathrm{Dangle}(\tau))$.

Proof. We may divide the analyses into two cases, depending on q.

Case 1: $q = 1$, *i.e.*, $\mathcal{T}_i(\tau)$ only includes a subtour C.

If $i = 0$, then τ includes r and $w(C) \leq 2w(\tau \cup \text{Dangle}(\tau))$. When $i > 0$, it is clear that τ has a vertex $u \in V_{<i}$. Based on the construction of C, we have $w(C) \leq 2(1 + \varepsilon)^{i-1}M + 2w(\tau \cup \text{Dangle}(\tau))$. Since $\tau \in \mathcal{S}_i^m$, we can assert $w(\tau) \geq \gamma \cdot (1 + \varepsilon)^i M$, implying $(1 + \varepsilon)^{i-1}M \leq \frac{2+\varepsilon}{\varepsilon} \cdot w(\tau)$. Thus, we obtain $w(C) \leq 2(1 + \varepsilon)^{i-1}M + 2w(\tau \cup \text{Dangle}(\tau)) \leq (2 \cdot \frac{2+\varepsilon}{\varepsilon} + 2) \cdot w(\tau \cup \text{Dangle}(\tau)) = (4 + \frac{4}{\varepsilon}) \cdot w(\tau \cup \text{Dangle}(\tau))$, which means $\sum_{C \in \mathcal{T}_i(\tau)} w(C) = w(C) \leq (4 + \frac{4}{\varepsilon}) \cdot w(\tau \cup \text{Dangle}(\tau))$.

Case 2: $q \geq 2$.

By Step 8, we have $2w(\tau \cup \text{Dangle}(\tau)) > (q - 1) \cdot 2(1 + \varepsilon)^{i+1}M$, namely, $(1 + \varepsilon)^{i+1}M < \frac{w(\tau \cup \text{Dangle}(\tau))}{q-1}$. Since $E(\tau \cup \text{Dangle}(\tau)) \subseteq E_{\leq i+1}$, it follows that $V(\tau \cup \text{Dangle}(\tau)) \subseteq V_{\leq i+1}$, implying that each shortest path added from r has total length at most $(1 + \varepsilon)^{i+1}M$. Hence, we have $\sum_{C \in \mathcal{T}_i(\tau)} w(C) \leq 2q \cdot (1 + \varepsilon)^{i+1}M + 2w(\tau \cup \text{Dangle}(\tau)) \leq (\frac{2q}{q-1} + 2) \cdot w(\tau \cup \text{Dangle}(\tau)) \leq 6 \cdot w(\tau \cup \text{Dangle}(\tau))$.

Combining the two preceding arguments in Cases 1–2, we obtain $\sum_{C \in \mathcal{T}_i(\tau)} w(C) \leq \max\{4 + \frac{4}{\varepsilon}, 6\} \cdot w(\tau \cup \text{Dangle}(\tau))$. $\qquad\square$

Using Lemmas 3–5, we obtain the following

Lemma 6. *If $w(E_{\geq i}) \leq M \cdot \Lambda((1 + \varepsilon)^{i-1})$, then the collection $\{\mathcal{T}_i\}_{i \geq 0}$ obtained from Step 8 is $(\alpha, \beta)_{M,\varepsilon}$-assignable, where $\alpha = 4 + 4\varepsilon$, $\beta = \max\{6, 4 + \frac{4}{\varepsilon}\}$ and $\mathcal{T}_i = \bigcup_{\tau \in \mathcal{S}_i^m} \mathcal{T}_i(\tau)$.*

Proof. We shall prove that the collection $\{\mathcal{T}_i\}_{i \geq 0}$ satisfies the two properties in Definition 1. By Lemma 4, it is clear that the property (1) in Definition 1 holds. Now, we only need to show that $\sum_{j \geq i} w(\mathcal{T}_j) \leq \max\{6, 4 + \frac{4}{\varepsilon}\} \cdot w(E_{\geq i})$.

In the HCP algorithm, we note that $\{\text{Dangle}(\tau) \mid \tau \in \mathcal{S}_j^m\}$ is a partition of $\mathcal{S}_{j+1}^u$, implying $\sum_{j \geq i}(w(\mathcal{S}_j^m) + w(\mathcal{S}_{j+1}^u)) \leq \sum_{j \geq i} w(\mathcal{S}_j) = w(E_{\geq i})$.

Using Lemma 3, we have $\sum_{\tau \in \mathcal{S}_j^m} w(\tau \cup \text{Dangle}(\tau)) = w(\mathcal{S}_j^m) + w(\mathcal{S}_{j+1}^u)$. Using Lemma 5, we can ensure $w(\mathcal{T}_j) = \sum_{\tau \in \mathcal{S}_j^m} w(\mathcal{T}_j(\tau)) = \sum_{\tau \in \mathcal{S}_j^m} \sum_{C \in \mathcal{T}_j(\tau)} w(C) \leq \max\{4 + \frac{4}{\varepsilon}, 6\} \cdot \sum_{\tau \in \mathcal{S}_j^m} w(\tau \cup \text{Dangle}(\tau))$, which gives $w(\mathcal{T}_j) \leq \max\{4 + \frac{4}{\varepsilon}, 6\} \cdot (w(\mathcal{S}_j^m) + w(\mathcal{S}_{j+1}^u))$. Thus, it follows that

$$\sum_{j \geq i} w(\mathcal{T}_j) \leq \max\{4 + \frac{4}{\varepsilon}, 6\} \cdot \sum_{j \geq i}(w(\mathcal{S}_j^m) + w(\mathcal{S}_{j+1}^u)) \leq \max\{4 + \frac{4}{\varepsilon}, 6\} \cdot w(E_{\geq i}).$$

This completes the proof of the lemma. $\qquad\square$

Using Lemmas 1–2 and 6, we obtain the following

Theorem 1. *The HCP algorithm is a constant factor approximation algorithm to solve the HCPP, and it runs in polynomial time.*

Proof. By Lemma 2, the decision condition in Step 3 does not hold whenever $M \geq OPT$. Based on the update rule for M, we deduce that Steps 4–10 are executed with $M \leq (1 + \delta) \cdot OPT$.

When fixing $\varepsilon = 0.5652$, using Lemma 6, we can obtain a $(6.2608, 11.07714)$-assignable collection $\{\mathcal{T}_i\}_{i\geq 0}$. By Lemma 1, Step 9 can assign in polynomial time $\{\mathcal{T}_i\}_{i\geq 0}$ to k vehicles, satisfying that the completion time on any vehicle is at most $20.8765 \cdot M$, which means $OUT \leq 20.8765 \cdot M \leq 20.8765(1 + \delta) \cdot OPT$.

Notice that every step in the HCP algorithm can be executed in polynomial time. We shall bound the number of iterations. By using the arguments mentioned-above, the HCP algorithm stops before $M > (1 + \delta)OPT$, where

$$(1 + \delta)OPT \leq (1 + \delta) \cdot \frac{\sum_{uv \in E}(d(r, u) + w(uv) + d(v, r))}{\lambda_{\max}}$$

$$\leq (1 + \delta)|E| \cdot \frac{\max_{uv \in E}(d(r, u) + w(uv) + d(v, r))}{\lambda_{\max}}$$

Since M is initialized at $\frac{\max_{uv \in E}(d(r,u)+w(uv)+d(v,r))}{\lambda_{\max}}$, and increased by an $(1 + \delta)$-factor for each iteration, we conclude that the number of iterations is at most $O(\frac{1}{\delta} \log |E|)$. Hence, the HCP algorithm can be implemented in polynomial time.

This completes the proof of the theorem.   □

4   Conclusion and Further Work

In this paper, we consider the heterogeneous Chinese postman problem (the HCPP), and design a $20.8765(1 + \delta)$-approximation algorithm in polynomial time to solve the HCPP for any small number $\delta > 0$.

In further work, a challenge is to design a constant factor approximation algorithm in polynomial time to solve the HCPP with added capacity constraint, and we shall study other versions of the routing problems with nonuniform speeds.

References

1. Christofides, N.: Worst-case analysis of a new heuristic for the traveling salesman problem, Report 388. Graduate School of Industrial Administration, Carnegie Mellon University (1976)
2. Edmonds, J.: Maximum matching and a polyhedron with $(0, 1)$-vertices. J. Res. Natl. Bureau Stand. B **69**, 125–130 (1965)
3. Edmonds, J., Johnson, E.L.: Matching, Euler tours and the Chinese postman. Math. Program. **5**, 88–124 (1973)
4. Frederickson, G.N., Hecht, M.S., Kim, C.E.: Approximation algorithms for some routing problems. SIAM J. Comput. **7**(2), 178–193 (1978)
5. Golden, B.L., Raghavan, S., Wasil, E.A. (eds.): The Vehicle Routing Problem: Latest Advances and New Challenges. Springer, New York (2008). https://doi.org/10.1007/978-0-387-77778-8
6. Gørtz, I.L., Molinaro, M., Nagarajan, V., Ravi, R.: Capacitated vehicle routing with nonuniform speeds. Math. Oper. Res. **41**(1), 318–331 (2016)
7. Guan, M.G.: Graphic programming using odd and even points. Acta Math. Sinica **10**, 263–266 (1960). (In Chinese). English translation: Chinese Mathematics 1, 273–277 (1962)

8. Korte, B., Vygen, J.: Combinatorial Optimization: Theory and Algorithms, 6th edn. Springer, Heidelberg (2018)
9. Laporte, G.: Fifty years of vehicle routing. Transp. Sci. **43**, 408–416 (2009)
10. Toth, P., Vigo, D.: Vehicle Routing: Problems, Methods and Applications. MOS-SIAM, Philadelphia (2014)
11. Vazirani, V.V.: Approximation Algorithms. Springer, Berlin (2001)

On Stochastic k-Facility Location

Yicheng Xu[1], Chunlin Hao[2], Chenchen Wu[3]($\boxtimes$), and Yong Zhang[1]

[1] Shenzhen Institutes of Advanced Technology, Chinese Academy of Sciences,
Beijing, People's Republic of China
{yc.xu,zhangyong}@siat.ac.cn
[2] Department of Operations Research and Information Engineering, Beijing University
of Technology, Beijing 100124, People's Republic of China
haochl@bjut.edu.cn
[3] College of Science, Tianjin University of Technology, Tianjin, People's Republic of China

Abstract. In the stochastic facility location problem, there are two-stage processes for the decision. A set of facilities may be opened without information for the demand of clients at the first stage; an additional set of facilities may further be opened at the second stage where the scenario of the clients is realized according to some given distribution. One has to take the risk into consideration and make decision on the open facilities in each stage and each scenario such that the total expected cost of the opening and service is minimized. In this paper, we consider a global cardinality constraint in this model, i.e., there is an upper bound k for the number of open facilities at the second stage. This model generalizes both stochastic facility location and the k-median. Our main result is a provable efficient approximation algorithm with a performance ratio of 6 based on primal-dual schema.

Keywords: Stochastic facility location · k-facility location · Approximation
algorithm · Primal-dual

Facility location is one of the most fundamental and classical models in the fields of applied mathematics, operations research and management science. Traditional facility location solves a subset from a given candidates of facilities to open in order to minimize the total cost of connecting each client to its nearest open facility plus the opening cost. This problem is somehow well studied in the literature. The state-of-the-art approximation algorithm hits a performance ratio of 1.488 by Li [12], which is very close to the inapproximability bound of 1.463 proved by Guha and Khuller [8] under the assumption of NP $\subseteq$ DTIME$[n^{O(\log \log n)}]$ and strengthened to P $\neq$ NP by Sviridenko (Personal Communication). Facility location problem has lots of variants, such as, robust facility location (c.f. [9]), capacitated facility location (c.f. [10]), facility location with penalties (c.f. [5]). In traditional facility location, all parameters are deterministic and known before making decisions. A new research trend focuses on the uncertainty study in facility location decisions. In stochastic facility location, uncertain parameters are governed by a certain probability distribution that are known and the decision maker should take the risk into consideration before make a decision. And a

© Springer Nature Switzerland AG 2021
W. Wu and H. Du (Eds.): AAIM 2021, LNCS 13153, pp. 47–56, 2021.
https://doi.org/10.1007/978-3-030-93176-6_5

common goal is to optimize the expected value of some cost or profit function in this stochastic situation.

Stochastic features are introduced to facility location by Ermoliev and Leonardi [6]. In the stochastic facility location, the decision maker is required to choose locations before the uncertain demands are known, and react once the uncertain scenario has been realized. Louveaux and Peeters [14] extends the approaches proposed for deterministic facility location and first present a heuristic dual-based procedure for this problem. One of the mainstream study for stochastic facility location is a scenario-based stochastic one that considers two-stage decision processes where at the first stage one has to choose the location before the uncertain demand of clients realized, and at the second stage one has to decide the location as well as the assignment for each scenario may occur. The facilities can be opened in either the first stage or the second with a fixed opening cost under the assumption that the opening cost is higher in the second stage. Gupta et al. [7] give a 8.45-approximation algorithm for this problem. Ravi and Sinha [18] improve this to reach a performance ratio of 8 based on LP rounding algorithm for deterministic facility location. Swamy and Shmoys [19] improve upon a family of optimization models including stochastic facility location using a convex program relaxation and specific rounding approach which achieves a $(3.04 + \varepsilon)$-approximation. Furthermore, the two-stage nature of stochastic facility location makes it a popular application of general stochastic programming methods [3]. Also, multi-stage decision processes in stochastic facility location is considered by Hochreiter and Pflug [11] who view this problem as multi-dimensional facility location and conclude that as least good heuristic algorithms exists. Basciftci et al. [2] consider distributioinally robust facility location problem in which we are just given the moment information of the distribution of the demand.

Another attractive problem in facility location family is the k-median. Traditional k-median problem seeks for at most k facilities in a given set to open so as to minimize the total distance from the clients to their nearest open facilities. The state-of-the-art approximation results for this problem is proposed by Byrka et al. [4] using a novel dependent rounding approach which gives a $(2.675 + \varepsilon)$-approximation. The k-median is extended to a more generalized model called k-facility location in which opening cost is also considered in the objective. For this problem, Zhang [21] proposes a $(2+\sqrt{3}+\varepsilon)$-approximation algorithm which is currently the state-of-the-art. Stochastic features are introduced to the k-median before which is well studied. However, the uncertainty in the k-median concerns mainly the costs [1,17,20], and a different stream of studies concerns the uncertainty of the locations [16].

To the best of our knowledge, the connection between stochastic facility location and stochastic k-median is first considered by Louveaux [13], who mainly study from the modeling instead of algorithmic perspective. In this paper, we propose the so-called stochastic k-facility location problem (SkFLP) which is a common generalization of stochastic facility location and k-median. Based on primal-dual schema and Lagrange multiplier preservation property, we give a provable efficient approximation algorithm for SkFLP with a performance ratio of 6.

1 Preliminaries

In the SkFLP, there are two stages in the location process. At the first stage, we are given the facility set F. Each facility can be opened to serve the client realized latter. The opening cost for facility at this stage is f_i^0. Moreover, we are also given the distribution of the demand to be served at the second stage. The client set D_t need to be served with the probability q_t (note that $\sum_{t=1}^{T} q_t = 1$). At the second stage, the client set is realized. We are allowed to open more facility as the supplement with higher opening cost. The opening cost in scenario s (that is, the client set D_s is realized) is f_i^s. The metric service cost is c_{ij} between the facility i and client j. There is also an upper bound k of the number of the opened facilities. The goal of the problem is to decide which facilities to open at both stages such that the expected total cost including the opening cost and connection cost is minimized. Then, the SkFLP

$$\min_{F_0, F_t \subseteq F} \left\{ \sum_{i \in F_0} f_i^0 + \sum_{t=1,2,\cdots,T} q_t \left[\sum_{i \in F_t} f_i^t + \sum_{j \in D_t} \min_{i \in F_0 \cup F_t} c_{ij} \right] \right\}.$$

For easy to describe the problem, we denote some new notations as follows.

- facility-scenario set $\mathcal{F} := \{(i, s) : i \in F, s = 0, 1, 2, \cdots, T\}$, if $s = 0$ means the facility is at the first stage;
- client-senario set $\mathcal{D} := \{(j, t) : j \in D_t, t = 1, 2, \cdots, T\}$;
- the connection cost between $(i, s) \in \mathcal{F}$ and $(j, t) \in \mathcal{D}$

$$c_{ij}^{st} := \begin{cases} c_{ij}^{st}, & \text{if } s = 0 \text{ or } t; \\ +\infty, & \text{otherwise.} \end{cases}$$

Therefore, the SkFLP is equivalent to the problem that we are given the facility-scenario set $\mathcal{F}$ and the client-scenario set $\mathcal{D}$. The facility-scenario pair $(i, s) \in \mathcal{F}$ is $q_s f_i^s$ (set $q_0 = 1$). The connection cost between facility pair $(i, s) \in \mathcal{F}$ and $(j, t) \in \mathcal{D}$ is c_{ij}^{st}. The demand of the client-scenario pair (j, t) is q_t. Then, we need to decide which facility-scenario pairs opened such that the total cost including the opening cost and connection cost is minimized. Based on the above description, we introduce the variables y_i^0 to show whether the facility i is opened at the first stage, y_i^s to show whether the facility i is opened in scenario s at the second stage, x_{ij}^{st} to show whether the client j in scenario t is severed by the facility i opened in scenario s (note that, $s = 0$ or $s = t$). Thus, we have the following formulation

$$\min \sum_{(i,s) \in \mathcal{F}} q_s f_i^s y_i^s + \sum_{(i,s) \in \mathcal{F}} \sum_{(j,t) \in \mathcal{D}} q_s c_{ij}^{st} x_{ij}^{st}$$

$$\text{s. t.} \sum_{(i,s) \in \mathcal{F}} x_{ij}^{st} \geq 1, \qquad \forall (j, t) \in \mathcal{D}, \tag{1}$$

$$x_{ij}^{st} \leq y_i^s, \qquad \forall (i, s) \in \mathcal{F}, (j, t) \in \mathcal{D},$$

$$\sum_{(i,s)} y_i^s \leq k,$$

$$x_{ij}^{st}, y_i^s \in \{0, 1\}, \qquad \forall (i, s) \in \mathcal{F}, (j, t) \in \mathcal{D}.$$

The first constraint shows that there must be connected to a facility-scenario pair for each client-scenario (j, t); The second constraint shows that if the client-scenario pair (j, t) is connected to facility-scenario pair (i, s), the facility-scenario pair (i, s) should be opened. The objective is the total expected total cost of the problem.

Relaxing the integer constraint of program (1), we obtain the linear program as follows.

$$\min \sum_{(i,s)\in\mathcal{F}} q_s f_i^s y_i^s + \sum_{(i,s)\in\mathcal{F}} \sum_{(j,t)\in\mathcal{D}} q_s c_{ij}^{st} x_{ij}^{st}$$

$$\text{s. t.} \sum_{(i,s)\in\mathcal{F}} x_{ij}^{st} \geq 1, \qquad \forall (j,t) \in \mathcal{D}, \tag{2}$$

$$x_{ij}^{st} \leq y_i^s, \qquad \forall (i,s) \in \mathcal{F}, (j,t) \in \mathcal{D},$$

$$\sum_{(i,s)} y_i^s \leq k,$$

$$x_{ij}^{st}, y_i^s \geq 0, \qquad \forall (i,s) \in \mathcal{F}, (j,t) \in \mathcal{D}.$$

Then, the dual program of the relaxation is

$$\max \sum_{(j,t)} \alpha_j^t - k \cdot \lambda$$

$$\text{s. t.} \alpha_j^t \leq \beta_{ij}^{st} + q_t c_{ij}^{st}, \qquad \forall (i,s) \in \mathcal{F}, (j,t) \in \mathcal{D}, \tag{3}$$

$$\sum_{(j,t)\in\mathcal{D}} \beta_{ij}^{st} \leq q_s f_i^s + \lambda, \qquad \forall (i,s) \in \mathcal{F},$$

$$\alpha_j^t, \beta_{ij}^{st} \geq 0, \qquad \forall (j,t) \in \mathcal{D}, (i,s) \in \mathcal{F},$$

In the above program, we can observe that α_j^t means the budget of (j, t) to connect to a facility-scenario pair, β_{ij}^{st} means the contribution of (j, t) to open (i, s). If λ is fixed, $k \cdot \lambda$ is a constant. Given a fixed λ, we propose the dual program of (3) after dropping the term of the constant $k \cdot \lambda$.

$$\min \sum_{(i,s)\in\mathcal{F}} (q_s f_i^s + \lambda) y_i^s + \sum_{(i,s)\in\mathcal{F}} \sum_{(j,t)\in\mathcal{D}} q_s c_{ij}^{st} x_{ij}^{st}$$

$$\text{s. t.} \sum_{(i,s)\in\mathcal{F}} x_{ij}^{st} \geq 1, \qquad \forall (j,t) \in \mathcal{D}, \tag{4}$$

$$x_{ij}^{st} \leq y_i^s, \qquad \forall (i,s) \in \mathcal{F}, (j,t) \in \mathcal{D},$$

$$x_{ij}^{st}, y_i^s \geq 0, \qquad \forall (i,s) \in \mathcal{F}, (j,t) \in \mathcal{D}.$$

The above program is a stochastic facility location problem without the constraint of k, where the facility cost is $q_s f_i^s + \lambda$. If λ is big enough, we just open few facilities (less than k) to obtain fewer opening cost. If λ is 0, we can open more faciities (more than k) to obtain fewer connection cost.

2   Main Results

In this section, we will propose our primal-dual algorithms for the stochastic k-facility location problem, along with their performance guarantee. Our algorithm is two-setps algorithm. Firstly, a dual ascending process will produce an integer solution for a fixed λ. We run the process twice for two different values of λ to obtain $k_1 > k$ facilities and $k_2 \leq k$ facilities to open, separately. Secondly, we construct an integer solution with exact k facilities.

Now, the primal-dual process for a fixed λ can be give as follows (c.f. Primal-dual (λ)).

In the Algorithm 2, we obtain a feasible solution for stochastic facility location problem with $f_i^s + \lambda$ as the facility-scenario pair opening cost. We estimate the cost of the solution with the constant λ. Firstly, we estimate the facility opening cost.

Lemma 1. *The facility cost $Cost_f(\hat{\mathcal{F}})$ of the solution obtained by the Algorithm 2 is* $\sum_{(i,s)\in\hat{\mathcal{F}}} \sum_{(j,t)\in\tilde{\mathcal{D}}^1_{(i,s)}} \beta_{ij}^{st}$.

Proof. *By the case 2 in Step 1.2 in the Algorithm 2, we have*

$$\sum_{(i,s)\in\hat{\mathcal{F}}} (f_i^s + \lambda) = \sum_{(i,j)\in\hat{\mathcal{F}}} \sum_{(j,t)\in\mathcal{D}} \beta_{ij}^{st}$$

$$= \sum_{(i,s)\in\hat{\mathcal{F}}} \sum_{(j,t)\in\tilde{\mathcal{D}}^1_{(i,s)}} \beta_{ij}^{st}$$

The second equality holds since $\tilde{\mathcal{D}}^1_{(i,s)}$ is disjoint for different (i, s).

Secondly, we give the estimation of the connection cost for the client-scenario pairs which have contribution to the facility-scenario pairs in $\hat{\mathcal{F}}$. Indeed, for these client-scenario pair, the connection cost is no more than $\frac{\alpha_j^t - \beta_{ij}^{st}}{q_t}$, where (i, s) is the facility-scenario pair in $\hat{\mathcal{F}}$ which (j, t) has positive contribution.

Thirdly, we give the estimation of the connection cost for the client-scenario pairs which does not have contribution to the facility-scenario pairs in $\hat{\mathcal{F}}$.

Lemma 2. *The connection cost $cost_c(\{j, t\})$ for the client-scenario pair which does not have contribution to is no more than $3\frac{\alpha_j^t}{q_t}$.*

Proof. We consider the facility-scenario pair (i, s) which is the fist such that the dual variables of (j, t) stop increasing. By the Step 2, there must be another facility-scenario pair $(i', s') \in \hat{\mathcal{F}}$ such that $\tilde{\mathcal{D}}^1_{(i,s)} \cap \tilde{\mathcal{D}}^1_{(i',s')} \neq \emptyset$. Assume (j', t') is a client-scenario pair in $\tilde{\mathcal{D}}^1_{(i,s)} \cap \tilde{\mathcal{D}}^1_{(i',s')}$. Note that $s' = s$ or 0. Then, the client-scenario pair (j, t) can be connected to (i', s'). Thus,

---

Algorithm 1. Primal-dual(λ)

Input: The facility-scenario pair set $\mathcal{F}$, the client-scenario pair set $\mathcal{D}$, the connection cost c_{ij}^{st}, the probability q_t for each scenario t, the opening cost f_i^s for each facility-scenario pair, and a fixed λ.

Step 1. Constructing a dual feasible solution.

Step 1.1 Initially, set all dual variables $\alpha_j^t := 0$ and $\beta_{ij}^{st} := 0$ for all $(j,t) \in \mathcal{D}$, $(i,s) \in \mathcal{F}$. All facility-scenario pairs are close, that is, set $\tilde{\mathcal{F}} = \emptyset$. All client-scenario pairs are not connected, that is, set $\tilde{\mathcal{D}} := \emptyset$.

Step 1.2 Increasing the dual variable α_j^t with the rate q_t, the following cases will happen.

case 1 There is a facility-scenario pair $(i,s) \in \mathcal{F} \backslash \tilde{\mathcal{F}}$ and a client-scenario pair $(j,t) \in \mathcal{D} \backslash \tilde{\mathcal{D}}$ such that
$$\alpha_j^t = q_t c_{ij}^{st}.$$

In this case, we start to increase the corresponding β_{ij}^{st} with the same rate of α_j^t.

case 2 There is a facility-scenario pair $(i,s) \in \mathcal{F} \backslash \tilde{\mathcal{F}}$ such that
$$\sum_{(j,t) \in \mathcal{D}} \beta_{ij}^{st} = q_s f_i^s + \lambda.$$

In this case, we temporarily open the pair (i,s), that is, set $\tilde{\mathcal{F}} := \tilde{\mathcal{F}} \cup \{(i,s)\}$. For the proof convenience, we denote $\tilde{D}_{(i,s)}^1$ and $\tilde{D}_{(i,s)}^2$ be the set in which the client-scenario pair in in $\mathcal{D}$ has positive contribution to (i,s) and the set in which the set in which the client-scenario pair in $\mathcal{D} \backslash \tilde{\mathcal{D}}$ has positive contribution to (i,s). That is,
$$\tilde{D}_{(i,s)}^1 := \{(j,t) \in \mathcal{D} : \beta_{ij}^{st} > 0\},$$
and
$$\tilde{D}_{(i,s)}^2 := \{(j,t) \in \mathcal{D} \backslash \tilde{\mathcal{D}} : \beta_{ij}^{st} > 0\}.$$

Moreover, we connect all client-scenario pairs in $\tilde{\mathcal{D}}$ to facility-scenario pair (i,s) who have positive contribution to (i,s), that is, $\tilde{\mathcal{D}} := \tilde{\mathcal{D}} \cup \{(j,t) \in \tilde{\mathcal{D}} | \beta_{ij}^{st} > 0\}$. The dual variables α_j^t and all β_{ij}^{st} stop increasing in $\tilde{\mathcal{D}}$.

case 3 There is a facility-scenario pair $(i,s) \in \tilde{\mathcal{F}}$ and a client-scenario pair $(j,t) \in \tilde{\mathcal{D}}$ such that
$$\alpha_j^t = q_t c_{ij}^{st}.$$

In this case, connec the client-scenario pair (j,t) to facility-scenario pair (i,s), that is, $\tilde{\mathcal{D}} := \tilde{\mathcal{D}} \cup \{(j,t)\}$.

Repeat the above until all client-scenario pairs are connected, that is, $\tilde{\mathcal{D}} = \mathcal{D}$.

Step 2. Constructing a primal integer feasible solution. Two facility-scenario pairs are dependent if there is a client-scenario pair such that it has positive contribution to both two facility-scenario pairs. Order the facility-scenario pair $(i,0)$ in $\tilde{\mathcal{F}}$ first, then, order the facility pair $(i,1)$ in $\tilde{\mathcal{F}}$, and go on. Find the maximal independent set $\hat{\mathcal{F}}$ in $\tilde{\mathcal{F}}$.

$$Cost_c(\{j,t\}) \leq c_{i'j}^{s't}$$
$$\leq c_{ij}^{st} + c_{ij'}^{st'} + c_{i'j'}^{s't'}$$
$$\leq \frac{\alpha_j^t}{q_t} + \frac{\alpha_{j'}^{t'}}{q_{t'}} + \frac{\alpha_{j'}^{t'}}{q_{t'}}$$
$$\leq 3\frac{\alpha_j^t}{q_t}.$$

Lastly, we can give the estimation of the algorithm of Primal-dual (λ).

Lemma 3. *The cost of the solution obtained by the algorithm of Primal-dual(λ) has*

$$\sum_{(i,s)\in\hat{\mathcal{F}}} (q_s f_i^s + \lambda) + \sum_{(j,t)\in\mathcal{D}} cost_c(\{j,t\}) \leq 3 \sum_{(i,t)\in\mathcal{D}} \alpha_j^t.$$

Proof. Combining Lemma 1 and Lemma 2, we have

$$\sum_{(i,s)\in\hat{\mathcal{F}}} (q_s f_i^s + \lambda) + \sum_{(j,t)\in\mathcal{D}} cost_c(\{j,t\})$$
$$= \sum_{(i,s)\in\hat{\mathcal{F}}} \sum_{(j,t)\in\tilde{\mathcal{D}}_{(i,s)}^1} \beta_{ij}^{st} + \sum_{(i,s)\in\hat{\mathcal{F}}} \sum_{(j,t)\in\cup_{(i,s)\in\hat{\mathcal{F}}}\tilde{\mathcal{D}}_{(i,s)}^1} q_t \frac{\alpha_j^t - \beta_{ij}^{st}}{q_t}$$
$$+ 3 \sum_{(i,s)\in\hat{\mathcal{F}}} \sum_{(j,t)\notin\cup_{(i,s)\in\hat{\mathcal{F}}}\tilde{\mathcal{D}}_{(i,s)}^1} q_t \frac{\alpha_j^t}{q_t}$$
$$\leq 3 \sum_{(j,t)\in\tilde{\mathcal{D}}} \alpha_j^t.$$

When we run the algorithm of Primal-dual(λ) with two different λ_1 and λ_2, we can obtain two solutions $\hat{\mathcal{F}}_1$ with cardinality $k_1 \leq k$ and $\hat{\mathcal{F}}_2$ with cardinality $k_2 > k$. Based on these two solutions, we construct an integer feasible solution with k opened facilities. We set constants a and b such that $ak_1 + bk_2 = k$ and $a + b = 1$. Then, Algorithm2 gives the process of constructing the integer feasible solution.

Algorithm 2. Integer feasible solution

Step 1   For each facility-scenario pair in $\hat{\mathcal{F}}_1$, find the closest facility in $\hat{\mathcal{F}}_2$. Let $\hat{\mathcal{F}}' \subseteq \hat{\mathcal{F}}_2$ be the facility-scenario pair set. If $|\hat{\mathcal{F}}'| < k_1$, pick the facility-scenario pair in $\hat{\mathcal{F}}_2\backslash\hat{\mathcal{F}}'$ arbitrarily adding to $\hat{\mathcal{F}}'$ such that $|\hat{\mathcal{F}}'| = k_1$.

Step 2   With probability a, open all facilities in $\hat{\mathcal{F}}_1$, and with probability b (note that $b = 1 - a$), open all facilities in $\hat{\mathcal{F}}'$. Moreover, open $k - k_1$ facilities randomly in randomly from $\hat{\mathcal{F}}_2\backslash\hat{\mathcal{F}}'$.

Theorem 1. *Algorithm 2 is a 6-approximation algorithm for the SkFLP.*

Proof. Firstly, we estimate the expected opening cost. All probability of the facility-scenario pair in $\hat{\mathcal{F}}_1$ and $\hat{\mathcal{F}}_2$ is a and b. Then, the expected opening cost is

$$a \sum_{(i,s)\in\hat{\mathcal{F}}_1} q_s f_i^s + b \sum_{(i,s)\in\hat{\mathcal{F}}_2} q_s f_i^s.$$

Secondly, we estimate the connection cost. For each (j,t), we denote (i_1, s_1) and (i_2, s_2). Then, we have the following cases.

Case 1. (i_1, s_1) is opened in the integer solution. The probability of the case is a. The connection cost of (j,t) is $c_{i_1 j}^{s_1 t}$.

Case 2. (i_1, s_1) is not opened, but (i_2, s_2) is opened in the integer solution. The probability of the case is $(1-a)b = b^2$. The connection cost of (j,t) is $c_{i_2 j}^{s_2 t}$.

Case 3. Both (i_1, s_1) and (i_2, s_2) are not opened in the integer solution. The probability of the case is $(1-a)(1-b) = ab$. We connect (j,t) to the nearest facility-scenario pair of (i_1, s_2) which is open in $\hat{\mathcal{F}}_2$. The connection cost of (j,t) is no more than $c_{i_1 j}^{s_1 t} + c_{i_2 j}^{s_2 t}$.

Thus, the expected connection cost of (j,t) is

$$a c_{i_1 j}^{s_1 t} + b^2 c_{i_2 j}^{s_2 t} + ab \left(c_{i_1 j}^{s_1 t} + c_{i_2 j}^{s_2 t}\right).$$

Then, we have that the expected connection cost is no more than

$$(1 + \max\{a, b\}) \left(a_{i_1 j}^{s_1 t} + b c_{i_2 j}^{s_2 t}\right).$$

Then, the expected integer feasible solution is no more than

$$a \sum_{(i,s)\in\hat{\mathcal{F}}_1} q_s f_i^s + b \sum_{(i,s)\in\hat{\mathcal{F}}_2} q_s f_i^s$$
$$+ (1 + \max\{a, b\}) \sum_{(j,t)\in\mathcal{D}} \left(a cost_c^1(\{j,t\}) + b cost_c^2(\{j,t\})\right),$$

where $cost_c^1(\{j,t\})$ and $cost_c^2(\{j,t\})$ is the connection cost of (j,t) in the solution $\hat{\mathcal{F}}_1$ and $\hat{\mathcal{F}}_2$, respectively. By Lemma 3, we have

$$a \sum_{(i,s)\in\hat{\mathcal{F}}_1} (q_s f_i^s + \lambda) + a \sum_{(j,t)\in\mathcal{D}} cost_c^1(\{j,t\})$$
$$+ b \sum_{(i,s)\in\hat{\mathcal{F}}_2} (q_s f_i^s + \lambda) + b \sum_{(j,t)\in\mathcal{D}} cost_c^2(\{j,t\})$$
$$\leq 3 \left(a\alpha_j^t(1) + b\alpha_j^t(2)\right),$$

where $\alpha_j^t(1)$ and $\alpha_j^t(2)$ is the dual variables in the solution $\hat{\mathcal{F}}_1$ and $\hat{\mathcal{F}}_1$, respectively. Thus, we obtain

$$
\begin{aligned}
& a \sum_{(i,s)\in\hat{\mathcal{F}}_1} q_s f_i^s + b \sum_{(i,s)\in\hat{\mathcal{F}}_2} q_s f_i^s \\
& + a \sum_{(j,t)\in\mathcal{D}} cost_c^1(\{j,t\}) + b \sum_{(i,s)\in\hat{\mathcal{F}}_2} cost_c^2(\{j,t\}) \\
& \leq 3\left(a\alpha_j^t(1) + b\alpha_j^t(2)\right) - k\lambda.
\end{aligned}
\tag{5}
$$

The RHS of (5) is a feasible dual solution of (3). Then

$$
\begin{aligned}
& a \sum_{(i,s)\in\hat{\mathcal{F}}_1} q_s f_i^s + b \sum_{(i,s)\in\hat{\mathcal{F}}_2} q_s f_i^s \\
& + (1+\max\{a,b\}) \sum_{(j,t)\in\mathcal{D}} \left(a cost_c^1(\{j,t\}) + b cost_c^2(\{j,t\})\right) \\
& \leq 6 OPT,
\end{aligned}
\tag{6}
$$

where OPT is the optimal solution of SkFLP.

Moreover, we view the stochastic k-median problem is a special case of the SkFLP if the opening cost if the opening costs are all 0.

Theorem 2. *There is a 6-approximation algorithm for the stochastic k-median problem.*

Acknowledgements. The research of the third author is supported by NSFC (No. 11971349). The research of the fourth author is supported by NSFC (No. 12071460).

References

1. Adibi, A., Razmi, J.: 2-stage stochastic programming approach for hub location problem under uncertainty: a case study of air network of Iran. J. Air Transp. Manag. **47**, 172–178 (2015)
2. Basciftci, B., Ahmed, S., Shen, S.: Distributionally robust facility location problem under decision-dependent stochastic demand. Eur. J. Oper. Res. **292**, 548–561 (2021)
3. Birge, J.R., Louveaux, F.: Introduction to Stochastic Programming. Springer, Heidelberg (2011). https://doi.org/10.1007/978-1-4614-0237-4
4. Byrka, J., Pensyl, T., Rybicki, B., Srinivasan, A., Trinh, K.: An improved approximation for k-median, and positive correlation in budgeted optimization. In: Proceedings of the 26th Annual ACM-SIAM Symposium on Discrete Algorithms, pp. 737–756 (2014)
5. Du, D., Lu, R., Xu, D.: A primal-dual approximation algorithm for the facility location problem with submodular penalties. Algorithmica **63**, 191–200 (2012)
6. Ermoliev, Y.M., Leonardi, G.: Some proposals for stochastic facility location models. Math. Modelling **3**(5), 407–420 (1982)
7. Gupta, A., Pál, M., Ravi, R., Sinha, A.: Boosted sampling: approximation algorithms for stochastic optimization. In: Proceedings of the 36th Annual ACM Symposium on Theory of Computing, pp. 417–426 (2004)

8. Guha, S., Khuller, S.: Greedy strikes back: improved facility location algorithms. J. Algorithms **31**(1), 228–248 (1999)
9. Han, L., Xu, D., Li, M., Zhang, D.: Approximation algorithms for the robust facility leasing problem. Optim. Lett. **12**(3), 625–637 (2018). https://doi.org/10.1007/s11590-018-1238-x
10. Han, L., Xu, D., Du, D., Zhang, D.: A local search approximation algorithm for the uniform capacitated k-facility location problem. J. Comb. Optim. **35**(2), 409–423 (2018)
11. Hochreiter, R., Pflug, G.C.: Financial scenario generation for stochastic multi-stage decision processes as facility location problems. Ann. Oper. Res. **152**(1), 257–272 (2007)
12. Li, S.: A 1.488 approximation algorithm for the uncapacitated facility location problem. Inf. Comput. **222**, 45–58 (2013)
13. Louveaux, F.V.: Discrete stochastic location models. Ann. Oper. Res. **6**(2), 21–34 (1986)
14. Louveaux, F.V., Peeters, D.: A dual-based procedure for stochastic facility location. Oper. Res. **40**(3), 564–573 (1992)
15. Mahdian, M.: Facility location and the analysis of algorithms through factor-revealing problems, Ph.D. thesis, Massachusetts Institute of Technology, Cambridge, MA (2004)
16. Mirchandani, P.B., Odoni, A.R.: Locations of medians on stochastic networks. Transp. Sci. **13**(2), 85–97 (1979)
17. Mirchandani, P.B., Oudjit, A., Wong, R.T.: Multidimensional extensions and a nested dual approach for the m-median problem. Eur. J. Oper. Res. **21**(1), 121–137 (1985)
18. Ravi, R., Sinha, A.: Hedging uncertainty: approximation algorithms for stochastic optimization problems. Math. Program. **108**(1), 97–114 (2006)
19. Swamy, C., Shmoys, D.B.: Approximation algorithms for 2-stage stochastic optimization problems. ACM SIGACT News **37**(1), 33–46 (2006)
20. Tadei, R., Ricciardi, N., Perboli, G.: The stochastic p-median problem with unknown cost probability distribution. Oper. Res. Lett. **37**(2), 135–141 (2009)
21. Zhang, P.: A new approximation algorithm for the k-facility location problem. Theoret. Comput. Sci. **384**(1), 126–135 (2007)

The Complexity of Finding a Broadcast Center

Hovhannes A. Harutyunyan[1] and Zhiyuan Li[2(✉)]

[1] Department of Computer Science and Software Engineering, Concordia University,
Montreal, QC H3G 1M8, Canada
haruty@cs.concordia.ca
[2] Computer Science and Technology, United International College,
Zhuhai 0086-519000, China
goliathli@uic.edu.cn

Abstract. Broadcasting in networks is one of the most important information dissemination processes. It is known that finding the optimal (minimum) broadcast time is NP-hard. Some of the follow-up researches focus on approximations/heuristics to find the broadcast center, a set of nodes from which the broadcast time in the network is minimum, in a given network by assuming the problem is hard without an actual proof. In this paper, we show that answering the questions "is a set of vertices a broadcast center to the given graph", and "does the given graph has a broadcast center of size smaller than k" are both NP-hard under Turing reductions.

1 Introduction

Broadcasting is an information dissemination process in an interconnected network and has been studied over the past decades. In this paper, we follow the basic assumptions (listed below), and model the network as a graph $G = (V, E)$. The vertex set and the edge set for a given graph G can be denoted by $V(G)$ and $E(G)$. The distance between two vertices v and u is denoted by $d(v, u)$.

1. The process of broadcasting is split into discrete time units.
2. Initially, only one vertex (*originator*) has the message.
3. In each time unit, a vertex with the message (*sender*) can *call* at most one uninformed neighbor (*receiver*).
4. All the calls are in parallel during the same time unit.
5. If all the vertices in the graph have the message, the process halts.

Each call can be represented by an ordered pair of two vertices (u, v), where u is the sender and v is the receiver. The set C_i consists of all calls in time unit i. And the sequence of calls $(C_1, C_2, \cdots, C_t)$ presents the process of broadcast, named *broadcast scheme*, where $C_i = \{(u, v)$—for all calls from informed vertex

Granted by the Guangdong Provincial Innovation and Enhancement Project of China under Grant No. R5201918.

W. Wu and H. Du (Eds.): AAIM 2021, LNCS 13153, pp. 57–70, 2021.
https://doi.org/10.1007/978-3-030-93176-6_6

u to uninformed vertex v in time unit i and the index $i = 1, 2, \cdots, t$ represents the time unit. A vertex v is *idle* in time unit t if v is informed before time t but does not make any call in time t.

Because all non-originating vertices can only be informed by exactly one vertex, the broadcast scheme forms a spanning tree (*broadcast tree*) rooted at the originator. We are also free to omit the direction of each call in the broadcast tree. The *broadcast time* for a vertex v in a given graph G, denoted by $b(G, v)$, is the minimum number of time units required by broadcasting from the vertex v in G, formally $b(G) = min_{v \in V(G)}(b(G, v))$. Consequently, an *optimal broadcast scheme*, denoted by $\mathcal{S}(G, v)$ uses $b(G, v)$ time units. From the assumption 3, the number of informed vertices are at most doubled in each time unit. So, the number of informed vertices up to time unit i is no greater than 2^i in general. By taking the inverse, $b(G, v) \geq \lceil \log n \rceil$, where n is the number of vertices in G. Obviously, $b(G) \geq \lceil \log n \rceil$. A graph G is a *broadcast graph* if $b(G) = \lceil \log n \rceil$. G is a minimum broadcast graph on n vertices if $|E(G)|$ is the minimum among all broadcast graphs on n vertices. Intuitively, a broadcast graph is the best topology structure in the sense of fast broadcasting, and a minimum broadcast graph has the lowest cost in terms of number of edges in such graphs.

We can distinguish two main lines of research on broadcasting in graphs. The first seeks to construct graphs (networks) with given broadcast times; the ongoing search for minimum broadcast graphs is an example. The second is given a graph and message originator and seeks to find the broadcast time; the work of this paper is an example.

The *broadcast time problem (BTP)* is defined as

Problem 1. BTP
Instance: (G, v, t), where $G = (V, E)$ is a graph, $v \in V$ is the originator, and t is a natural number.
Output: "Yes" if $b(G, v) \leq t$; "No" otherwise.

The broadcast center $BC_G = \{v | \forall u \in V(G), b(G, v) \leq b(G, u)\}$ refers to the set of vertices whose broadcast time is minimum in the graph G. We define the *broadcast center deciding problem (BCD)* and the *broadcast center size problem (BCS)* as follows.

Problem 2. BCD
Instance: (G, U), where $G = (V, E)$ is a graph and U is a subset of V.
Output: "Yes" if U is a subset of BC_G; "No" otherwise.

Problem 3. BCS
Instance: (G, x), where $G = (V, E)$ is a graph and x is a natural number.
Output: "Yes" if $|BC_G| \geq x$; "No" otherwise.

The study of broadcast center can be traced back to [29]. Slater, Cockayne, and Hedetniemi defined the broadcast center and proposed a linear algorithm to find the broadcast center of a tree. In the same paper, the authors also proved that

Theorem 1. *BTP is NP-hard.*

As a consequence, the follow-up research assumed that finding broadcast centers was also not easy and only focused on certain classes of graphs, even the complexity of finding broadcast centers remains open.

Harutyunyan, Liestman, and Shao designed a linear algorithm for k-broadcast center of trees [22]. Maraachlian studied the broadcast center of unicyclic graphs [24] and tree cluster networks [23]. In [30], Su, Lin, and Lee presented a linear algorithm for finding the broadcast time and broadcast centers in heterogeneous trees. More recently, Cevni and Zerovnik studied the broadcast center for cactus graphs [27].

On the opposite of the broadcast time problem, researchers also show a lot of interests in the constructions of broadcast graphs. Many construction methods have been proposed in the past 40 years; direct construction from binomial trees [14,19], vertex addition/deletion methods [5,15,19,20], and compounding multiple existing broadcat graphs [1,4,17–19]. However, similar to the BCS problem, the complexity of deciding whether a given graph is a broadcast graph or not remains open. For more algorithms of the broadcast time problem, see [2,3,7,8,15]. For more broadcast graph constructions see [6,9,10,13,16,28,31]. And for more broadcasting in general, see [11,21,25,26]. For the details of NP-completeness and more complexity classes, readers are referred to [12].

In this paper, we prove that BCD and BCS are both NP-hard under polynomial-time Turing reductions, denoted by $\leq_p$ from BTP. Note that Turing reductions are different from Karp reductions, which are commonly used in NP-complete proofs.

2   Preliminaries

This section presents some well-studied graphs which are usefull to our reduction. Similar discussions can be found in [15].

Definition 1. *A path on n vertices is a graph $L = (V, G)$, where $V = \{v_1, v_2, \cdots, v_n\}$ and $E = \{(v_i, v_{i+1})|1 \leq i \leq n-1\}$. The vertex v_1 and v_n are the two end vertices (leaves) of L. The length of the path L is $length(L) = n-1$.*

Observation 1. *In a path L as it is defined above,*

$$b(L, v_i) = \begin{cases} max(i-1, n-i) & \text{if } i-1 \neq n-i \\ i & \text{otherwise} \end{cases}$$

The proof of the observation is simple. The optimal broadcast scheme originated from one of the two end vertices simply make calls along the path, which takes $d(v_1, v_n) = length(L)$ time units. For the originator v_i other than the two end vertices, it needs to call along the direction to the farther end vertex first, and the other direction second, which is $max(d(v_1, v_i), d(v_i, v_n)) = max(i-1, n-i)$. If the originator is right in the middle of the path, calling which direction first makes no difference. Broadcasting on one direction takes $d(v_1, v_i) = d(v_i, v_n)$ time units. But the broadcast time $b(L, v_i)$ requires one extra time unit. So it is equal to $d(v_1, v_i) + 1 = i$.

Observation 2. *The broadcast center BC_L of the path L on n vertices is*

- $\{v_{\frac{n}{2}}, v_{\frac{n}{2}+1}\}$ *if n is even;*
- $\{v_{\frac{n+1}{2}-1}, v_{\frac{n+1}{2}}, v_{\frac{n+1}{2}+1}\}$ *if n is odd.*

Proof. The proof is done by illustrating the broadcast time from different origi-nators for all cases. Assume n is even.

- If the originator is $v_{\frac{n}{2}}$, $\frac{n}{2}-1 \neq n-\frac{n}{2}$. By Observation 1, $b(L, v_{\frac{n}{2}}) = max(\frac{n}{2}-1, n-\frac{n}{2}) = \frac{n}{2}$.
- Similarly, for the originator $v_{\frac{n}{2}+1}$, $b(L, v_{\frac{n}{2}+1}) = \frac{n}{2}$.
- If the originator v_i is neither $v_{\frac{n}{2}}$ nor $v_{\frac{n}{2}+1}$, assume $i < \frac{n}{2}$ without lose of generality. Since $i-1 < n-i$ and by Observation 1 again, $b(L, v_i) = max(i-1, n-i) = n-i > \frac{n}{2}$.

Therefore, $BC_L = \{v_{\frac{n}{2}}, v_{\frac{n}{2}+1}\}$ when n is even.

When n is odd, the proof is similar. The only difference is that $v_{\frac{n+1}{2}}$ is the vertex in the middle because $\frac{n+1}{2}-1 = n-\frac{n+1}{2}$. □

Definition 2. *A star on n vertices is a graph $S = (V, G)$, where $V = \{v_1, \cdots, v_n\}$ and $E = \{(v_1, v_i)|2 \leq i \leq n\}$. v_1 is the center while $v_2, \cdots, v_n$ are the leaves.*

Observation 3. *For any vertex v_i in S, $b(S, v_i) = deg(v_1) = n-1$.*

Proof. If the originator is the center vertex v_1, all of v_1's neighbors are pendent vertices which are only adjacent to v_1. To finish broadcasting, v_1 needs to inform them one by one. So, the broadcast time $b(S, v_1) = n-1$.

If the originator is a leaf v_i, v_i informs the center v_1 in the first time unit. Then, v_1 calls other neighbors one by one. Thus, $b(S, v_1) = n-1$. □

From Observation 3, we directly get

Observation 4. *For a star S, the broadcast center $BC_S = V(S)$.*

3   NP-Hardness of BCD

Our proof of NP-hardness of BCD is simple. We assume that there is an oracle B which solves BCD in polynomial time and try to construct an algorithm to solve BTP in polynomial time by calling the oracle as a subroutine.

The sketch of the reduction is as follows. Assume that an instance for BTP is (G, v, t). Directly calling the oracle B on the instance $(G, \{v\})$ does not work because it can only tell whether $\{v\}$ is a subset of the broadcast center or not. That is, whether broadcasting from the vertex v uses shorter time than from other vertices, but not the broadcast time from v in the graph G. Thus, we construct a new graph G' by attaching a path L to the originator v in the graph G (see Fig. 1 for example). Broadcasting from v in G' can be regarded as two separated broadcastings, one from v in the graph G and the other one from v in L.

Now, let us consider broadcasting from an arbitrary vertex u in G'. By the construction of G', u can be either a vertex in G or a vertex in L. If u is a vertex in G, u has to inform v before informing all other vertices on the path L. There is no other way to call the vertices on the path without going through v. Thus, if the broadcast time from v to all vertices on the path L is large, say more than $b(G, u)$ for example, the optimal broadcast scheme $\mathcal{S}(G', u)$ needs to inform v as quickly as possible and let v call other vertices in L immediately. So, the broadcast time $b(G', u)$ depends on the time it takes informing v from the originator u, plus the time it takes informing all other vertices of L from v. This time is obviously larger than the time it takes from v to other vertices in L. In addition, if the optimal broadcast scheme $\mathcal{S}(G', v)$, originating from v, lets v call the vertices in L first when $b(L, v) > b(G, v)$, we can trivially conclude that $b(G', u) > b(G', v)$. In other words, if $b(L, v)$ is large enough, broadcasting from v takes a shorter time than from another vertex u, such that u is in the graph G. And by Observation 1, $b(L, v) = length(L)$ because v is an end of L.

A similar analysis can be done on the vertices in L. And the conclusion is that when the length of L is small enough, broadcasting from v takes a shorter time than from those on the path. After combining the results from the above discussions, when the length of L is carefully selected, broadcasting from v in G' requires the minimum time among all vertices. Then, v is in the broadcast center.

To find the correct length of L, our reduction loops on the length of L from a small value. Each iteration calls the oracle B, which solves BCD problem in polynomial time, to check whether the vertex v is in the broadcast center. Note that in our real reduction, some additional conditions need to be verified, even when $B(G', v)$ returns a positive answer. This condition checking is for handling some exceptions.

The reduction is formally presented as in Algorithm 1. Lines 2 and 3 give the initial construction of G'. Suppose L denotes the path $\{(v, v_1), (v_1, v_2)\}$. G' is defined as graph G and attached path L at the vertex v. The length of L is represented by i, and is equal to 2 initially. In each iteration, if v and v_2 are not both in $BC_{G'}$, the algorithm extends L by attaching one vertex to the end of L on the opposite side of v.

The initialization and each iteration of the algorithm obviously takes a constant time. Next, if the "While" loop excutes in polynomial time, then the algorithm is a polynomial-time reduction. The poof is split into several lemmas. At the same time, we prove the reduction is correct.

Lemma 1. *If $length(L) > b(G, v) + 2$, then $v \notin BC_{G'}$.*

Proof. To prove $v \notin BC_{G'}$, we will show that $b(G', v_1) < b(G', v)$. Assume v_i is the last vertex on the path L. If the originator is v, then the broadcasting from originator v takes at least $d(v, v_i) = length(L)$ time units to broadcast along the path. Thus, $b(G', v) \geq length(L)$.

Algorithm 1: $BTP \leq_p BCD$

Input : Graph $G = (V, E)$ on $n \geq 3$ vertices and a vertex v.

Output: The broadcast time originated from v in G

1 **begin**

2   Construct a new graph $G' = (V', E')$;

3   Let $V' = V \cup \{v_1, v_2\}$ and $E' = E \cup \{(v, v_1), (v_1, v_2)\}$;

4      $i \longleftarrow 2$;

5   **while** $\text{B}(G', \{v, v_2\}) = 0$ **do**

6          $V(G') \longleftarrow V(G') \cup \{v_i\}$;

7          $E(G') \longleftarrow E(G') \cup \{(v_{i-1}, v_i)\}$;

8          $i++$;

9   **end**

10   Return $i - 2$;

11 **end**

Then, we consider the broadcasting from v_1. Since v_1 is on the path, it has only two neighbors, v and v_2. So, broadcasting from v_1 has only two choices.

1. v_1 informs v_2 in the first time unit. Then, v_2 makes calls along the path. In parallel, v_1 informs v in the second time unit. Then, v broadcasts in G. This broadcast scheme takes $t_1 = max\{length(L) - 1, b(G, v) + 2\}$ time units. $length(L) - 1$ is given by the distance from the vertex v_1 to v_i. $b(G, v) + 2$ appears here because v is informed in the second time unit, and broadcasting from the originator v to all other vertices in G takes $b(G, v)$ time.
2. v_1 informs v in the first time unit and v_2 in the second time unit. The rest of broadcasting are the same as in case 1. Similarly, this broadcast scheme takes $t_2 = max\{length(L), b(G, v) + 1\}$ time units.

Therefore, the optimal broadcast scheme from the originator v_1 depends on the values of t_1 and t_2. From the assumption of Lemma 1, $length(L) > b(G, v) + 2$ and $length(L) - 1 > b(G, v) + 1$. Thus, $t_1 = length(L) - 1 < t_2 = length(L)$. Then, $b(G', v_1) = length(L) - 1$.

Comparing with $b(G', v) \geq length(L)$, $b(G', v_1)$ is smaller. Therefore, v is not in the broadcast center of G', $v \notin BC_{G'}$. $\square$

Lemma 2. *If* $length(L) < b(G, v) + 2$, *then* $v_2 \notin BC_{G'}$.

Proof. Similar to Lemma 1, we can show that $b(G', v_1) < b(G', v_2)$. As a consequence, $v_2 \notin BC_{G'}$.

If the originator is v_2, then v_2 needs at least 2 time units to let v get the information. Then, v needs at least $b(G, v)$ time units to broadcast within the graph G. Thus, $b(G', v_2) \geq b(G, v) + 2$.

If the originator is v_1, one can easily derive

$$b(G', v_1) = min\{max\{length(L) - 1, b(G, v) + 2\}, max\{length(L), b(G, v) + 1\}\}$$

From the assumption, $length(L) < b(G, v) + 2$. Thus, $b(G', v_1) = min\{b(G, v) + 2, b(G, v) + 1\} = b(G, v) + 1$, which is smaller than $b(G', v_2) \geq b(G, v) + 2$.

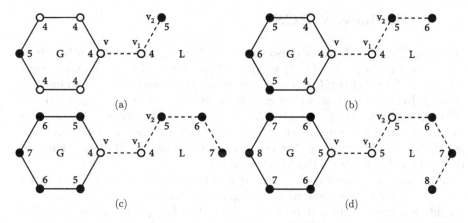

Fig. 1. This is an example to show how the reduction works. Each G' consists of the original graph G and a path L. The solid edges are the edges in G. The dashed edges are the ones on the path L. And the hollow vertices are the ones in the broadcast center. The number beside each vertex is the broadcast time. Initially (a), L contains only three vertices v, v_1, and v_2. If v and v_2 are not both in the broadcast center (a–c), $length(L) < b(G, v) + 2$ and the algorithm attaches one extra vertex to L at the other end. When v and v_2 are both in the broadcast center (d), $length(L) = b(G, v) + 2$ and the algorithm stops.

Therefore, broadcasting from v_1 in G' takes a shorter time than from v_2. Hence, v_2 is not in the broadcast center of $BC_{G'}$, $v_i \notin BC_{G'}$. $\square$

When we summarize Lemma 1 and Lemma 2, we get that

1. if $v \in BC_{G'}$, then $length(L) \leq b(G, v) + 2$;
2. if $v_2 \in BC_{G'}$, then $length(L) \geq b(G, v) + 2$.

Therefore,

Lemma 3. *If $v, v_2 \in BC_{G'}$, $length(L) = b(G, v) + 2$.*

So, the "While" loop in Algorithm 1 stops when $\{v, v_2\}$ is a subset of $BC_{G'}$ and $B(G', \{v, v_2\}) = 1$. Then, the algorithm obtains $b(G, v)$ by returning $i - 2 = length(L) - 2$.

Furthermore, the length of L is initialized as 3 since the reduction needs to make sure that $length(L) < b(G, v) + 2$ initially. So, v_2 cannot be in $BC_{G'}$ when the algorithm starts. Then, as $length(L)$ increases, v and v_2 are both in $BC_{G'}$ when $length(L) = b(G, v) + 2$. The assumption of the reduction ensures that $length(L) < b(G, v) + 2$ at the begining. The input graph G is on $n \geq 3$ vertices. Thus, $b(G, v) \geq \lceil \log(3) \rceil = 2$, which is larger than $length(L) - 2 = 3 - 2 = 1$.

Therefore, the reduction is correct. At the same time, the "While" loop runs $b(G, v) + 2 - 2 = b(G, v)$ rounds. Thus, the reduction takes polynomial time. Figure 1 shows an example when G has 6 vertices and $b(G, v) = 3$.

Theorem 2. *BCD is NP-hard.*

4   NP-Hardness of BCS

Following Theorem 2, we can also show that finding the size of the broadcast center for a given graph (BCS) is NP-hard in general. The reduction is again from BTP. Recall that (G, v, t) is an instance of BTP, and (G'', x) is an instance of BCS. In the text below, we restrict the input graph of BCS being denoted by G'' to distinguish between graphs G or G'. However, the difference in notation should not cause misunderstanding because we will consider our initial problems of BTP and BCS. Suppose that the oracle $\mathtt{C}(G'', x)$ solves BCS in polynomial time. The reduction is similar to Algorithm 1.

First note that, the direct construction of the graph G'' by attaching a path L to G similar to the construction of graph G' for BCD problem, does not provide a correct reduction $BTP \leq_p BCS$. The key point for the reduction $BTP \leq_p BCD$ is to let $\{(v, v_1), (v_1, v_2)\}$ form a bridge between G and the rest of part L. When v and v_2 are both in the broadcast center, then the two sides of the bridge are "balanced" in the sense of broadcasting. This works for BCD because the oracle $\mathtt{B}$ can varify whether some specific vertices are in the broadcast center or not. But for BCS, the oracle $\mathtt{C}$ can only tell the size of the broadcast center but cannot decide if some vertices are in the broadcast center. If G'' is constructed by attaching the path, then the function call of $\mathtt{C}$ needs to be $\mathtt{C}(G'', 2)$ or $\mathtt{C}(G'', 3)$ since by Observation 2, the broadcast center of a path has only 2 or 3 vertices. Then, the main issue is that when the function call $\mathtt{C}(G'', 2)$ returns positive, we cannot guarantee that v and v_2 are in the broadcast center. It is possible that some totally different vertices in G or on the path L are in $BC_{G''}$. As a consequence, the broadcast time cannot be derived by the length of L when $\mathtt{C}(G'', 3)$ is positive.

In fact, following Observation 4, every vertex in a star is in the broadcast center of the star graph. This property helps the reduction from BTP to BCS because here we gain the control on the size of the broadcast center. Therefore, the graph G'' is constructed by attaching a star S on $i+1$ vertices to the vertex v in the graph G. Formally, $V(G'') = V(G) \cup \{v_1, v_2, \cdots, v_i\}$ and $E(G'') = E(G) \cup E(S)$ where $E(S) = \{(v, v_1), (v_1, v_2), (v_1, v_3), \cdots, (v_1, v_i)\}$. And the vertex set of the star S is $V(S) = \{v, v_1, v_2, \cdots, v_i\}$. So, the center of S is v_1 of degree $deg(v_1) = i$ and the size of S is $size(S) = deg(v_1) + 1 = i + 1$.

From the construction the edge (v, v_1) is a bridge between G and S. If S is heavier in the sense of broadcasting, then the center v_1 is in the broadcast center, because it is the only vertex that can call other vertices in S. At the same time, the other leaves are also in the broadcast center because it makes no difference which leaf is called by v_1 first, as long as $size(S)$ is large enough. On the other hand, if G is heavier, then the leaves of S cannot be in the broadcast center. Thus, the reduction starts from $size(S) = n + 2$ (which is greater than $b(G, v)$); make function calls $\mathtt{C}(G'', size(S))$; reduce $size(S)$ by one if the subroutine returns "Yes"; and keep doing this until the $\mathtt{C}(G'', size(S))$ returns "No". The above two cases are formally proved later to ensure correctness of the reduction. Formally, the reduction is as follows.

Algorithm 2: $BTP \leq pBCS$

Input : Graph $G = (V, E)$ on $n \geq 3$ vertices and a vertex $v \in V$.
Output: The broadcast time originated from v in G

1 **begin**
2   Construct a new graph $G'' = (V'', E'')$;
3   Let $V'' = V \cup \{v_1, v_2, \cdots, v_{n+1}\}$ and
      $E'' = E \cup \{(v, v_1), (v_1, v_2), (v_1, v_3), \cdots, (v_1, v_{n+1})\}$;
4     $i \longleftarrow n + 3$;
5   **while** $\mathsf{C}(G'', i)$ *and* $i \geq 1$ **do**
6         $V(G'') = V(G'') \backslash \{v_{i-1}\}$;
7         $E(G'') = E(G'') \backslash \{(v_1, v_{i-1})\}$;
8         $i - -$;
9   **end**
10   **return** $i - 2$;
11 **end**

The variable i on Line 4 is equal to $size(S)$, which is initialized to be $n + 3$. In iteration, if $\mathsf{C}(G'', i)$ returns positive, the leaves are in the broadcast center. Line 6 and 7 removes one leaf from the star. Then, the algorithm continues the loop.

The construction of G'' takes $O(n)$ time to attach the star S; and the "While" loop runs in at most $n + 3$ iterations. Thus, the algorithm is a polynomial time reduction. Next, we only need to prove the correctness of our reduction.

Lemma 4. *If* $size(S) \geq b(G, v) + 3$, $V(S) = BC_{G''}$.

Proof. We prove the lemma by showing

1. $b(G'', v_2) = b(G'', v_3) = \cdots = b(G'', v_i) = size(S) - 1$;
2. $b(G'', v_1) = size(S) - 1$;
3. $b(G'', v) = size(S) - 1$;
4. For all $u \in V(G) \backslash \{v\}$, $b(G'', u) > size(S) - 1$

Case 1: Since all of $v_2, v_3, \cdots, v_i$ are leaves of the star S. Thus, without loss of generality, the proof considers only the case when v_2 is the originator.

To inform all vertices in S, v_2 needs 1 time unit1 to call the center v_1. Then, v_1 needs another $deg(v_1) - 1 = size(S) - 2$ time units to inform all other leaves in S. Thus,

$$b(G'', v_2) \geq size(S) - 1 \tag{1}$$

Next, consider the following broadcast scheme. v_2 informs v_1 in the first time unit. v_1 informs v in the second time unit. Then, in parallel v broadcasts in G in the next $b(G, v)$ time units; and v_1 calls other leaves in S in the next $deg(v_1) - 2$ time units. By counting the time units, $b(G'', v_2) \leq max\{b(G, v) + 2, deg(v_1)\} =$

$max\{b(G, v)+2, size(S)-1\}$. Sine the lemma assumes that $size(S) \geq b(G, v)+3$, we obtain

$$b(G'', v_2) \leq size(S) - 1 \tag{2}$$

Combining Eq. 1 and 2, we get that $b(G'', v_2) = size(S) - 1$.

Case 2: When the broadcasting originates from v_1, v_1 needs at least one time unit to inform v. Then, v can broadcast in G. At the same time, v_1 can call other leaves. Thus, $b(G'', v_1) = max\{deg(v_1), b(G, v) + 1\} = max\{size(S) - 1, b(G, v) + 1\}$. In fact, $size(S) - 1 \geq b(G, v) + 1$ because Lemma 4 has assumed $size(S) \geq b(G, v) + 3$. So, $b(G'', v_1) = size(S) - 1$.

Case 3: If the originator is v, we further assume v_1 is informed by v in time unit t in the optimal broadcast scheme. After v_1 is informed, it needs another $deg(v_1) - 1$ time units to call other leaves. So, the broadcasting in S finishes in time unit $deg(v_1) - 1 + t = size(S) - 2 + t$, for $t \geq 1$. Then, consider the broadcasting on the side of the graph G. v has to waste one time unit to call v_1. So, the broadcasting in G is delayed by one time unit and finishes in time unit $b(G, v) + 1$. Thus, $b(G'', v) = max\{size(S) - 2 + t, b(G, v) + 1\}$ for $t \geq 1$. Further, we know that $size(S) \geq b(G, v) + 3$ by the assumption, which implies the minimum of $size(S) - 2 + t$ is $size(S) - 1 \geq b(G, v) + 1$. Therefore, $b(G'', v) = size(S) - 1$.

Case 4: If the originator is an arbitrary vertex $u \in V(G)\backslash\{v\}$, u needs at least $d(u, v)$ time units to call v. Then, v needs at least $size(S) - 1$ time units to call every vertex in S. Thus, $b(G'', v) \geq d(u, v) + size(S) - 1 > size(S) - 1$.

By summarizing the above 4 cases, the lemma is proved. □

Lemma 5. *If $size(S) < b(G, v) + 3$, $v_2, v_3, \cdots, v_i \notin BC_{G''}$.*

Proof. Without loss of generality, we pick the leaf v_2 and show that $b(G'', v_2) > b(G'', v_1)$. If broadcasting from v_1 takes a shorter time than from v_2, then v_2 is not in the broadcast center.

If the originator is v_2, a broadcast scheme requires at least 2 time units from v_2 to v. And v needs $b(G, v)$ time units to broadcast in G. Thus, $b(G'', v_2) \geq b(G, v) + 2$.

If the originator is v_1, consider the following broadcast scheme. v_1 informs v in the first time unit. v start the broadcasting in G, which finishes in time unit $b(G, v) + 1$. Then, v_1 informs other leaves one by one and this part of calls finish in time unit $size(S) - 1$. This broadcasting takes $max\{b(G, v) + 1, size(S) - 1\}$ time units, which is equal to $b(G, v) + 1$ because $size(S) < b(G, v) + 3$ from the assumption. Thus, $b(G'', v_1) \leq b(G, v) + 1$ which is smaller than $b(G'', v_2)$. □

Lemma 6. *For any vertex $u \in V(G)\backslash\{v\}$, if $size(S) \geq b(G, v) + 2$, then $u \notin BC_{G''}$.*

Proof. This proof is the same as the one for Lemma 5. By following the same analysis, we can derive $b(G'', u) \geq size(S)$ and $b(G'', v_1) = size(S) - 1$. Thus, u cannot be in the broadcast center. □

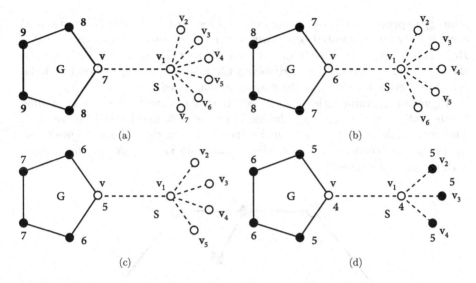

Fig. 2. This is an example to show how the reduction $BTP \leq_p BCS$ works. Each G'' consists of the original graph G and a star S. The solid edges are the edges in G. The dashed edges are the ones on the path S. And the hollow vertices are the ones in the broadcast center. The number beside each vertex is the broadcast time.

Now, we are ready to claim the correctness of our reduction. If $size(S) \geq b(G, v) + 3$, Lemma 4 states $V(S) = BC_{G''}$, which directly imples that $\mathsf{C}(G'', size(S)) = 1$. If $size(S) = b(G, v) + 2$, Lemma 5 and Lemma 6 together show that $BC_{G''}$ contains at most two vertices v and v_1. Thus, $\mathsf{C}(G'', size(S)) = 0$. So, when $\mathsf{C}(G'', size(S))$ returns negative the first time, the "While" loop in the algorithm terminates and we know that $i = size(S) = b(G, v) + 2$. Therefore, the reduction finds the correct solution for the broadcast time, and we claim the final result.

Theorem 3. *BCS is NP-hard.*

Figure 2 gives an example of the reduction when G is on 5 vertices and $b(G, v) = 3$. Initially (Fig. 2a), S contains 8 vertices $v, v_1, \cdots, v_7$. If $\mathsf{C}(G'', size(S)) = 1$ (Fig. 2a–2c), one leaf is removed from S in each iteration. When $\mathsf{C}(G'', size(S)) = 0$ (Fig. 2d), $size(L) = b(G, v) + 2$ and the algorithm stops.

5   Properties of BC

The above section shows that finding the broadcast center is NP-complete. Then, one may try to find some good approximation or heuristic algorithms to find local optimal solutions for the problem. One natural approach is starting from an arbitrary vertex in the graph, then moving to its neighbor whose broadcast time is approximately smaller. Eventually, the algorithm stops at a vertex or a group of vertices whose broadcast time is minimal. The performance of this strategy depends on the convexness of the broadcast time over the vertices. If there is only one connected broadcast center ("connected" means that $\forall u, v \in BC$, there exists a path P from u to u such that $\forall w$ on P, $w \in BC$), then the algorithm works effectively. However, the following example shows that the broadcast center for some graphs may be disconnected.

Figure 3 is a graph with two isolated broadcast centers. The broadcast time for the hollow vertices is 5, for the solid vertices is 6, and for the squares is 7. The two broadcast centers are split by the vertices in the middle. Figure 3 also shows that the vertices in the broadcast center are not necessarily the vertices with the highest degrees.

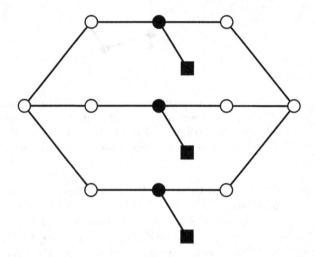

Fig. 3. An example of multiple broadcast centers

6   Conclusion and Future Works

This paper shows that "whether a given set of vertices is a broadcast center to the graph?" and "does a graph have a broadcast center of size no smaller than x?" are both NP-hard problems. These results provide solid theoretical foundations and motivations for designing algorithms to find broadcast centers only for special classes of graphs.

In the future, the new results can also potentially assist the studies on the broadcast graph problem, which is another problem that have been studied for decades while the complexity remains open.

References

1. Averbuch, A., Shabtai, R.H., Roditty, Y.: Efficient construction of broadcast graphs. Discret. Appl. Math. **171**, 9–14 (2014)

2. Bar-Noy, A., Guha, S., Naor, J.S., Schieber, B.: Multicasting in heterogeneous networks. In: Proceedings of the Thirtieth Annual ACM Symposium on Theory of Computing, (STOC), pp. 448–453. ACM (1998)
3. Beier, R., Sibeyn, J.F.: A powerful heuristic for telephone gossiping. In: Proceedings of the 7th Colloquium on Structural Information and Communication Complexity (SIROCCO), pp. 17–35. Carleton Scientific (2000)
4. Bermond, J.-C., Fraigniaud, P., Peters, J.G.: Antepenultimate broadcasting. Networks 26(3), 125–137 (1995)
5. Bermond, J.-C., Hell, P., Liestman, A.L., Peters, J.G.: Sparse broadcast graphs. Discret. Appl. Math. 36(2), 97–130 (1992)
6. Dinneen, M.J., Ventura, J.A., Wilson, M.C., Zakeri, G.: Compound constructions of broadcast networks. Discret. Appl. Math. 93(2), 205–232 (1999)
7. Elkin, M., Kortsarz, G.: Sublogarithmic approximation for telephone multicast: path out of jungle (extended abstract). In: Proceedings of the Fourteenth Annual ACM-SIAM Symposium on Discrete Algorithms (SODA), pp. 76–85. Society for Industrial and Applied Mathematics (2003)
8. Elkin, M., Kortsarz, G.: A combinatorial logarithmic approximation algorithm for the directed telephone broadcast problem. SIAM J. Comput. 35(3), 672–689 (2005)
9. Farley, A.M.: Minimal broadcast networks. Networks 9(4), 313–332 (1979)
10. Farley, A.M., Hedetniemi, S., Mitchell, S., Proskurowski, A.: Minimum broadcast graphs. Discret. Math. 25(2), 189–193 (1979)
11. Fraigniaud, P., Lazard, E.: Methods and problems of communication in usual networks. Discret. Appl. Math. 53(1–3), 79–133 (1994)
12. Garey, M.R., Johnson, D.S.: Computers and Intractability: A Guide to the Theory of NP-Completeness. W.H. Freeman, New York (1979)
13. Gargano, L., Vaccaro, U.: On the construction of minimal broadcast networks. Networks 19(6), 673–689 (1989)
14. Grigni, M., Peleg, D.: Tight bounds on mimimum broadcast networks. SIAM J. Discret. Math. 4(2), 207–222 (1991)
15. Grigoryan, H., Harutyunyan, H.A.: New lower bounds on broadcast function. In: Gu, Q., Hell, P., Yang, B. (eds.) AAIM 2014. LNCS, vol. 8546, pp. 174–184. Springer, Cham (2014). https://doi.org/10.1007/978-3-319-07956-1_16
16. Harutyunyan, H.A.: Broadcast networks with near optimal cost. In: Gu, Q., Hell, P., Yang, B. (eds.) AAIM 2014. LNCS, vol. 8546, pp. 312–322. Springer, Cham (2014). https://doi.org/10.1007/978-3-319-07956-1_28
17. Harutyunyan, H.A., Li, Z.: A simple construction of broadcast graphs. In: Du, D.-Z., Duan, Z., Tian, C. (eds.) COCOON 2019. LNCS, vol. 11653, pp. 240–253. Springer, Cham (2019). https://doi.org/10.1007/978-3-030-26176-4_20
18. Harutyunyan, H.A., Li, Z.: A new construction of broadcast graphs. Discret. Appl. Math. 280, 144–155 (2020)
19. Harutyunyan, H.A., Liestman, A.L.: More broadcast graphs. Discret. Appl. Math. 98(1), 81–102 (1999)
20. Harutyunyan, H.A., Liestman, A.L.: Upper bounds on the broadcast function using minimum dominating sets. Discret. Math. 312(20), 2992–2996 (2012)
21. Harutyunyan, H.A., Liestman, A.L., Peters, J.G., Richards, D.: Broadcasting and gossiping. In: Handbook of Graph Theory, pp. 1477–1494. Chapman and Hall (2013)
22. Harutyunyan, H.A., Liestman, A.L., Shao, B.: A linear algorithm for finding the k-broadcast center of a tree. Networks 53(3), 287–292 (2009)

23. Harutyunyan, H., Maraachlian, E.: Linear algorithm for broadcasting in unicyclic graphs. In: Lin, G. (ed.) COCOON 2007. LNCS, vol. 4598, pp. 372–382. Springer, Heidelberg (2007). https://doi.org/10.1007/978-3-540-73545-8_37

24. Harutyunyan, H.A., Maraachlian, E.: On broadcasting in unicyclic graphs. J. Comb. Optim. **16**(3), 307–322 (2008)

25. Hedetniemi, S.M., Hedetniemi, S.T., Liestman, A.L.: A survey of gossiping and broadcasting in communication networks. Networks **18**(4), 319–349 (1988)

26. Hromkovič, J., Klasing, R., Monien, B., Peine, R.: Dissemination of Information in Interconnection Networks (Broadcasting & Gossiping). In: Du, D.Z., Hsu, D.F. (eds.) Combinatorial Network Theory. Applied Optimization, vol. 1, pp. 125–212. Springer, Boston (1996). https://doi.org/10.1007/978-1-4757-2491-2_5

27. Čevnik, M., Žerovnik, J.: Broadcasting on cactus graphs. J. Comb. Optim. **33**(1), 292–316 (2015). https://doi.org/10.1007/s10878-015-9957-8

28. Maheo, M., Saclé, J.-F.: Some minimum broadcast graphs. Discret. Appl. Math. **53**(1–3), 275–285 (1994)

29. Slater, P.J., Cockayne, E.J., Hedetniemi, S.T.: Information dissemination in trees. SIAM J. Comput. **10**(4), 692–701 (1981)

30. Su, Y.-H., Lin, C.-C., Lee, D.T.: Broadcasting in heterogeneous tree networks. In: Thai, M.T., Sahni, S. (eds.) COCOON 2010. LNCS, vol. 6196, pp. 368–377. Springer, Heidelberg (2010). https://doi.org/10.1007/978-3-642-14031-0_40

31. Zhou, J., Zhang, K.: A minimum broadcast graph on 26 vertices. Appl. Math. Lett. **14**(8), 1023–1026 (2001)

An Online Algorithm for Data Caching Problem in Edge Computing

Xinxin Han[1,2], Guichen Gao[1], Yang Wang[1], and Yong Zhang[1(✉)]

[1] Shenzhen Institute of Advanced Technology, Chinese Academy of Sciences, Beijing, China
{xx.han,gc.gao,yang.wang1,zhangyong}@siat.ac.cn
[2] University of Chinese Academy of Sciences, Beijing, China

Abstract. In edge computing, the edge server can provide highly accessible computing and storage resources for nearby application users. Caching data on the edge server can retrieve application data to ensure rapid completion of users' requests with low latency. However, edge servers usually have limited resources and sometimes have to transfer data from other servers or buy data from the data center. In this paper, we study data caching problem in edge computing in order to minimize the system cost, including data caching cost, data transmission cost and data purchase cost. The traditional data center is only the choice to purchase data, the data is always available after purchasing. However, we consider that data has an expiration date. It only provides a finite number of data invoking to buyers for free after purchasing data. Based on this assumption, when the future arrival requests are unknown, we give an algorithm to minimize the total cost for satisfying users' requests. We prove that the asymptotic performance ratio of the algorithm can be guaranteed with 3.

Keywords: Edge computing · Data caching · Transmission · Online algorithm · Competitive ratio

1 Introduction

Over the past decade, with the rapid increase of mobile devices in daily life, internet-based mobile data has grown dramatically. It is increasingly difficult for traditional cloud computing to support massive, diverse and rapidly growing mobile services or applications. As an effective way, edge computing is a distributed computing paradigm with advantages of reducing latency [1,2], congestion and cost. This pattern is achieved by deploying a common edge server at a base station or access point near the mobile user. Edge servers provide services such as caching, computing, and communication to service agents. Data caching on the edge server can reduce the delay of mobile users' accessing popular data and reduce the data traffic between mobile users and centralized cloud [3] and [4].

Data caching in edge computing is a common technique that allows a copy of the data to be stored somewhere in order to quickly retrieve the data for

© Springer Nature Switzerland AG 2021
W. Wu and H. Du (Eds.): AAIM 2021, LNCS 13153, pp. 71–80, 2021.
https://doi.org/10.1007/978-3-030-93176-6_7

future requests. Data caching is widely used in networks [5], databases [6], Web servers [7], multi-processors [8] and storage [9] to improve information retrieval efficiency and reduce data access delay. This is especially important for delay-sensitive applications, such as interactive games, real-time navigation, augmented reality, and more. How to cache data, especially popular data, to optimize system performance in these research areas [9]. In addition, caching data on edge servers can also reduce the traffic burden on the backbone of the Internet, as the amount of mobile traffic data transferred between the cloud and the mobile devices of application users is significantly reduced [10].

Many works studied data caching with a data center in their network structure [14,15]. However, in these studies, data center only provides an alternative to download data, which is always valid once purchased. They ignore the validity period of the data and the number of valid invoking. However, as far as we know, new data pricing is being considered based on the number of invoking. This hypothesis is more practical. In addition, the methods in these articles are based on future requests' information (distribution known or predicted) [11–13], which is not always feasible when predicting the type of request in the future is inaccurate.

In this paper, we consider a high-speed, fully connected network with a data center. 1) Servers are connected to each other; 2) When the required data exists on the target server, the user's requests can be satisfied immediately; 3) The time of data transmission between servers and the invocation time from the data center can be ignored, but the cost must be considered; Assuming that user requests can be met by the same type of data, when the requests of surrounding users arrive the edge server, the server needs to retrieve the required data to quickly fulfill the requests. The required data can be provided in three ways: stored on the server, transferred from other servers, or purchased and invoked from the data center. Each of the three data provisioning options has a corresponding cost, and we need to find a strategy to minimize the total cost for the user's online request streams. Our main results in this paper are summarized as follows.

- Consider a network with data centers, and the data is not always available for the server after buying data from the data center. Assume that one purchase of data can be invoked k times from the data center for free. We are the first to consider this data price model.
- We present an online algorithm and prove that the performance ratio of the algorithm is guaranteed by constant 3 under the worst case.

In the remaining part, the paper is organized as follows. Related work of data caching problem are shown in Sect. 2. In Sect. 3, we analyze the system model and give an online competitive algorithm. Then, we prove the performance ratio of the algorithm in Sect. 4. Finally, Sect. 5 concludes our main results and future work.

2   Related Work

Data caching is not a new problem, and there has been extensive research on its various forms and research objectives [19,20,23,26]. Based on the assumption that single data item can satisfy all the task, although it is not reasonable in reality, [26] provided relevant research about data caching with single data. About the homogeneous data caching model associated with single data in a high-speed network, that is, the storage cost of servers and the transfer cost between servers are both identical, [23] presented an optimal caching strategy in polynomial time minimizing the total cost of requests. Meanwhile, they gave a dynamic programming technique to minimize the total cost in fully connected topology about the heterogeneous data caching model with constraint which is NP-hard [16,22]. In [26], they proposed an offline optimal algorithm based on DP (dynamic programming) technique in polynomial time to minimize the total cost. Meanwhile, they also studied the online form of the problem and proved a 3-approximate algorithm by pre-caching. Following network architecture and models in [23], Wang et al. consider the data caching problem with multiple type of data contents [24]. Under their homogeneous model, they discuss different situation and improved the time complexity to $O(m^2n)$. Concretely, when a factor of transmission cost divided by storage cost is small, a single copy of every data item is enough to serve all the task. Moreover, they consider the multicopy situation. In addition, about the NP-hard heterogeneous model, they present a DP-based algorithm to solve the heterogeneous model with constrains and give a approximation algorithm about the general model with the performance ratio 2. In [21], they proposed a greedy heuristic strategy to decide when, where, and how long to cache about minimizing the total cost of data caching.

There exist rich research works focus on online data caching problem. When the popularities of the content item isn't known exactly, [17] gave a constant competition online algorithm in content centric networking to minimize the total cost. Based on the identical target function, if they didn't know any knowledge about the popularities of contents, [18] presented an online algorithm with the competitive ratio of $O(\log n)$ for the collaborative caching model in multiple cluster collaborative systems. Considering the idea of co-migration, Wang et al. presented a random competition algorithm which is a parallel dynamic programming algorithm based on the combination of branch and bound strategy and sampling technique [25]. Wang et al. studied a fresh homogeneous cache model, i.e. the transimission cost of the content between servers and caching cost of every edge node are fixed respectively [26]. In their paper, they presented a $O(mn)$ polynomial time algorithm upon dynamic programming to minimize total cost, where m and n are the number of nodes and the number of requests respectively. In addition, they utilize the idea of data caching in advance to minimize the total cost for the online model, and the given algorithm can achieve 3 competition ratio. Subsequently, they extended their analysis to semi-homogeneous model, the transmission cost of the data is identical, but the storage cost of servers depends on the storage capacity of the server itself. [27] gave an offline algorithm to gain the optimal caching strategy with time complexity $O(mn \log(mn))$ and presented a 2-competitive algorithm by tradeoff the transimission cost and the caching cost for the online model of this problem. Han et al. studied data caching problem with mutiple data types [28]. They gave corresponding online algorithms for three different data caching models.

3 System Model

In this section, we propose our new data caching model, which is different from the previous model. To better define the model, some basic parameters and symbols are shown in Table 1.

In the fully connected network, there is a data center which stores the data all the way and server set $S = \{1, 2, \cdots, m\}$ for our system model. Servers can be connected with each other by wireless network. Meanwhile, servers connect the data center by links. When users' requests arrive at the server, the target server needs to provide data for request. At the beginning, there is no data stored at servers and servers have to buy data from the data center. The server buys data from the data center once for cost B, and then it can call this data k times for free. After that, there are three options for the target server to supply the data for the request. (1) The server can store the data for the request and the caching cost per unit time is μ_j; Or (2) the server can transfer the data to the target server for the transmission cost λ ($\lambda < B$); Or (3) the server can call the data from the data center for free. The objective is minimizing the total cost of online request stream to satisfy all users' requests.

Table 1. Parameters definition

Parameter	Paraphrase/Significance
$S = \{1, 2, \cdots, m\}$	The server set which contains m servers
$r_j^i = (j, t_j^i)$	The $i - th$ request arrives server j at time $t = t_j^i$
λ	The transmission cost of the data among servers
μ_j	The pet unit time cost of storing data on server j
B	The cost of buying the data from the data center
N_j	A variable
Δt_j	The retention time of the data on server j when $N_j = k \mid N_j = 0$
Δt_j^B	The retention time of the data on server j when $0 < N_j < k$
Δ	Time interval
$C_A(r_j^i)$	The cost of request r_j^i of the algorithm A
$C_{OPT}(r_j^i)$	The optimal cost for request r_j^i
C_A	The total implementation cost of arbitrary deterministic algorithm $\mathcal{A}$
OPT	The optimal total cost of the first n requests
c	A positive constant
k	A positive constant, the number of calling time for free from the data center

4 Model Analysis

After purchasing data from the data center, assume that the average invocation cost is lower than the transmission cost from other servers, but not too small in practical. i.e. $\frac{\lambda}{2} \leq \frac{B}{k} \leq \lambda$. Upon this assumption, we propose the following balance between caching and transmission online algorithm (called BBCTO Algorithm) for any online request stream.

Without losing generality, we assume $\mu_1 \leq \mu_2 \leq \cdots \leq \mu_m$. In a centralized system, in order to minimize the total system cost, if servers have to buy data from the data center, server 1 is the most cost-effective. Because the storage price per unit time of server 1 is the cheapest. Use a variable N_j to record the number of data invocation after server j buys data from the data center, initialized to 0. $N_j = 0$ indicates that server j does not purchase data from the data center. And $N_j = k$ indicates that server j has used up the number of free invocation for the purchased data. Define $\Delta t_j = \frac{\lambda}{\mu_j}$, $\Delta t_j^B = \frac{B/k}{\mu_j}$.

1. Server j that does not purchase data from data center or ran out of free data calls

- Cache: The retention period of the data is Δt_j;
- Transfer/Copy: When a new request arrives at server j, if there exists data at the target server, then the request will be satisfied; else if there exist data copies among other servers, then the request can be satisfied by data transmission from other servers; else if there is no data among servers and $0 < N_1 < k$, call data from the data center to server 1, then transfer the data to the target server.

2. Server 1 that purchases data from data center and has free data calls times $(N_1 < k)$

- Cache: The retention period of the data is Δt_1^B.
- Transfer/Copy: When a new request arrives at server 1, if there exists data at the target server, then the request will be satisfied; else if there exist data copies among other servers, then the request can be satisfied by data transmission from other servers; else call data from the data center.

3. In general, when is data deleted from the server

- Delete: After the holding time, if there exist data copies among servers or $0 < N_1 < k$, then delete the data from the current server. Otherwise, transfer data to server 1 and delete the data from the current server, then go on holding data on server 1 with the duration $\frac{B}{\mu_1}$. When $t = \frac{B}{\mu_1}$, data on the server 1 will be deleted. Buy data from the data center until next new request arrives, set $N_1 = 1$.

Observation 1. *After purchasing data from the data center, the free invocation is caused when there is no data cached on all the servers.*

Proof. After purchasing data from the data center, while data invocation is free, as long as there is data between servers, it is valuable to avoid call data from the data center. When the next request doesn't arrive too soon and there is no data on the server, calling data from the data center will play a great role.

Observation 2. *For the optimal cost, the previous increase in data caching time does not reduce the number of purchases.*

Proof. After buying data from the data center, according to the BBCTO Algorithm, data caching time on the server is Δt_j^B or Δt_j. If new request arrives at the server beyond the caching time, the strategy of algorithm is to call data from the data center, then the cost is $\frac{B}{k}$. While the strategy of optimal offline algorithm is calling data from the data center or caching data on the server. Suppose that in order to reduce the number of purchases, OPT chooses to store the data to meet the requests rather than invoke the data from the data center, then the cost is more than $\frac{B}{k}$, written as $\frac{B}{k}+$.

When the number of free calls runs out and there is no new request arriving at other servers, data is cached on server 1. When next new request arrives, the BBCTO Algorithm will buy data from the data center after the data on server 1 has been stored for $t = \frac{B}{\mu_1}$. While the strategy of optimal offline algorithm is to call data from the data center, then the cost is $\frac{B}{k}$. As the number of future requests increases, so calling data from the data center is the optimal strategy for the optimal cost.

Therefore, the ratio between BBCTO Algorithm and OPT is

$$\frac{C()}{OPT()} \leq \frac{\frac{B}{\mu_1}\mu_1 + B + k(\mu_1 \Delta t_1^B)}{B}.$$

Theorem 1. *After server 1 buys data from the data center, according to the BBCTO Algorithm, the total cost of k data calls caused by requests does not exceed 3 times the optimal cost.*

Proof. According to Observation 1, the free call is caused because the data is not cached on the server. There are two cases in an algorithm that can cause a free call: requests at the other servers which need data and there is no data copy among servers (case 1), or the request arrives the server after caching data Δt_1^B (case 2) and there is no data copy among servers. For k data calls from the data center, assume that the number of case 1 and case 2 are respectively l_1 and l_2. We prove the ratio by taking k calls after purchasing the data as a whole. Thus, the total cost of k calls is $\frac{B}{\mu_1}\mu_1 + B + l_2(\mu_1 \Delta t_1^B)$.

$$\frac{C_k()}{OPT_k()} \leq \frac{\frac{B}{\mu_1}\mu_1 + B + l_2(\mu_1 \Delta t_1^B)}{B} < 3.$$

Theorem 2. *The BBCTO online algorithm is a 3-competitive algorithm.*

Proof. We prove the ratio in detail from the following three cases.

- server j that does not purchase data from data center or has no free calling time

– a. The new request arrives at the server when $t \leq t_j^{i-1} + \Delta t_j$, according to the algorithm, the request will be satisfied directly.

$$\frac{C(r_j^i)}{OPT(r_j^i)} = \frac{\mu_j(t - t_j^{i-1})}{\mu_j(t - t_j^{i-1})} = 1.$$

– b. The new request arrives at the server when $t > t_j^{i-1} + \Delta t_j$ and there exists data copy at other servers, according to the algorithm, the request will be satisfied by transmission from other servers.

$$\frac{C(r_j^i)}{OPT(r_j^i)} = \frac{\mu_j \Delta t_j + \lambda}{\lambda} = 2.$$

– c. The new request arrives at the server when $t > t_j^{i-1} + \Delta t_j$ and there is no data copy at other servers, but $0 < N_1 < k$, according to the algorithm, the request will be satisfied by transmission from server 1 which calls data from the data center first.

$$\frac{C(r_j^i)}{OPT(r_j^i)} = \frac{\mu_j \Delta t_j + \lambda}{\lambda} = 2.$$

– d. The new request arrives at the server when $t > t_j^{i-1} + \Delta t_j$ and there exists data copy at other servers (there is no free data calls and data has been transferred to server 1), according to the algorithm, the request will be satisfied by transimission from server 1.

$$\frac{C(r_j^i)}{OPT(r_j^i)} = \frac{\mu_j \Delta t_j + \lambda + \mu_1 \Delta + \lambda}{\mu_j (\Delta t_j + \Delta)} < 3.$$

- server 1 which buys data from the data center and its holding time is Δt_j^B

– a. The new request arrives at server 1 when $t \leq t_1^{i-1} + \Delta t_1^B$, according to the algorithm, the request will be satisfied directly.

$$\frac{C(r_1^i)}{OPT(r_1^i)} = \frac{\mu_1(t - t_1^{i-1})}{\mu_1(t - t_1^{i-1})} = 1.$$

– b. The new request arrives at server 1 when $t > t_1^{i-1} + \Delta t_1^B$ and there exists data copy at other servers, according to the algorithm, the request will be satisfied by transimission from other servers.

$$\frac{C(r_1^i)}{OPT(r_1^i)} = \frac{\mu_1 \Delta t_1^B + \lambda}{\min\{\lambda, \mu_1(\Delta t_1^B + \epsilon), \frac{B}{k}\}} < 3.$$

– c. The new request arrives at server 1 when $t > t_1^{i-1} + \Delta t_1^B$ and there is no data copy at other servers, but $0 < N_1 < k$, according to the algorithm, the request will be satisfied by calling from the data center. In this case, as shown Theorem 1, we prove the ratio by taking k calls after purchasing the data as a whole.

- new requests have not arrived for a long time and there are no free calls for every server, then server 1 has to cache data

- a. The new request arrives at server 1 during holding time $t < \frac{B}{\mu_1}$, according to the algorithm, data will be stored at server 1 untill the time period is $\frac{B}{\mu_1}$. The optimal cost maybe caches data or calls data from the data center (for some free calling of previous purchase, OPT chooses to cache more to reduce a calling). If the request flow stops in this case, the cost of the BBCTO Algorithm will increases a constant multiple of B, but it does not affect the asymptotic performance ratio.
- b. The new request arrives the server at $\frac{B}{\mu_1}$, according to the algorithm, the request will be satisfied by buying data from the data center. As requests arrive in the future, when the cost of OPT increases B, it must also buy data from the data center. In this case, as shown Theorem 1, we have proved the ratio by taking k calls after purchasing the data as a whole.

Suppose that C_{BO} is the total cost of an online requests stream including n requests gained by the BBCTO algorithm, OPT is the optimal total cost of the requests stream which can be obtained by an offline algorithm. Then

$$C_{BO}/OPT = (\sum_{1 \leq j \leq n} C_A(r_j) + cB)/\sum_{1 \leq j \leq n} C_{OPT}(r_j)$$

$$< (\sum_{1 \leq j \leq n} 3C_{OPT}(r_j^i) + cB)/\sum_{1 \leq j \leq n} C_{OPT}(r_j^i)),$$

$$\lim_{n \to \infty} C_{BO}/OPT = 3.$$

When $\frac{B}{k} < \frac{\lambda}{2}$, we modify the second part of the online BBCOT Algorithm.

2. Server j that purchases data from data center and has free data calls times $(N_j < k)$

- Cache: The retention period of the data is Δt_j^B.
- Transfer: When a new request arrives at server S_j, if there exists data at the target server, then the request will be satisfied; call data from data center, otherwise.

Theorem 3. *The competitive ratio of the modified BBCTO Algorithm is 3.*

Proof. The proof is the same as Theorem 2 without the case b in case 2 (server 1 which buys data from the data center and its holding time is Δt_1^B).

5   Conclusion and Future Work

In this paper, we studied data caching problem with data center in edge computing. In previous studies, data center was used only to purchase data. We consider that the data purchasing from the data center is not always valid, i.e., the number of invoking data after purchasing data is limited. In edge computing, it is the first time such a pay-per-call data pricing has been considered.

We design an algorithm to minimize the total cost of completing users' online requests when future requests arriving at the server are unknown. And we prove that the competitive ratio of the proposed algorithm is 3. In our future work, we will consider a distributed game system, where each server is selfish and wants to minimize its own costs, it will be more challenging.

References

1. Hu, Y.C., Patel, M., Sabella, D., Sprecher, N., Young, V.: Mobile edge computinga key technology towards 5G. ETSI White Pap. **11**(11), 1–16 (2015)
2. Wang, S., Zhao, Y., Xu, J., Yuan, J., Hsu, C.-H.: Edge server placement in mobile edge computing. J. Parallel Distrib. Comput. **127**, 160–168 (2019)
3. Shi, W., Cao, J., Zhang, Q., Li, Y., Xu, L.: Edge computing: vision and challenges. IEEE Internet Things J. **3**(5), 637–646 (2016)
4. Mao, Y., You, C., Zhang, J., Huang, K., Letaief, K.B.: A survey on mobile edge computing: the communication perspective. IEEE Commun. Surv. Tutor. **19**(4), 2322–2358 (2017)
5. Tang, B., Gupta, H., Das, S.R.: Benefit-based data caching in ad hoc networks. IEEE Trans. Mob. Comput. **7**(3), 289–304 (2008)
6. Dar, S., Franklin, M.J., Jonsson, B.T., Srivastava, D., Tan, M.: Semantic data caching and replacement. VLDB **96**, 330–341 (1996)
7. Wang, J.: A survey of web caching schemes for the internet. ACM SIGCOMM Comput. Commun. Rev. **29**(5), 36–46 (1999)
8. Stenstrom, P.: A survey of cache coherence schemes for multiprocessors. Computer **23**(6), 12–24 (1990)
9. Oh, Y., Choi, J., Lee, D., Noh, S.H.: Caching less for better performance: balancing cache size and update cost of flash memory cache in hybrid storage systems. In: FAST, vol. 12 (2012)
10. Patel, M., et al.: Mobile edge computing-introductory technical white paper. Mobile-Edge Computing (MEC) Industry Initiative, White Paper, pp. 1089–7801 (2014)
11. Wang, X., Chen, M., Taleb, T., Ksentini, A., Leung, V.C.M.: Cache in the air: exploiting content caching and delivery techniques for 5G systems. IEEE Commun. Mag. **52**(2), 131–139 (2014)
12. Amble, M.M., Parag, P., Shakkottai, S., Ying, L.: Content-aware caching and traffic management in content distribution networks. In: 2011 Proceedings IEEE INFO-COM, pp. 2858–2866 (2011)
13. Urgaonkar, R., Wang, S., He, T., Zafer, M., Chan, K., Leung, K.K.: Dynamic service migration and workload scheduling in edge-clouds. Perform. Eval. **91**, 205–228 (2015)
14. Ostovari, P., Wu, J., Khreishah, A.: Efficient online collaborative caching in cellular networks with multiple base stations. In: 2016 IEEE 13th International Conference on Mobile Ad Hoc and Sensor Systems (MASS), pp. 136–144 (2016)
15. Tan, H., Jiang, S.H.-C., Han, Z., Li, M.: Asymptotically optimal online caching on multiple caches with relaying and bypassing. IEEE/ACM Trans. Netw. **29**(4), 1841–1852 (2021)
16. Charikar, M., Halperin, D., Motwani, R.: The dynamic servers problem. In: Proceedings of Ninth Annual ACM-SIAM Symposium on Discrete Algorithms (SODA 1998), vol. 98, pp. 410–419 (1998)

17. Gharaibeh, A., Khreishah, A., Khalil, I.: An O(1)-competitive online caching algorithm for content centric networking. In: IEEE INFOCOM 2016–35th Annual IEEE International Conference on Computer Communications, pp. 1–9 (2016)
18. Gharaibeh, A., Khreishah, A., Ji, B., Ayyash, M.: A provably efficient online collaborative caching algorithm for multicell-coordinated systems. IEEE Trans. Mob. Comput. **15**(8), 1863–1876 (2016)
19. Huang, G., Luo, C., Wu, K., Ma, Y., Zhang, Y., Liu, X.: Software-defined infrastructure for decentralized data lifecycle governance: principled design and open challenges. In: Proceedings of the IEEE 39th International Conference on Distributed Computing Systems (ICDCS), pp. 1674–1683 (2019)
20. Karger, D., et al.: Web caching with consistent hashing. In: Proceedings of 8th International Conference on World Wide Web, pp. 1203–1213 (1999)
21. Papadimitriou, C., Ramanathan, S., Rangan, P., Sampathkumar, S.: Multimedia information caching for personalized video-on-demand. Comput. Commun. **18**(3), 204–216 (1995)
22. Papadimitriou, C., Ramanathan, S., Rangan, P.: Optimal information delivery. In: Proceedings of Sixth International Symposium on Algorithms and Computation (ISAAC 1995), pp. 181–187 (1995)
23. Veeravalli, B.: Network caching strategies for a shared data distribution for a predefined service demand sequence. IEEE Trans. Knowl. Data Eng. **15**(6), 1487–1497 (2003)
24. Wang, Y., Veeravalli, B., Tham, C.: On data staging algorithms for shared data accesses in clouds. IEEE Trans. Parallel Distrib. Syst. **24**(4), 825–838 (2013)
25. Wang, Y., Shi, W., Hu, M.: Virtual servers co-migration for mobile accesses: online versus off-line. IEEE Trans. Mob. Comput. **14**(12), 2576–2589 (2015)
26. Wang, Y., He, S., Fan, X., Xu, C., Sun, X.: On cost-driven collaborative data caching: a new model approach. IEEE Trans. Parallel Distrib. Syst. **30**(3), 662–676 (2019)
27. Wang, et al.: Cost-driven data caching in the cloud: an algorithmic approach. In: INFOCOM 2021 (2021)
28. Han, X., Gao, G., Wang, Y., Ting, H.F., You, I., Zhang, Y.: Online data caching in edge computing. Concurr. Comput.: Pract. Exp. (2021)

Scheduling

Scheduling on Multiple Two-Stage Flowshops with a Deadline

Jianer Chen[1,2(✉)], Minjie Huang[1], and Yin Guo[1]

[1] Guangzhou University, Guangzhou 510006, People's Republic of China
[2] Texas A&M University, College Station, TX 77843, USA
chen@cse.tamu.edu

Abstract. Motivated by applications in cloud computing, we study approximation algorithms for scheduling two-stage jobs on multiple two-stage flowshops with a deadline, aiming at maximizing the profit. For the case where the number of flowshops is part of the input, we present a fast approximation algorithm with a constant ratio. The ratio is improved via a study of the relationship between the problem and the multiple knapsack problem, combined with a recently developed approximation algorithm for the multiple knapsack problem. By integrating techniques in the study of the classical KNAPSACK problem and the MAKESPAN problem on multiple processors, plus additional new techniques, a polynomial-time approximation algorithm with a further improved ratio is developed for the case where the number of flowshops is a fixed constant.

Keywords: Scheduling · Two-stage flowshop · Approximation algorithm

1 Introduction

Recently, there have been increasing interests in the study of scheduling multiple two-stage flowshops [4,5,11–16]. The study was partially motivated by the research in cloud computing and data centers [17]. In certain applications of cloud computing, a client request can be regarded as a two-stage job, consisting of a disk-reading stage and a network-transformation stage, where the network-transformation will not start until the disk-reading brings the required data from disks into the main memory. A server in the cloud can be regarded as a two-stage flowshop that can handle both the disk-reading and network-transformation for a client request [12]. It has been observed [17] that the costs of the two stages of a client request can be comparable and are not necessarily closely correlated. Moreover, they depend on different cloud services, the involved servers in the cloud, and bandwidth of local I/O (disk or flash) and of network.

Most research on scheduling multiple two-stage flowshops has been focused on minimizing the scheduling makespan. The problem is NP-hard [12].

This work was supported in part by the National Natural Science Foundation of China under grant 61872097.

Fully polynomial-time approximation schemes have been developed for the problem of scheduling on a fixed number of two-stage flowshops [4,11,12]. For the case where the number of flowshops is part of the input, a sequence of improved approximation algorithms have been developed [5,14,15]. In particular, a polynomial-time approximation scheme for the problem has been developed very recently [5].

The current paper initiates the study of scheduling multiple two-stage flowshops with a given deadline. Thus, we are given a deadline D and a set $\mathcal{G}$ of two-stage jobs, where each job is associated with a profit. The objective is to select a subset $\mathcal{G}'$ of jobs from $\mathcal{G}$ such that $\mathcal{G}'$ can be scheduled on m two-stage flowshops with makespan bounded by D and that $\mathcal{G}'$ maximizes the profit. The problem is again motivated by applications in cloud computing: in certain applications, cloud computing can be modeled by running two-stage jobs on multiple two-stage flowshops. In particular, scheduling multiple two-stage flowshops with deadlines models the practice of buying cloud services in which the cloud service provider companies are aimed at maximizing the profit under time constraints.

To authors' best acknowledge, the problem has not been systematically studied in the scheduling literature. The problem on a single flowshop was studied by Dawande et al. [3], where approximation algorithms and heuristic algorithms were studied and experimental results were presented. Another related research line is the study of approximation algorithms for the Multiple-KNAPSACK problem. Indeed, an item (s, p) of size s and profit p in an instance of KNAPSACK can be regarded as a one-stage job of time s and profit p, while a knapsack of capacity D can be regarded as a one-stage flowshop with deadline D. Therefore, Multiple-KNAPSACK can be regarded as scheduling one-stage jobs on multiple one-stage flowshops with a deadline. The problem has drawn recent interests and has led to some deep and significant results [1,2,7,10]. In particular, polynomial-time approximation schemes have been developed for the problem [1,10].

The current paper studies approximation algorithms for scheduling multiple two-stage flowshops with a deadline. Our first result is a fast approximation algorithm of ratio 4 for the problem in which the number of flowshops is part of the input. This result significantly extends a result in [3], which presented an approximation algorithm of the same ratio for the problem on a single flowshop. Note that extending approximation algorithms from one flowshop to multiple flowshops is not always routine and straightforward. For example, it is well-known that the classical KNAPSACK problem has fully polynomial-time approximation schemes [6], while Multiple-KNAPSACK with two knapsacks has no fully polynomial-time approximation scheme unless P = NP [1].

We then study connections between the Multiple-KNAPSACK problem and the problem of scheduling multiple two-stage flowshops with a deadline. This study plus the approximation algorithms for Multiple-KNAPSACK developed in [1] gives a polynomial-time approximation algorithm of ratio $3 + \epsilon$, where $\epsilon > 0$ is any constant, for the problem of scheduling multiple two-stage flowshops with a deadline, in which the number of flowshops is given as part of the input. This again extends a result in [3] that presented an approximation algorithm of the same ratio for scheduling a single two-stage flowshop with a deadline.

Finally, we study scheduling two-stage flowshops with a deadline in which the number of flowshops is a fixed constant. By integrating techniques in approximation algorithms for the classical KNAPSACK problem and that for the classical MAKESPAN problem on multiple processors, we classify two-stage jobs in terms of their work (i.e., the total time required by the two-stage processing), their profits, and their profit ratios. This classification enables us to focus on the set of jobs with small work and small profits, in which each job's inclusion in the final solution would not significantly impact the makespan and the profit of the solution. Our study suggests a new approximation algorithm for the problem. By thorough analysis of the algorithm, we show that the proposed new algorithm has an approximation ratio bounded by $2 + \epsilon$ for any constant $\epsilon > 0$. The new algorithm significantly improves previous known results. In particular, the best known approximation algorithm for the problem of scheduling a single two-stage flowshop with a deadline has a ratio $3 + \epsilon$. Our algorithm not only improves the ratio to $2 + \epsilon$, but also is applicable to multiple two-stage flowshops.

2   Preliminaries and Simple Facts

For scheduling on two-stage flowshops, we make the following assumptions:

1. a *two-stage job* consists of an *R-operation*, a *T-operation*, and a *profit*;
2. a *two-stage flowshop* has an *R-processor* and a *T-processor* that run in parallel and process the R- and T-operations, respectively, of the assigned jobs;
3. the R- and T-operations of a two-stage job must be executed in the R- and T-processors, respectively, of the *same* two-stage flowshop, in such a way that the T-operation cannot start unless the R-operation is completed; and
4. there are no job precedence constraints, and preemption is not allowed.

We will remove the prefix "two-stage" in our discussion. Thus, a "job" or a "flowshop" is always interpreted as being two-stage.

A job J is given as a triple $J = (r, t; p)$ of non-negative integers, where r and t are, respectively, the *R-time* and *T-time* of J (i.e., the time units required for processing the R- and T-operations, respectively, of J), and p is the *profit*. Thus, if the job J can be completed by the given deadline D, then the profit gained is p. The *profit* of a job set is equal to the sum of the profits of the jobs in the set.

A *schedule* of a job set $\mathcal{G}$ on a set $\mathcal{F}$ of flowshops is an assignment of the jobs in $\mathcal{G}$ to the flowshops in $\mathcal{F}$, plus a schedule of the job subset assigned to each flowshop in $\mathcal{F}$. The *makespan* of a schedule is the time that is needed to complete all jobs under the schedule. When the deadline D is given, we say that a schedule is *feasible* if the makespan of the schedule is bounded by D.

Now we are ready to define our problem:

MULTIPLE TWO-STAGE FLOWSHOP PACKING (MFL$_2$-PACKING):
Given a set $\mathcal{G} = \{J_i = (r_i, t_i; p_i) \mid 1 \le i \le n\}$ of jobs, a *deadline* D, and the number m of flowshops, find a subset $\mathcal{G}'$ of $\mathcal{G}$ and a feasible schedule of $\mathcal{G}'$ on the m flowshops such that $\mathcal{G}'$ maximizes the profit over all subsets of jobs in $\mathcal{G}$ that have feasible schedules on the m flowshops.

We will denote the problem by mMFL$_2$-PACKING if the number m of flow-shops in the problem is a fixed constant.

The problem of scheduling on a single two-stage flowshop to minimize the makespan is the classical TWO-STAGE FLOWSHOP problem, which can be solved *optimally* in time $O(n \log n)$ by *Johnson's algorithm* that schedules the jobs in *Johnson order* [9]. Thus, if an assignment of a set of jobs to the flowshops is given, we can easily determine if there is a feasible schedule under the assignment, and, if yes, construct the feasible schedule. Therefore, the main difficulty of scheduling multiple flowshops is the assignment of the jobs to the flowshops.

There are some simple facts on the MFL$_2$-PACKING problem that can be easily derived based on known results for the classical KNAPSACK problem and the MAKESPAN problems on multiple processors, which are listed below.

The KNAPSACK problem can be easily reduced to the MFL$_2$-PACKING problem on a single flowshop, i.e., the 1MFL$_2$-PACKING problem: given an instance $(\mathcal{S}, D)$ of KNAPSACK, where $\mathcal{S} = \{(s_i, p_i) \mid 1 \le i \le n\}$ is a set of items, in which s_i and p_i are the size and profit, respectively, of the i-th item, and D is the knapsack capacity, we construct an instance $(\mathcal{G}, D)$ for 1MFL$_2$-PACKING, where $\mathcal{G} = \{(s_i, 0; p_i) \mid 1 \le i \le n\}$, i.e., the i-th job has R-time s_i, T-time 0, and profit p_i, and D is the deadline. It is easy to see that an optimal solution to $(\mathcal{G}, D)$ gives an optimal solution to $(\mathcal{S}, D)$. Thus, the NP-hardness of KNAPSACK gives

Theorem 1. *The* 1MFL$_2$-PACKING *problem, thus* MFL$_2$-PACKING, *is NP-hard.*

In the study of the Multiple-KNAPSACK problem, Chekuri and Khanna [1] proved that Multiple-KNAPSACK with two knapsacks has no fully polynomial-time approximation scheme unless P = NP. Based on the relationship between KNAPSACK and the MFL$_2$-PACKING problem as described above, we have

Theorem 2. *The* 2MFL$_2$-PACKING *problem, thus* MFL$_2$-PACKING, *has no fully polynomial-time approximation scheme unless $P = NP$.*

On the other hand, we give below a pseudo-polynomial time algorithm for the mMFL$_2$-PACKING problem, when the number m of flowshops is a fixed constant.

The algorithm uses the standard techniques of developing pseudo-polynomial time algorithms. Let $\mathcal{G}$ be a set of n jobs, and let D be the given deadline. Without loss of generality, assume that the job set $\mathcal{G}$ is already sorted into Johnson order (otherwise we can simply spend additional $O(n \log n)$ time to sort $\mathcal{G}$). For each i, let $\mathcal{G}_i$ be the set of the first i jobs in $\mathcal{G}$. We maintain a $(2m + 1)$-dimensional array $H[0..n, 0..D, 0..D, \ldots, 0..D, 0..D]$ such that the element $H[i, \rho_1, \tau_1, \ldots, \rho_m, \tau_m]$ of the array records the assignment of a subset $\mathcal{G}'_i$ of $\mathcal{G}_i$ to the m flowshops such that for each h, $1 \le h \le m$, the job set $\mathcal{G}'_{i,h}$ of the jobs assigned to the h-th flowshop F_h satisfies the conditions: (1) $\sum_{(r_i, t_i; p_i) \in \mathcal{G}'_{i,h}} r_i = \rho_h$, and (2) the minimum-makespan schedule for the job set $\mathcal{G}'_{i,h}$ on F_h has its makespan equal to τ_h. Moreover, the job set $\mathcal{G}'_i$ recorded in the element $H[i, \rho_1, \tau_1, \ldots, \rho_m, \tau_m]$ has the maximum profit over all assignments of subsets of $\mathcal{G}_i$ to the m flowshops whose corresponding schedule satisfies the conditions (1) and (2) on all flowshops F_h. Since the job set $\mathcal{G}$ is in Johnson order, the jobs assigned to each flowshop in the above process also follow Johnson order. Therefore, when a new job is added to the flowshop, the minimum-makespan schedule of the new job set in the flowshop can be constructed from the minimum-makespan schedule of the old job set in constant time [12]. Now using the standard techniques of dynamic programming, we can construct an optimal solution to the instance $(\mathcal{G}, D)$ of $m\mathrm{MFL}_2$-PACKING. It is not difficult to see that the above dynamic programming algorithm runs in time $O(nD^{2m})$.

Theorem 3. *The $m\mathrm{MFL}_2$-PACKING problem is solvable in pseudo-polynomial time. More specifically, the problem can be solved in time $O(nD^{2m})$, where n is the number of jobs in the input and D is the deadline.*

We remark that Theorem 3 is a bit surprising. In many cases in the study of approximation algorithms, fully polynomial-time approximation schemes for an optimization problem can be derived naturally from pseudo-polynomial time algorithms for the problem, using the techniques of scaling [6]. On the other hand, although Theorem 3 offers a pseudo-polynomial time algorithm for the $m\mathrm{MFL}_2$-PACKING problem for any fixed constant $m \geq 1$, as shown in Theorem 2, the problem has no fully polynomial-time approximation schemes for any $m \geq 2$.

3   A Fast 4-Approximation Algorithm for MFL_2-PACKING

In this section, we present an approximation algorithm for the MFL_2-PACKING problem, in which the number m of flowshops is given as part of the input. Thus, our instance is of the form $(\mathcal{G}, D, m)$, where $\mathcal{G}$ is a set of n jobs, D is the deadline, and m is the number of flowshops. Without loss of generality, we assume that each job $J' = (r', t', p')$ given in the input satisfies $r + t \leq D$. A job set sometimes is given as a *job sequence*, in which the jobs are given in a specified order. Of course, a job sequence uniquely defines a job set.

We introduce some notations. For a job $J' = (r', t'; p')$, we define $r(J') = r'$, $t(J') = t'$, $p(J') = p'$, $w(J') = r' + t'$, and $\sigma(J') = p(J')/w(J')$, where $p(J')$, $w(J')$, and $\sigma(J')$ are called the *profit*, the *work*, and the *σ-value* of the job J', respectively. The notations of R-time, T-time, profit, and work can be extended to a set or a sequence $\mathcal{G}'$ of jobs. Thus, $r(\mathcal{G}') = \sum_{J \in \mathcal{G}'} r(J)$, $t(\mathcal{G}') = \sum_{J \in \mathcal{G}'} t(J)$, $p(\mathcal{G}') = \sum_{J \in \mathcal{G}'} p(J)$ and $w(\mathcal{G}') = \sum_{J \in \mathcal{G}'} w(J)$.

Consider the algorithm given in Fig. 1. Note that though the output of the algorithm is an assignment $(\mathcal{G}_1, \mathcal{G}_2, \ldots, \mathcal{G}_m)$ of the subset $\mathcal{G}' = \mathcal{G}_1 \cup \cdots \cup \mathcal{G}_m$ of the input job set $\mathcal{G}$, the corresponding feasible schedule of $\mathcal{G}'$ on the m flowshops can be constructed by Johnson's algorithm in additional time $O(n \log n)$.

Inductively for each h, $1 \leq h \leq m$, we initially assign the jobs in the set $\mathcal{H}$ to the flowshop F_h, assuming that the jobs $J_1, J_2, \ldots, J_{i-1}$ have been assigned to the first h flowshops. Now we continue assigning the jobs $J_i, J_{i+1}, \ldots, J_n$, in that order, to the flowshop F_h until we encounter the first job J_k such that the jobs in $\mathcal{H} \cup \{J_i, J_{i+1}, \ldots, J_{k-1}, J_k\}$ cannot be processed by the flowshop F_h without exceeding the deadline D. Note that k could be i, and that we can use Johnson's

---

Algorithm MFL$_2$Apx($\mathcal{G}, D, m$)
INPUT: a set $\mathcal{G}$ of n jobs, a makespan deadline D, and the number m of flowshops
OUTPUT: a feasible schedule of a subset $\mathcal{G}'$ of $\mathcal{G}$ on m flowshops.
1.   sort the jobs in $\mathcal{G}$ in non-increasing order of their σ-values: $J_1, J_2, \ldots, J_n$;
2. $\mathcal{H} = \{J_1\}$; $i = 2$;
3.   **for** $h = 1$ **to** m **do**
3.1   let k be the index such that the job set $\mathcal{H} \cup \{J_i, \ldots, J_{k-1}\}$ can be completed
         on the flowshop F_h by the deadline D, but the job set $\mathcal{H} \cup \{J_i, \ldots, J_{k-1}, J_k\}$
         cannot be completed on F_h by the deadline D;
3.2   **if** $(p(\mathcal{H} \cup \{J_i, \ldots, J_{k-1}\}) > p(J_k))$
3.2.1   **then** { assign $\mathcal{G}_h = \mathcal{H} \cup \{J_i, \ldots, J_{k-1}\}$ to the flowshop F_h; $\mathcal{H} = \{J_k\}$; }
3.2.2   **else** { assign $\mathcal{G}_h = \{J_k\}$ to the flowshop F_h; $\mathcal{H} = \mathcal{H} \cup \{J_i, \ldots, J_{k-1}\}$; }
3.3 $i = k + 1$;
4.   **return** $(\mathcal{G}_1, \mathcal{G}_2, \ldots, \mathcal{G}_m)$.

Fig. 1. An approximation algorithm for MFL$_2$-PACKING

algorithm to determine if a job set can be processed by the flowshop F_h within the deadline D. Then we compare the profits of the two job sets $\mathcal{H} \cup \{J_i, \ldots, J_{k-1}\}$ and $\{J_k\}$, leaving the one with larger profit (i.e., $\mathcal{G}_h$) in the h-th flowshop F_h, and making the one with smaller profit as the new job set $\mathcal{H}$ (which will be assigned to the flowshop F_{h+1} if $h < m$). The index i is then set to $k + 1$, which gives the first job in the sequence that has not been assigned to a flowshop, yet. At this point, the first h flowshops have been assigned, respectively, with the job sets $\mathcal{G}_1, \mathcal{G}_2, \ldots, \mathcal{G}_h$, and the union $(\bigcup_{k=1}^{h} \mathcal{G}_k) \cup \mathcal{H}$ is equal to $\{J_1, J_2, \ldots, J_{i-1}\}$, where $J_{i-1} = J_k$. Then we work on a subset schedule of the jobs in the set $\mathcal{H} \cup \{J_i, J_{i+1}, \ldots, J_n\}$ on the rest $m - h$ flowshops.

In the h-th execution of the **for**-loop in step 3 of the algorithm, the job set $\mathcal{G}_h$ is set to be either $\mathcal{H} \cup \{J_i, \ldots, J_{k-1}\}$ or $\{J_k\}$. By the definition of the index k and by our assumption on the processing time of a single job, the job set $\mathcal{G}_h$ can always be processed by the flowshop F_h and completed by the deadline D. Therefore, the output $(\mathcal{G}_1, \mathcal{G}_2, \ldots, \mathcal{G}_m)$ of the algorithm **MFL$_2$Apx** is a feasible schedule of the job subset $\mathcal{G}' = \mathcal{G}_1 \cup \cdots \cup \mathcal{G}_m$ of the input job set $\mathcal{G}$ on the m flowshops, thus, a valid solution to the instance $(\mathcal{G}, D, m)$ of MFL$_2$-PACKING.

The algorithm **MFL$_2$Apx** uses ideas that are similar to that used in approximation algorithms for the classical KNAPSACK problem [6]. However, since the MFL$_2$-PACKING problem is a significant extension of the KNAPSACK problem: the machine model is extended from one-stage to two-stage, and the number of machines is extended from 1 to any given number m, the problem becomes much more difficult and solving it requires new techniques. The following theorem summarizes our results, whose formal proof will be given in the complete version of this paper.

Theorem 4. *Algorithm* **MFL$_2$Apx** *is an $O(n^2)$-time approximation algorithm for the* MFL$_2$-PACKING *problem, whose approximation ratio is bounded by 4.*

Theorem 4 extends a result in [3], where an approximation algorithm of ratio 4 on *one* flowshop is given. However, as shown in the proof of Theorem 4, which will be given in the complete version of this paper, extending the result to multiple flowshops requires significant additional analysis and new techniques.

4   MFL$_2$-PACKING and Multiple-KNAPSACK

As shown in the previous sections, there seems a close connection between the MFL$_2$-PACKING problem and the Multiple-KNAPSACK problem. The latter has drawn notable attentions recently and has seen some significant progresses [1, 2, 7, 10]. In this section, we study the relationship between the two problems, which leads to a new approximation algorithm for the MFL$_2$-PACKING problem.

An instance of Multiple-KNAPSACK is a triple $(\mathcal{S}, B, m)$, where $\mathcal{S}$ is a set of *items*, each item is given as a pair (s, p) of non-negative integers, where s and p are the *size* and *profit* of the item, respectively, B is the *knapsack capacity*, and m is the number of knapsacks. Similar to what we did in the study of MFL$_2$-PACKING, we can extend the concepts of size $s(I)$ and profit $p(I)$ of an item I to that of a set $\mathcal{S}'$ of items. The objective of $(\mathcal{S}, B, m)$ is to select a subset $\mathcal{S}'$ of items in $\mathcal{S}$ that can be packed into the m knapsacks of capacity B and maximizes the profit. We will denote by $opt(\mathcal{S}, B, m)$ the profit of an optimal solution to the instance $(\mathcal{S}, B, m)$ of Multiple-KNAPSACK. Without loss of generality, we assume $s(I) \leq B$ for all items $I \in \mathcal{S}$ in an instance $(\mathcal{S}, B, m)$ of Multiple-KNAPSACK.

Lemma 1. *Let $\mathcal{S}$ be a set of items. Then for the two instances $(\mathcal{S}, B, m)$ and $(\mathcal{S}, 2B, m)$ of Multiple-KNAPSACK, we have $opt(\mathcal{S}, B, m) \geq opt(\mathcal{S}, 2B, m)/3$.*

We remark that the bound given in Lemma 1 is tight: consider a set $\mathcal{S}$ of n items of size 2 and profit 1. Let $B = 3$. Then for any m, an optimal solution to the instance $(\mathcal{S}, B, m)$ of Multiple-KNAPSACK has profit m, while an optimal solution to the instance $(\mathcal{S}, 2B, m)$ of Multiple-KNAPSACK has profit $3m$.

Theorem 5. *If the Multiple-KNAPSACK problem has a polynomial-time approximation algorithm of ratio γ, then the MFL$_2$-PACKING problem has a polynomial-time approximation algorithm of ratio 3γ.*

Proof. Let A_K be an approximation algorithm of ratio γ for Multiple-KNAPSACK. We develop an approximation algorithm A_F for MFL$_2$-PACKING as follows:

> Input: an instance $(\mathcal{G}, D, m)$ for MFL$_2$-PACKING, where $\mathcal{G} = (J_1, J_2, \ldots, J_n)$,
>   and for each i, $J_i = (r_i, t_i; p_i)$
> 1. construct an instance $(\mathcal{S}, D, m)$ for Multiple-KNAPSACK, where
> $\mathcal{S} = (I_1, I_2, \ldots, I_n)$, and for each i, $I_i = (r_i + t_i, p_i)$;
> 2. call the algorithm A_K on $(\mathcal{S}, D, m)$ to construct a solution to $(\mathcal{S}, D, m)$,
>   which is a subset $\mathcal{S}'$ of $\mathcal{S}$ with a partition $\mathcal{S}' = \mathcal{S}'_1 \cup \cdots \cup \mathcal{S}'_m$;
> 3. for each $1 \leq h \leq m$, let $\mathcal{G}'_h = \{J_i \mid I_i \in \mathcal{S}'_h\}$. Then $\mathcal{G}' = \mathcal{G}'_1 \cup \cdots \cup \mathcal{G}'_m$ is
>   the solution to the instance $(\mathcal{G}, D, m)$ of MFL$_2$-PACKING, where for each
> h, $\mathcal{G}'_h$ is the subset of jobs assigned to the h-th flowshop.

We first show that the schedule, i.e., the job subset $\mathcal{G}'$ and its partition, constructed by the algorithm A_F is a feasible schedule of the instance $(\mathcal{G}, D, m)$ of MFL$_2$-PACKING. Consider the subset $\mathcal{G}'_h$ in $\mathcal{G}'$. By the algorithm, the corresponding subset $\mathcal{S}'_h$ of items in the solution $\mathcal{S}'$ to the instance $(\mathcal{S}, D, m)$ of

Multiple-KNAPSACK constructed in step 2 can be packed into a knapsack of capacity D. Since the size of an item in S'_h is equal to the work of the corresponding job in G'_h, we get $w(G'_h) = s(S'_h) \leq D$. Therefore, it is clear that the job subset G'_h can be completed in a flowshop with completion time bounded by D. As a result, the subset G' of jobs in G and the partition of G' constructed by the algorithm A_F is a feasible schedule of the instance (G, D, m) of MFL$_2$-PACKING.

The algorithm A_F obviously runs in polynomial time under the assumption that the algorithm A_K runs in polynomial time.

Now consider the approximation ratio of the algorithm A_F. Observe that for the instance (G, D, m) of MFL$_2$-PACKING and the instance $(S, 2D, m)$ of Multiple-KNAPSACK, $opt(G, D, m) \leq opt(S, 2D, m)$. This is because for every feasible schedule $G' = G'_1 \cup \cdots \cup G'_m$ of (G, D, m), the corresponding $S' = S'_1 \cup \cdots \cup S'_m$ is a valid solution to $(S, 2D, m)$, where for each h, $S'_h = \{I_i \mid J_i \in G'_h\}$. This combined with Lemma 1 gives $opt(G, D, m) \leq opt(S, 2D, m) \leq 3 \cdot opt(S, D, m)$.

By the assumption, in step 3 of the algorithm A_F, the called algorithm A_K produces a solution $S' = S'_1 \cup \cdots \cup S'_m$ to the instance (S, D, m) of Multiple-KNAPSACK satisfying $opt(S, D, m)/p(S') \leq \gamma$. By the algorithm, $p(S') = p(G')$, where G' is the solution produced by the algorithm A_F to the instance (G, D, m), which thus satisfies: $opt(G, D, m)/p(G') = 3 \cdot opt(S, D, m)/p(S') \leq 3\gamma$. Therefore, the approximation ratio of the algorithm A_F for the problem MFL$_2$-PACKING is bounded by 3γ. This completes the proof of the theorem.   □

Chekuri and Khanna [1] developed a polynomial-time approximation scheme for the Multiple-KNAPSACK problem, which is an approximation algorithm of ratio $1 + \epsilon$ for any $\epsilon > 0$ and runs in polynomial time when ϵ is a fixed constant. This result combined with Theorem 5 gives directly the following corollary:

Corollary 1. *There is an approximation algorithm for the* MFL$_2$-PACKING *problem that has approximation ratio bounded by $3 + \epsilon$ for any $\epsilon > 0$ and runs in polynomial time when ϵ is a fixed constant.*

Corollary 1 extends a result in [3], where a polynomial-time approximation algorithm of the same ratio was presented for the problem 1MFL$_2$-PACKING.

5   A $(2 + \epsilon)$-Approximation for mMFL$_2$-PACKING

In this section, we consider the problem mMFL$_2$-PACKING in which the number m of flowshops is a fixed constant, whose instances are given as (G, D) with G being a job set and D being the deadline. The profit of an optimal solution to the instance (G, D) will be denoted by $opt(G, D)$.

In our discussion below, we will need to order jobs using either their profits or their σ-values. In order to ensure a unique ordering, we will use the job indices given in the input job set $G = \{J_1, \ldots, J_n\}$ to resolve ties. Thus, for two jobs J_x and J_y in G, by "the profit of J_x is larger than that of J_y", we really mean that either $p(J_x) > p(J_y)$, or $p(J_x) \leq p(J_y)$ and $x > y$. Similar rule is applied when we order jobs by their σ-values. Under these rules of ordering, a job set will have a unique ordering when it is sorted by either the profits or the σ-values, and the partition of a job set based on the profit or σ-value of a job is unique.

Let $\mathcal{G}' = \langle J_1', \ldots, J_h' \rangle$ be a job sequence. By "greedily add the jobs in the sequence $\mathcal{G}'$ to the m flowshops", we refer to the following procedure: following the order of the sequence $\mathcal{G}'$, repeatedly add the next job J in $\mathcal{G}'$ to a flowshop F in which all assigned jobs plus J have a schedule in F with completion time bounded by D (note that this can be determined in polynomial time using Johnson's algorithm). The procedure ends either when all jobs in $\mathcal{G}'$ are added to the flowshops, or it encounters a job J in the sequence such that adding J to any of the flowshops will result in a job subset that has no schedule in the flowshop with completion time bounded by D. Note that the m flowshops are allowed to be non-empty before the procedure adds the jobs in $\mathcal{G}'$ to the flowshops.

Let k_0 be a fixed integer whose value will be determined later. We say that a job J' is a *large-job* if $w(J') \geq 2mD/k_0$, and a *small-job* otherwise. Note that if a job subset $\mathcal{G}'$ has a feasible schedule on the m flowshops, then $\mathcal{G}'$ contains at most k_0 large-jobs. Consider the algorithm given in Fig. 2.

Algorithm mMFL$_2$Apx$(\mathcal{G}, D, k_0)$
INPUT: a set $\mathcal{G}$ of n jobs, a makespan deadline D, and an integer $k_0 > 0$
OUTPUT: a feasible schedule of a subset $\mathcal{G}'$ of the job set $\mathcal{G}$ on m flowshops.

1. $\mathcal{G}_0 = \emptyset$;
2. **for** (each subset $\mathcal{G}_1'$ of at most k_0 large-jobs plus at most k_0 small-jobs)
3.   **if** (the subset $\mathcal{G}_1'$ has a feasible schedule in the m flowshops) **then**
3.1   let $\mathcal{G}_2'$ be the sequence of the small-jobs in $\mathcal{G} \setminus \mathcal{G}_1'$ whose profits are smaller than that of all small-jobs in $\mathcal{G}_1'$, sorted decreasingly by their σ-values;
3.2   make a feasible schedule of the job subset $\mathcal{G}_1'$ to the m flowshops;
3.3   greedily add the small-jobs in $\mathcal{G}_2'$ to the m flowshops;
3.4   let the resulting schedule be $\mathcal{G}'$; **if** $(p(\mathcal{G}') > p(\mathcal{G}_0))$ **then** $\mathcal{G}_0 = \mathcal{G}'$;
4. return($\mathcal{G}_0$).

Fig. 2. An approximation algorithm for mMFL$_2$-PACKING

We study the approximation ratio of the algorithm m**MFL$_2$Apx**$(\mathcal{G}, D, k_0)$. Let $\mathcal{G}^* = \{J_1^*, J_2^*, \ldots, J_v^*\}$ be an optimal solution for the instance $(\mathcal{G}, D)$. Thus, $p(\mathcal{G}^*) = opt(\mathcal{G}, D)$, $w(\mathcal{G}^*) \leq 2mD$. If $\mathcal{G}^*$ contains at most k_0 small-jobs, then, since $\mathcal{G}^*$ can neither contain more than k_0 large-jobs, the set $\mathcal{G}^*$ will be examined in step 2 of the algorithm and assigned to the m flowshops in step 3.2. By step 3.4, the output of the algorithm in this case will be an optimal solution to $(\mathcal{G}, D)$.

Thus, we can assume that the set $\mathcal{G}^*$ contains more than k_0 small-jobs. Let $\mathcal{G}_1 = \{J_1^*, \ldots, J_q^*\}$ be the set of the large-jobs plus the k_0 small-jobs with the largest profits in $\mathcal{G}^*$. Since $\mathcal{G}^*$ cannot contain more than k_0 large-jobs and the job set $\mathcal{G}_1$ certainly has a feasible schedule on the m flowshops, steps 3.1–3.4 will be applied on the set $\mathcal{G}_1$. Let $\mathcal{G}_2$ be the sequence of the rest small-jobs in $\mathcal{G}^*$, sorted decreasingly by σ-values. Then, $\mathcal{G}^* = \mathcal{G}_1 \cup \mathcal{G}_2$.

Case 1. $p(\mathcal{G}_1) \geq p(\mathcal{G}^*)/2$.

Since $\mathcal{G}_1$ is examined in step 2 and has feasible schedules, step 3.2 makes a feasible schedule for $\mathcal{G}_1$ so that the corresponding schedule $\mathcal{G}'$ in step 3.4, thus the output $\mathcal{G}_0$ of the algorithm, has profit at least $p(\mathcal{G}_1) \geq p(\mathcal{G}^*)/2 = opt(\mathcal{G}, D)/2$.

Case 2. $p(\mathcal{G}_1) < p(\mathcal{G}^*)/2$, thus $p(\mathcal{G}_2) > p(\mathcal{G}^*)/2$.

We divide the case into two subcases.

Case 2.1. There is a postfix $\mathcal{G}_{2'}$ of $\mathcal{G}_2$ such that $p(\mathcal{G}_{2'}) \geq p(\mathcal{G}^*)/2$, $w(\mathcal{G}_{2'}) \leq mD$.

Again, if $\mathcal{G}_{2'}$ contains no more than k_0 jobs, then $\mathcal{G}_{2'}$ will be examined in step 2, for which step 3.4 will produce a schedule $\mathcal{G}'$ whose profit is at least $p(\mathcal{G}_{2'}) \geq p(\mathcal{G}^*)/2 = opt(\mathcal{G}, D)/2$.

Thus, we assume that $\mathcal{G}_{2'}$ contains more than k_0 jobs. In this case, let $\mathcal{G}_1^*$ be the set consisting of the k_0 jobs in $\mathcal{G}_{2'}$ that have the largest k_0 profits in $\mathcal{G}_{2'}$, $\mathcal{G}_2^* = \mathcal{G}_{2'} \backslash \mathcal{G}_1^*$, and suppose $\mathcal{G}_2^*$ is sorted decreasingly by the σ-values. Thus, in this case, we get two job subsets $\mathcal{G}_1^*$ and $\mathcal{G}_2^*$ that satisfy the following conditions:

Condition-1.
(1.1) $\mathcal{G}_1^* \cup \mathcal{G}_2^* \subseteq \mathcal{G}^*$, $p(\mathcal{G}_1^* \cup \mathcal{G}_2^*) \geq p(\mathcal{G}^*)/2$, $w(\mathcal{G}_1^* \cup \mathcal{G}_2^*) \leq mD$;
(1.2) $\mathcal{G}_1^*$ contains at most k_0 large-jobs plus k_0 small-jobs;
(1.3) $\mathcal{G}_2^*$ contains only small-jobs whose profit < the profit of all small-jobs in $\mathcal{G}_1^*$, and the jobs in $\mathcal{G}_2^*$ are sorted decreasingly by σ-values.

Case 2.2. No postfix $\mathcal{G}_{2'}$ of $\mathcal{G}_2$ satisfies both $p(\mathcal{G}_{2'}) \geq p(\mathcal{G}^*)/2$ and $w(\mathcal{G}_{2'}) \leq mD$.

In Case 2.2, since $p(\mathcal{G}_2) > p(\mathcal{G}^*)/2$, we must have $w(\mathcal{G}_2) > mD$. This combined with $w(\mathcal{G}^*) = w(\mathcal{G}_1 \cup \mathcal{G}_2) \leq 2mD$ gives $w(\mathcal{G}_1) < mD$.

Without loss of generality, suppose $\mathcal{G}_2 = \langle J_{q+1}^*, \ldots, J_v^* \rangle$ is sorted decreasingly by the σ-values. Let z be the first index in $\mathcal{G}_2$ such that $w(\{J_{z+1}^*, \ldots, J_v^*\}) \leq mD$. The index z satisfies $q + 1 \leq z < v$ because $w(\mathcal{G}_2) > mD$ and the jobs in $\mathcal{G}_2$ are small-jobs whose work is bounded by $2mD/k_0$ (assuming $k_0 \geq 10$). In Case 2.2, we must have $p(\{J_{z+1}^*, \ldots, J_v^*\}) < p(\mathcal{G}^*)/2$. Then $p(\{J_1^*, \ldots, J_q^*, J_{q+1}^*, \ldots, J_z^*\}) > p(\mathcal{G}^*)/2$. In this case, we let $\mathcal{G}_1^* = \mathcal{G}_1 = \{J_1^*, \ldots, J_q^*\}$, and $\mathcal{G}_2^* = \langle J_{q+1}^*, \ldots, J_{z-1}^* \rangle$, where $\mathcal{G}_2^*$ is sorted decreasingly by the σ-values. Thus, in this case, we get two job subsets $\mathcal{G}_1^*$ and $\mathcal{G}_2^*$ that satisfy the following conditions:

Condition-2.
(2.1) $\mathcal{G}_1^* \cup \mathcal{G}_2^* \subseteq \mathcal{G}^*$, $p(\mathcal{G}_1^* \cup \mathcal{G}_2^*) + p(\mathcal{G}^*)/k_0 > p(\mathcal{G}^*)/2$, $w(\mathcal{G}_1^* \cup \mathcal{G}_2^*) \leq mD$;
(2.2) $\mathcal{G}_1^*$ contains at most k_0 large-jobs plus k_0 small-jobs;
(2.3) $\mathcal{G}_2^*$ contains only small-jobs whose profit < the profit of all small-jobs in $\mathcal{G}_1^*$, and the jobs in $\mathcal{G}_2^*$ are sorted decreasingly by σ-values.

The first inequality in (2.1) is because $p(\{J_1^*, \ldots, J_q^*, J_{q+1}^*, \ldots, J_z^*\}) > p(\mathcal{G}^*)/2$ and J_z^* is a small-job in $\mathcal{G}_2$ whose profit is not large than that of any of the k_0 small-jobs in $\mathcal{G}_1^* = \mathcal{G}_1 \subseteq \mathcal{G}^*$, and the second inequality in (2.1) is by the definition of the index z.

Note that Condition-1 implies Condition-2. Thus, in both Cases 2.1 and 2.2, we can always construct the subsets $\mathcal{G}_1^*$ and $\mathcal{G}_2^*$ that satisfy Condition-2.

Lemma 2. *Under the condition of Case 2, the solution $\mathcal{G}_0$ returned by the algorithm $m\textbf{MFL}_2\textbf{Apx}(\mathcal{G}, D, k_0)$ has profit at least $opt(\mathcal{G}, D)(1/2 - (m+1)/k_0)$.*

Proof. By the above discussion, from the optimal solution $\mathcal{G}^*$, we can always construct the subsets $\mathcal{G}_1^*$ and $\mathcal{G}_2^*$ that satisfy Condition-2.

By (2.2) of Condition-2, the subset $\mathcal{G}_1^*$ is examined in step 2 as the set $\mathcal{G}_1'$ in the algorithm. Since $\mathcal{G}_1^*$ clearly has feasible schedules on the m flowshops, the algorithm proceeds to steps 3.1–3.4, and greedily adds the jobs in the job sequence $\mathcal{G}_2'$ to the flowshops in step 3.3. Note that $\mathcal{G}_2' = \langle J_1', J_2', \ldots, J_u' \rangle$ is the job sequence consisting of all small-jobs in $\mathcal{G} \backslash \mathcal{G}_1^*$ whose profit is smaller than that of the small-jobs in $\mathcal{G}_1^*$. By (2.3) of Condition-2 and the uniqueness of our ordering rules, $\mathcal{G}_2^*$ is a subsequence of $\mathcal{G}_2'$.

Let J_y' be the first job in the sequence $\mathcal{G}_2'$ that cannot be added by step 3.3 to any of the m flowshops. Let $\mathcal{G}_{y-1}' = \mathcal{G}_1^* \cup \{J_1', \ldots, J_{y-1}'\}$. Note that $\mathcal{G}_{y-1}'$ is the set $\mathcal{G}'$ obtained in step 3.4 when the job set $\mathcal{G}_1'$ examined in step 2 is $\mathcal{G}_1^*$. Therefore, the profit of the subset $\mathcal{G}_0$ returned by algorithm $m\textbf{MFL}_2\textbf{Apx}$ is at least as large as $p(\mathcal{G}_{y-1}')$. Thus, it suffices to show $p(\mathcal{G}_{y-1}') \geq opt(\mathcal{G}, D)(1/2 - (m+1)/k_0)$.

If $\mathcal{G}_{y-1}' = \mathcal{G}_1^* \cup \mathcal{G}_2'$, then since $\mathcal{G}_2^*$ is a subsequence of $\mathcal{G}_2'$, we have

$$p(\mathcal{G}_{y-1}') \geq p(\mathcal{G}_1^* \cup \mathcal{G}_2^*) \geq p(\mathcal{G}^*)(1/2 - 1/k_0) = opt(\mathcal{G}, D)(1/2 - 1/k_0),$$

where we have used (2.1) in Condition-2. Thus, in this case, the lemma is proved.

If $\mathcal{G}_{y-1}' \neq \mathcal{G}_1^* \cup \mathcal{G}_2'$, then job J_y' in $\mathcal{G}_2'$ exists and is not in $\mathcal{G}_{y-1}'$. Let $\mathcal{G}_y' = \mathcal{G}_{y-1}' \cup \{J_y'\}$ and let $\mathcal{G}_{1+2}^* = \mathcal{G}_1^* \cup \mathcal{G}_2^*$. Let $\mathcal{I}_y' = \mathcal{G}_y' \backslash (\mathcal{G}_y' \cap \mathcal{G}_{1+2}^*)$ and $\mathcal{I}_{1+2}^* = \mathcal{G}_{1+2}^* \backslash (\mathcal{G}_y' \cap \mathcal{G}_{1+2}^*)$. Since $\mathcal{G}_1^*$ is a subset of the intersection $\mathcal{G}_y' \cap \mathcal{G}_{1+2}^*$, for each job J_i' in $\mathcal{I}_y'$ and each job J_h^* in $\mathcal{I}_{1+2}^*$, we always have $p(J_i')/w(J_i') = \sigma(J_i') \geq \sigma(J_h^*) = p(J_h^*)/w(J_h^*)$. Moreover, for the job J_y', we also have $p(J_y')/w(J_y') \geq p(J_h^*)/w(J_h^*)$ for all J_h^* in $\mathcal{I}_{1+2}^*$. Therefore,

$$\frac{p(\mathcal{I}_y') + (m-1)p(J_y')}{w(\mathcal{I}_y') + (m-1)w(J_y')} = \frac{\sum_{J_i' \in \mathcal{I}_y'} p(J_i') + (m-1)p(J_y')}{\sum_{J_i' \in \mathcal{I}_y'} w(J_i') + (m-1)w(J_y')}$$

$$\geq \frac{\sum_{J_h^* \in \mathcal{I}_{1+2}^*} p(J_h^*)}{\sum_{J_h^* \in \mathcal{I}_{1+2}^*} w(J_h^*)} = \frac{p(\mathcal{I}_{1+2}^*)}{w(\mathcal{I}_{1+2}^*)}. \tag{1}$$

By the definition of the index y, with the jobs in $\mathcal{G}_{y-1}'$ assigned to the m flowshops, adding the job J_y' to any flowshop F would result in a subset of jobs in F that has no schedule of completion time bounded by D in the flowshop F. Therefore, the total work of the jobs in $\mathcal{G}_{y-1}'$ assigned to the flowshop F plus $w(J_y')$ must be larger than D. As a result, the total work of the jobs in $\mathcal{G}_{y-1}'$ plus $m \cdot w(J_y')$, i.e., the value $w(\mathcal{G}_y') + (m-1)w(J_y')$, is larger than mD. On the other hand, by (2.1) of Condition-2, we have $w(\mathcal{G}_{1+2}^*) \leq mD$. Therefore,

$$w(\mathcal{I}_y') + (m-1)w(J_y') = w(\mathcal{G}_y') + (m-1)w(J_y') - w(\mathcal{G}_y' \cap \mathcal{G}_{1+2}^*)$$
$$> w(\mathcal{G}_{1+2}^*) - w(\mathcal{G}_y' \cap \mathcal{G}_{1+2}^*) = w(\mathcal{I}_{1+2}^*).$$

Combining this with (1), we get $p(\mathcal{I}_y') + (m-1)p(J_y') \geq p(\mathcal{I}_{1+2}^*)$. Thus,

$$
\begin{aligned}
p(\mathcal{G}_{y-1}') = p(\mathcal{G}_y') - p(J_y') &= p(\mathcal{I}_y') + p(\mathcal{G}_y' \cap \mathcal{G}_{1+2}^*) - p(J_y') \\
&\geq p(\mathcal{I}_{1+2}^*) - (m-1)p(J_y') + p(\mathcal{G}_y' \cap \mathcal{G}_{1+2}^*) - p(J_y') \\
&= p(\mathcal{G}_{1+2}^*) - mp(J_y').
\end{aligned} \tag{2}
$$

From (2.1) of Condition-2, we have $p(\mathcal{G}_{1+2}^*) = p(\mathcal{G}_1^* \cup \mathcal{G}_2^*) > p(\mathcal{G}^*)(1/2 - 1/k_0)$. Moreover, since the job J_y' in $\mathcal{G}_2'$ has its profit not larger than that of any of the k_0 small-jobs in $\mathcal{G}_1^*$, and $\mathcal{G}_1^*$ is a subset of the optimal solution $\mathcal{G}^*$, we have $p(J_y') \leq p(\mathcal{G}^*)/k_0$. Bringing all these into (2) completes the proof of the lemma:

$$
\begin{aligned}
p(\mathcal{G}_{y-1}') &\geq p(\mathcal{G}^*)(1/2 - 1/k_0) - mp(\mathcal{G}^*)/k_0 \\
&= p(\mathcal{G}^*)(1/2 - (m+1)/k_0) = opt(\mathcal{G}, D)(1/2 - (m+1)/k_0).
\end{aligned}
$$

□

It is easy to see that the **for**-loop in step 2 of the algorithm $m\mathbf{MFL}_2\mathbf{Apx}$ is executed at most $O(n^{2k_0})$ times, and each execution takes polynomial time. Now if we set $k_0 = \lceil 2(m+1)(2+\epsilon)/\epsilon \rceil$, then by Lemma 2 and noticing that all m, ϵ and k_0 are constants, we get our main theorem for this section:

Theorem 6. *For any fixed constant $\epsilon > 0$, there is a polynomial-time algorithm that on an instance $(\mathcal{G}, D)$ of the $m\mathrm{MFL}_2$-PACKING problem, constructs a feasible schedule of a subset $\mathcal{G}_0$ of $\mathcal{G}$ on m flowshops satisfying $opt(\mathcal{G}, D)/p(\mathcal{G}_0) \leq 2+\epsilon$.*

Theorem 6 significantly extends and improves the results in [3]. It not only extends the results in [3] from one flowshop to multiple flowshops, but also improves the approximation ratio from $3 + \epsilon$ to $2 + \epsilon$, where $\epsilon > 0$ is any constant.

References

1. Chekuri, C., Khanna, S.: A polynomial time approximation scheme for the multiple knapsack problem. SIAM J. Comput. **35**(3), 713–728 (2005)
2. Chen, L., Zhang, G.: Packing groups of items into multiple knapsacks. ACM Trans. Algorithms **14**(4), 1–24 (2018)
3. Dawande, M., Gavirneni, S., Rachamadugu, R.: Scheduling a two-stage flowshop under makespan constraint. Math. Comput. Model. **44**, 73–84 (2006)
4. Dong, J., et al.: An FPTAS for the parallel two-stage flowshop problem. Theor. Comput. Sci. **657**, 64–72 (2017)
5. Dong, J., Jin, R., Luo, T., Tong, W.: A polynomial-time approximation scheme for an arbitrary number of parallel two-stage flow-shops. Eur. J. Oper. Res. **281**, 16–24 (2020)
6. Garey, M.R., Johnson, D.S.: Computers and Intractability: A Guide to the Theory of NP-Completeness. W.H. Freeman and Company, New York (1979)
7. Jansen, K.: Parameterized approximation scheme for the multiple knapsack problem. SIAM J. Comput. **39**(4), 1392–1412 (2010)

8. Jansen, K.: A fast approximation scheme for the multiple knapsack problem. In: Bieliková, M., Friedrich, G., Gottlob, G., Katzenbeisser, S., Turán, G. (eds.) SOF-SEM 2012. LNCS, vol. 7147, pp. 313–324. Springer, Heidelberg (2012). https://doi.org/10.1007/978-3-642-27660-6_26

9. Johnson, S.M.: Optimal two-and three-stage production schedules with setup times included. Naval Res. Logist. Q. 1(1), 61–68 (1954)

10. Kellerer, H.: A polynomial time approximation scheme for the multiple knapsack problem. In: Hochbaum, D.S., Jansen, K., Rolim, J.D.P., Sinclair, A. (eds.) APPROX/RANDOM -1999. LNCS, vol. 1671, pp. 51–62. Springer, Heidelberg (1999). https://doi.org/10.1007/978-3-540-48413-4_6

11. Kovalyov, M.Y.: Efficient epsilon-approximation algorithm for minimizing the makespan in a parallel two-stage system. Vesti Acad. navuk Belaruskai SSR Ser. Phiz.-Mat. Navuk 3, 119 (1985). (in Russian)

12. Wu, G., Chen, J., Wang, J.: Scheduling two-stage jobs on multiple flowshops. Theor. Comput. Sci. 776, 117–124 (2019)

13. Wu, G., Chen, J., Wang, J.: On scheduling inclined jobs on multiple two-stage flowshops. Theor. Comput. Sci. 786, 67–77 (2019)

14. Wu, G., Chen, J., Wang, J.: Improved approximation algorithms for two-stage flowshops scheduling problem. Theor. Comput. Sci. 806, 509–515 (2020)

15. Wu, G., Chen, J., Wang, J.: On scheduling multiple two-stage flowshops. Theor. Comput. Sci. 818, 74–82 (2020)

16. Zhang, X., van de Velde, S.: Approximation algorithms for the parallel flow shop problem. Eur. J. Oper. Res. 216(3), 544–552 (2012)

17. Zhang, Y., Zhou, Y.: TransOS: a transparent computing-based operating system for the cloud. Int. J. Cloud Comut. 4(1), 287–301 (2012)

Single Machine Scheduling with Rejection to Minimize the Weighted Makespan

Lingfa Lu[1(✉)], Liqi Zhang[2], and Jinwen Ou[3]

[1] School of Mathematics and Statistics, Zhengzhou University, Zhengzhou, Henan, People's Republic of China
lulingfa@zzu.edu.cn

[2] College of Information and Management Science, Henan Agricultural University, Zhengzhou, Henan, People's Republic of China

[3] Department of Administrative Management, Jinan University, Guangzhou, Guangdong, People's Republic of China
toujinwen@jnu.edu.cn

Abstract. In this paper, we consider the single machine scheduling problem with rejection to minimize the weighted makespan. In this problem, each job is either accepted and processed on the single machine, or is rejected by paying a rejection cost. The objective is to minimize the sum of the weighted makespan (the maximum weighted completion time) of accepted jobs and the total rejection cost of rejected jobs. We first show that this problem is binary NP-hard and then propose a pseudo-polynomial dynamic programming algorithm. Furthermore, based on the relaxed integral programming, we propose a 2-approximation algorithm for this problem. Finally, based on the dynamic programming algorithm and the vector trimming technique, we also obtain a fully polynomial-time approximation scheme (FPTAS) for this problem.

Keywords: Scheduling with rejection · Weighted makespan · Binary NP-hard · Dynamic programming · FPTAS

1 Introduction

In real life situations, a manufacturer often receive a great deal of orders (jobs) from the customers. Due to the lack of enough resources such as machines, operators or warehouses, processing all jobs may occur high inventory or tardiness costs. Thus, sometimes the manufacturer has to reject some jobs or outsource some jobs to a third-party manufacturer (an outsourcer). When a job is rejected or outsourced, a corresponding rejection cost or outsourcing cost is required. The decision-maker needs to determine which jobs should be accepted (and a feasible schedule for the accepted jobs), and which jobs should be rejected or outsourced, such that the total cost (including the production cost and the total rejection cost) is minimized. Thus, from the practical point of view, rejecting or outsourcing some jobs can save time and reduce costs. If the outsourcing cost is treated as the rejection cost, scheduling with rejection and scheduling with outsourcing are in fact equivalent.

© Springer Nature Switzerland AG 2021
W. Wu and H. Du (Eds.): AAIM 2021, LNCS 13153, pp. 96–110, 2021.
https://doi.org/10.1007/978-3-030-93176-6_9

1.1   Scheduling with Rejection

Bartal et al. [1] first studied the multi-processor scheduling problem with rejection which includes the on-line version and the off-line version. The objective is to minimize the makespan of the accepted jobs plus the total rejection penalty of the rejected jobs. They provided an on-line algorithm with best-possible competitive ratio $\frac{\sqrt{5}+3}{2} \approx 2.618$ for the on-line version and a polynomial-time approximation scheme (PTAS) for the off-line version, respectively. Since then, scheduling with rejection has received more and more attention in recent two decades. Hoogeveen et al. [7] considered the off-line multi-processor scheduling problem with rejection, where preemption is allowed. They provided some effective approximation algorithms and some FPTASs for their problems. Zhang et al. [21] studied the single machine scheduling problem with release date and rejection. They showed that this problem is NP-hard and then presented an FPTAS for this problem. For more results on scheduling with rejection, we refer the readers to the surveys provided by Shatbay et al. [17] and Zhang [22], respectively. More recent papers dealing scheduling with rejection are [3,6,9–13,15].

1.2   Scheduling to Minimize the Weighted Makespan

To the best of our knowledge, Feng and Yuan [4] first introduced the weighted makespan $WC_{\max}$ (the maximum weighted completion time) in a single-machine scheduling problem. Li [8] studied the online single machine scheduling problem to minimize the weighted makespan $WC_{\max}$. In this problem, all jobs arrive over time, and all information about a job is unknown until the job arrives. For this problem, they showed that the competition ratio of any online algorithm is at least 2, and they also obtained an online algorithm with the competition ratio 3. For the case where all jobs have the same processing time, they also presented an online algorithm with the best-possible competition ratio $\frac{\sqrt{5}+1}{2} \approx 1.618$. For the general online problem in [8], Chai et al. [2] provided two on-line algorithms with the best-possible competitive ratio 2.

2   Problem Formulation

Single machine scheduling with rejection to minimize the weighted makespan can be described as follows: There are n jobs $J_1, J_2, \cdots, J_n$ and a single machine. Each job has a processing time p_j, a weight w_j and a rejection cost e_j. Without loss of generality, we assume that all p_j, w_j and e_j values are positive integers. Each job J_j is either accepted and processed on the single machine, or is rejected by paying the corresponding rejection cost e_j. Let A and R be the set of the accepted jobs and the set of the rejected jobs, respectively. Furthermore, let π be a feasible schedule for the jobs in A and let C_j be the completion time of J_j in π, where $J_j \in A$. We define $WC_{\max} = \max\{w_j C_j : J_j \in A\}$ as the weighted makespan of the accepted jobs. The objective is to minimize the sum of the weighted makespan of the accepted jobs and the total rejection cost of the rejected jobs. Using the general notation for scheduling problems, our problem can be denoted by $1||WC_{\max} + \sum_{J_j \in R} e_j$.

In order to design an effective approximation algorithm, we introduce the concept of "split". That is, if a job J_j is split into two parts J_j^A and J_j^R, then we have $p_j^A + p_j^R = p_j$ and $\frac{e_j^A}{p_j^A} = \frac{e_j^R}{p_j^R} = \frac{e_j}{p_j}$. Furthermore, J_j^A is accepted and processed on the machine, and J_R is rejected by paying the rejection cost e_j^R. The corresponding problem can be denoted by $1|split|WC_{\max} + \sum_{J_j \in R} e_j$. If we treat p_j^A, p_j^R and e_j^R as the actual processing time t_j, the compression amount u_j and the compression cost $x_j u_j$ of job J_j, i.e., $p_j^A = t_j = p_j - u_j$, $p_j^R = u_j$ and $e_j^R = x_j u_j = \frac{e_j}{p_j} u_j$, then each scheduling problem with split and rejection is in fact equivalent to the corresponding scheduling problem with controllable processing times. Therefore, problem $1|split|WC_{\max} + \sum_{J_j \in R} e_j$ is equivalent to $1|t_j = p_j - u_j|WC_{\max} + \sum_{j=1}^{n} x_j u_j$, respectively. For the more general problem $1|t_j = p_j - u_j|f_{\max} + \sum_{j=1}^{n} x_j u_j$, van Wassenhove and Baker [19] presented an $O(n^2)$-time algorithm for finding an optimal schedule (and all Pareto optimal points) when the function f_j satisfies the condition that there exists a permutation π such that $f_{\pi(1)}(t) \le f_{\pi(2)}(t) \le \cdots \le f_{\pi(n)}(t)$ for all t. Clearly, in our problem, $f_j(t) = w_j t$ satisfies the above condition. Thus, problem $1|t_j = p_j - u_j|WC_{\max} + \sum_{j=1}^{n} x_j u_j$ (and also $1|split|WC_{\max} + \sum_{J_j \in R} e_j$) can be solved in $O(n^2)$ time. For the more results on scheduling with the controllable processing times, we refer the readers to two surveys presented by Nowicki and Zdrzalka [14], Shabtay and Steiner [18].

In this paper, we show that problem $1||WC_{\max} + \sum_{J_j \in R} e_j$ are binary NP-hard and propose a pseudo-polynomial dynamic programming algorithm. Furthermore, based on the optimal schedule for problem $1|split|WC_{\max} + \sum_{J_j \in R} e_j$, we presented an effective 2-approximation algorithm for problem $1||WC_{\max} + \sum_{J_j \in R} e_j$. Finally, based on the dynamic programming algorithm and the vector trimming technique, we also obtain a fully polynomial-time approximation scheme (FPTAS) for this problem.

3   NP-Hardness Proof

In this section, we show that problem $1||WC_{\max} + \sum_{J_j \in R} e_j$ is binary NP-hard.

Theorem 1. *Problem $1||WC_{\max} + \sum_{J_j \in R} e_j$ is NP-hard.*

Proof. We use the NP-complete Partition problem (see Garey and Johnson [5]) for the reduction.

Partition Problem: Given $t + 1$ positive integers $a_1, a_2, \cdots, a_t, B$ with $\sum_{i=1}^{t} a_i = 2B$, is there a partition $(S, \overline{S})$ of $\{1, \cdots, t\}$ such that $\sum_{i \in S} a_i = \sum_{i \in \overline{S}} a_i = B$?

For a given instance of the Partition problem, we construct an instance of the decision version of $1||WC_{\max} + \sum_{J_j \in R} e_j$ as follows.

- $n = t + 1$ jobs.
- $w_j = 2, p_j = e_j = a_j$ for $1 \leq j \leq t$.
- $w_j = 2B + 2, p_j = 1, e_j = 3B + 3$ for $j = t + 1$.
- The threshold value is defined by $Y = 3B + 2$.
- The decision version asks whether there is a schedule

π such that $WC_{\max} + \sum_{J_j \in R} e_j \leq Y$.

Assume that Partition problem has a solution, i.e., there exists a partition $(S, \overline{S})$ such that $\sum_{j \in S} a_j = \sum_{j \in \overline{S}} a_j = B$. We construct a schedule in the following way: Schedule the job J_{t+1} as the first processed job on the machine, and then schedule all jobs J_j with $j \in S$ consecutively after J_{t+1} on the machine. It can be seen that, for the job J_{t+1}, we have $w_{t+1}C_{t+1} = (2B + 2) \times 1 = 2B + 2$; for the job J_j with $j \in S$, we have $w_j C_j \leq (B + 1) \times 2 = 2B + 2$. Thus, we have $WC_{\max} = 2B + 2$. We further reject all jobs J_j with $j \in \overline{S}$. Thus, we have $\sum_{J_j \in R} e_j = \sum_{j \in \overline{S}} a_j = B$. It follows that $WC_{\max} + \sum_{J_j \in R} e_j = 2B + 2 + B = 3B + 2 = Y$.

Next, we assume there exists a schedule π such that $WC_{\max} + \sum_{J_j \in R} e_j \leq Y$. We will prove that Partition problem has a solution $(S, \overline{S})$. Let A and R be the set of accepted jobs and the set of rejected jobs in π, respectively. We have the following two claims.

Claim 1. Job J_{t+1} must be accepted and J_{t+1} is the first processed job in π.

Proof. If job J_{t+1} is rejected, then we have $WC_{\max} + \sum_{J_j \in R} e_j \geq e_{t+1} = 3B + 3 > Y$, a contradiction. Furthermore, if job J_{t+1} is not the first processed job in π, then we have $C_{t+1}(\pi) \geq 2$. Thus, we have $WC_{\max} + \sum_{J_j \in R} e_j \geq (2B + 2) \times 2 = 4B + 4 > Y$, a contradiction again. Claim 1 follows.

Claim 2. $\sum_{J_j \in R} a_j = B$.

Proof. By Claim 1, we have $WC_{\max} \geq w_{t+1}C_{t+1}(\pi) = 2B + 2$. Furthermore, we have $\sum_{J_j \in R} a_j = \sum_{J_j \in R} e_j \leq B$ since $WC_{\max} + \sum_{J_j \in R} e_j \leq Y = 3B + 2$. Next, we assume that $\sum_{J_j \in R} a_j < B$. Thus, we have

$$WC_{\max} + \sum_{J_j \in R} e_j \geq 2\Big(\sum_{J_j \in A \setminus J_{t+1}} p_j + 1 \Big) + \sum_{J_j \in R} e_j$$

$$= 2\Big(\sum_{J_j \in A \setminus J_{t+1}} a_j + 1 \Big) + \sum_{J_j \in R} a_j$$

$$= 2 + 2\Big(\sum_{J_j \in A} a_j + \sum_{J_j \in R} a_j \Big) - \sum_{J_j \in R} a_j$$

$$= 2 + 4B - \sum_{J_j \in R} a_j$$

$$> 4B + 2 - B$$
$$= 3B + 2$$
$$= Y,$$

a contradiction. Thus, we have $\sum_{J_j \in R} a_j = B$.

Set $S = \{j : 1 \le j \le t \text{ and } J_j \in R\}$ and $\overline{S} = \{j : 1 \le j \le t \text{ and } J_j \in A\}$. Clearly, $(S, \overline{S})$ is a solution of Partition problem. Theorem 1 follows.

4   Dynamic Programming Algorithm

When job rejection is not allowed, the corresponding scheduling problem can be denoted by $1||WC_{\max}$. For problem $1||WC_{\max}$, Li [8] showed that the LW rule yields an optimal schedule. The LW rule can be stated as follows:

LW Rule: All jobs are processed on the machine by the LW (Largest Weight first) rule. That is, whenever the machine is idle, among all unfinished jobs, the job with the largest weight is scheduled.

Lemma 1. *For problem* $1||WC_{\max} + \sum_{J_j \in R} e_j$, *there is an optimal schedule such that all accepted jobs are processed in the LW rule.*

Sort all jobs such that $w_1 \ge \cdots \ge w_n$. Let $f_j(t, E)$ be the minimum $WC_{\max}$ value when (1) the jobs in consideration are $J_1, \cdots, J_j$; (2) the current makespan of the accepted jobs among $J_1, \cdots, J_j$ is exactly t; and (3) the total rejected cost of the rejected jobs among $J_1, \cdots, J_j$ is exactly E. In any such schedule, there are two possible cases: either job J_j is rejected or job J_j is accepted.

Case 1: If job J_j is rejected, then the makespan of $J_1, \cdots, J_{j-1}$ is still t, the total rejection cost of $J_1, \cdots, J_{j-1}$ is $E - e_j$. Thus, we have $f_j(t, E) = f_{j-1}(t, E - e_j)$.

Case 2: If job J_j is accepted, then the makespan of $J_1, \cdots, J_{j-1}$ is $t - p_j$, the total rejection cost of $J_1, \cdots, J_{j-1}$ is still E. Note that the minimum $WC_{\max}$ value of $J_1, \cdots, J_{j-1}$ is $f_{j-1}(t - p_j, E)$ and $w_j C_j = w_j \cdot t$. Thus, we have $f_j(t, E) = \max\{f_{j-1}(t - p_j, E), w_j \cdot t\}$.

Combining the above two cases, we have the following dynamic programming algorithm DP1.

Dynamic Programming Algorithm DP1

The Boundary Conditions:

$$f_1(t, E) = \begin{cases} 0, & \text{if } t = 0 \text{ and } E = e_1; \\ w_1 p_1, & \text{if } t = p_1 \text{ and } E = 0; \\ +\infty, & \text{otherwise.} \end{cases}$$

The Recursive Function:

$$f_j(t, E) = \min\{\max\{f_{j-1}(t - p_j, E), w_j \cdot t\}, f_{j-1}(t, E - e_j)\}.$$

The Optimal Value:

$$\min\{f_n(t, E) + E : 0 \leq t \leq \sum_{j=1}^{n} p_j, 0 \leq E \leq \sum_{j=1}^{n} e_j\}.$$

Theorem 2. *For problem $1||WC_{\max} + \sum_{J_j \in R} e_j$, algorithm DP1 yields an optimal schedule in $O(n \cdot \sum_{j=1}^{n} p_j \cdot \sum_{j=1}^{n} e_j)$ time.*

Proof. The correctness of algorithm DP1 is guaranteed by the above discussion. The recursive function has at most $O(n \cdot \sum_{j=1}^{n} p_j \cdot \sum_{j=1}^{n} e_j)$ states and each iteration costs a constant time. Hence, the total running time is bounded by $O(n \cdot \sum_{j=1}^{n} p_j \cdot \sum_{j=1}^{n} e_j)$.

5 Approximation Algorithms

5.1 A 2-Approximation Algorithm

In this subsection, we provide a 2-approximation algorithm for problem $1||WC_{\max} + \sum_{J_j \in R} e_j$. Sort all jobs such that $w_1 \geq \cdots \geq w_n$. Let $x_j = 1$ if job J_j is accepted and $x_j = 0$ if job J_j is rejected. Thus, problem $1||WC_{\max} + \sum_{J_j \in R} e_j$ is equivalent to the following Integer Linear Programming (**ILP**).

$$\min \ WC_{\max} + \sum_{j=1}^{n} (1 - x_j) e_j$$

$$w_j \sum_{k=1}^{j} x_k p_k \leq WC_{\max} \text{ for each } k = 1, \cdots, n.$$

$$x_j \in \{0, 1\} \text{ for each } j = 1, \cdots, n.$$

For each rejected job J_j, if $\sum_{k=1}^{j} x_k p_k > 0$, then there is some i with $i < j$ such that $C_i = \sum_{k=1}^{i} x_k p_k = \sum_{k=1}^{j} x_k p_k$ and $w_i \geq w_j$. Note that $w_i \sum_{k=1}^{i} x_k p_k \leq WC_{\max}$. Thus, $w_j \sum_{k=1}^{j} x_k p_k \leq WC_{\max}$ holds for each rejected job J_j. If we replace $x_j \in \{0,1\}$ by $0 \le x_j \le 1$ for each $j = 1, \cdots, n$, we can obtain a Relaxed Linear Programming (**RLP**).

$$\min \; WC_{\max} + \sum_{j=1}^{n} (1 - x_j)e_j$$

$$w_j \sum_{k=1}^{j} x_k p_k \le WC_{\max} \text{ for each } k = 1, \cdots, n.$$

$$0 \le x_j \le 1 \text{ for each } j = 1, \cdots, n.$$

Algorithm A_1

Step 1: Solve the **RLP**. Let $(x_1^*, \cdots, x_n^*, WC_{\max}^*)$ be an optimal solution of **RLP**. If $x_j^* \ge \frac{1}{2}$, then we set $x_j = 1$; otherwise, we set $x_j = 0$.

Step 2: Accept all jobs with $x_j = 1$ and reject all jobs with $x_j = 0$. Process all accepted jobs in the LW rule.

Note that the RLP is equivalent to problem $1|\text{split}|WC_{\max} + \sum_{J_j \in R} e_j$. Furthermore, both of them are equivalent to problem $1|t_j = p_j - u_j|WC_{\max} + \sum_{j=1}^{n} x_j u_j$. For the latter problem, van Wassenhove and Baker [19] presented an $O(n^2)$-time algorithm, which is faster than some well-known algorithms for the RLP. Thus, we also use van Wassenhove and Baker's algorithm to obtain the optimal solution of the RLP.

Let π be the schedule obtained from algorithm A_1. Furthermore, we let Z and Z^* be the corresponding objective values of π and an optimal schedule π^*, respectively.

Theorem 3. $Z \le 2Z^*$.

Proof. Note that $w_j C_j = w_j \sum_{k=1}^{j} x_k p_k \le 2w_j \sum_{k=1}^{j} x_k^* p_k \le 2WC_{\max}^*$ for each accepted job J_j. That is, $WC_{\max} = \max\{w_j C_j : 1 \le j \le n\} \le 2WC_{\max}^*$. Furthermore, we also have $\sum_{J_j \in R} e_j = \sum_{j=1}^{n} (1 - x_j)e_j \le 2\sum_{j=1}^{n} (1 - x_j^*)e_j$. Thus, we have

$$Z = WC_{\max} + \sum_{J_j \in R} e_j \le 2WC_{\max}^* + 2\sum_{j=1}^{n} (1 - x_j^*)e_j = 2Z^*.$$

This completes the proof of Theorem 3.

5.2   A Fully Polynomial-Time Approximation Scheme

In order to design a fully polynomial-time approximation scheme (FPTAS) for an NP-hard problem, a common way is to transform a optimal but slow algorithm (a pseudo-polynomial-time dynamic programming algorithm) into a near optimal but faster algorithm (a fully polynomial-time approximation scheme).

The main idea of this way is to remove some yielded data during the execution of the algorithm and clean up the algorithm's memory. As a result, the algorithm becomes faster and the outputted solution is nearly optimal. This approach is called "vector trimming" in the literature (see [16,20]). Before we propose our FPTAS for problem $1||WC_{\max} + \sum_{J_j \in R} e_j$, we first introduce the following notation.

Assume that the jobs have been sorted in the LW rule such that $w_1 \geq \cdots \geq w_n$. Let $I_j = \{J_1, \cdots, J_j\}$ be the set of the first j jobs. Assume that π is a feasible schedule for the jobs in I_j. Furthermore, let $t(\pi)$, $E(\pi)$ and $WC_{\max}(\pi)$ be the the makespan of the accepted jobs, the total rejected cost of the rejected jobs and the corresponding $WC_{\max}$ value in π, respectively. In this situation, we use a 3-dimensional vector $v(\pi) = [t(\pi), E(\pi), WC_{\max}(\pi)]$ to encode the schedule π. Clearly, if there are two feasible schedules π_1 and π_2 such that $t(\pi_1) = t(\pi_2)$, $E(\pi_1) = E(\pi_2)$ and $WC_{\max}(\pi_1) < WC_{\max}(\pi_2)$, then π_2 and $v(\pi_2)$ are **dominated** by π_1 and $v(\pi_1)$, respectively. Thus, if there are multiple schedules for the jobs in I_j with the same makespan t and the same total rejection cost E, we can only reserve the schedule and the corresponding vector with the minimum $WC_{\max}$ value. It follows that

$$\min\{WC_{\max}(\pi) : t(\pi) = t \text{ and } E(\pi) = E\} = f_j(t, E),$$

where $f_j(t, E)$ is the minimum $WC_{\max}$ value which is defined in DP1. Consequently, we also can use the vector $v = [t, E, f_j(t, E)]$ to represent a non-dominated schedule. Let VS_j be the set of all vectors with respect to all non-dominated schedules for the jobs in I_j. Therefore, we can obtain the following algorithm DP2.

Algorithm DP2

Initialization: $VS_1 = \{(0, e_1, 0), (p_1, 0, w_1 p_1)\}$.

Phase j: For each vector $[t, E, f_{j-1}(t, E)] \in VS_{j-1}$, we add two vectors $[t, E + e_j, f_{j-1}(t, E)]$ and $[t + p_j, E, \max\{f_{j-1}(t, E), w_j(t + p_j)\}]$ into the set VS_j. If there are multiple vectors in VS_j with the same makespan t and the same total rejection cost E, we can only reserve the corresponding vector with the minimum $WC_{\max}$ value in VS_j.

Outputs:

$$\min\{f_n(t, E) + E : [t, E, f_n(t, E)] \in VS_n, 0 \leq t \leq \sum_{j=1}^{n} p_j \text{ and } 0 \leq E \leq \sum_{j=1}^{n} e_j\}.$$

It seems that algorithm DP2 is slower than DP1. However, algorithm DP2 is more suitable to obtain an FPTAS by the vector trimming technology. Let Z and Z^* be the corresponding objective values obtained by the 2-approximation algorithm A_1 and an optimal schedule π^*, respectively. Furthermore, let $t^*, E^*, WC_{\max}^*$ be the corresponding values with respect to π^*.

That is, $t^* = t(\pi^*)$, $E^* = E(\pi^*)$ and $WC^*_{\max} = WC_{\max}(\pi^*)$. From Theorem 5.1, we have $Z^* \leq Z \leq 2Z^*$. Without loss of generality, we assume that all weights w_j are positive integers. That is, $w_j \geq 1$ holds for each $j = 1, \cdots, n$. Otherwise, if $w_j = 0$ holds for some job J_j, then J_j is always accepted and processed in the end of any optimal schedule. Thus, we can remove all the jobs with zero weights. As a result, we have $t^* \leq WC^*_{\max} \leq Z^* \leq 2Z$ and $E^* \leq Z^* \leq 2Z$. That is, all vectors with respect to all optimal schedules fall in the cube $[0, 2Z]^3$. Set $L = (1 + \epsilon)Z$. Thus, to obtain an FPTAS for our problem, we only need to consider those vectors with $t \leq E + WC_{\max} \leq (1 + \epsilon)Z^* \leq (1 + \epsilon)Z = L$. That is, all considered vectors in our FPTAS fall in the cube $[0, L]^3$. Furthermore, we set $\Delta = \frac{\epsilon Z}{4n}$. Consequently, we subdivide the cube $[0, L]^3$ into $m = (\frac{L}{\Delta})^3 = O(\frac{n^3}{\epsilon^3})$ boxes. In this situation, if a vector $[x, y, z]$ with $(i - 1)\Delta \leq x < i\Delta$, $(j - 1)\Delta \leq y < j\Delta$ and $(k - 1)\Delta \leq z < k\Delta$, then we say that vector $[x, y, z]$ falls into the $(i \times j \times k)$-th box.

If two vector $[x, y, z]$ and $[x', y', z']$ fall into the same box, then we have $0 \leq |x - x'| \leq \Delta$, $0 \leq |y - y'| \leq \Delta$ and $0 \leq |z - z'| \leq \Delta$. Since Δ is very small, all vectors in the same box are very close to each other. Thus, if there are multiple vectors in the same box, we only reserve the vector with the minimum t value. Such a procedure is called "vector trimming" in the literature. Clearly, for any vector set $S \subseteq [0, L]^3$, we have $|S^\#| = O(\frac{n^3}{\epsilon^3})$, where $S^\#$ is the set obtained by the trimming procedure from set S. Based on this idea, we present an FPTAS for problem $1||WC_{\max} + \sum_{J_j \in R} e_j$.

Algorithm A_ϵ

Initialization: Let $VS_1^\#$ be the trimmed set obtained from VS_1, where $VS_1 = \{(0, e_1, 0), (p_1, 0, w_1 p_1)\}$.

Phase j: For each vector $[t, E, f_{j-1}(t, E)] \in VS_{j-1}^\#$, we add two vectors $[t, E + e_j, f_{j-1}(t, E)]$ and $[t + p_j, E, \max\{f_{j-1}(t, E), w_j(t + p_j)\}]$ into the set VS_j'. Furthermore, Let $VS_j^\#$ be the trimmed set obtained from VS_j'.

Outputs:

$$\min\{f_n(t, E) + E : [t, E, f_n(t, E)] \in VS_n^\# \text{ and } \max\{t, E, f_n(t, E)\} \leq L\}.$$

To show that algorithm A_ϵ is an FPTAS for problem $1||WC_{\max} + \sum_{J_j \in R} e_j$, we have the following lemma.

Lemma 2. *For each vector $[x, y, z] \in VS_j$, there is a vector $[x^\#, y^\#, z^\#] \in VS_j^\#$ such that $x^\# \leq x$, $y^\# \leq y + j\Delta$ and $z^\# \leq z + j\Delta$.*

Proof. We prove this lemma by induction on j. Clearly, Lemma 2 holds when $j = 1$. Now, we assume that the lemma holds for $j - 1$ and consider any vector $[x, y, z]$ with $[x, y, z] \in VS_j$. From algorithm DP2, there is a vector $[a, b, c] \in VS_{j-1}$ such that $[x, y, z]$ is generated by $[a, b, c]$. Thus, we have two possibilities: either

$[x, y, z] = [a, b + e_j, c]$ if J_j is rejected or $[x, y, z] = [a + p_j, b, \max\{c, w_j(a + p_j)\}]$ if J_j is accepted. Note that $[a, b, c] \in VS_{j-1}$. By the inductive assumption, there is a vector $[a^\#, b^\#, c^\#] \in VS_{j-1}^\#$ such that $a^\# \leq a$, $b^\# \leq b + (j - 1)\Delta$ and $c^\# \leq c + (j - 1)\Delta$. We distinguish two cases in the following discussion.

Case 1. $[x, y, z] = [a, b + e_j, c]$.

Note that $[a^\#, b^\#, c^\#] \in VS_{j-1}^\#$. By Phase j of A_ϵ, we have $[a^\#, b^\# + e_j, c^\#] \in VS_j'$. Let $[x^\#, y^\#, z^\#] \in VS_j^\#$ be the selected vector which is in the same box as $[a^\#, b^\# + e_j, c^\#]$. Thus, by the trimming rule, we have

$$x^\# \leq a^\# \leq a = x,$$

$$y^\# \leq b^\# + e_j + \Delta \leq b + (j - 1)\Delta + e_j + \Delta = y + j\Delta,$$

and

$$z^\# \leq c^\# + \Delta \leq c + (j - 1)\Delta + \Delta = z + j\Delta.$$

Case 2. $[x, y, z] = [a + p_j, b, \max\{c, w_j(a + p_j)\}]$.

Note that $[a^\#, b^\#, c^\#] \in VS_{j-1}^\#$. By Phase j of A_ϵ, we have

$$[a^\# + p_j, b^\#, \max\{c^\#, w_j(a^\# + p_j)\}] \in VS_j'.$$

Let $[x^\#, y^\#, z^\#] \in VS_j^\#$ be the selected vector which is in the same box as $[a^\# + p_j, b^\#, \max\{c^\#, w_j(a^\# + p_j)\}]$. Thus, by the trimming rule, we have

$$x^\# \leq a^\# + p_j \leq a + p_j = x,$$

$$y^\# \leq b^\# + \Delta \leq b + (j - 1)\Delta + \Delta = y + j\Delta,$$

and

$$\begin{aligned}
z^\# &\leq \max\{c^\#, w_j(a^\# + p_j)\} + \Delta \\
&\leq \max\{c + (j - 1)\Delta, w_j(a + p_j)\} + \Delta \\
&\leq \max\{c + (j - 1)\Delta, w_j(a + p_j) + (j - 1)\Delta\} + \Delta \\
&= \max\{c, w_j(a + p_j)\} + (j - 1)\Delta + \Delta \\
&= z + j\Delta.
\end{aligned}$$

Combine the above discussion in two cases, we can conclude that Lemma 2 holds by induction on j.

Note that $[t^*, E^*, WC_{\max}^*] \in VS_n$. By Lemma 2, there is a vector $[t^\#, E^\#, WC_{\max}^\#] \in VS_n^\#$ such that $t^\# \leq t^*$, $E^\# \leq E^* + n\Delta$ and $WC_{\max}^\# \leq WC_{\max}^* + n\Delta$. Let

$$Z_\epsilon = \min\{f_n(t, E) + E : [t, E, f_n(t, E)] \in VS_n^\# \text{ and } \max\{t, E, f_n(t, E)\} \leq L\}$$

be the objective value obtained from algorithm A_ϵ. Thus, we have

$$Z \le WC^\#_{\max} + E^\# \le WC^*_{\max} + E^* + 2n\Delta \le Z^* + \epsilon\frac{Z}{2} \le (1+\epsilon)Z^*.$$

Now, we consider the time complexity of A_ϵ. Note that there are n phases and in each phase j, we have

$$|VS^\#_j| \le |VS'_j| \le 2|VS^\#_{j-1}| = O(\frac{n^3}{\epsilon^3}).$$

Thus, the time complexity of A_ϵ is $\sum_{j=1}^n |VS^\#_j| = O(\frac{n^4}{\epsilon^3})$. Therefore, we have the following theorem.

Theorem 4. *Algorithm A_ϵ is an FPTAS for problem $1||WC_{\max} + \sum_{J_j \in R} e_j$ and its time complexity is $O(\frac{n^4}{\epsilon^3})$.*

6   Discussions on Some Special Cases

For any instance $I = \{J_1, \cdots, J_n\}$, let n_p, n_w and n_e be the numbers of distinct processing times, distinct weights and distinct rejection costs, respectively. In this section, we consider some special cases with $n_p = k$, or $n_w = k$ or $n_e = k$, where k is a fixed constant. The corresponding problems are denoted $1|n_p = k|WC_{\max} + \sum_{J_j \in R} e_j$, $1|n_w = k|WC_{\max} + \sum_{J_j \in R} e_j$ and $1|n_e = k|WC_{\max} + \sum_{J_j \in R} e_j$, respectively.

6.1   Problem $1|n_p = k|WC_{\max} + \sum_{J_j \in R} e_j$

Suppose that $a_1, a_2, \cdots, a_k$ are k distinct processing times for instance $I = \{J_1, \cdots, J_n\}$. Furthermore, we write $S_i = \{J_j : p_j = a_i\}$ and $|S_i| = m_i$. In this subsection, we will provide a polynomial-time algorithm for this problem.

First, similar to algorithm DP1, we can obtain a new dynamic programming algorithm DP3 for the general problem $1||WC_{\max} + \sum_{J_j \in R} e_j$. Let $f_j(t, WC_{\max})$ be the minimum total rejection cost when (1) the jobs in consideration are $J_1, \cdots, J_j$; (2) the makespan of the accepted jobs among $J_1, \cdots, J_j$ is exactly t; and (3) the weighted makespan of the accepted jobs among $J_1, \cdots, J_j$ is exactly $WC_{\max}$.

Case 1: If job J_j is rejected, then we have $f_j(t, WC_{\max}) = f_{j-1}(t, WC_{\max}) + e_j$. For convenience, we write $V_R = f_{j-1}(t, WC_{\max}) + e_j$ if J_j is rejected.
Case 2: If job J_j is accepted, then the makespan of $J_1, \cdots, J_{j-1}$ is $t - p_j$ and the completion time of J_j is t. Thus, we have $WC_{\max} \ge w_j \cdot t$. If $WC_{\max} > w_j \cdot t$, then the weighted makespan of the accepted jobs among $J_1, \cdots, J_{j-1}$ is still $WC_{\max}$. Furthermore, we have $f_j(t, WC_{\max}) = f_{j-1}(t - p_j, WC_{\max})$.

If $WC_{\max} = w_j \cdot t$, let $WC'_{\max}$ be the weighted makespan of the accepted jobs among $J_1, \cdots, J_{j-1}$. Thus, we have $WC'_{\max} \le w_j \cdot t$. Furthermore, we have $f_j(t, WC_{\max}) = f_{j-1}(t - p_j, WC'_{\max})$. For convenience, we write $V_A^1 = f_{j-1}(t - p_j, WC'_{\max})$ if J_j is accepted and $WC_{\max} > w_j \cdot t$. Furthermore, we also write $V_A^2 = \min\{f_{j-1}(t - p_j, WC'_{\max}) : WC'_{\max} \le w_j \cdot t\}$ if J_j is accepted and $WC_{\max} = w_j \cdot t$.

Combining the above two cases, we have the following dynamic programming algorithm DP3.

Dynamic Programming Algorithm DP3

The Boundary Conditions:

$$f_1(t, WC_{\max}) = \begin{cases} 0, & \text{if } t = p_1 \text{ and } WC_{\max} = w_1 p_1; \\ e_1, & \text{if } t = 0 \text{ and } WC_{\max} = 0; \\ +\infty, & \text{otherwise.} \end{cases}$$

The Recursive Function:

$$f_j(t, WC_{\max}) = \begin{cases} V_R, & \text{if } t < p_j \text{ or } WC_{\max} < w_j t; \\ \min\{V_R, V_A^1\}, & \text{if } t \ge p_j \text{ and } WC_{\max} > w_j t; \\ \min\{V_R, V_A^2\}, & \text{if } t \ge p_j \text{ and } WC_{\max} = w_j t. \end{cases}$$

The Optimal Value:

$$\min\{f_n(t, WC_{\max}) + WC_{\max} : 0 \le t \le \sum_{j=1}^n p_j, 0 \le WC_{\max} \le w_1 \cdot \sum_{j=1}^n p_j\}.$$

Theorem 5. *For problem* $1||WC_{\max} + \sum_{J_j \in R} e_j$, *algorithm DP3 yields an optimal schedule in* $O(n \cdot w_1 \cdot (\sum_{j=1}^n p_j)^2)$ *time.*

Proof. The correctness of algorithm DP3 is also guaranteed by the above discussion. The recursive function has at most $O(n \cdot w_1 \cdot (\sum_{j=1}^n p_j)^2)$ states. When $WC_{\max} < w_j \cdot t$ or $WC_{\max} > w_j \cdot t$, each iteration costs a constant time; when $WC_{\max} = w_j \cdot t$, the recursive function has at most $O(n \cdot \sum_{j=1}^n p_j)$ states and each iteration costs an $O(w_1 \cdot \sum_{j=1}^n p_j)$ time. Hence, the time complexity of algorithm DP3 is bounded by $O(n \cdot w_1 \cdot (\sum_{j=1}^n p_j)^2)$.

Specially, if all jobs have k distinct processing times $a_1, \cdots, a_k$, then we have $t \in \{x_1 a_1 + \cdots + x_k a_k : 0 \le x_k \le m_k\}$ and $WC_{\max} \in \{w_j \cdot (x_1 a_1 + \cdots + x_k a_k) : 1 \le j \le n \text{ and } 0 \le x_k \le m_k\}$. Thus, we have $O(\Pi_{i=1}^k m_k) = O(n^k)$ choices for each t and $O(n^{k+1})$ choices for each $WC_{\max}$ or $WC'_{\max}$. As a result, we have the following corollary.

Corollary 1. *Algorithm DP3 solves* $1|n_p = k|WC_{\max} + \sum_{J_j \in R} e_j$ *in* $O(n^{2k+2})$ *time.*

6.2   Problem $1|n_w = k|WC_{\max} + \sum_{J_j \in R} e_j$

Clearly, when $k = 1$, i.e., $w_j = w$ for $j = 1, \cdots, n$, problem $1|n_w = 1|WC_{\max} + \sum_{J_j \in R} e_j$ is equivalent to problem $1||w \cdot C_{\max} + \sum_{J_j \in R} e_j$. Thus, it is trivial to obtain an optimal schedule by accepting the jobs with $wp_j < e_j$ and rejecting other jobs. Note that in the NP-hard proof of Theorem 1, there are only $k = 2$ distinct weights. Thus, when $k \geq 2$, problem $1|n_w = k|WC_{\max} + \sum_{J_j \in R} e_j$ becomes NP-hard.

6.3   Problem $1|n_e = k|WC_{\max} + \sum_{J_j \in R} e_j$

Note that in all algorithms DP1, DP2 and DP3, the current makespan t is always used to compute the $w_j C_j$ value if J_j is accepted. Thus, it seems to be difficult to design an algorithm which does not include t as a parameter. Thus, we conjecture that problem $1|n_e = k|WC_{\max} + \sum_{J_j \in R} e_j$ is also NP-hard even when $n_e = 1$. It might be a challenging problem to determine its exact computational complexity.

6.4   Problem $1|n_w = k_1, n_e = k_2|WC_{\max} + \sum_{J_j \in R} e_j$

Note that problem $1|n_w = k|WC_{\max} + \sum_{J_j \in R} e_j$ is NP-hard when $k \geq 2$ and we conjecture that problem $1|n_e = k|WC_{\max} + \sum_{J_j \in R} e_j$ is also NP-hard. Thus, we consider a more special case with k_1 distinct weights and k_2 distinct rejection costs. The corresponding problem is denoted by $1|n_w = k_1, n_e = k_2|WC_{\max} + \sum_{J_j \in R} e_j$. In this subsection, we will show that problem $1|n_w = k_1, n_e = k_2|WC_{\max} + \sum_{J_j \in R} e_j$ can be solved in polynomial time.

Let $b_1, \cdots, b_{k_1}$ and $c_1, \cdots, c_{k_2}$ be the distinct weights and rejection costs, respectively. Furthermore, we set $S_{xy} = \{J_j : w_j = b_x$ and $e_j = c_y\}$ for each $x = 1, \cdots, k_1$ and $y = 1, \cdots, k_2$. Resort all jobs in S_{xy} in the SPT (Shortest Processing Time-first) rule. Note that the jobs in $S_{x,y}$ have the same weight and rejection cost. Thus, if there are l_{xy} jobs are accepted in an optimal schedule π^*, we can assume that the first l_{xy} jobs in S_{xy} are accepted. Since $0 \leq l_{xy} \leq |S_{xy}| \leq n$, by enumerating all possibilities about the distinct l_{xy} values and selecting the best schedule, we can find an optimal schedule in $O(n^{k_1 \cdot k_2})$ time. That is, problem $1|n_w = k_1, n_e = k_2|WC_{\max} + \sum_{J_j \in R} e_j$ can solved optimally in $O(n^{k_1 \cdot k_2})$ time.

7   Conclusions and Future Research

In this paper we consider the single machine scheduling problem with rejection to minimize the weighted makespan. The objective is to minimize the sum of the weighted makespan (the maximum weighted completion time) of accepted jobs and the total rejection cost of rejected jobs. We first show that this problem is binary NP-hard and then propose a pseudo-polynomial dynamic programming algorithm. Furthermore, based on the relaxed integral programming, we propose a 2-approximation algorithm for this problem. Finally, based on the dynamic programming algorithm and the vector trimming technique, we also obtain a fully polynomial-time approximation scheme (FPTAS) for this problem. In additions, we also discuss some special cases and provide two polynomial-time algorithms for them.

Note that it is still open whether problem $1|n_e = k|WC_{\max} + \sum_{J_j \in R} e_j$ is NP-hard or not even when $n_e = 1$. Thus, an interesting direction is to consider its computational complexity of problem $1|n_e = k|WC_{\max} + \sum_{J_j \in R} e_j$. Moreover, it is also interesting to consider the online or semi-online versions of this problem. Finally, we will extend this problem to parallel machine setting in the future.

Acknowledgments. This research was supported by NSFCs (11901168, 11971443 and 11771406).

References

1. Bartal, Y., Leonardi, S., Spaccamela, A.M., Stougie, J.: Multi-processor scheduling with rejection. SIAM J. Discret. Math. **13**, 64–78 (2000)
2. Chai, X., Lu, L.F., Li, W.H., Zhang, L.Q.: Best-possible online algorithms for single machine scheduling to minimize the maximum weighted completion time. Asia-Pacific J. Oper. Res. **35**, 1850048 (2018)
3. Chen, R.-X., Li, S.-S.: Minimizing maximum delivery completion time for order scheduling with rejection. J. Comb. Optim. **40**(4), 1044–1064 (2020). https://doi.org/10.1007/s10878-020-00649-2
4. Feng, Q., Yuan, J.J.: NP-hardness of a multicriteria scheduling on two families of jobs. OR Trans. **11**, 121–126 (2007)
5. Garey, M.R., Johnson, D.S.: Computers and Intractablity: A Guide to the Theory of NP-Completeness. Freeman, San Francisco, CA (1979)
6. Hermelin, D., Pinedo, M., Shabtay, D., Talmon, N.: On the parameterized tractability of a single machine scheduling with rejection. Eur. J. Oper. Res. **273**, 67–73 (2019)
7. Hoogeveen, H., Skutella, M., Woeginger, G.J.: Preemptive scheduling with rejection. Math. Program. **94**, 361–374 (2003)
8. Li, W.J.: A best possible online algorithm for the parallel-machine scheduling to minimize the maximum weighted completed time. Asia-Pacific J. Oper. Res. **32**, 1550030 (2015)
9. Liu, P., Lu, X.: New approximation algorithms for machine scheduling with rejection on single and parallel machine. J. Comb. Optim. **40**(4), 929–952 (2020). https://doi.org/10.1007/s10878-020-00642-9
10. Liu, Z.X.: Scheduling with partial rejection. Oper. Res. Lett. **48**, 524–529 (2020)
11. Ma, R., Guo, S.N.: Applying Peeling Onion approach for competitive analysis in online scheduling with rejection. Eur. J. Oper. Res. **290**, 57–67 (2021)
12. Mor, B., Mosheiov, G., Shapira, D.: Flowshop scheduling with learning effect and job rejection. J. Sched. **23**(6), 631–641 (2019). https://doi.org/10.1007/s10951-019-00612-y
13. Mor, B., Mosheiov, G., Shabtay, D.: Minimizing the total tardiness and job rejection cost in a proportionate flow shop with generalized due dates. J. Sched. **24**(6), 553–567 (2021). https://doi.org/10.1007/s10951-021-00697-4

14. Nowicki, E., Zdrzalka, S.: A survey of results for sequencing problems with controllable processing times. Discret. Appl. Math. **26**, 271–287 (1990)
15. Oron, D.: Two-agent scheduling problems under rejection budget constraints. Omega **102**, 102313 (2021)
16. Schuurman, P., Woeginger, G.J.: Approximation Schemes - A Tutorial. Lectures on Scheduling, LA Wolsey (2009)
17. Shabtay, D., Gaspar, N., Kaspi, M.: A survey on off-line scheduling with rejection. J. Sched. **16**, 3–28 (2013)
18. Shabtay, D., Steiner, G.: A survey of scheduling with controllable processing times. Discret. Appl. Math. **155**, 1643–1666 (2007)
19. Van Wassenhove, L.N., Baker, K.R.: A bicriterion approach to time/cost trade-offs in sequencing. Eur. J. Oper. Res. **11**, 48–54 (1982)
20. Woeginger, G.J.: When does a dynamic programming formulation guarantee the exitence of an FPTAS? INFORMS J. Comput. **12**, 57–74 (2000)
21. Zhang, L.Q., Lu, L.F., Yuan, J.J.: Single machine scheduling with release dates and rejection. Eur. J. Oper. Res. **198**, 975–978 (2009)
22. Zhang, Y.Z.: A survey on job scheduling with rejection (in Chinese). OR Trans. **24**(2), 111–130 (2020)

Maximizing Energy Efficiency for Charger Scheduling of WRSNs

Yi Hong[1,2], Chuanwen Luo[1,2(✉)], Zhibo Chen[1,2], Xiyun Wang[1], and Xiao Li[3]

[1] School of Information Science and Technology, Beijing Forestry University,
Beijing 100083, China
{hongyi,chuanwenluo,zhibo}@bjfu.edu.cn
[2] Engineering Research Center for Forestry-oriented Intelligent Information Processing
of National Forestry and Grassland Administration, Beijing 100083, China
[3] Department of Computer Science, University of Texas at Dallas, Richardson, USA
xiao.li@utdallas.edu

Abstract. Wireless Rechargeable Sensor Networks (WRSNs) has emerged with the advantages of high charging efficiency, which can guarantee the timeliness of charging and the service quality of network coverage. To guarantee the continuous coverage of the rechargeable sensors, continuous power supply for sensors becomes more important. In this paper, we focus on the Charging Scheduling problem with Maximized Energy Efficiency in WRSNs (CS-MEE Problem), in which a mobile charger is used to charge the low energy sensors in WRSN. The problem aims to optimize travelling path of the mobile charger for maximizing the charging energy efficiency of the charging process. We firstly give the mathematical model and NP-hardness proof of the problem. Then we propose an heterogeneous-weighted-graph algorithm, called CS-HWG, to solve the problem. To evaluate the performance of the proposed algorithm, the extensive simulation experiments are conducted under four influencing factors in terms of the energy efficiency of the mobile charger to verify the effectiveness of the algorithm.

Keywords: WRSNs · Charging scheduling · Energy efficiency · Heterogeneous-weighted-graph

1 Introduction

The most applications of Wireless Sensor Networks (WSNs) have the common requirement of continuous monitoring [1], which poses challenges to the battery-powered sensors and brings the energy efficiency problems in virtual backbone construction [2,3] and broadcast and multicast routing [4]. To solve the energy problems of the WSNs, most researchers proposed two kinds of strategies, i.e., one is the wake-sleep batch scheduling of sensors, and another one is collecting energy from external environment based on energy transformation module of sensors. However, the former strategy may cause the reduction of data reliability and the latter one has low efficiency of energy

Supported by the National Natural Science Foundation of China under Grant (62002022) and the Fundamental Research Funds for the Central Universities (No. BLX201921, No. 2021ZY88).

© Springer Nature Switzerland AG 2021
W. Wu and H. Du (Eds.): AAIM 2021, LNCS 13153, pp. 111–122, 2021.
https://doi.org/10.1007/978-3-030-93176-6_10

transformation. To this end, Wireless Rechargeable Sensor Networks (WRSNs) has emerged with the advantages of high charging efficiency via static charging stations or mobile charging vehicles, which can guarantee the timeliness of charging and the service quality of network coverage.

The most important problem of the WRSN with mobile chargers is to design the charging plans which mainly focuses on the charging pattern, charging order arrangement and charging amount assignment. This paper studies the charging planning problem of a mobile charger for charging sensors from the perspectives of charging amount assignment and charging path planning, which is to maximize the charging efficiency of the charger for guaranteeing the continuously works of the WRSN.

The existing research on the charing planning of mobile chargers focused on two aspects, demand-driven charging strategies and periodic charging ones. For the demand-driven charging strategies [5] proposed a path planning algorithm to choose the sensors in low-power status and satisfy their charging requirement based on a threshold value β on the remaining energy. The authors in [6] predicted the energy consumption of sensors and transformed the charging cost as a monotone submodular function, then introduced a $(1 - \frac{1}{e})/4$-ratio algorithm for the problem. The authors in [7] proposed a spatial-and-temporal optimization algorithm for real-time charging for eliminating the exhausted sensors and adding the powered new ones. The studies in [8,9] aimed at designing the algorithm of path planning and charging assignment to maximize the network lifetime and minimize the charging consumption.

For periodic charging strategies, the authors in [10] designed a constant-ratio approximation algorithm for charging path planning problem under the powering limitation model. And the authors in [11] applied the region-separation and charging-discretion into the charing solution and proposed a $\frac{1-\xi}{4(1-1/e)}$-ratio algorithm. The authors in [12] considered the one-to-many charging model and designed a constant-ratio algorithm. Recently, the new charging technology has drawn attentions of researchers like the $(1 - \xi)(1 - e)/e$-ratio algorithm based on the energy transferring depending on the obstacles in [13], the $(3 + \xi)$-ratio algorithm for multiple-chargers in one-vehicle model in [14] and the periodic charging algorithm with the optimal movement speed in [15].

However, the existing literature mentioned above did not consider the energy efficiency. In this paper, we consider the demand-driven charging planning for a single mobile charger in WRSNs, which includes the charging path planning and the charging energy assignment to maximize the energy efficiency of the charger. The contributions of this paper are shown as below.

(1) We propose a single-charger charging planning problem for WRSNs, called the Charging Scheduling problem with Maximized Energy Efficiency in WRSNs (CS-MEE Problem) based on the energy consumption model. The goal of the problem is to maximize the charging energy efficiency of the charging process in a period. The mathematical model and NP-hardness proof of problem are both given.

(2) To solve the CS-MEE problem, we propose an heterogeneous-weighted-graph algorithm, CS-HWG Algorithm, which is composed of Charging Energy Assignment and Charging Path Planning. And we analyze the time complexity of the algorithm.

(3) The extensive simulations are performed to verify the effectiveness of the proposed algorithm for the CS-MEE problem.

This paper is organized as follows. Section 2 introduces the network model, energy consumption model and problem formulation. In Sect. 3, we propose a heuristic algorithm to solve the problem and analyze the approximation ratio of the algorithm. Simulations are shown in Sect. 4. Section 5 concludes this paper.

2 System Model and Problem Formulation

2.1 Network Model

We consider a WRSN composed of n stationary rechargeable sensors deployed in a two-dimensional plane, which are denoted by set $S = \{s_1, s_2, ..., s_n\}$. Each sensor s_i is deployed at the position $(x[s_i], y[s_i])$ and powered by a rechargeable battery with the maximum energy capacity E_i^0. These sensors perform the area coverage task collaboratively and the current battery energy of s_i is denoted as E_i^{cur}. We assume that the coverage strategy is determined or periodically adjusted depending on their initial energies. There are three kinds of status of sensors depending on the charging requirements based on two thresholds, E^{low} and E^{min}: (1) **Working Status.** $E^{low} < E_i^{cur} \le E_i^0$; (2) **Low-power Status.** $E^{min} < E_i^{cur} \le E^{low}$; (3) **Charging Status.** $0 < E_i^{cur} \le E^{min}$.

There is one mobile charger to charge sensor nodes with low remaining energy, which is denoted as node c. Charger c starts the charging task from its service station located at $c_0 = (x[c], y[c])$ and ends the task back to its station. And the charger has the initial energy E_{max} in the assumption that E_{max} can satisfy the charging requirement of all the sensors or the charging amount for sensors cannot be less than $\theta \cdot E_{max}$, where θ is a parameter closed to 1. Since the sensors with lower remaining energy may cause exhaustion and monitoring failure, the charging task firstly guarantee the impletion of the sensors in **Charging Status**. If there is the remaining energy for the charger, the charging for sensors in **Low-power Status** will be considered and the charging for sensors in **Working Status** is in a similar way.

2.2 Energy Consumption Model

In the cooperative coverage task, the static sensors are in charge of covering the target area and the mobile charger is responsible for charging the sensors into Working Status. The energy consumption of chargers includes two aspects:

(1) Charging Energy Consumption. Due to the determined coverage strategy, the maximum charging energy for sensor s_i with the energy capacity E_i^0 and the remaining energy E_i^{cur} in the current coverage mission. Since the coverage scheduling is assumed to be determined in advance, E_i^{cur} has been known before charging scheduling. We denote the scheduled charging energy for s_i as $C(s_i)$. Furthermore, we consider the inevitable energy loss in the process of charging and the charging energy consumption is regarded as α time of the required amount, i.e. $E_i^{charging} = \alpha \cdot C(s_i) \cdot g_i$, where g_i are defined as follows:

$$g_i = \begin{cases} 1, & \text{if } c \text{ has been scheduled to charge sensor } s_i, \\ 0, & \text{otherwise.} \end{cases} \tag{1}$$

(2) Moving Energy Consumption. Considering the charging model in close range, the mobile charger should move to the position of the sensor with low power for charging. Thus the moving distance is the Euclidean distance and calculated based on the locations of the charger and sensor, which is denoted as $dist(c, s_i)$. Here we denote the energy consumption rate as β, thus the moving energy consumption is $E_i^{moving} = \beta \cdot dist(c, s_i) \cdot g_i$.

Based on the two kinds of energy consumption of chargers, we define the energy efficiency of the charging process as follows.

Definition 1 (Charging Energy Efficiency). *Based on a charging scheduling, the energy efficiency is the proportion of the energy consumption on charging in the overall energy cost, which is denoted as* $EE = \dfrac{\sum_{s_i \in S} E_i^{charging}}{\sum_{s_i \in S} (E_i^{charging} + E_i^{moving})}$.

2.3 Problem Formulation

We study the charging planning problem to realize the goal of maximizing the charging energy efficiency. The charging scheduling is composed of two parts, the charging energy assignment denoted as $EA = \{C(s_i) | 1 \leq i \leq n\}$ (where $C(s_i)$ is the scheduled charging energy on s_i) and the charging path $path_c$ denoted as a sequence of locations passing by c. Based on the above preliminaries, we refer to the problem as the Charging Scheduling problem with Maximized Energy Efficiency in WRSNs (CS-MEE Problem), whose detailed definition is shown as follows.

Definition 2 (CS-MEE Problem)
*Given a set $S = \{s_1, s_2, \cdots, s_n\}$ of n rechargeable sensors where each sensor s_i has the battery capacity E_i^0 and the initial energy E_i^{cur}, one mobile charger c with its starting service station c_0 and the initial energy E_{max}, Charging Scheduling problem with Maximized Energy Efficiency in WRSNs (CS-MEE Problem) is to find **a charging scheduling strategy denoted as two-tuples** $(EA, path)$, such that*

(1) the $path_c$ starts from c_0 and ends at c_0,
(2) for each sensor $s_i \in S$, $E_i^{charging} + E_i^{cur} \leq E_i^0$,
(3) the Charging Priorities(CP) for the sensors are increased according to their status:
     *CP(**Working Status**)<CP(**Low-power Status**)<CP(**Charging Status**);*
(4) $\theta \cdot E_{max} \leq \sum_{s_i \in S} (E_i^{charging} + E_i^{moving}) \leq E_{max}$, or all sensors can be charged by c, where θ is a parameter closed to 1,

(5) the charging energy efficiency $EE = \dfrac{\sum_{s_i \in S} E_i^{charging}}{\sum_{s_i \in S} (E_i^{charging} + E_i^{moving})}$ is maximized.

In the following, we will introduce the mathematical formulation of the CS-MEE Problem.

$$Maximize \quad \frac{\sum_{s_i \in S} \left(\alpha \cdot C(s_i) \cdot g_i \right)}{\sum_{s_i \in S} \left(\alpha \cdot C(s_i) \cdot g_i + \beta \cdot dist(c, s_i) \cdot g_i \right)} \quad (2)$$

s.t.

$$E_i^{cur} + \alpha \cdot C(s_i) \cdot g_i \leq E_i^0 \qquad i = 1, 2, \cdots, n \qquad (3)$$

$$\theta \cdot E_{max} \leq \sum_{s_i \in S} \left(C(s_i) \cdot g_i + \beta \cdot dist(c, s_i) \cdot g_i \right) \leq E_{max} \qquad (4)$$

$$g_i \in \{0, 1\} \qquad i = 1, 2, \cdots, n \qquad (5)$$

The function (2) is to maximize the charging energy efficiency. Constraint (3) express that the charged energy amount of each sensor cannot beyond the sensor's battery capacity. Constraint (4) is the charging energy consumption constraint which ensures that the charging energy amount of the charger will not exceed the initial energy of the charger. Constraints (5) defines the domain of the variable g_i.

In the following theorem, we will give the NP-hardness proof of the problem.

Theorem 1. *CS-MEE Problem is NP-hard.*

Proof. To prove the NP-hardness of CS-MEE Problem, we consider a special case of it: all the sensors are in Charging Status ($g_i = 1$ for $1 \leq i \leq n$) and they have the same current energies E_i^{cur}s. In this case, the charging energy for sensor s_i, $C(s_i)$ is the maximum amount $E_i^0 - E_i^{cur}$, which is unified represented as C.

Thus the objective of the problem is driven to be maximizing $\frac{\sum_{s_i \in S} \left(\alpha \cdot C(s_i) \right)}{\sum_{s_i \in S} \left(\alpha \cdot C(s_i) + \beta \cdot dist(c, s_i) \right)}$. Based on equivalent conversion, the objective can be rewritten into maximizing $\frac{1}{1 + \frac{\beta}{\alpha \cdot C} \sum_{s_i \in S} dist(c, s_i)}$. Note that α, β and C are predefined or can be calculated. By denoting $\frac{\beta}{\alpha \cdot C}$ as a constant *const* the objective becomes from maximizing $\frac{1}{1 + const \cdot \sum_{s_i \in S} dist(c, s_i)}$ to minimizing $\sum_{s_i \in S} dist(c, s_i)$.

It can be easily found that the problem in this special case is equivalent to the Travelling Salesman Problem (TSP), which has been proved NP-hard [16]. Since a special case of CS-MEE problem is NP-hard, CS-MEE problem is also NP-hard, which completes the proof. □

3   Algorithms for CS-MEE Problem

In this section, we propose an heterogeneous-weighted-graph algorithm, CS-HWG Algorithm, which is composed of two phases, **Charging Energy Assignment** and **Charging Path Planning**. And we will analyze the time complexity of CS-HWG Algorithm.

We firstly give the preliminaries in Lines 1–7 of Algorithm 1: since the sensors with the higher charging requirements have larger priorities, we give a baseline value according to the divergence indicator among the sensors' battery capacities, i.e.,

$DI = \lceil \max_{1 \le i,j \le n} \frac{E_i^0}{E_j^0} \rceil$. We assign the priorities for the three kinds of sensors' status respectively: (1) For the ones in Charging Status, its priority $pri(s_i) = DI^2$; (2) For the sensors in Low-Power Status, $pri(s_i) = DI$; (3) For the sensors in Working Status, the charging priority is assigned as $pri(s_i) = \frac{1}{DI}$. This priority assignment measure can guarantee that the gap between the pairs of the priorities belonged to different requirements can be widen. Furthermore, it also consider the charging demands and the energy capacity of sensors: for the sensors in Charging Status, the charging requirement is greatest and the maximum charging energy $(E_i^0 - E_i^{cur})$ could be satisfied. Thus $pri(s_i) = DI^2$ which is larger than those in other two statuses.

Phase 1: Charging Energy Assignment

Phase 1.1: Heterogeneous-weighted Graph Construction
Constructing the auxiliary graph is a classic method to model the practical problem, and the auxiliary graph is either a node-weighted graph or an edge-weighted graph. In CS-MEE Problem, we construct a particular auxiliary graph with node-weights and edge-weights, heterogeneous-weighted graph, as shown in Lines 10–19 of Algorithm 1. The node set is composed of the positions of sensors and a charger, $V = S \bigcup C$. With the consideration of the sensor deployment density, we give the assumptions for spares graphs and dense ones. For spares graphs, we introduce a limitation value l_0 of the moving distance between twice of charging, which can avoid excess consumption of the chargers' energy for some single charging. Thus $E = \{(s_i, s_j) | \forall s_i, s_j \in V \text{ and } dist(s_i, s_j) \le l_0\}$. For dense graphs, l_0 can be regarded as infinity.

The weight assignment is with the consideration of charging cost and moving cost: When considering the charging cost, it is decided by each sensor's maximum charging requirement or the charged energy amount. Thus the node weight is denoted as $weight(s_i) = \alpha \cdot (E_i^0 - E_i^{cur})$ and the node weight set $VW = \{weight(s_i) | \forall s_i \in S\}$. When considering the moving cost, it is determined by the Euclidean distances between the pairs of nodes in the network, i.e., the edge weight is calculated by $weight(s_i, s_j) = \beta \cdot dist(s_i, s_j)$. And the edge weight set $EW = \{weight(s_i, s_j) | \forall (s_i, s_j) \in E\}$. Then we complete the construction of the heterogeneous-weighted graph, $G = (V, E, VW, EW)$.

Phase 1.2: Charged Node Filtering
Considering high charging efficiency and limitation of the charger's initial energy E_{max}, we filter the nodes with necessary charging requirements like those in Charging Status. Since E_{max} is limited to satisfy the charging requirements for part of sensors, we firstly reserve the consumption on charging movement E_{moving}^{res}, which is calculated in Step 21. And the calculation is based on the length of the Minimum Hamilton Cycle which can guarantee to pass across all the sensors in Charging Status. Then the remaining energy $E_{max} - E_{moving}^{res}$ can be assign for charging sensors.

Based on the new E_{max}, we assign the charging amount according to the sensors' charging priorities and filter the sensors with necessary charging requirements. The assignment is realized in three loops as shown in Lines 24–28: firstly the charging requirement of the sensors in Charging Status can be satisfied and the assigned charging amount is $C(s_i) = \frac{pri(s_i)}{DI^2} \cdot (E_i^0 - E_i^{cur})$. If the charger has the remaining energy, the sensors in Low-Power Status can be charged. The charging for sensors in Working Status is in the similar way.

The filtering is based on the assigned charging energy $C(s_i)$ as shown in Lines 29–31: if $C(s_i) = 0$, s_i will be out of the consideration later and eliminated from V and E. The we obtain the filtered node set V' and node set E'.

Phase 1.3: Edge-weighted Graph Transformation

To solve the problem on the constructed heterogeneous-weighted graph, the weights' distribution on both nodes and edges is not beneficial to global optimization. In other words, the energy consumption of chargers is composed of charging cost and moving cost, which cannot exceed the maximum limitation E_{max}. Thus the two kinds of energy cost should be measured by uniform standard, and we adopt the edge weight as the measurement. Here we introduce an equivalent transformation method of blending node weights into edge weights, as shown in Lines 34–40 of Algorithm 1:

For each node in the filtered set V', we revalue the node weight with the consideration of the charging priority and the uniform magnitude of node weights and edge weights, i.e., $weight'(s_i) = \frac{1}{pri(s_i)} \cdot \beta \cdot avrdist \cdot \frac{weight(s_i)}{E_i^0}$, where $avrdist = \frac{\sum_{1 \le i,j \le n} dist(s_i, s_j)}{|E'|}$ is the average distance among all the pairs of sensors. Note that $avrdist$ is a normalization factor for modifying the node weight into the similar magnitude with those of the edge weight. And $\frac{1}{pri(s_i)}$ indicates that the node with higher charging priority has smaller node weight, which is consistent with that the node pair with low moving cost has smaller edge weight.

Since the sensor's charging can be finished by the charger's only one pass, we equally divide the node weight into two parts, e.g. $\frac{1}{2}weight'(s_i)$. And then we distribute the divided node weight to the weight of the node's associated edges, i.e., $weight'(s_i, s_j) = weight(s_i, s_j) + \frac{1}{2}weight'(s_i) + \frac{1}{2}weight'(s_j)$, which updates the edge weight set. Then we will perform charging planning based on the transformed edge-weighted graph $G' = (V', E', EW')$.

Phase 2: Charging Path Planning

Based on the auxiliary graph $G' = (V', E', EW')$, we perform the algorithm for TSP Problem and the charging path $path_c$ of the charger c can be obtained. The detailed description is shown in Algorithm 1.

Theorem 1. *The time complexity of CS-HWG Algorithm is $O(n^3)$, where n is the number of sensors.*

Proof. According to the description of Algorithm 1, there are three parts as shown in Algorithm 1, the preliminaries, Phase 1 and Phase 2. We analyze the time complexities for these parts as follows: For the preliminaries in Lines 1–7, the charging priority assignment has the time complexity of $O(n)$. For Phase 1, Phase 1.1 (Heterogeneous-weighted Graph Construction) and Phase 1.3 (Edge-weighted Graph Transformation) both perform for all the nodes and edges, whose time consumptions are directly related to the number of nodes and that of edges. Thus their time complexities are both $O(n^2)$. For the node filtering (Lines 23–32) in Phase 1.2 (Charged Node Filtering), its time complexity is $O(n)$. Furthermore, TSP Algorithm is applied in Phase 1.2 and Phase 2, which has a larger time complexity of $O(n^3)$ [17].

To sum up, the time complexity of CS-HWG Algorithm is $O(n^3)$, which completes the proof.   □

Algorithm 1. CS-HWG Algorithm for CS-MEE Problem

Input: $S = \{s_1, s_2, \cdots, s_n\}$, $\{E_i^0 | 1 \leq i \leq n\}$, $\{C(s_i) | 1 \leq i \leq n\}$, a set $C = \{c\}$ and E_{max}
Output: $EA = \{C(s_i) | 1 \leq i \leq n\}$ and $path_c = \{c, s_{i_1}, s_{i_2}, ..., s_{i_k}, c | 1 \leq i_1, i_2, ..., i_k \leq n\}$

1: Set the divergence indicator $DI = \lceil \max_{1 \leq i,j \leq n} \frac{E_i^0}{E_j^0} \rceil$.

2: **for** $\forall s_i \in S$ **do**

3:     $C(s_i) = 0$;

4:     Case1: If s_i in Charging Status, $pri(s_i) = DI^2$;

5:     Case2: If s_i in Low-Power Status, $pri(s_i) = DI$;

6:     Case3: If s_i in Working Status, $pri(s_i) = \frac{1}{DI}$;

7: **end for**

8: //**Phase 1: Charging Energy Assignment**

9: //**Phase 1.1: Heterogeneous-weighted Graph Construction**

10: Set $V, E, VW, EW \leftarrow \emptyset$

11: $V \leftarrow S \cup C$, $E = \{(s_i, s_j) | \forall s_i, s_j \in V \text{ and } dist(s_i, s_j) \leq l_0)\}$

12: **for** $\forall s_i \in V$ **do**

13:     $weight(s_i) = \alpha \cdot (E_i^0 - E_i^{cur})$

14: **end for**

15: $VW = \{weight(s_i) | \forall s_i \in S\}$

16: **for** $\forall (s_i, s_j) \in E$ **do**

17:     $weight(s_i, s_j) = \beta \cdot dist(s_i, s_j)$

18: **end for**

19: $EW = \{weight(s_i, s_j) | \forall (s_i, s_j) \in E\}$

20: //**Phase 1.2: Charged Node Filtering**

21: Perform TSP Algorithm on $G[\{s_i | \forall s_i \text{ in Charging Status}\}]$ and obtain a Hamilton Cycle with edge weight E_{moving}^{res}

22: $E_{max} = E_{max} - E_{moving}^{res}$

23: Set $V' = V, E' = E, EW' \leftarrow \emptyset$

24: **while** $E_{max} > 0$ **do**

25:     For each s_i in Charging Status, $C(s_i) = \frac{pri(s_i)}{DI^2} \cdot (E_i^0 - E_i^{cur})$, $E_{max} = E_{max} - C(s_i)$//**Case1**

26:     For each s_i in Low-Power Status, $C(s_i) = \frac{pri(s_i)}{DI^2} \cdot (E_i^0 - E_i^{cur})$, $E_{max} = E_{max} - C(s_i)$ //**Case2**

27:     For each s_i in Working Status, $C(s_i) = \frac{pri(s_i)}{DI^2} \cdot (E_i^0 - E_i^{cur})$, $E_{max} = E_{max} - C(s_i)$ //**Case3**

28: **end while**

29: **for** $\forall s_i \in V$ **do**

30:     If $C(s_i) = 0$, $V' = V' \setminus \{s_i\}$, $E' = E' \setminus \{(s_i, s_j) | \forall s_j \in V\}$

31: **end for**

32: $EA = \{C(s_i) | \forall s_i \in V'\}$

33: //**Phase 1.3: Edge-weighted Graph Transformation**

34: **for** $\forall s_i \in V'$ **do**

35:     $weight'(s_i) = \beta \cdot avrdist \cdot \frac{1}{pri(s_i)} \cdot \frac{weight(s_i)}{\alpha \cdot E_i^0}$, where $avrdist = \frac{\sum_{1 \leq i,j \leq n} dist(s_i, s_j)}{|E'|}$

36: **end for**

37: **for** $\forall (s_i, s_j) \in E'$ **do**

38:     $weight'(s_i, s_j) = weight(s_i, s_j) + \frac{1}{2} weight'(s_i) + \frac{1}{2} weight'(s_j)$

39: **end for**

40: $EW' = \{weight'(s_i, s_j) | \forall (s_i, s_j) \in E'\}$

41: //**Phase 2: Charging Path Planning**

42: Perform TSP Algorithm on $G' = (V', E', EW')$ and obtain the charging $path_c$ of c

---

4 Simulation Results

The simulation experiments are performed in a two-dimension planar with the size of $M * M$. On the plane, there are n sensors randomly deployed; for each sensor s_i, there is a uniform parameter E^0 denoting the maximum battery capacity. And s_i's current battery energy E_i^{cur} is valued in the range of $[0, \frac{4}{5} \cdot E^0]$ and it battery capacity E_i^0 is set in $[E_i^{cur}, E^0]$. The two indictors for the sensors' status, E^{low} and E^{min} are assigned $\frac{3}{5} \cdot E^0$ and $\frac{3}{10} \cdot E^0$. The moving distance limitation between any pair sensors l_0 is set as 50.

For the optimization goal of MVB-GRC Problem, we evaluate the proposed algorithm in terms of the energy efficiency, which is denoted as **Energy Efficiency**. The four parameters, the side length of the region M, the number of sensors n, the initial energy of the charger E_{max}, and the maximum energy capacity of sensors E^0 are considered as the potential factors on performance of charing scheduling. And we consider the following four groups of parameter settings and we repeat the experiment 100 times and adopt the average values for each setting: (1) M varies from 40 to 160 by the step of 20 with fixed n, E^0 and E_{max}; (2) n varies from 60 to 160 by the step of 20 with fixed M, E^0 and E_{max}; (3) E_{max} varies from 2500 to 5500 by the step of 500 with fixed M, E^0 and n; (4) E^0 varies from 40 to 100 by the step of 20 with fixed M, n and E_{max}.

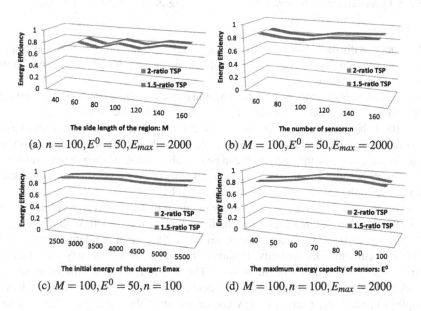

(a) $n = 100, E^0 = 50, E_{max} = 2000$ (b) $M = 100, E^0 = 50, E_{max} = 2000$

(c) $M = 100, E^0 = 50, n = 100$ (d) $M = 100, n = 100, E_{max} = 2000$

Fig. 1. Energy efficiency by comparing two TSP algorithms

Firstly we apply two TSP Algorithms with approximation ratios of 2 and 1.5 [17] (denoted as **2-ratio TSP** and **1.5-ratio TSP**) in Charging Path Planning Phase and evaluate their performance measure. As shown in Fig. 1, with the changes of the four parameters, **1.5-ratio TSP** outperforms **2-ratio TSP** on energy efficiency. The reason is that the former algorithm can construct a better Hamilton Cycle which is closer to the optimal one, i.e., **1.5-ratio TSP** constructs a charging path with less length than that generated from **2-ratio TSP**. Then the scheme with less charging path length can enhance the whole charging efficiency. Moreover with the growth of M as shown in (a), the energy efficiency fluctuates in the range $[0.65, 0.87]$ and gets stable when $M > 120$; with the increasing of n, E_{max} or E^0 in (b)-(d), two algorithms both enter a smooth status with little fluctuations on the energy efficiency. It is because that the region scale directly influences the maximum length of charging paths, which determines the results obtained by TSP algorithms.

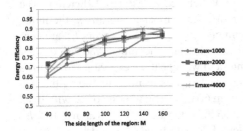

Fig. 2. Energy efficiency by varing M

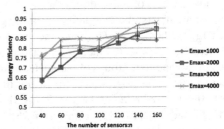

Fig. 3. Energy efficiency by varing n

Secondly, with the advantages of **1.5-ratio TSP** Algorithm, we continue to perform the simulations via applying it and evaluate the algorithm's performance with the change of four parameters.

As shown in Fig. 2, with the growth of M, the energy efficiency presents upward tendency and moderate fluctuation when $E_{max} = 1000, 2000, 3000, 4000$. Compared with $E_{max} = 1000$, the influence of M on the energy efficiency becomes smaller with a larger $E_{max} = 4000$, which shows that the region scale has little impact on the algorithm with a larger initial energy of the charger. It can be explained by that when the initial energy of the charger is sufficient, the charging requirements of all the sensors can be satisfied which can keep the energy efficiency on a high level.

As shown in Fig. 3, the energy efficiency obtained by CS-HWG Algorithm remains upward trend with the increasing of n at the fixed $E_{max} = 1000, 2000, 3000, 4000$. Especially when $n > 100$, the results among different E_{max} enter a relative steady state $[0.80, 0.95]$ with little fluctuations. It shows the algorithm can satisfy the charging requirements of the majority of the sensors. The reason is that with the increasing of network scale with a fixed region scale, the deployment density becomes higher which is helpful to reduce the moving energy consumption; at the same time, the increased initial energy of the charger can meet more charing requirements for the sensors.

As shown in Fig. 4, the energy efficiency fluctuates up and down with the increasing of E_{max}, which is especially apparent when $E_{max} \in [3500, 5500]$ with a fixed $n = 100$. The amplitude of fluctuation becomes unapparent with the growth of n, i.e., the results when $n = 200$ remain in $[0.87, 0.93]$. It can be explained by that when the network scale gets larger, the different between each pair of sensors' charing requirement becomes relative smaller, which is benefit for improve the charger's charging efficiency. Thus

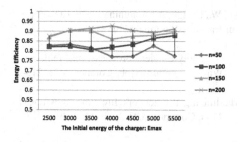

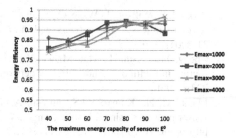

Fig. 4. Energy efficiency by varing E_{max} **Fig. 5.** Energy efficiency by varing E^0

with the increased initial energy of the charger, the change of energy efficiency can enter a smooth status.

As shown in Fig. 5, comparing the results when $E_{max} = 1000, 2000, 3000, 4000$, the changing of E^0 has a steady impact on increasing the energy efficiency with different E_{max}, i.e. the results remain $[0.85, 0.95]$ when $E^0 > 80$. It shows that the maximum battery capacity of sensors E^0 has limited influence on the performance of our algorithm. It is because that E^0 determines the maximum charging requirements for sensors, which cannot decide the actual charging amount. The algorithm is designed to meet the most necessary charging requirements first and perform selective fully-charged-mode charging to sensors in different status, which is for enhancing the whole energy efficiency.

Finally, we can draw the conclusion that the network scale n and the initial energy of the charger E_{max} has more influence than the region scale M and the energy capacity of the sensors E^0 in terms of the energy efficiency of the whole charging process.

5 Conclusion

In this paper, we investigate the maximum energy efficiency charing planning problem for one mobile charger in WRSNs. We formally define the problem and propose a heuristic algorithm composed of charing energy assignment and path planning, which is based on heterogeneous-weighted graph construction and edge-weighted graph transformation. Furthermore, we apply two approximation algorithms in the charging path phase and perform the simulation experiments to evaluate the algorithm's performance. In the future, we have great interest on investigating the maximum energy efficiency charging planning problem for multiple chargers in WRSNs.

References

1. Luo, C., Satpute, M.N., Li, D., Wang, Y., Chen, W., Wu, W.: Fine-grained trajectory optimization of multiple UAVs for efficient data gathering from WSNs. IEEE/ACM Trans. Netw. **29**(1), 162–175 (2021)
2. Wang, W., et al.: On construction of quality fault-tolerant virtual backbone in wireless networks. IEEE/ACM Trans. Netw. **21**(5), 1499–1510 (2013)

3. Park, M.A., Willson, J., Wang, C., Thai, M., Wu, W., Farago, A.: A dominating and absorbent set in a wireless ad-hoc network with different transmission ranges. In: MobiHoc, pp. 22–31(2007)
4. Cheng, M.X., Sun, J., Min, M., Li, Y., Wu, W.: Energy-efficient broadcast and multicast routing in multihop ad hoc wireless networks. Wirel. Commun. Mob. Comput. **6**(2), 213–223 (2006)
5. Magadevi, N., Kumar, V.J.S., Suresh, A.: Maximizing the network life time of wireless sensor networks using a mobile charger. Wirel. Pers. Commun. **102**(2), 1029–1039 (2017). https://doi.org/10.1007/s11277-017-5131-1
6. Wu, T., Yang, P., Dai, H., Xu, W., Xu, M.: Collaborated tasks-driven mobile charging and scheduling: a near optimal result. In: Proceedings of IEEE INFOCOM, pp. 1810–1818 (2019)
7. Lin, C., Zhou, J., Guo, C., Song, H., Guowei, W., Obaidat, M.S.: TSCA: a temporal-spatial real-time charging scheduling algorithm for on-demand architecture in wireless rechargeable sensor networks. IEEE Trans. Mob. Comput. **17**(1), 211–224 (2018)
8. Xu, W., Liang, W., Jia, X., Xu, Z.: Maximizing sensor lifetime in a rechargeable sensor network via partial energy charging on sensors. In: Proceedings of IEEE SECON, pp. 1–9 (2016)
9. Lin, C., Zhou, Y., Dai, H., Deng, J., Wu, G.: MPF: prolonging network lifetime of wireless rechargeable sensor networks by mixing partial charge and full charge. In: Proceedings of IEEE SECON, pp. 379–387 (2018)
10. Liang, W., Zichuan, X., Wenzheng, X., Shi, J., Mao, G., Das, S.K.: Approximation algorithms for charging reward maximization in rechargeable sensor networks via a mobile charger. IEEE/ACM Trans. Netw. **25**(5), 3161–3174 (2017)
11. Wu, T., Yang, P., Dai, H., Xu, W., Xu, M.: Charging oriented sensor placement and flexible scheduling in rechargeable WSNs. In: Proceedings of IEEE INFOCOM, pp. 73–81 (2019)
12. Ma, Yu., Liang, W., Wenzheng, X.: Charging utility maximization in wireless rechargeable sensor networks by charging multiple sensors simultaneously. IEEE/ACM Trans. Netw. **26**(4), 1591–1604 (2018)
13. Lin, C., Gao, F., Dai, H., Wang, L., Wu, G.: When wireless charging meets fresnel zones: even obstacles can enhance charging efficiency. In: Proceedings of IEEE SECON, pp. 1–9 (2019)
14. Zou, T., Wenzheng, X., Liang, W., Peng, J., Cai, Y., Wang, T.: Improving charging capacity for wireless sensor networks by deploying one mobile vehicle with multiple removable chargers. Ad Hoc Netw. **63**, 79–90 (2017)
15. Shu, Y., et al.: Near-optimal velocity control for mobile charging in wireless rechargeable sensor networks. IEEE Trans. Mob. Comput. **15**(7), 1699–1713 (2016)
16. Karp, R.M.: Reducibility among combinatorial problems. In: Miller, R.E., Thatcher, J.W., Bohlinger, J.D. (eds.) Complexity of Computer Computations. The IBM Research Symposia Series, pp. 85–103. Springer, Boston, MA (1972). https://doi.org/10.1007/978-1-4684-2001-2_9
17. Christofides, N.: Worst-case analysis of a new heuristic for the travelling salesman problem, Technical report, 388. Carnegie Mellon University, Graduate School of Industrial Administration (1976)

A New Branch-and-Price Algorithm for Daily Aircraft Routing and Scheduling Problem

Yu Si, Suixiang Gao, and Wenguo Yang[✉]

School of Mathematical Sciences, University of Chinese Academy of Sciences,
Beijing 100049, China
siyu191@mails.ucas.ac.cn, {sxgao,yangwg}@ucas.edu.cn

Abstract. The problem of aircraft routing and scheduling is one of the essential problems in airline industry. Airlines need to identify multiple routes that may be profitable and allocate their fleets to these routes. To describe this problem, we present two equivalent models and analyze the advantages and disadvantages of them. The first one has less variables but often yields weak linear relaxation, and the second model is just the opposite. We propose a branch-and-price algorithm with a new branching rule to solve the second model. In addition, we describe the process of column generation algorithm involved in branch-and-price, and design a labeling algorithm to solve the subproblem in column generation. Finally we report computational results obtained on data provided by airlines. These results indicate that our approach significantly reduces the amount of CPU time compared to the basic method.

Keywords: Airline routing · Fleet scheduling · Branching · Column generation · Labeling algorithm

1 Introduction

In this paper, we consider a problem faced by airlines. Given the airline's fleet plan, which determines the availability of aircraft with different capacities and ranges characteristics, airlines need to construct daily schedules for heterogeneous fleets. An aircraft schedule consists of a sequence of flight legs (the routing aspect) and a aircraft to carry out the sequence (the scheduling aspect). Therefore, this problem involves two issues, one is to plan aircraft routes network and the other is to assign fleets to those routes [1]. In general, the problem of fleet routing and scheduling is affected by many factors, and various types of constraints can be considered when constructing such a schedule. Consequently the definition of daily aircraft routing and scheduling problem (DARSP) varies across the literature. Levin [2] was the first to have proposed a model for the DARSP with variable departure times which could only take from several values. Then, Abara [3] and Hane [4] presented two multi-commodity network flow models for fleet scheduling problem with fixed departure time. Desaulniers [5] addressed

© Springer Nature Switzerland AG 2021
W. Wu and H. Du (Eds.): AAIM 2021, LNCS 13153, pp. 123–133, 2021.
https://doi.org/10.1007/978-3-030-93176-6_11

DARSP for a heterogeneous fleet with time windows and used a branch-and-price algorithm to solve it. In recent years, people have begun to consider scheduling problem in more practical scenarios. Kenan [6] and Cadarso [7] considered flight scheduling under uncertainty. Cur [8] proposed models for aircraft routing problem with consideration of remaining time and robustness. Zheng [9] considered aircraft scheduling with parking problem and combined two heuristic algorithms for solving. Most of these problems can be considered as extended versions of basic fleet scheduling problem which will be considered in this paper. We described how to build this problem into a multi-commodity flow model and proposed a Branch-and-Price algorithm with new branching rule.

The following paper is arranged as below. Section 2 presents the problem definition. Meanwhile, two equivalent mathematical models are established. We describe our algorithm in Sect. 3, and the experimental results are shown in Sect. 4. Finally, our work is concluded in Sect. 5.

2   Problem Description for DARSP

Given the airline's fleet plan, which determines the availability of aircraft with different capacity and range characteristics, the next major step in the airline planning process is to determine the specific routes to be flown. The DARSP requires that the fleet be allocated to flight legs to maximize total revenue. Each leg has a fixed departure time and duration. Two legs can be connected if the destination of one is same as the departure airport of the other and the interval time between two legs is greater than a certain range. The total duration of each aircraft on the route shall not exceed a certain amount. Each aircraft can only take off from the station every day and fly back to the station at the end.

We will convert this problem into multi-commodity network flow model and give two different formulations: the `arc-flow` or `conventional` formulation and the `path-flow` or `column-generation` formulation. Let $\mathscr{V}$ be the node set consisting of all operational flight legs. Define $\mathscr{V}_o$ as the node set consisting of all legs taking off from the station, and $\mathscr{V}_d$ as the node set consisting of all legs landing on the station. For each leg l, it has a departure time a_l and duration d_l. For every two flight legs i, j, if $a_i + d_i + \Delta \leqslant a_j$, where Δ is the minimum flight connection time, we can draw an arc e_{ij} from i to j which means those two flight legs can be flown consecutively by the same aircraft. The station is represented by two virtual nodes o and d which represent source node and sink node, respectively. In addition, o is the predecessor node for $\mathscr{V}_o$ and d is the successor node for $\mathscr{V}_d$. All arcs form a set $\mathscr{A}$. Let $\widetilde{\mathscr{V}} = \mathscr{V} \cup \{o, d\}$ represents the set of all nodes. Then we can build a directed graph $G(\widetilde{\mathscr{V}}, \mathscr{A})$ as shown in Fig. 1.

The arc-flow formulation is based on the graph G. We present the following notation that based on a standard multi-commodity network.

Notations

$\mathcal{K}$: Set of all aircraft.

$\mathcal{A}$: Set of all arcs.

$\mathcal{V}$: Set of all legs represented by nodes.

$\widetilde{\mathcal{V}}$: $\mathcal{V} \cup \{o, d\}$. Set of all nodes.

$\mathcal{V}_o$: Set of all legs whose origin is the station.

$\mathcal{V}_d$: Set of all legs whose destination is the station.

$\mathcal{V}^+(i)$: $\{j \in \widetilde{\mathcal{V}} \mid (i, j) \in \mathcal{A}\}$. Set of successor nodes of node i.

$\mathcal{V}^-(i)$: $\{j \in \widetilde{\mathcal{V}} \mid (j, i) \in \mathcal{A}\}$. Set of predecessor nodes of node i.

o, d : Constant. Indicating source and sink that both represent the station.

t_i : Constant. Indicating duration required for leg i, $t_o = t_d = 0$ for notational conciseness.

w_i^k : Constant. Indicating revenue of aircraft k flying leg i, $w_o^k = w_d^k = 0$ for notational conciseness.

b_i : Constant. Equal to 1 at source, equal to -1 at sink, equal to zero at remaining points.

T^k : Constant. Indicating the total length of time allowed for the kth aircraft to fly.

x_{ij}^k : Binary variable. Indicating whether the aircraft k uses the arc (i, j).

Accordingly, we formulate this problem in the arc-flow form as follows:

$$\max \sum_{k \in \mathcal{K}} \sum_{(i,j) \in \mathcal{A}} w_j^k x_{ij}^k \qquad \text{(arc-flow)}$$

$$\text{s.t.} \sum_{k \in \mathcal{K}} \sum_{j \in \mathcal{V}^+(i)} x_{ij}^k \leqslant 1, \qquad \forall i \in \widetilde{\mathcal{V}}, \qquad (1)$$

$$\sum_{j \in \mathcal{V}^+(i)} x_{ij}^k - \sum_{j \in \mathcal{V}^-(i)} x_{ji}^k = b_i \qquad \forall i \in \widetilde{\mathcal{V}}, \forall k \in \mathcal{K}, \qquad (2)$$

$$\sum_{(i,j) \in \mathcal{A}} t_j x_{ij}^k \leqslant T^k, \qquad \forall k \in \mathcal{K}, \qquad (3)$$

$$x_{ij}^k \in \{0, 1\}, \qquad \forall k \in \mathcal{K}, \forall (i, j) \in \mathcal{A}. \quad (4)$$

The objective function is to maximize the profits made up of revenue of each selected flight leg. The degree constraints (1) require each flight leg at most be carried out once. Constraints (2) are the flow conservation constraints. Constraints (3) limit the maximum time that each aircraft can work per day.

The arc-flow model often yields a very weak linear relaxation. In order to fix this disadvantage, by applying Dantzig-Wolfe decomposition, path-flow formulation can be obtained. The additional notations are as follows:

Notations

$\mathcal{P}$: Set of o-d paths.

$\mathcal{P}(k)$: $\{p \in \mathcal{P} \mid \sum_{(i,j)\in p} t_j \leqslant T^k\}$. Set of paths which is available for aricraft k.

w_p^k : Constant. Equaling to $\sum_{(i,j)\in p} w_{ij}^k$.

δ_i^p : Binary variable. Indicating whether the path p contains the node i.

y_i^p : Binary variable. Indicating whether aircraft k carry out path i.

We can get the path-flow formulation:

$$\max \quad \sum_{k\in\mathcal{K}} \sum_{p\in\mathcal{P}(k)} w_p^k y_p^k \qquad \text{(path-flow)}$$

$$\text{s.t.} \quad \sum_{k\in\mathcal{K}} \sum_{p\in\mathcal{P}(k)} \delta_i^p y_p^k \leqslant 1, \qquad\qquad \forall i \in \mathcal{V}, \qquad\qquad (5)$$

$$\sum_{p\in\mathcal{P}(k)} y_p^k = 1 \qquad\qquad \forall k \in \mathcal{K}, \qquad\qquad (6)$$

$$y_p^k \in \{0,1\}, \qquad\qquad \forall k \in \mathcal{K}, \forall p \in \mathcal{P}(k). \qquad (7)$$

The objective function of path-flow is as same as arc-flow's. The path-flow model only has two kinds of constraints. Constraints (5) represent each leg can only be allocated to at most one aircraft. Constraints (6) require that each aircraft must be assigned to one route. Given aircraft type, the feasibility of a path is tested in a shortest path algorithm with constrained capacity representing the duration along path. To be feasible, the duration must be less than allowed amount for the given type.

In practice, path-flow model often yields better upper bounds than arc-flow model and has less symmetry than arc-flow model with three-index arc-flow variables [10]. However, it has a huge number of variables, which is far more than the arc-flow model's, which means that it is almost impossible to solve this model directly. We chose a branch-and-price approach to solve this problem.

3 Branch-and-Price Strategies

Branch-and-price approach, which is often used in Vehicle Routing Problem, is designed to solve large-scale integer linear programming (ILP). Branch-and-price is a combination of column generation and branch-and-bound algorithm which are used to solve the linear relaxation of ILP and to eliminate fraction in the solutions of linear relaxations, respectively.

3.1 Column Generation

Considering the path-flow model with $y_p^k \in [0,1]$, which called the *master problem* or MP, the general idea is that not all columns (variables) in the constraint

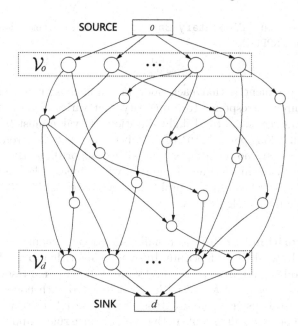

Fig. 1. The nodes and the arcs of network G. Rectangles and circles represent station and flight legs, respectively, and the larger circles represent the flight legs connected to the station.

matrix will be included in the optimal solutions to master problem. In fact, only a very small subset of all columns will be in optimal solutions with all other columns equal to zeros. We can start with solving a *restrict master problem* consisting of a small subset $\bar{P}$ of columns in MP. After finding the solution to RMP, we determine whether there are any columns not included in the RMP with positive reduced cost. If no columns are found, the current solution will also be the optimal solution for MP. Otherwise, we need to add these columns to RMP and repeat the above process.

Finding columns with largest reduced cost is called *pricing problem*. Let π_i and μ_k be the dual variables associated with constraints (5) and (6), respectively. π_i is positive and μ_i is unrestricted. The reduced cost of variable y_p^k can be calculated by following formula:

$$\overline{w}_k^p = w_p^k - \sum_{i \in \mathcal{V}} \delta_i^p \pi_i - \mu_k = -\left(\mu_k + \sum_{i \in p \setminus \{o,d\}} (\pi_i - w_i)\right)$$

For every arc in $\mathscr{A}$, let the cost of arc (i,j) for aircraft k, which is denoted by $\bar{c}_{ij}^k$, equals to $\pi_j - w_j$ if j is not the sink, otherwise equals to μ_k. So, finding a variable with the largest reduced cost

$$\underset{k \in \mathcal{K}, p \in \mathcal{P}(k)}{\arg\max} \quad \overline{w}_k^p$$

corresponds to an **Elementary Shortest Path Problem with Resource Constraints** or **ESPPRC**.

$$\underset{k \in \mathcal{K},\, p \in \mathcal{P}(k)}{\arg\min} \quad \bar{c}_k^p$$

Elementary requirement is that each node in graph must be covered as most once. The resource corresponds to a quantity that varies along a path. In our background, resource is the total flight duration. Its value must be less than a prespecified value along path. ESPPRC have been proved to be strongly NP-hard which means there are no pseudo-polynomial time algorithms. But in our problem, the network is a directed acyclic graph, which means elementary requirement already be satisfied, so the problem degenerates to **SPPRC**, which have pseudo-polynomial time algorithms.

Labeling Algorithm. SPPRC is usually solved by dynamic programming, more precisely, by a labeling algorithm. In such an algorithm, partial paths start at the source node o and are represented by multi-dimensional resource vectors, called labels. Starting from $|\mathcal{K}|$ initial labels associated with node o, labels are propagated forwardly using resource extension functions(REFs) through network G. To avoid enumerating all feasible paths, a dominance rule is applied to discard unpromising labels.

A label $E_{pi}^k = (Z_{pi}^k, T_{pi}^k)$ representing a partial path p for aircraft k from node o to node i contains two components. Z_{pi}^k represents the cost of p and T_{pi}^k represents the duration time accumulated along p. E_{pi}^k is feasible if $T_{pi}^k \in [0, T^k]$. Path p (label E_{pi}^k) can be extended by appending arc $(i,j) \in \mathscr{A}$. Resulting path is represented by a label $E_{pj}^k = (Z_{pj}^k, T_{pj}^k)$ whose components are computed using the following REFs:

$$Z_{pj}^k = Z_{pi}^k + \bar{c}_{ij}$$
$$T_{pj}^k = T_{pi}^k + t_j$$

Therefore, E_{pj}^k is feasible if $T_{pj}^k \leqslant T^k$. Consider two labels (E_1 and E_2) representing two feasible partial paths ending at the same node. Label E_1 dominates label E_2 if $E_1 \leqslant E_2$ component-wise (that is, $Z_1 \leqslant Z_2$ and $T_1 \leqslant T_2$). In this case, E^2 can be discarded. When all components are equal, keep one of the two labels.

At first, for each aircraft, we initialize one label at the source node. Every next step we select one label with the minimum Z and extend it to all subsequent nodes. Then we use the dominance rule to remove unpromising labels. Repeat until all the labels extend to the sink node. We select the label with the minimum Z at sink, so that we can get the feasible shortest path and the corresponding aircraft.

The detailed algorithm is shown as below. Let $\mathcal{U}_i$ represents the set of unprocessed labels at node i, and $\mathcal{L}_i$ represents the set of processed labels at node i. Let $i(E)$ represents the last node of the path associated with E. Let $DOM(\mathcal{U}_j, \mathcal{L}_j)$ represents the dominance algorithm applied to labels in $\mathcal{U}_j$ and $\mathcal{L}_j$ that returns a possibly reduced set $\mathcal{U}_j$.

Algorithm 1 : Labeling algorithm

1: Set $\mathcal{U}_i = \mathcal{L}_i = \emptyset$ for all node in $\widetilde{\mathcal{V}}$ except $\mathcal{U}_o = \{E^1_{1o}, E^2_{2o}, \cdots, E^{|\mathcal{K}|}_{|\mathcal{K}|o}\}$ where
   $E^k_{po} = (0,0)$ for all p and k.
2: **while** $\bigcup_{i \in \widetilde{\mathcal{V}}} \mathcal{U}_i \neq \emptyset$ **do**
3:     Choose a label $E \in \bigcup_{i \in \widetilde{\mathcal{V}}} \mathcal{U}_i$ and remove E from $\mathcal{U}_{i(E)}$.
4:     **for all** arcs $(i(E), j) \in \mathscr{A}$ **do**
5:         Extend E along $(i(E), j)$ using the REFs to create a new label E'
6:         **if** E' is feasible **then**
7:             Add E' to $\mathcal{U}_j$.
8:             $\mathcal{U}_j = DOM(\mathcal{U}_j, \mathcal{L}_j)$.
9:         Add E to $\mathcal{L}_{i(E)}$.
10:    Filter $\mathcal{L}_d$ to find a shortest o-d path and corresponding aircraft.
11: **return** The shortest path and corresponding aircraft and its cost

---

Then we can get the column generation algorithm. It will output a solution of MP.

Algorithm 2 : Column generation algorithm

1: Initial: Use DFS to find feasible o-d path for each aircraft and build RMP according
   to the corresponding variables $\{y^1_1, y^2_2, \cdots, y^{|\mathcal{K}|}_{|\mathcal{K}|}\}$.
2: **while** True **do**
3:     Solve RMP and get the optimal solution y^* and the corresponding dual variable
       π and μ.
4:     Modify the cost of arc for each aircraft.
5:     Use *labeling algorithm* to get the shortest path p and aircraft k and cost c^k_p.
6:     **if** Reduced cost $RC = -c^k_p \leqslant 0$ **then**
7:         **return** the optimal solution y^*
8:     **else**
9:         Add y^k_p to RMP.

3.2   A New Branch Strategy

The solution obtained by the column generation may not conform to integer constraints. If the solution is not integral, we need to use a branching approach to eliminate the fractional variables. In order to incorporate column generation with branching strategies, the conventional integer programming branching on variables may not be feasible because fixing variables can destroy the structure of the pricing problem. For example, suppose there is a fractional variable $y^k_p = 0.5$, we need to divide it into two branches, $y^k_p = 1$ and $y^k_p = 0$. The first branch is easy to enforce because we can just let aircraft k covers path p and remove variables related to p. However, the latter branch is almost impossible to enforce because we cannot restrict aircraft k from choosing path p. In fact, it is likely the shortest path for k is indeed path p. In this case, the pricing problem solution must be achieved using a n-th *shortest path procedure* which is almost impossible for large-scale problems. Barnhart [11] gave a branching strategy for multi-commodity flow problem. It is based on the variables in arc-flow formulation and compatible with the pricing problem solution procedure. Based on Barnhart's strategy, we proposed a new branch strategy for DARSP that can achieve faster convergence.

The basic idea for Barnhart's branching rule is that each aircraft can only be assigned to one o-d path. If aircraft k' has been assigned to more than one path, then we can select two distinct paths p_1 and p_2 from them. Considering node o is in both p_1 and p_2, we can extend along paths from o until the two paths separate. We call the node at which two paths split as *divergence node s* and two arcs in p_1 and p_2 whose origin node is s as a_1 and a_2. Let $A(s, a_1)$ and $A(s, a_2)$ represent some partition of the set of arcs originating at s such that the subset $A(s, a_1)$ contains a_1 and the subset $A(s, a_2)$ contains a_2. Then we can creates two branches. The first requires

$$\sum_{p \cap A(s,a_1) \neq 0} y_p^{k'} = 0,$$

and the second requires

$$\sum_{p \cap A(s,a_2) \neq 0} y_p^{k'} = 0.$$

When faced with problems with high symmetry, this branching rule may become inefficient. For example, if there are more than one aircraft of same type as k' and we only disallow one aircraft k' from paths p_1 and p_2, there will be other aircraft assigned to p_1 and p_2 after branching and have the same value. Our remedy is to forbid more than one aircraft from the subset of arcs to weaken symmetry.

The key of our rule is that the constraints (5) restrict each node (except source) can only be covered by at most one aircraft, which means that there will be no two arcs in any integer solution that have the same origin node (except source). Our new rule is that if divergence node s is not source, we can create two new branches, one of which forbids **all** aircraft in $\mathcal{K}$ from using arcs in $A(s, a_1)$,

$$\sum_{k \in \mathcal{K}} \sum_{p \cap A(s,a_1) \neq 0} y_p^k = 0,$$

and the other forbids **all** aircraft in $\mathcal{K}$ from using arcs in $A(s, a_2)$,

$$\sum_{k \in \mathcal{K}} \sum_{p \cap A(s,a_2) \neq 0} y_p^k = 0.$$

Therefore, in the new two branches, we not only eliminate the emergence of symmetric solutions, but also remove a large number of infeasible solutions which use two distinct arcs having the same origin s. The branching rule is valid because all possible integer solutions are preserved after branching.

The branching rule will be more complicated if divergence node is the source. Since the source node will be used by all aircraft, we cannot prohibit all aircraft from using arcs in $A(s, a_1)$ or $A(s, a_2)$. Aircraft k_1 uses arc a_1 and aircraft k_2 uses arc a_2 may be part of the optimal solution, but this situation is not available in any of the two branches mentioned in the above rule. To fix this problem, let $\mathcal{K}(k')$ represents the set containing all other aircraft of the same type as k' (k' is not in $\mathcal{K}(k')$), we modify the branching rule if the divergence node is the source. For the first branch, we do not allow aircraft k' to use any of the arcs in $A(s, a_1)$ AND do not allow aircraft in $\mathcal{K}(k')$ to use arc a_2,

$$\sum_{p \cap A(s,a_1) \neq 0} y_p^{k'} + \sum_{k \in \mathcal{K}(k')} \sum_{p:a_2 \in p} y_p^k = 0,$$

and the second branch dose not allow aircraft k' to use any of the arcs in $A(s, a_2)$ AND does not allow aircraft in $\mathcal{K}(k')$ to use arc a_1,

$$\sum_{p \cap A(s,a_2) \neq 0} y_p^{k'} + \sum_{k \in \mathcal{K}(k')} \sum_{p:a_1 \in p} y_p^k = 0.$$

This is valid because each pair of arcs (i, j) can be branched at most once and after each branch, we retain at least one feasible solution of each symmetry. This means that at least one optimal solution remains in the left or right branch.

A major benefit of our branching rule is that it mitigates the effect of symmetry and reduces domains of problems after branching. The experiments in Sect. 4 show that our algorithm has better performance than basic branching rule.

4 Computational Results

We use 8 sets of data corresponding to several short and medium haul routes in China. These datasets consist of fleets of heterogeneous aircraft and legs between some major cities. Their characteristics are given in Table 1.

Table 1. Problem characteristics

Problem	Fleets		Legs		
	Types	Number of aircraft	Airports	Number of legs	Arcs
1	6	41	8	82	549
2	10	63	6	96	856
3	12	70	16	107	487
4	14	74	16	213	1543
5	11	105	45	247	972
6	16	96	25	211	1197
7	4	23	27	358	2595
8	9	104	60	187	781

We converted these problems into path-flow models and solved by column generation. To evaluate our branching rule, we compared the performance of using our branching rule and the basic branching rule in the same column generation framework. The LP in column generation was solved by Gurobi 9.1.1. The results are given in Table 2. The table gives the number of branch-and-bound nodes searched, the gap between lower bound and upper bound and the CPU time in seconds on an Intel Xeon platinum 8280. Solution procedure is terminated when either a provably optimal integer solution is found or the run time exceeds one hour. Solution times of less than one hour indicate that optimal solutions are obtained. We observe that most of test problems can be solved in reasonable times using our branching rule. Compared to the basic branching rule, our branching strategy require less time and calculate fewer branch nodes. In problem 2, our solution time is significantly lower than the basic branching rule. In problems 6 and 7, the basic approach could not find the optimal solution within a limited time, but our algorithm found the optimal solution in half an hour.

Table 2. Improved vs. basic branching

Problem	Improved branching			Basic branching		
	Nodes	Time	Gap	Nodes	Time	Gap
1	13	123.04	0.00%	19	125.77	0.00%
2	27	92.95	0.00%	2645	1922.84	0.00%
3	9	40.41	0.00%	15	35.99	0.00%
4	229	3600	0.48%	308	3600	0.48%
5	2	798.16	0.00%	5	886.59	0.00%
6	87	1419.56	0.00%	789	3600	0.05%
7	13	545.71	0.00%	202	3600	0.06%
8	17	662.42	0.00%	24	669.91	0.00%

5   Conclusions

In this paper, we focus on the aircraft routing and scheduling problem. We proposed two equivalent models to describe this problem and analyze the differences between the two models. Since there were too many variables in the second model, we used the algorithms of column generation and branch-and-bound to get the optimal solutions. In the process of column generation, we analyzed the properties of the subproblems and gave a labeling algorithm to solve it. Finally, we gave a new branching rule according to the special constraints of the model. The experimental results on the real routes networks indicate that our proposed methods can efficiently provide a fleet routing and scheduling scheme.

References

1. Belobaba, P., Odoni, A., Barnhart, C.: The Global Airline Industry. John Wiley & Sons, Hoboken (2015)
2. Levin, A.: Scheduling and fleet routing models for transportation systems. Transp. Sci. **5**(3), 232–255 (1971)
3. Abara, J.: Applying integer linear programming to the fleet assignment problem. Interfaces **19**(4), 20–28 (1989)
4. Hane, C.A., Barnhart, C., Johnson, E.L., et al.: The fleet assignment problem: solving a large-scale integer program. Math. Program. **70**(1–3), 211–232 (1995)
5. Desaulniers, G., Desrosiers, J., Dumas, Y., et al.: Daily aircraft routing and scheduling. Manag. Sci. **43**(6), 841–855 (1997)
6. Cadarso, L., de Celis, R.: Integrated airline planning: robust update of scheduling and fleet balancing under demand uncertainty. Transp. Res. Part C Emerg. Technol. **81**, 227–245 (2017)
7. Kenan, N., Jebali, A., Diabat, A.: An integrated flight scheduling and fleet assignment problem under uncertainty. Comput. Oper. Res. **100**, 333–342 (2018)
8. Cui, R., Dong, X., Lin, Y.: Models for aircraft maintenance routing problem with consideration of remaining time and robustness. Comput. Ind. Eng. **137**, 106045 (2019)
9. Zheng, S., Yang, Z., He, Z., et al.: Hybrid simulated annealing and reduced variable neighbourhood search for an aircraft scheduling and parking problem. Int. J. Prod. Res. **58**(9), 2626–2646 (2020)
10. Costa, L., Contardo, C., Desaulniers, G.: Exact branch-price-and-cut algorithms for vehicle routing. Transp. Sci. **53**(4), 946–985 (2019)
11. Barnhart, C., Hane, C.A., Vance, P.H.: Using branch-and-price-and-cut to solve origin-destination integer multicommodity flow problems. Oper. Res. **48**(2), 318–326 (2000)

Optimizing Mobile Charger Scheduling for Task-Based Sensor Networks

Xiangguang Meng[1], Jianxiong Guo[2], Xingjian Ding[3]([⊠]), and Xiujuan Zhang[4]

[1] School of Information, Beijing Forestry University, Beijing 100083, China
`mengxg@cuc.edu.cn`
[2] BNU-UIC Institute of Artificial Intelligence and Future Networks,
Beijing Normal University at Zhuhai, Zhuhai, Guangdong 519087, China
`jianxiongguo@bnu.edu.cn`
[3] School of Software Engineering, Beijing University of Technology,
Beijing 100124, China
`dxj@bjut.edu.cn`
[4] School of Information, Renmin University of China, Beijing 100872, China
`xiujuanzhang@qfnu.edu.cn`

Abstract. Replenishing energy to wireless sensor networks is always a crucial problem as the energy capacity of sensor nodes is very limited. Scheduling mobile chargers to charge sensor nodes has been widely studied due to its efficiency and flexibility. However, most existing works focus on maximizing the charging utility or charging efficiency, which ignores the task performing function of sensor nodes. In this paper, we study the mobile charger scheduling problem with the objective to maximize the task utility achieved by sensor nodes. We consider two different scenarios where sensor nodes are deployed on a line and a ring, respectively. We prove the NP-Hardness of our problems and design two approximation algorithms with guaranteed performance. We prove the approximation ratio of our algorithms through theoretical analysis, and conduct extensive simulations to validate the performance of our algorithms. Simulation results show that our algorithms always outperform the baselines, which demonstrates the effectiveness of our algorithms.

Keywords: Mobile charger · Wireless sensor network · Wireless power transfer

1 Introduction

Wireless Sensor Networks (WSNs) are widely used to monitor the physical world [8,9], where the sensor nodes in WSNs are usually powered by built-in batteries. As the energy capacity of the built-in battery is limited, how to efficiently replenish energy to sensor nodes is an important problem to be addressed. The breakthrough of the wireless power transfer technology brings an efficient and flexible way to supply energy to sensor nodes [4], where end-users can schedule mobile chargers to charge sensor nodes periodically.

© Springer Nature Switzerland AG 2021
W. Wu and H. Du (Eds.): AAIM 2021, LNCS 13153, pp. 134–145, 2021.
https://doi.org/10.1007/978-3-030-93176-6_12

Recently, there are lots of researchers study the mobile charger scheduling problem to achieve continuous operations of wireless sensor networks. Existing works mainly focus on improving the charging efficiency or charging utility by scheduling the charging routes of mobile chargers [1,5,10,12]. That is, their goal is to make the mobile charger spend more energy on charging sensor nodes rather than on moving, or to make sensor nodes receive as much energy as possible in a charging scheduling period. However, these works ignore the tasks performed by sensor nodes. In many practical WSNs applications, sensor nodes may performing different tasks with different energy requirements [2,13], and will achieve different task utility according to the type of the performed tasks. Obviously, focus on maximizing task utility is a more practical and meaningful problem compared to maximizing charging efficiency or charging utility.

In this paper, therefore, we investigate the task utility maximization problem by scheduling a mobile charger. Wireless sensor networks are widely used in many scenarios, such as forests, parks, fields, etc. In our work, we consider two different network scenarios. In the first scenario, a wireless sensor network is deployed to monitor a part of a road, where sensor nodes are deployed on the side of the road, and the mobile charger could move freely on the road to provide energy to sensor nodes. In the second scenario, a wireless sensor network is deployed on a ring road in a park, where the mobile charger can travel on the ring road in any direction (clockwise or anticlockwise). Thus we address our problem in both line scenario and ring scenario. As the energy capacity of the mobile charger is also limited, and is used for both moving and charging, how to efficiently select sensor nodes to charge is the core issue of our problem. The main contributions of this work are list as follows.

- We consider the task utility maximization problem by scheduling a mobile charger in both line and ring scenarios, and prove that the problem to be addressed is NP-Hard in both scenarios.
- We design two approximation algorithms for our problems, and prove the approximation ratio through theoretical analysis.
- We validate the effectiveness of our algorithms by conducting extensive simulations.

The remainder of this paper is structured as follows. In Sect. 2, we introduce the related works of this paper. In Sect. 3, we formally define our problems to be addressed in two scenarios, and prove the NP-hardness of our problems. In Sect. 4, we design two approximation algorithms for our problems and analyze the ratio through theorems. In Sect. 5, we conduct extensive simulations to evaluate our algorithms. And finally, we conclude this paper in Sect. 6.

2   Related Works

Scheduling mobile chargers to charge wireless sensor networks has been widely studied these years. Lin et al. [6] study the charging delay minimization problem for WSNs under the directional charging model, in which they determine the charging orientation for the mobile charger and charging time at each charging location, their objective is to minimize the total charging time while fully charge all the sensor nodes. Xu et al. [11] study how to maximize the sum of sensor node lifetimes by scheduling a mobile charger, in which they assume that each sensor node can be partially charged so as to charge more sensor nodes before they exhaust their energy. Liu et al. [7] consider that mobile chargers could charge each other, and study how to schedule minimum mobile chargers to charge a large-scale wireless sensor networks. Jiang et al. [3] jointly consider location selection of depots and charging routes planning of mobile chargers for large-scale WSNs, they aim to minimize the number of mobile chargers and improve the energy efficiency of mobile chargers. Zhang et al. [14] consider the scenario that there are a set of candidate charging itineraries for mobile chargers, they aim to select itineraries and determine a corresponding charging association so as to maximize the charging efficiency, while making sure that all sensor nodes are fully charged.

3   Model and Problem Formulation

3.1   Models and Assumptions

Consider there are a set of m rechargeable sensor nodes deployed on a line or a ring, the position of each sensor node is known in advance and fixed. These rechargeable sensor nodes are denoted by $\mathcal{S} = \{s_1, s_2, \ldots, s_m\}$. We assume that each sensor node can perform a specific task by consuming a certain amount of energy. For the sensor node $s_i \in \mathcal{S}$, the energy it needed to perform the task is denoted by E_i, and the corresponding achieved task utility is denoted by ω_i. We also assume that the energy of each sensor node is supplied by a Mobile Charger (MC) which is located at the *depot* r. The depot will periodically dispatch the MC to replenish energy to sensor nodes. For the line scenario, the depot r is located at the left end side of the line, and without loss of generality, we assume that sensor nodes are sequentially deployed from left to right according to their subscripts, as shown in Fig. 1(a). For the ring scenario, sensor nodes are sequentially deployed in the anticlockwise direction of the ring based on their subscripts, and the depot r is located between sensors s_1 and s_m, as shown in Fig. 1(b).

In our study, in order to ensure the charging efficiency, we consider that the MC needs to move to the site of a sensor node before it begins charging the sensor node. The energy capacity of the MC is denoted by EI, which is limited and used for both moving and charging sensors. We use η to denote the energy consumption rate of the MC for moving per unit distance. During each charging scheduling, the MC moves along the line or the ring to charge sensor nodes, and then it needs to move back to the depot before its energy is exhausted.

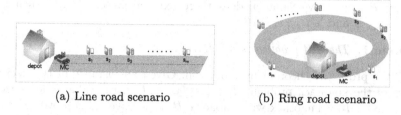

(a) Line road scenario   (b) Ring road scenario

Fig. 1. Illustration of the two network scenarios.

3.2 Problem Formulation

In this work, we aim to schedule an MC with limited energy to charge sensor nodes on a line or a ring, so as to maximize the total task utility achieved by all sensor nodes. As mentioned before, both moving and charging sensors will consume the energy of MC, so we need to find a charging route and determine which sensor nodes to be charged, while the energy consumption of the MC is limited by its energy capacity. We define our problem as follows.

Problem 1. Task utility Maximization problem by scheduling a mobile charger (TM). Given a set $\mathcal{S}$ of rechargeable sensor nodes deployed on a line or a ring road, the energy requirement E_i of each sensor $s_i \in \mathcal{S}$ for performing tasks, the corresponding achieved task utility ω_i of each sensor s_i, a mobile charger with energy capacity EI located at the depot r, and the energy consumption rate η of the MC. The TM problem aims to find a charging route and determine the set of sensor nodes to be charged, such that the total achieved task utility is maximized, while the energy consumption of the MC is limited by EI.

Note that we assume the MC can find a route to fully charge any sensor node without exhausts its energy, otherwise, we can drop the sensor node from our problem.

Next, we will prove that the TM problems is NP-hard. We first consider a special case of the problem, in which we assume that $\eta = 0$, that is, the MC moving does not consume any energy. In such a special case, we only need to consider which sensor nodes to be charged. For clarity, the special case is termed as the STM problem. If we can prove the NP-Hardness of the STM problem, then we can conclude that the general TM problem is also NP-hard. In the following, we introduce some problems to help us prove the NP-Hardness of our problems.

The Decision Version of the STM Problem: Given a set $\mathcal{S}$ of rechargeable sensor nodes, the energy requirement E_i and corresponding task utility ω_i of each sensor s_i, the energy capacity EI of the MC, a positive number l, does there exist a subset $\mathcal{S}' \subseteq \mathcal{S}$ such that the total charging energy for nodes in $\mathcal{S}'$ is not larger than EI (i.e., $\sum_{s_i \in \mathcal{S}'} E_i \leq EI$) and the total achieved task utility of $\mathcal{S}'$ is at least l (i.e., $\sum_{s_i \in \mathcal{S}'} \omega_i \geq l$)?

The Decision Version of the Knapsack Problem: Given a set of m items $I = \{I_1, I_2, \ldots, I_m\}$, the value v_i and size s_i for each item I_i, a knapsack with capacity B, a positive value p, does there exist a subset $I' \subseteq I$ such that $\sum_{I_i \in I'} s_i \leq B$ and $\sum_{I_i \in I'} v_i \geq q$?

Theorem 1. *The TM problem is NP-Hard.*

Proof. We prove the theorem by reduction. Given an instance of the decision version of the knapsack problem, $I = \{I_1, I_2, \ldots, I_m\}$, the value v_i and size s_i for each item I_i, a knapsack with capacity B, a positive value p, we construct an instance of the decision version of the STM problem as follows. Let $\mathcal{S} = I$, $E_i = s_i$ and $\omega_i = v_i$ for each sensor $s_i \in \mathcal{S}$, $EI = B$, and $l = p$. Then, if the decision version of the knapsack problem has a "Yes" answer, the decision version of the STM problem also has a "Yes" answer, and vice versa. As the knapsack problem is a typical NP-Hard problem, we can conclude that the STM problem is also NP-Hard. The STM problem is a special case of the TM problem, so we can easily know that problem TM is NP-Hard.

4 Approximation Algorithms

To distinguish the two scenarios, we term our problem as the TML problem for the line road scenario, and the TMR problem for the ring road scenario. Next, we will describe the proposed algorithms for problems TML and TMR, and give the performance analysis for our algorithms.

4.1 Algorithm for the TML Problem

We name the proposed algorithm for the TML problem as ATML. As the sensor nodes are deployed on a line, we let the mobile charger moves along the line and selects some sensor nodes to charge, and then move back to the depot before its energy is exhausted. The ATML algorithm mainly has two phases. In the first phase, we iteratively determine the charging route for the mobile charger, and then select sensor nodes to be charged under each charging route. In the second phase, we choose the best strategy got in the first phase as our solution. In the following, we will describe our algorithm in detail.

In the first phase, we consider m possible charging routes for the mobile charger. For each $1 \leq i \leq m$, we let the MC moves to the i-th sensor node (s_i) and then back to the depot r. For clarity, we let $\mathcal{S}_i = \{s_1, s_2, \ldots, s_i\}$ for each $1 \leq i \leq m$. As the MC will pass through sensor nodes in $\mathcal{S}_i$, we then select sensor nodes from $\mathcal{S}_i$ to charge limited to the energy capacity of the MC. We use L_i to denote the distance from depot r to sensor node s_i, then in the i-th iteration, the energy of the MC that can be used for charging sensors is $EI_i = EI - 2 * \eta L_i$. We use N_i to denote the selected sensor node set to be charged in the i-th iteration, and use $U(N_i)$ to denote the total task utility achieved by sensor nodes in N_i, i.e., $U(N_i) = \sum_{s_i \in N_i} \omega_i$. In each iteration, we find two candidate solutions which are denoted by N_i^1 and N_i^2, respectively. N_i^1 contains the sensor node with the largest task utility selected from $\mathcal{S}_i$, i.e., $N_i^1 = argmax\{U(s_j) | s_j \in \mathcal{S}_i\}$. The second candidate solution N_i^2 is initialized as $\emptyset$. Then we iteratively select sensor node s_j from $\mathcal{S}_i$ with largest $\frac{\omega_j}{E_j}$, if s_j can be added into N_i^2 without exceeding the energy budget EI_i, then move s_j from $\mathcal{S}_i$ to N_i^2, otherwise, delete s_j from $\mathcal{S}_i$. This process terminates until all sensor nodes in $\mathcal{S}_i$ are checked (i.e., $\mathcal{S}_i$ is empty). If $U(N_i^1) \geq U(N_i^2)$, we let $N_i = N_i^1$, otherwise, $N_i = N_i^2$.

In the second phase, we select the best strategy got by the first phase as our final solution. Let N denote the final solution of our algorithm, then, $N = argmax\{U(N_i)|1 \leq i \leq m\}$.

The pseudo-code of the ATML algorithm is shown in Algorithm 1.

Algorithm 1. Algorithm for the TML problem (ATML)

Input: $\mathcal{S}$, depot r, EI, η, the needed energy E_i and task utility ω_i for each $s_i \in \mathcal{S}$.
Output: A charging strategy N for the MC.
1: **for** $1 \leq i \leq m$ **do**
2: Compute the distance L_i between r and s_i;
3: $EI_i = EI - 2 * \eta L_i$, $\mathcal{S}_i = \{s_1, s_2, \ldots, s_i\}$;
4: $N_i^1 \leftarrow argmax\{U(s_j)|s_j \in \mathcal{S}_i\}$;
5: $N_i^2 \leftarrow \emptyset$, $N_i \leftarrow \emptyset$;
6: **while** $\mathcal{S}_i \neq \emptyset$ **do**
7:   Select $s_j \in \mathcal{S}_i$ that with maximum $\frac{\omega_j}{E_j}$;
8:   **if** $E_j + \sum_{s_k \in N_i^2} E_k \leq EI_i$ **then**
9:    $N_i^2 \leftarrow N_i^2 \cup \{s_j\}$;
10:   **end if**
11:   $\mathcal{S}_i \leftarrow \mathcal{S}_i \backslash \{s_j\}$;
12: **end while**
13: $N_i = argmax\{U(N_i^k)|k = 1, 2\}$;
14: **end for**
15: $N \leftarrow argmax\{U(N_i)|1 \leq i \leq m\}$;
16: **return** the charging strategy N;

Next, we analyze the performance of the ATML algorithm.

Theorem 2. *The ATML algorithm achieves a 2-approximation ratio for the TML problem.*

Proof. We assume that the last sensor node in the optimal solution N^* is s_k. Then the moving distance of the MC in the optimal solution is $2 * L_k$, and the energy used for charging is at most $EI_k = EI - 2 * \eta L_k$. In the k-th iteration of the for-loop, we greedily select sensor nodes from $\mathcal{S}_k$ according to the value of $\frac{\omega_i}{E_i}$ of each sensor $s_i \in \mathcal{S}_k$, and add them into N_k^2. Without loss of generality, we suppose that $\frac{\omega_1}{E_1} \geq \frac{\omega_2}{E_2} \geq \cdots \geq \frac{\omega_k}{E_k}$. Assume that s_q is the first sensor node that does not satisfy the condition in line 8, then we have the following two inequalities.

$$U(N^*) \leq \sum_{1 \leq i \leq q} \omega_i + (EI_i - \sum_{1 \leq i \leq q} E_i)\frac{\omega_{q+1}}{E_{q+1}}. \tag{1}$$

$$0 \leq EI_i - \sum_{1 \leq i \leq q} E_i \leq E_{q+1}. \tag{2}$$

Combining inequalities (1) and (2), we have the following inequality.

$$U(N^*) \leq \sum_{1 \leq i \leq q} \omega_i + \omega_{q+1}. \tag{3}$$

As we have checked all the sensor nodes in S_k, we can easily know that $U(N_k^2) \geq \sum_{1 \leq i \leq q} \omega_i$. N_k^1 contains the sensor node with the largest task utility that is selected from S_k, then we know that $U(N_k^1) \geq \omega_{q+1}$. We choose the better one of N_k^1 and N_k^2 as the solution of the k-th iteration, so we have

$$U(k) \geq \frac{1}{2}(U(N_k^1) + U(N_k^2)). \tag{4}$$

As the solution of our algorithm is the best one selected from m iterations, we have

$$U(N) \geq U(N_k) \geq \frac{1}{2}(U(N_k^1) + U(N_k^2)) \geq \frac{1}{2}(\sum_{1 \leq i \leq q} \omega_i + \omega_{q+1}) \geq \frac{1}{2}U(N^*). \tag{5}$$

Thus the theorem holds.

Theorem 3. *The time complexity of algorithm ATML is $\mathcal{O}(m^3)$.*

Proof. In the first phase of the ATML algorithm, there are m iterations in the for-loop. In each iteration, the calculation of N_i^1 takes $\mathcal{O}(m)$ time as we need to find the sensor node that provides the maximum task utility from S_i. There are $\mathcal{O}(m)$ iterations in the while-loop, and in each iteration of the while-loop, it takes $\mathcal{O}(m)$ time to find $s_j \in S_i$ that with maximum $\frac{\omega_j}{E_j}$. Summarily, the time complexity of the first phase is $\mathcal{O}(m * (m + m^2)) = \mathcal{O}(m^3)$. The second phase of the ATML algorithm takes $\mathcal{O}(m)$ time as we only need to find the best strategy from m solutions. Therefore, the time complexity of algorithm ATML is $\mathcal{O}(m^3) + \mathcal{O}(m) = \mathcal{O}(m^3)$.

4.2   Algorithm for the TMR Problem

We name the proposed algorithm for the TMR problem as ATMR. Similar to the ATML algorithm, the ATMR algorithm also has two phases. In the first phase, we iteratively determine the charging route for the mobile charger and then select sensor nodes to be charged. In the second phase, we choose the best strategy got in the first phase as our final solution. The main difference between algorithms ATMR and ATML is that the charging route of the MC is different. In the following, we will give the detailed description of the ATMR algorithm.

Algorithm 2. Algorithm for the TMR problem (ATMR)

Input: $\mathcal{S}, depot$ r,EI,η, the needed energy E_i and task utility ω_i for each $s_i \in \mathcal{S}$.
Output: A charging strategy N for the MC.
 1: Compute the length L of the ring, and the shortest distance L_i between each sensor $s_i \in \mathcal{S}$ and the depot;
 2: Find the last sensor node s_p that doesn't exceed the middle point of the ring along the anticlockwise direction;
 3: **for** $1 \leq i \leq m + 1$ **do**
 4:   **if** $1 \leq i \leq p$ **then**
 5:     $\mathcal{S}_i = \{s_1, s_2, \ldots, s_i\}$;
 6:   **else if** $p < i \leq m$ **then**
 7:     $\mathcal{S}_i = \{s_m, s_{m-1}, \ldots, s_i\}$;
 8:   **else**
 9:     $\mathcal{S}_i = \mathcal{S}$;
10:   **end if**
11:   $EI_i = EI - 2 * \eta L_i$;
12:   $N_i^1 \leftarrow argmax\{U(s_j) | s_j \in \mathcal{S}_i\}$;
13:   $N_i^2 \leftarrow \emptyset, N_i \leftarrow \emptyset$;
14:   **while** $\mathcal{S}_i \neq \emptyset$ **do**
15:     Select $s_j \in \mathcal{S}_i$ that with maximum $\frac{\omega_j}{E_j}$;
16:     **if** $E_j + \sum_{s_k \in N_i^2} E_k \leq EI_i$ **then**
17:       $N_i^2 \leftarrow N_i^2 \cup \{s_j\}$;
18:     **end if**
19:     $\mathcal{S}_i \leftarrow \mathcal{S}_i \backslash \{s_j\}$;
20:   **end while**
21:   $N_i = argmax\{U(N_i^k) | k = 1, 2\}$;
22: **end for**
23: $N \leftarrow argmax\{U(N_i) | 1 \leq i \leq m + 1\}$;
24: **return** the charging strategy N;

In the first phase, we need to find possible charging routes for the MC. In the TMR problem, sensor nodes are deployed on a ring, there are three possible moving ways for the MC. The first way is to move along the ring in the anticlockwise direction, then moves back to the depot on the same path before it reaches the middle point of the ring. The second way is to move along the ring in the clockwise direction, and back to the depot on the same path before it reaches the middle point of the ring. The last way is to move around the ring (in the anticlockwise or clockwise direction). Next, we will consider $m + 1$ possible charging routes for the MC. Assume that s_p is the last sensor node that doesn't exceed the middle point of the ring along the anticlockwise direction. For each $1 \leq i \leq m + 1$, if $i \leq p$, we define $\mathcal{S}_i = \{s_1, s_2, \ldots, s_i\}$, and let the MC moves to the sensor node s_i in the anticlockwise direction, and then moves back to the depot on the same path. If $p < i \leq m$, we define $\mathcal{S}_i = \{s_m, s_{m-1}, \ldots, s_i\}$, and let the MC moves to the sensor node s_i in the clockwise direction, and then moves back to the depot on the same path. If $i = m + 1$, we define $\mathcal{S}_i = \mathcal{S}$, and let the MC moves around the ring in the anticlockwise direction. We use L_i

$(1 \leq i \leq m)$ to denote the shortest distance between sensor node s_i and depot r, and L_{m+1} is equal to the length L of the ring. In each iteration, the charging route of the MC is determined, and the energy of the MC that can be used for charging sensors is $EI_i = EI - 2 * \eta L_i$. We use N_i to denote the selected sensor nodes to be charged in the i-th iteration. The calculation of N_i is the same as that of the ATML algorithm, here we omit the detailed descriptions.

In the second phase, we choose the best strategy got by the first phase as our final solution, that is, $N = argmax\{U(N_i)|1 \leq i \leq m+1\}$.

The pseudo-code of the ATNR algorithm is described in Algorithm 2.

Theorem 4. *The ATMR algorithm achieves a 2-approximation ratio with time complexity* $\mathcal{O}(m^3)$ *for the TMR problem.*

Proof. As described before, we have tried three moving ways of the MC in our algorithm, it is obvious that the charging route of the optimal solution must be the same as one case in the $m+1$ iterations of our algorithm. Denote the optimal solution by N^*, and assume that the optimal charging route is the same as the i-th iteration, according to the analysis in Theorem 2, we have $U(N) \geq U(N_i) \geq \frac{1}{2}U(N^*)$. Thus ATMR is a 2-ratio algorithm. The process of the ATMR algorithm is almost the same as the ATML algorithm, besides the charging routes in iterations are different. Obviously, the time complexity of the ATMR algorithm is the same as that of the ATML algorithm, i.e., $\mathcal{O}(m^3)$.

5　Simulations

In this section, we validate the performance of our algorithms by conducting extensive simulations.

5.1　Simulation Settings

The basic parameters of our simulations are set as follows. We use the same default parameters for both TML and TMR problems. The number of sensor nodes is set as 100, the required energy of each sensor node is randomly chosen in $[3, 10]$ kJ, the task utility of each sensor node is randomly chosen in $[10, 60]$. Both the length of the line for the TML problem and the length of the ring for the TMR problem are set as 1000 m. The energy capacity of the MC is set as 400 kJ, and the energy consumption rate η of the MC is set as 200 J/m .

5.2　Results

To evaluate the performance of our algorithms, we design two baseline algorithms for problems TML and TMR. The baseline algorithm for the TML problem is named FirstK, in which we let the MC move along the line to charge every sensor node it passing through, and let the MC return to the depot if it can't charge the next sensor node without exhaust its energy. The baseline algorithm for the TMR problem is named FirstK-Ring, which is similar to the FirstK algorithm. In FirstK-Ring algorithm, we first let the MC move along the ring in the anticlockwise direction to charge every sensor node it passing through, and let the MC return to the depot if it can't charge the next sensor node without exhaust its energy. Then we let the MC move along the ring in the clockwise direction to do the same thing. We choose the better one of the above strategies as the final solution. In the following, we will compare our algorithms with the two baselines under different parameters.

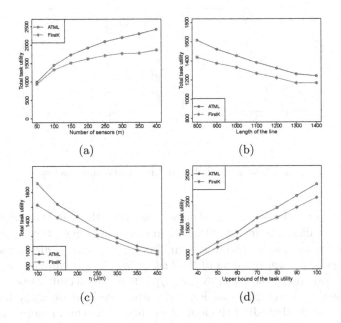

Fig. 2. Comparison between ATML and the baseline.

1) Impact of the number of sensors (m): As shown in Fig. 2(a) and Fig. 3(a), with the number of sensor nodes increasing from 50 to 400, both algorithms ATML and ATMR perform better than their baselines, especially when there are more sensor nodes in the network. The reason is that when there are more sensor nodes, our algorithms have more chances to select "better" sensor nodes that could achieve more task utility.

2) Impact of the length of the line or the ring: Fig. 2(b) and Fig. 3(b) show the performance comparison of our algorithms and baselines under different length settings. We can see that when the length of the line or the ring increases, all of the four algorithms will achieve little task utility, as the MC will spend more energy on moving. Our algorithms always outperform the baselines especially when the length of the line or the ring is small.

3) Impact of the energy consumption rate η of the MC: From Fig. 2(c) and Fig. 3(c) we can see that when the energy consumption rate of the MC for

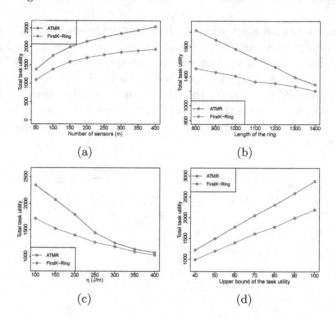

Fig. 3. Comparison between ATMR and the baseline.

moving becomes larger, the gap between our algorithms and the two baselines becomes smaller. The reason is that when η becomes larger, the MC will spend more energy on moving, and thus spend little energy on charging sensor nodes.

4) Impact of the upper bound of the task utility: In our simulations, the task utility is randomly chosen in $[30, \omega_{max}]$, where the default value of ω_{max} is set to be 60. In Fig. 2(d) and Fig. 3(d), when we increase ω_{max} from 40 to 100, we can see that all of the four algorithms will achieve more task utility, as the average task utility of sensor nodes is increased. We can also see that our algorithms always outperform the baselines under any network settings, which validate the performance of our algorithms.

6   Conclusions

In this paper, we study the task utility maximization problem under the line and ring scenarios by scheduling a mobile charger. We prove that the problem to be addressed is NP-Hard, and design two approximation algorithms for the problem under the two different scenarios. We analyze the approximation ratio of our algorithms through theorems, and validate the performance of our designs by extensive simulations. The results show the effectiveness of our algorithms. In the future, we will expand our research to the 2-D scenario, and study the scheduling problem of multiple mobile chargers.

References

1. Ding, X., Chen, W., Wang, Y., Li, D., Hong, Y.: Efficient scheduling of a mobile charger in large-scale sensor networks. Theor. Comput. Sci. **840**, 219–233 (2020)
2. Ding, X., Guo, J., Wang, Y., Li, D., Wu, W.: Task-driven charger placement and power allocation for wireless sensor networks. Ad Hoc Networks, p. 102556 (2021)
3. Jiang, G., Lam, S.K., Sun, Y., Tu, L., Wu, J.: Joint charging tour planning and depot positioning for wireless sensor networks using mobile chargers. IEEE/ACM Trans. Netw. **25**(4), 2250–2266 (2017)
4. Kurs, A., Karalis, A., Moffatt, R., Joannopoulos, J.D., Fisher, P., Soljačić, M.: Wireless power transfer via strongly coupled magnetic resonances. Science **317**(5834), 83–86 (2007)
5. Liang, W., Xu, Z., Xu, W., Shi, J., Mao, G., Das, S.K.: Approximation algorithms for charging reward maximization in rechargeable sensor networks via a mobile charger. IEEE/ACM Trans. Netw. **25**(5), 3161–3174 (2017)
6. Lin, C., Zhou, Y., Ma, F., Deng, J., Wang, L., Wu, G.: Minimizing charging delay for directional charging in wireless rechargeable sensor networks. In: IEEE INFO-COM 2019-IEEE Conference on Computer Communications, pp. 1819–1827. IEEE (2019)
7. Liu, T., Wu, B., Wu, H., Peng, J.: Low-cost collaborative mobile charging for large-scale wireless sensor networks. IEEE Trans. Mob. Comput. **16**(8), 2213–2227 (2017)
8. Luo, C., Hong, Y., Li, D., Wang, Y., Chen, W., Hu, Q.: Maximizing network lifetime using coverage sets scheduling in wireless sensor networks. Ad Hoc Netw. **98**, 102037 (2020)
9. Luo, C., Satpute, M.N., Li, D., Wang, Y., Chen, W., Wu, W.: Fine-grained trajectory optimization of multiple uavs for efficient data gathering from wsns. IEEE/ACM Trans. Netw. **29**(1), 162–175 (2020)
10. Ma, Y., Liang, W., Xu, W.: Charging utility maximization in wireless rechargeable sensor networks by charging multiple sensors simultaneously. IEEE/ACM Trans. Netw. **26**(4), 1591–1604 (2018)
11. Xu, W., Liang, W., Jia, X., Xu, Z., Li, Z., Liu, Y.: Maximizing sensor lifetime with the minimal service cost of a mobile charger in wireless sensor networks. IEEE Trans. Mob. Comput. **17**(11), 2564–2577 (2018)
12. Ye, X., Liang, W.: Charging utility maximization in wireless rechargeable sensor networks. Wirel. Netw. **23**(7), 2069–2081 (2016). https://doi.org/10.1007/s11276-016-1271-6
13. Zeng, D., Li, P., Guo, S., Miyazaki, T., Hu, J., Xiang, Y.: Energy minimization in multi-task software-defined sensor networks. IEEE Trans. Comput. **64**(11), 3128–3139 (2015)
14. Zhang, S., Qian, Z., Wu, J., Kong, F., Lu, S.: Optimizing itinerary selection and charging association for mobile chargers. IEEE Trans. Mob. Comput. **16**(10), 2833–2846 (2016)

Semi-online Early Work Maximization Problem on Two Hierarchical Machines with Partial Information of Processing Time

Man Xiao, Xiaoqiao Liu, and Weidong Li[✉]

School of Mathematics and Statistics, Yunnan University,
Kunming 650504, People's Republic of China

Abstract. In this paper, we study three semi-online early work maximization problems on two hierarchical machines. When the total processing time of low or high hierarchy is known, we propose an optimal algorithm with a competitive ratio of $\sqrt{5} - 1$. When the total processing times of low and high hierarchy are known, we propose an optimal algorithm with a competitive ratio of $\frac{6}{5}$.

Keywords: Semi-online · Early work · Hierarchy · Competitive ratio

1 Introduction

The online hierarchical scheduling problem has been widely studied, and many online and semi-online algorithms for different objectives have been derived. For the (semi-)online scheduling problem, the jobs arrive one by one. The performance of the (semi-)online algorithm is measured by the competitive ratio. For a maximization problem and any job instance I, the competitive ratio of algorithm A is defined as the minimum ρ that satisfies the $C^{OPT}(I) \leq \rho C^A(I)$, where $C^A(I)$(abbreviated as C^A) denotes the output value of online algorithm A, and $C^{OPT}(I)$(abbreviated as C^{OPT}) denotes the offline optimal value. For an (semi-)online problem, if there is no algorithm with competitive ratio less than ρ, then ρ is a lower bound of the problem. If there is an algorithm whose competitive ratio is equal to ρ, this algorithm is called an optimal online algorithm.

Given a feasible solution for the online hierarchical scheduling problem on two machines, let L_i be the load of machine M_i for $i = 1, 2$. There are three typical objectives, makespan minimization, machine covering and early work maximization. The makespan minimization problem [7] is to find a feasible schedule such that $\max\{L_1, L_2\}$ is minimized. The machine covering problem [1] is to find a feasible schedule such that $\min\{L_1, L_2\}$ is maximized. The early work maximization problem [3] is to find a feasible schedule such that $\sum_{i=1}^{2} \min\{L_i, d\}$ is maximized, where d is the common due date.

For the makespan minimization problem, Park et al. [7] and Jiang [5] independently gave an optimal algorithm with competitive ratio of $\frac{5}{3}$. When the

W. Wu and H. Du (Eds.): AAIM 2021, LNCS 13153, pp. 146–156, 2021.
https://doi.org/10.1007/978-3-030-93176-6_13

total processing time of all jobs is known in advance, Park et al. [7] proposed an optimal online algorithm with competitive ratio of $\frac{3}{2}$. When the total processing time of low-hierarchy jobs is known in advance, Chen et al. [2] and Luo and Xu [6] independently presented an optimal online algorithm with competitive ratio of $\frac{3}{2}$. When the total processing time of high-hierarchy jobs is known in advance, Luo and Xu [6] presented an optimal online algorithm with competitive ratio of $\frac{20}{13}$. When both the total processing times of low-hierarchy and high-hierarchy jobs are known in advance, Chen et al. [2] and Luo and Xu [6] independently presented an optimal online algorithm with competitive ratio of $\frac{4}{3}$.

For the machine covering problem, Chassid and Epstein [1] proved that the competitive ratio of any online algorithm is unbounded. When the total processing time of all jobs is known in advance, Chassid and Epstein [1] proposed an optimal online algorithm with competitive ratio of 2. When the total processing time of low-hierarchy jobs is known in advance, Xiao et al. [8] presented an optimal online algorithm with competitive ratio of 2. When both the total processing times of low-hierarchy and high-hierarchy jobs are known in advance, Xiao et al. [8] presented an optimal online algorithm with competitive ratio of $\frac{3}{2}$.

For the early work maximization problem on two identical machines, Chen et al. [4] designed an optimal online algorithm with competitive ratio of $\sqrt{5}-1$. When the total processing time of all jobs is known in advance, Chen et al. [3] designed an optimal online algorithm with competitive ratio of $\frac{6}{5}$. For the early work maximization problem on two hierarchical machines, Xiao et al. [9] designed an optimal online algorithm with competitive ratio of $\sqrt{2}$. When the total processing time of all jobs is known in advance, Xiao et al. [9] designed an optimal online algorithm with competitive ratio of $\frac{4}{3}$.

In this paper, we consider the early work maximization problem with a common due date on two hierarchical machines, and presented some optimal semi-online algorithms. The remainder of this paper is organized as follows. In Sect. 2, we describe some preliminaries. In Sect. 3, we design an optimal online algorithm with competitive ratio of $\sqrt{5}-1$ when the total processing time of low-hierarchy jobs is known in advance. In Sect. 4, we design an optimal online algorithm with competitive ratio of $\sqrt{5}-1$ when the total processing time of high-hierarchy jobs is known in advance. In Sect. 5, we design an optimal online algorithm with competitive ratio of $\frac{6}{5}$ when both the total processing times of low-hierarchy and high-hierarchy jobs are known in advance. We present some conclusions and possible directions for future research in the last section.

2   Preliminaries

We are given two machines M_1 and M_2, and a series of jobs arriving online which are to be scheduled irrevocably at the time of their arrivals. The arrival of a new job occurs only after the current job is scheduled. Let $J = \{J_1, J_2, ..., J_n\}$ be the set of all jobs arranged in the order of arrival. Denote the j-th job as $J_j = (p_j, g_j)$, where the p_j is the processing time (also called size) of the job J_j, and $g_j \in \{1, 2\}$ is the hierarchy of the job J_j. If $g_j = k$, we call J_j as a job of hierarchy k, $k \in \{1, 2\}$. For the hierarchical scheduling problem, M_1 can process all jobs, and M_2 can only process jobs of hierarchy 2.

As in [3,4], assume that all jobs have the same due date $d > 0$, and the job processing time

$$p_j \le d, \text{ for } j = 1, 2, \ldots, n.$$

The early work of job J_j is denoted by $X_j \in [0, p_j]$. If job J_j is completed before the due date d, the job is called totally early, and $X_j = p_j$. If the job J_j starts at the time of $S_j < d$, but finishes after the due date d, the job is called partially early, and $X_j = d - S_j$. If the job J_j starts at the time of $S_j \ge d$, the job is called totally late, and $X_j = 0$.

A schedule is actually a partition (S_1, S_2) of the job set J, such that $S_1 \cup S_2 = J$ and $S_1 \cap S_2 = \emptyset$. Let $L_i = \sum_{J_j \in S_i} p_j$ be the load of M_i, $i \in \{1, 2\}$. The object-ive is to find a schedule such that total early work

$$X = \sum_{j=1}^{n} X_j = \sum_{i=1}^{2} \min\{L_i, d\}$$

is maximized. For convenience, this problem is denoted as the $P2|GoS, online, d_j = d|\max(X)$.

For $i, k \in \{1, 2\}$ and $j \in \{1, 2, \ldots, n\}$, let T_k be the total size of the jobs with hierarchy k, and L_i^j be the load of M_i after job J_j is scheduled. Let T be the total size of the jobs. Clearly, we have

$$L_i^n = L_i \text{ and } T_1 + T_2 = T.$$

Consider the example given in Fig. 1, where the load of M_1 is less than d and the load of M_2 is more than d. Hence, the total early work is $X = \min\{L_1, d\} + \min\{L_2, d\} = L_1 + d$.

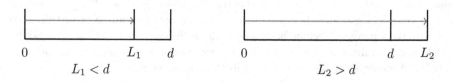

$$L_1 < d \qquad\qquad\qquad L_2 > d$$

Fig. 1. The total early work is $X = L_1 + d$

By the definition of X, we have

Lemma 1. The optimal value C^{OPT} satisfies that

$$C^{OPT} \le \min\{T_1 + T_2, 2d\}.$$

3   The Total Processing Time of Low-Hierarchy Jobs Is Known

In this section, we consider the case where the total processing time of the low-hierarchy jobs T_1 is known in advance. For convenience, this problem is denoted as $P2|GoS, online, d_j = d, T_1| \max(X)$. For this problem, we will give a lower bound of $\sqrt{5}-1$, and design an online algorithm with competitive ratio of $\sqrt{5}-1$.

Theorem 2. Any online algorithm A for $P2|GoS, online, d_j = d, T_1| \max(X)$ has a competitive ratio at least $\sqrt{5} - 1$.

Proof. Let $d = \frac{\sqrt{5}+1}{2}$ and $T_1 = 1$. The first two jobs are $J_1 = (1,1)$ and $J_2 = (1,2)$. J_1 can only be assigned to M_1. If J_2 is assigned to M_1, then no more jobs arrive. Hence, $C^{OPT} = 2$ and $C^A = \frac{\sqrt{5}+1}{2}$, implying that

$$\frac{C^{OPT}}{C^A} = \frac{2}{\frac{\sqrt{5}+1}{2}} = \frac{4}{\sqrt{5}+1} = \sqrt{5} - 1.$$

If J_2 is assigned to M_2, the last job $J_3 = (\frac{\sqrt{5}+1}{2}, 2)$ arrives. Regardless of how job J_3 is allocated, $C^{OPT} = \sqrt{5} + 1$ and $C^A = \frac{\sqrt{5}+1}{2} + 1$, implying that

$$\frac{C^{OPT}}{C^A} = \frac{\sqrt{5}+1}{\frac{\sqrt{5}+1}{2} + 1} = \frac{2(\sqrt{5}+1)}{\sqrt{5}+3} = \frac{2(2\sqrt{5}-2)}{4} = \sqrt{5} - 1.$$

Therefore, in any case, we have $\frac{C^{OPT}}{C^A} \geq \sqrt{5} - 1$, implying that the theorem holds. ∎

For $j = 1, 2, \ldots, n$, let $L_{1,2}^j$ be the total processing time of high-hierarchy jobs assigned to machine M_1, after job J_j is scheduled. The details of our algorithm are as follows.

Theorem 3. The competitive ratio of Algorithm 1 is at most $\sqrt{5} - 1$.

Proof. Based on Lemma 1, if $\min\{L_1, L_2\} \geq d$, we have $C^A = 2d \geq C^{OPT}$. If $\max\{L_1, L_2\} \leq d$, we have $C^A = T_1 + T_2 \geq C^{OPT}$. It implies that we only need to consider the case $\min\{L_1, L_2\} < d < \max\{L_1, L_2\}$, which implies that

$$C^A = d + \min\{L_1, L_2\}.$$

We distinguish the following two cases.

Case 1. $\max\{L_1, L_2\} = L_1 > d$.
  In this case, we have $C^A = d + L_2$. If there is no job of hierarchy 2 assigned to M_1, we have $L_1 = T_1 > d$ and $L_2 = T_2 < d$, implying that   Algorithm 1 reaches the optimal. Else, let $J_l = (p_l, 2)$ be the last job of hierarchy 2 assigned to M_1. There are three possibilities in Algorithm 1 to allocate J_l to the M_1.

Algorithm 1:

---

1 Initially, let $L_{1,2}^0 = 0$.

2 When a new job $J_j = (p_j, g_j)$ arrives,

3 **if** $g_j = 1$ **then**

4 $\quad\lfloor$ Assign job J_j to M_1, and let $L_{1,2}^j = L_{1,2}^{j-1}$.

5 **else**

6 $\quad$ **if** $T_1 \geq \frac{(\sqrt{5}-1)d}{2}$ **then**

7 $\quad\quad\lfloor$ Assign job J_j to M_2.

8 $\quad$ **else**

9 $\quad\quad$ **if** $L_{1,2}^{j-1} + T_1 + p_j \leq \frac{(\sqrt{5}-1)d}{2}$ **then**

10 $\quad\quad\quad\lfloor$ Assign J_j to M_1.

11 $\quad\quad$ **else**

12 $\quad\quad\quad$ **if** $\frac{(\sqrt{5}-1)d}{2} < L_{1,2}^{j-1} + T_1 + p_j \leq (\sqrt{5}-1)d$ **then**

13 $\quad\quad\quad\quad$ Assign J_j to M_1, and assign the remaining hierarchy 2 jobs to M_2(if there are jobs after J_j).

14 $\quad\quad\quad$ **else**

15 $\quad\quad\quad\quad$ Assign J_j to M_2, and assign the remaining hierarchy 2 jobs to M_1(if there are jobs after J_j).

16 If there is another job, $j == j + 1$, go to **step 2**. Otherwise, stop.

If J_l is assigned to M_1 at Line 10, from the choice of Algorithm 1, we have $L_1 = L_{1,2}^{l-1} + T_1 + p_l \leq \frac{(\sqrt{5}-1)d}{2}$, which contradictions the fact $L_1 > d$.

If J_l is assigned to M_1 at Line 13, from the choice of Algorithm 1, we have $L_1 = L_{1,2}^{l-1} + T_1 + p_l \leq (\sqrt{5}-1)d$. By Lemma 1, we have

$$\frac{C^{OPT}}{C^A} \leq \frac{T_1 + T_2}{d + L_2} = \frac{L_1 + L_2}{d + L_2} \leq \frac{(\sqrt{5}-1)d + L_2}{d + L_2} \leq \sqrt{5} - 1.$$

If J_l is assigned to M_1 at Line 15, let $J_t = (p_t, 2)$ be the job assigned to M_2 at Line 15. From the choice of Algorithm 1, we have $L_{1,2}^{t-1} + T_1 + p_t > (\sqrt{5}-1)d$ and $T_1 < \frac{(\sqrt{5}-1)d}{2}$. If $L_{1,2}^{t-1} = 0$, then $L_{1,2}^{t-1} + T_1 \leq \frac{(\sqrt{5}-1)d}{2}$. If $L_{1,2}^{t-1} > 0$, let $J_k = (p_k, 2)$ be the last job assigned to M_1 when the J_t arrives, according to the choice of Algorithm 1, we known the J_k be assigned to M_1 at Line 10. Thus, we have $L_{1,2}^{t-1} + T_1 = L_{1,2}^{k-1} + T_1 + p_k \leq \frac{(\sqrt{5}-1)d}{2}$. Therefore, $L_2 \geq p_t > \frac{(\sqrt{5}-1)d}{2}$. By Lemma 1, we have

$$\frac{C^{OPT}}{C^A} \leq \frac{2d}{d + L_2} \leq \frac{2d}{d + \frac{(\sqrt{5}-1)d}{2}} = \sqrt{5} - 1.$$

Case 2. $\max\{L_1, L_2\} = L_2 > d$.

In this case, we have $C^A = d + L_1$. Let $J_l = (p_l, 2)$ be the last job assigned to M_2. There are three possibilities in Algorithm 1 to allocate J_l to the M_2.

If J_l is assigned to M_2 at Line 7, from the choice of Algorithm 1, we have $L_1 = T_1 \geq \frac{(\sqrt{5}-1)d}{2}$. Based on Lemma 1, we have

$$\frac{C^{OPT}}{C^A} \leq \frac{2d}{d+L_1} \leq \frac{2d}{d + \frac{(\sqrt{5}-1)d}{2}} = \sqrt{5} - 1.$$

If J_l is assigned to M_2 at Line 13, let $J_t = (p_t, 2)$ be the job assigned to M_1 at Line 13. From the choice of Algorithm 1, we have $L_1 = L_{1,2}^{t-1} + T_1 + p_t > \frac{(\sqrt{5}-1)d}{2}$. Based on Lemma 1, we have

$$\frac{C^{OPT}}{C^A} \leq \frac{2d}{d+L_1} \leq \frac{2d}{d + \frac{(\sqrt{5}-1)d}{2}} = \sqrt{5} - 1.$$

If J_l is assigned to M_2 at Line 15, then Algorithm 1 does not run the Lines 7 and 13. Because only the job J_l is assigned to M_2 at Line 15, we have $L_2 = p_l \leq d$, which contradictions with the assumption $L_2 > d$. ∎

4 The Total Processing Time of High-Hierarchy Jobs Is Known

In this section, we consider the case where T_2 is known in advance, which is denoted as $P2|GoS, online, d_j = d, T_2| \max(X)$. For this problem, we will give a lower bound of $\sqrt{5} - 1$, and design an online algorithm with competitive ratio of $\sqrt{5} - 1$.

Theorem 4. Any online algorithm A for $P2|GoS, online, d_j = d, T_2| \max(X)$ has a competitive ratio at least $\sqrt{5} - 1$.

Proof. Let $d = \sqrt{5} + 1$ and $T_2 = 4$. The first job is $J_1 = (2, 2)$. If J_1 is assigned to M_1, the last two jobs $J_2 = (2, 2)$ and $J_3 = (\sqrt{5} + 1, 1)$ arrive. Therefore, $C^{OPT} = 2(\sqrt{5} + 1)$ and $C^A \leq \sqrt{5} + 3$, implying that

$$\frac{C^{OPT}}{C^A} \geq \frac{2(\sqrt{5}+1)}{\sqrt{5}+3} = \frac{2(2\sqrt{5}-2)}{4} = \sqrt{5} - 1.$$

If J_1 is assigned to M_2, the next job $J_2 = (2, 2)$ arrives. If J_2 is assigned to M_2, then there is no more jobs arrive. Therefore, $C^{OPT} = 4$ and $C^A = \sqrt{5} + 1$, implying that

$$\frac{C^{OPT}}{C^A} = \frac{4}{\sqrt{5}+1} = \sqrt{5} - 1.$$

If J_2 is assigned to M_1, the last job $J_3 = (\sqrt{5} + 1, 1)$ arrives. Therefore, $C^{OPT} = 2(\sqrt{5} + 1)$ and $C^A = \sqrt{5} + 3$, implying that

$$\frac{C^{OPT}}{C^A} = \frac{2(\sqrt{5}+1)}{\sqrt{5}+3} = \frac{2(2\sqrt{5}-2)}{4} = \sqrt{5} - 1.$$

Hence, in any case, we have $\frac{C^{OPT}}{C^A} \geq \sqrt{5} - 1$.

The details of our algorithm are as follows.

Algorithm 2:

---

1 Initially, let $L_{1,2}^0 = 0$.

2 When a new job $J_j = (p_j, g_j)$ arrives,

3 **if** $g_j = 1$ **then**

4 $\quad$ $\lfloor$ Assign the job J_j to M_1, and set $L_{1,2}^j = L_{1,2}^{j-1}$.

5 **else**

6 $\quad$ **if** $T_2 \leq (\sqrt{5} - 1)d$ **then**

7 $\quad\quad$ $\lfloor$ Assign the J_j to M_2.

8 $\quad$ **else**

9 $\quad\quad$ **if** $T_2 - L_{1,2}^{j-1} - p_j \geq \frac{(\sqrt{5}-1)d}{2}$ **then**

10 $\quad\quad\quad$ $\lfloor$ Assign the J_j to M_1.

11 $\quad\quad$ **else**

12 $\quad\quad\quad$ **if** $T_2 - L_{1,2}^{j-1} \leq (\sqrt{5} - 1)d$ **then**

13 $\quad\quad\quad\quad$ $\lfloor$ Assign the J_j to M_2, and assign the remaining jobs to M_2(if there are jobs after J_j).

14 $\quad\quad\quad$ **else**

15 $\quad\quad\quad\quad$ $\lfloor$ Assign the J_j to M_2, and assign the remaining jobs to M_1(if there are jobs after J_j).

16 If there is another job, $j == j + 1$, go to **step 2**. Otherwise, stop.

Theorem 5. The competitive ratio of Algorithm 2 is at most $\sqrt{5} - 1$.

Proof. As before, if $\min\{L_1, L_2\} \geq d$ or $\max\{L_1, L_2\} \leq d$, Algorithm 2 reaches the optimal. We only need to consider the case $\min\{L_1, L_2\} < d < \max\{L_1, L_2\}$, which implies that

$$C^A = d + \min\{L_1, L_2\}.$$

We distinguish the following two cases.

Case 1. $\max\{L_1, L_2\} = L_1 > d$.

In this case, we have $C^A = d + L_2$. If there is no job of hierarchy 2 assigned to M_1, we have $L_1 = T_1 > d$ and $L_2 = T_2 < d$, which implies that Algorithm 2 reaches the optimal. Else, let $J_l = (p_l, 2)$ be the last job of hierarchy 2 assigned to M_1. There are two possibilities in Algorithm 2 to allocate J_l to the M_1.

If J_l is assigned to M_1 at Line 10, from the choice of Algorithm 2, we have $L_2 = T_2 - L_{1,2}^{l-1} - p_l \geq \frac{(\sqrt{5}-1)d}{2}$. Based on Lemma 1, we have

$$\frac{C^{OPT}}{C^A} \leq \frac{2d}{d + L_2} \leq \frac{2d}{d + \frac{(\sqrt{5}-1)d}{2}} = \sqrt{5} - 1.$$

If J_l is assigned to M_1 at Line 15, let $J_t = (p_t, 2)$ is the job assigned to M_2 at Line 15, from the choice of Algorithm 2, we have $T_2 - L_{1,2}^{t-1} - p_t < \frac{(\sqrt{5}-1)d}{2}$

and $T_2 - L_{1,2}^{t-1} > (\sqrt{5} - 1)d$. It implies $L_2 \geq p_t > \frac{(\sqrt{5}-1)d}{2}$. Based on Lemma 1, we have

$$\frac{C^{OPT}}{C^A} \leq \frac{2d}{d + L_2} \leq \frac{2d}{d + \frac{(\sqrt{5}-1)d}{2}} = \sqrt{5} - 1.$$

Case 2. $\max\{L_1, L_2\} = L_2 > d$.

In this case, we have $C^A = d + L_1$. Let $J_l = (p_l, 2)$ be the last job assigned to M_2, there are three possibilities in Algorithm 2 to allocate J_l to the M_2.

If J_l is assigned to M_2 at Line 7, from the choice of Algorithm 2, we have $L_2 = T_2 \leq (\sqrt{5} - 1)d$. Based on Lemma 1, we have

$$\frac{C^{OPT}}{C^A} \leq \frac{T_1 + T_2}{d + L_1} = \frac{L_1 + L_2}{L_1 + d} \leq \frac{L_1 + (\sqrt{5} - 1)d}{L_1 + d} \leq \sqrt{5} - 1.$$

If J_l is assigned to M_2 at Line 13, let $J_t = (p_t, 2)$ be the first job assigned to M_2 at Line 13. From the choice of Algorithm 2, we have $L_2 = T_2 - L_{1,2}^{t-1} \leq (\sqrt{5} - 1)d$. Based on Lemma 1, we have

$$\frac{C^{OPT}}{C^A} \leq \frac{T_1 + T_2}{d + L_1} \leq \frac{L_1 + L_2}{L_1 + d} \leq \frac{L_1 + (\sqrt{5} - 1)d}{L_1 + d} = \sqrt{5} - 1.$$

If J_l is assigned to M_2 at Line 15, then Algorithm 2 does not run the Lines 7 and 13. Because only the job J_l is assigned to M_2 at Line 15, we have $L_2 = p_l \leq d$, which contradictions with the assumption $L_2 > d$. ∎

5   The Total Processing Times of Low-Hierarchy and High-Hierarchy Jobs Are Known

In this section, we consider the case where T_1 and T_2 are known in advance, which is denoted as $P2|GoS, online, d_j = d, T_1 \& T_2| \max(X)$. We will give a lower bound $\frac{6}{5}$, and design an online algorithm with competitive ratio of $\frac{6}{5}$.

Theorem 6. Any online algorithm A for the $P2|GoS, online, d_j = d, T_1 \& T_2| \max(X)$ has a competitive ratio at least $\frac{6}{5}$.

Proof. Let $d = 3n$, $T_1 = 1$ and $T_2 = 6n$. The first three jobs are $J_1 = (1, 1)$, $J_2 = (n, 2)$ and $J_3 = (n, 2)$. Job J_1 can only be assigned to M_1. If J_2 and J_3 are assigned to the same machine, the last two jobs $J_4 = (2n, 2)$ and $J_5 = (2n, 2)$ arrive, which implies that $C^{OPT} = 6n$ and $C^A \leq 5n + 1$. Therefore,

$$\frac{C^{OPT}}{C^A} \geq \frac{6n}{5n + 1} = \frac{6}{5}(n \to \infty).$$

Otherwise, the last job $J_4 = (4n, 2)$ arrives. Therefore, $C^{OPT} = 5n + 1$ and $C^A \leq 4n + 1$, then

$$\frac{C^{OPT}}{C^A} \geq \frac{5n + 1}{4n + 1} = \frac{5}{4}(n \to \infty).$$

Thus, $\frac{C^{OPT}}{C^A} \geq \frac{6}{5}$ in any case. ∎

Algorithm 3:

1 Initially, let $L_{1,2}^0 = 0$.
2 When a new job $J_j = (p_j, g_j)$ arrives,
3 **if** $g_j = 1$ **then**
4     $\lfloor$ Assign job J_j to M_1, and set $L_{1,2}^j = L_{1,2}^{j-1}$.

5 **else**
6   **if** $T_1 + L_{1,2}^{j-1} + p_j \leq \frac{6d}{5}$ **then**
7       $\lfloor$ Assign job J_j to M_1.

8   **else**
9   **if** $T_1 + L_{1,2}^{j-1} \geq \max\{p_j, T_2 - (L_{1,2}^{j-1} + p_j)\}$ **then**
10        $\lfloor$ Assign job J_j to M_2, and assign the remaining jobs to M_2(if there are jobs after J_j).

11   **else**
12   **if** $p_j \geq \max\{T_1 + L_{1,2}^{j-1}, T_2 - (L_{1,2}^{j-1} + p_j)\}$ **then**
13         $\lfloor$ Assign job J_j to M_2, and assign the remaining jobs to M_1(if there are jobs after J_j).

14   **else**
15         $\lfloor$ Assign job J_j to M_1, and assign the remaining jobs to M_2(if there are jobs after J_j).

16 If there is another job, $j == j + 1$, go to **step 2**. Otherwise, stop.

The details of our algorithm are as follows.

Theorem 7. The competitive ratio of Algorithm 3 is at most $\frac{6}{5}$.

Proof. As before, if $\min\{L_1, L_2\} \geq d$ or $\max\{L_1, L_2\} \leq d$, Algorithm 3 reaches the optimal. We only need to consider the case $\min\{L_1, L_2\} < d < \max\{L_1, L_2\}$, which implies that

$$C^A = d + \min\{L_1, L_2\}.$$

We distinguish the following two cases.

Case 1. $\max\{L_1, L_2\} = L_1 > d$.

In this case, we have $C^A = d + L_2$. If there is no job of hierarchy 2 assigned to M_1, we have $L_1 = T_1 > d$ and $L_2 = T_2 < d$, which implies that Algorithm 3 reaches the optimal. Else, let $J_l = (p_l, 2)$ be the last job of hierarchy 2 assigned to M_1, there are three possibilities in Algorithm 3 to allocate J_l to the M_1.

If J_l is assigned to M_1 at Line 7, from the choice of Algorithm 3, we have $L_1 = T_1 + L_{1,2}^{l-1} + p_l \leq \frac{6d}{5}$. Based on Lemma 1, we have

$$\frac{C^{OPT}}{C^A} \leq \frac{T_1 + T_2}{d + L_2} = \frac{L_1 + L_2}{d + L_2} \leq \frac{\frac{6d}{5} + L_2}{d + L_2} \leq \frac{6}{5}.$$

If J_l is assigned to M_1 at Line 13, let $J_t = (p_t, 2)$ be the job assigned to M_2 at Line 13. From the choice of Algorithm 3, we have $L_2 \geq p_t \geq \max\{T_1 + L_{1,2}^{t-1}, T_2 - (L_{1,2}^{t-1} + p_t)\}$, which implies that $L_2 \geq \frac{T_1 + T_2}{3}$.

If J_l is assigned to M_1 at Line 15, from the choice of Algorithm 3, we have $L_2 = T_2 - (L_{1,2}^{l-1} + p_l) \geq \max\{p_l, T_1 + L_{1,2}^{l-1}\}$, which implies that $L_2 \geq \frac{T_1 + T_2}{3}$.

When $T_1 + T_2 \leq 2d$, based on Lemma 1, we have

$$\frac{C^{OPT}}{C^A} \leq \frac{T_1 + T_2}{d + L_2} \leq \frac{T_1 + T_2}{\frac{T_1 + T_2}{2} + \frac{T_1 + T_2}{3}} = \frac{6}{5}.$$

When $T_1 + T_2 > 2d$, based on Lemma 1, we have

$$\frac{C^{OPT}}{C^A} \leq \frac{2d}{d + L_2} \leq \frac{2d}{d + \frac{2d}{3}} = \frac{6}{5}.$$

Case 2. $\max\{L_1, L_2\} = L_2 > d$.

In this case, we have $C^A = d + L_1$. Let $J_l = (p_l, 2)$ be the last job assigned to M_2. There are three possibilities in Algorithm 3 to allocate J_l to the M_2.

If J_l is assigned to M_2 at Line 10, let $J_t = (p_t, 2)$ is the first job assigned to M_2 at Line 10. From the choice of Algorithm 3, we have $L_1 = T_1 + L_{1,2}^{t-1} \geq \max\{p_t, T_2 - (L_{1,2}^{t-1} + p_t)\}$, this implies $L_1 \geq \frac{T_1 + T_2}{3}$.

When $T_1 + T_2 \leq 2d$, based on Lemma 1, we have

$$\frac{C^{OPT}}{C^A} \leq \frac{T_1 + T_2}{d + L_1} \leq \frac{T_1 + T_2}{\frac{T_1 + T_2}{2} + \frac{T_1 + T_2}{3}} = \frac{6}{5}.$$

When $T_1 + T_2 > 2d$, based on Lemma 1, we have

$$\frac{C^{OPT}}{C^A} \leq \frac{2d}{d + L_1} \leq \frac{2d}{d + \frac{2d}{3}} = \frac{6}{5}.$$

If J_l is assigned to M_2 at Line 13, then Algorithm 3 does not run the Lines 10 and 15. Because only the job J_l is assigned to M_2 at Line 13, we have $L_2 = p_l \leq d$, which contradictions with the assumption $L_2 > d$.

If J_l is assigned to M_2 at Line 15, let $J_t = (p_t, 2)$ be the job assigned to M_1 at Line 15, from the choice of Algorithm 3, we have $L_1 = T_1 + L_{1,2}^{t-1} + p_t > \frac{6d}{5}$, which contradictions with the assumption $L_1 < d$. ■

6   Discussion

In this paper, we considered the early work maximization problem on two hierarchical machines. We present some optimal online algorithms when the total processing time of low or (and) high hierarchy is known. It is interesting to design optimal online algorithms for the early work maximization problem on three or more hierarchical machines.

Acknowledgement. The work is supported in part by the National Natural Science Foundation of China [No. 12071417], and Project for Innovation Team (Cultivation) of Yunnan Province [No. 202005AE160006].

References

1. Chassid, O., Epstein, L.: The hierarchical model for load balancing on two machines. J. Comb. Optim. **15**, 305–314 (2008)
2. Chen, X., Ding, N., Dosa, G., Han, X.: Online hierarchical scheduling on two machines with known total size of low-hierarchy jobs. Int. J. Comput. Math. **92**(5), 873–881 (2015)
3. Chen, X., Kovalev, S., Liu, Y.Q., Sterna, M., Chalamon, I., Błażewicz, J.: Semi-online scheduling on two identical machines with a common due date to maximize total early work. Discret. Appl. Math. **290**, 71–78 (2021)
4. Chen, X., Sterna, M., Han, X., Błażewicz, J.: Scheduling on parallel identical machines with late work criterion: offline and online cases. J. Sched. **19**(6), 729–736 (2016)
5. Jiang, Y.W., He, Y., Tang, C.M.: Optimal online algorithms for scheduling on two identical machines under a grade of service. J. Zhejiang Univ. Sci. A **7**, 309–314 (2006)
6. Luo, T.B., Xu, Y.F.: Semi-online scheduling on two machines with GoS levels and partial information of processing time. Sci. World J. **2**, 576234 (2014)
7. Park, J., Chang, S.Y., Lee, K.: Online and semi-online scheduling of two machines under a grade of service provision. Oper. Res. Lett. **34**(6), 692–696 (2006)
8. Xiao, M., Wu, G., Li, W.: Semi-online machine covering on two hierarchical machines with known total size of low-hierarchy jobs. In: Sun, X., He, K., Chen, X. (eds.) NCTCS 2019. CCIS, vol. 1069, pp. 95–108. Springer, Singapore (2019). https://doi.org/10.1007/978-981-15-0105-0_7
9. Xiao, M., Liu, X.Q., Li, W.D.: Online and semi-online scheduling on two hierarchical machines with a common due date to maximize total early work, working paper (2021)

Nonlinear Combinatorial Optimization

Streaming Algorithms for Maximizing DR-Submodular Functions with *d*-Knapsack Constraints

Bin Liu[1]($\boxtimes$)(ID), Zihan Chen[1], and Hongmin W. Du[2]

[1] School of Mathematical Sciences, Ocean University of China, Qingdao 266100, People's Republic of China
binliu@ouc.edu.cn
[2] Accounting and Information Systems Department, Rutgers University, Newark, NJ, USA

Abstract. The problem of maximizing submodular functions has received considerable attention in the last few years. However, most of the submodular functions are defined on set. But recently some progress has been made on the integer lattice. In this paper, we study streaming algorithms for the problem of maximizing DR-submodular functions with *d*-knapsack constraints on the integer lattice. We first propose a one pass streaming algorithm that achieves a $\frac{1-\theta}{1+d}$-approximation with $O(\frac{\log(d\beta^{-1})}{\beta\epsilon})$ memory complexity and $O(\frac{\log(d\beta^{-1})}{\epsilon} \log \|\boldsymbol{b}\|_\infty)$ update time per element, where $\theta = \min(\alpha + \epsilon, 0.5 + \epsilon)$ and α, β are the upper and lower bounds for the cost of each item in the stream. Then we devise an improved streaming algorithm to reduce the memory complexity to $O(\frac{d}{\beta\epsilon})$ with unchanged approximation ratio and query complexity. As far as we know, this is the first streaming algorithm on the integer lattice under this constraint.

Keywords: Streaming algorithm · *d*-Knapsack constraints · Integer lattice · DR-submodular maximization

1 Introduction

Submodular functions play a significant role in combinatorial optimization. Suppose E is the ground set, a function $f : 2^E \rightarrow \mathbf{R}$ is called *submodular* if $f(A) + f(B) \geq f(A \cup B) + f(A \cap B)$ for any $A, B \subseteq E$. The main reason why submodular functions are widely used is that submodularity is equivalent to diminishing returns property, i.e. for any $S \subseteq T \subseteq E$ and $e \in E \setminus T$, it holds $f(S \cup \{e\}) - f(S) \geq f(T \cup \{e\}) - f(T)$. A set function f is called *monotone* if $f(S) \leq f(T)$ for any $S \subseteq T \subseteq E$. There are many applications to the problem

This work was supported in part by the National Natural Science Foundation of China (11971447, 11871442), and the Fundamental Research Funds for the Central Universities.

W. Wu and H. Du (Eds.): AAIM 2021, LNCS 13153, pp. 159–169, 2021.
https://doi.org/10.1007/978-3-030-93176-6_14

of submodular maximization, such as the submodular welfare problem [7], the influence maximization problem [3,4] and the sensor placement problem [9].

In the current big data environment, the input data of many applications is much larger than the storage capacity of individual computer. In this case, we need to process data by using streaming model, which is described as follows. When each element in set $E = \{e_1, e_2, ..., e_n\}$ arrives in order, the streaming algorithm must decide whether to keep it before the next element arrives. Obviously, streaming algorithm is different from other algorithms. Besides the approximation ratio and query complexity, the criteria to measure the quality of streaming algorithm include memory complexity and the number of passes to scan all data.

There have been many results in the study of one pass streaming algorithm. For the problem of maximizing submodular functions with a cardinality constraint, Badanidiyuru et al. [1] first obtained a 1/2-approximation streaming algorithm with memory $O(k \log k/\epsilon)$ and update time $O(\log k/\epsilon)$ per element. Later, Buchbinder et al. [2] devised a 1/4-approximation streaming algorithm with an improved memory $O(k)$. Norouzi-Fard et al. [10] proved that if the memory is $O(n/k)$, the approximation ratio of the streaming algorithm for this problem will not be better than 1/2 unless $P = NP$. Based on [1], Kazemi et al. [8] described a 1/2-approximation streaming algorithm with an improved memory $O(k/\epsilon)$. For maximizing submodular functions under a knapsack constraint, Wolsey et al. [14] proposed a $(1 - 1/e^\beta) \approx 0.35$-approximation streaming algorithm, where β is the unique solution of equation $e^x = 2 - x$. Yu et al. [15] designed a one pass $(1/3 - \epsilon)$-approximation streaming algorithm. Huang et al. [6] obtained a $(0.4 - \epsilon)$-approximation algorithm.

Submodular set functions are useful for solving problems with variable selection. But they can not solve problems that an element in the ground set is repeatedly selected. To address this situation, many papers have studied submodular functions on multiset, also known as submodular functions on the integer lattice. Different from set functions, submodularity defined on the integer lattice is not equivalent to diminishing return property. Therefore, submodular functions on the integer lattice can be divided into two types. A function $f : \mathbf{Z}_+^E \to \mathbf{R}$ is called *lattice submodular function* if $f(\boldsymbol{x}) + f(\boldsymbol{y}) \geq f(\boldsymbol{x} \vee \boldsymbol{y}) + f(\boldsymbol{x} \wedge \boldsymbol{y})$, for any $\boldsymbol{x}, \boldsymbol{y} \in \mathbf{Z}^E$, where $\vee$ and $\wedge$ are coordinate-wise max and min. A function $f : \mathbf{Z}_+^E \to \mathbf{R}$ is said to be *diminishing return submodular (DR-submodular)* if $f(\boldsymbol{x} + \mathcal{X}_e) - f(\boldsymbol{x}) \geq f(\boldsymbol{y} + \mathcal{X}_e) - f(\boldsymbol{y})$, for any $\boldsymbol{x}, \boldsymbol{y} \in \mathbf{Z}^E$ with $\boldsymbol{x} \leq \boldsymbol{y}$ and $e \in E$, where $\mathcal{X}_e$ denotes the unit vector with coordinate e being 1 and other components are 0.

Maximizing DR-submodular functions on the integer lattice arises from many applications, such as submodular welfare problem [5,7] and the budget allocation problem with decreasing influence probabilities [11]. For maximizing a monotone submodular function over the integer lattice subject to a knapsack constraint, Soma et al. [11] introduced a pseudo-polynomial-time algorithm whose approximation ratio is $1 - 1/e$. Later, Soma et al. [12] maintained the approximation ratio to $1 - 1/e$ and proposed polynomial time algorithms with a cardinality constraint, a polymatroid constraint and a knapsack constraint on the integer lattice, respectively.

In this paper, we study the streaming algorithm for maximizing DR-submodular functions with d-knapsack constraints on the integer lattice. Let $b \in \mathbf{Z}_+^E$ be an n-dimension integer vector, and $[b] = \{x \in \mathbf{Z}_+^n \mid 0 \leq x(e_i) \leq b(e_i), \forall 1 \leq i \leq n\}$ be the integer lattice domain. Let $f : \mathbf{Z}_+^E \to \mathbf{R}$ be a monotone non-negative DR-submodular function defined on the integer lattice. Then the constraint can be written as $c_j^T x \leq K_j, \forall j \in [d]$ and $x \leq b$. Where $c \in \mathbf{Z}_+^E$, $c_j^T x = \sum_{e \in \{x\}} c_j(e)$, $\{x\}$ is a multiset in which item e is repeated $x(e)$ times and K_j is the budget of j-th knapsack.

Many literatures have studied the streaming algorithm of maximizing submodular functions under d-knapsack constraints. Yu et al. [15] introduced a one pass $(\frac{1}{1+2d} - \epsilon)$-approximation streaming algorithm with memory complexity $O(\frac{b \log b}{d\epsilon})$ and update time $O(\frac{\log b}{\epsilon})$ per element, where b is the standardized d-knapsack capacity. Later, Wang et al. [13] obtained a one pass $\frac{1-\epsilon}{1+d}$-approximation streaming algorithm.

Our Contribution. In this paper, several streaming algorithms for maximizing DR-submodular functions on the integer lattice with d-knapsack constraints are proposed. When item e and its $b(e)$ copies arrive, we utilize binary search to decide the number of items kept. First, we give a one pass streaming algorithm with approximate ratio $\frac{1-\theta}{1+d}$, memory complexity $O(\frac{\log(d\beta^{-1})}{\beta\epsilon})$ and query complexity $O(\frac{\log(d\beta^{-1})}{\epsilon} \log \|b\|_\infty)$ per element. Later, on the basis of this algorithm, a new streaming algorithm named Accelerated-Streaming Algorithm is obtained, which retains the approximation ratio and query complexity, but improves the memory complexity to $O(\frac{d}{\beta\epsilon})$.

2   Preliminaries

In this section we will introduce the notations and definitions to be used later. Let $E = \{e_1, e_2, ..., e_n\}$ be the ground set, and e_i be the i-th item in the stream. Denote $b \in \mathbf{Z}_+^E$ be an n-dimensional integer vector, and $[b] = \{x \in \mathbf{Z}_+^n \mid 0 \leq x(e_i) \leq b(e_i), \forall 1 \leq i \leq n\}$ be the integer lattice domain. For an integer k, $[k]$ denotes the set $\{1, 2, ..., k\}$. For an n-dimensional vector x, $x(e_i)$ is the component value corresponding to the element e_i. $\{x\}$ denotes the multiset corresponding to x, where the element e_i repeats $x(e_i)$ times. Define $supp^+(x) = \{e \in E \mid x(e) > 0\}$. $\mathcal{X}_{e_i}$ is the unit vector, where e_i corresponds to component 1 and the other components are 0. $\mathbf{0}$ is a vector with all zero components. For vectors x, y, we define $f(x \mid y) = f(x + y) - f(y)$. $x \vee y$ and $x \wedge y$ are coordinate-wise max and min, respectively.

For $j \in [d]$, $c_j(e_i)$ which can be abbreviated as c_{ij} is the cost of element e_i in the j-th knapsack and $c_j^T x = \sum_{e \in \{x\}} c_j(e)$. $\lambda(e) \in [d]$ is the index of element e. Let $\alpha = \max_{j \in [d], i \in [n]} c_{ij}$ and $\beta = \min_{j \in [d], i \in [n]} c_{ij}$ be the upper and lower bounds of all element costs respectively, i.e. $\beta \leq c_{ij} \leq \alpha$ for $j \in [d]$, $i \in [n]$. Denote K_j be the budget of j-th knapsack. For convenience, let $K_j = 1$ for $j \in [d]$, and the cost of any item is less than 1.

In this paper, we study the maximization of DR-submodular functions under d-knapsack constraints. Using the above notations, we give the specific expression of the problem as follows.

$$\max f(\boldsymbol{x})$$
$$\text{s. t. } \boldsymbol{c}_j^T \boldsymbol{x} \le K_j, \forall j \in [d], \tag{1}$$
$$\boldsymbol{x} \le \boldsymbol{b}.$$

3 Threshold-Streaming Algorithm for DR-Submodular Maximization

In this section, we propose a one pass streaming algorithm for Problem (1). In order to get a better output, we first design a streaming algorithm on the premise that the optimal value of the problem is known. Then we use the traditional method to estimate the optimal value, and propose the final one pass streaming algorithm named Threshold-Streaming Algorithm.

3.1 A Streaming Algorithm with Known OPT

We assume that the optimal value OPT of Problem (1) is known. Then the following algorithm is constructed to select vector $\boldsymbol{x}$ by using the optimal value of Problem (1).

Overview of Algorithm. Suppose that we know a value v which satisfies $\lambda OPT \le v \le OPT$ for some $0 < \lambda \le 1$. Algorithm 1 works in two parts. First it records the current best singleton of the stream data. Next, when element e_i and its $\boldsymbol{b}(e_i)$ copies arrive, we utilize the binary search with threshold $\tau = \frac{v}{1+d}$ to get a appropriate l_i which satisfies

$$\frac{f(l_i \mathcal{X}_{e_i} | \boldsymbol{x})}{\boldsymbol{c}_j^T (l_i \mathcal{X}_{e_i})} \ge \frac{v}{1+d}, \text{ for any } j \in [d],$$

and

$$\frac{f((l_i + 1)\mathcal{X}_{e_i} | \boldsymbol{x})}{\boldsymbol{c}_j^T ((l_i + 1)\mathcal{X}_{e_i})} < \frac{v}{1+d}, \text{ for some } j \in [d].$$

If it satisfies the d-knapsack constraints, we add $l_i \mathcal{X}_{e_i}$ to the current solution $\boldsymbol{x}$. Finally, it outputs the best value of the above two parts.

Lemma 1. *Given vectors $\boldsymbol{x}, \boldsymbol{y}$ and $\boldsymbol{x}_i \in \boldsymbol{Z}_+^E, i = 1, 2, ..., n$ with $\boldsymbol{y} \le \boldsymbol{x} \le \boldsymbol{b}$, then we have*

$$f(\boldsymbol{x} + \sum_{i=1}^n \boldsymbol{x}_i) - f(\boldsymbol{x}) \le \sum_{i=1}^n [f(\boldsymbol{y} + \boldsymbol{x}_i) - f(\boldsymbol{y})]. \tag{2}$$

Algorithm 1. Streaming-Know-OPT

Require: function f, cost c_j, ground set E and v such that $\lambda OPT \leq v \leq OPT$
1: $\boldsymbol{x} \leftarrow 0, e_{max} \leftarrow \emptyset, \boldsymbol{b}(e_{max}) \leftarrow 0$
2: **for** $i = 1, 2, ..., n$ **do**
3:     **if** $f(\boldsymbol{b}(e_i)\mathcal{X}_{e_i}) > f(\boldsymbol{b}(e_{max})\mathcal{X}_{e_{max}})$ **then**
4:         $e_{max} \leftarrow e_i, \boldsymbol{b}(e_{max}) \leftarrow \boldsymbol{b}(e_i)$
5:     **end if**
6:     **if** $c_j^T \boldsymbol{x} \leq 1, j \in [d]$ **then**
7:         $l_i \leftarrow$ Binary Search $(f, \boldsymbol{x}, \boldsymbol{b}, e_i, c_j, d, \frac{v}{1+d})$
8:         **if** $c_j^T (\boldsymbol{x} + l_i\mathcal{X}_{e_i}) \leq 1, j \in [d]$ **then**
9:             $\boldsymbol{x} \leftarrow \boldsymbol{x} + l_i\mathcal{X}_{e_i}$
10:        **end if**
11:    **end if**
12: **end for**
13: **return** $\hat{\boldsymbol{x}} \leftarrow \arg\max(f(\boldsymbol{b}(e_{max})\mathcal{X}_{e_{max}}), f(\boldsymbol{x}))$

Algorithm 2. Binary Search

Require: function f, $\boldsymbol{x}$, $\boldsymbol{b}$, e_i, c_j, d, τ
1: $l_s \leftarrow 1, l_t \leftarrow (\boldsymbol{b}(e) - \boldsymbol{x}(e))$
2: **if** $\frac{f(l_t\mathcal{X}_{e_i}|\boldsymbol{x})}{c_j^T(l_t\mathcal{X}_{e_i})} \geq \tau$ **then**
3:     **return** l_t
4: **end if**
5: **if** $\frac{f(\mathcal{X}_{e_i}|\boldsymbol{x})}{c_j(e_i)} < \tau$ **then**
6:     **return** 0
7: **end if**
8: **while** $l_t > l_s + 1$ **do**
9:     $m = \lfloor \frac{l_s+l_t}{2} \rfloor$
10:    **if** $\frac{f(m\mathcal{X}_{e_i}|\boldsymbol{x})}{c_j^T(m\mathcal{X}_{e_i})} \geq \tau$ **then**
11:        $l_s = m$
12:    **else**
13:        $l_t = m$
14:    **end if**
15: **end while**
16: **return** l_s

Lemma 2. *Let $\boldsymbol{x}$ be the vector returned by line 9 of Algorithm 1 in the last iteration and $\boldsymbol{x}^*$ be the optimal solution, then there must exist $\lambda(e) \in [d]$ such that*

$$f(\mathcal{X}_e \mid \boldsymbol{x}) < \frac{c_{\lambda(e)}(e)v}{1+d}, \text{ for } e \in \{\boldsymbol{x}^*\} \setminus \{\boldsymbol{x}\}.$$

Lemma 3. *Denote $\boldsymbol{x}$ be the vector returned by line 9 of Algorithm 1 in the last iteration, then we have*

$$f(\boldsymbol{x}) \geq \frac{\lambda(1-\alpha)}{1+d}OPT,$$

where α is the upper bound for the cost of each item in the stream.

Proof. Let x_i be the vector x at the end of iteration i of Algorithm 1 with an initial solution $x_0 = 0$. We assume that $supp^+\{x_i\} = \{e_1, e_2, ..., e_i\}$. Let l_i be the output of Algorithm 2 when e_i and its copies arrive. Then x_i can be denoted as $x_i = l_1 \mathcal{X}_{e_1} + ... + l_i \mathcal{X}_{e_i}$. According to Algorithm 1, we have

$$\frac{f(l_i \mathcal{X}_{e_i} \mid x_{i-1})}{c_j^T(l_i \mathcal{X}_{e_i})} \geq \frac{v}{1+d}, \text{ for } j \in [d].$$

Then for any $j \in [d]$, we can obtain

$$
\begin{aligned}
f(x) &= f(l_1 \mathcal{X}_{e_1} + ... + l_n \mathcal{X}_{e_n}) \\
&= f(x_n) - f(x_{n-1}) + f(x_{n-1}) - f(x_{n-2}) + ... + f(x_1) - f(x_0) \\
&= \sum_{t=1}^{n} f(l_t \mathcal{X}_{e_t} \mid x_{t-1}) \\
&\geq \sum_{t=1}^{n} c_j^T(l_t \mathcal{X}_{e_t}) \cdot \frac{v}{1+d} \\
&= c_j^T x \frac{v}{1+d}.
\end{aligned}
$$

Denote $\gamma = \max_{j \in [d]} c_j^T x$ be the maximal cost of x in d knapsacks. Thus we have $f(x) \geq \frac{v}{1+d}\gamma$.

Next, we split the proof into two cases according to the value of γ.

Case 1. We consider the case $\gamma \geq (1 - \alpha)$. Then we have

$$f(x) \geq \frac{\gamma v}{1+d} \geq \frac{(1-\alpha)v}{1+d} \geq \frac{\lambda(1-\alpha)}{1+d}OPT.$$

Case 2. We consider the case $\gamma < (1 - \alpha)$. Denote x^* be the optimal solution of Problem (1). It is clear that for any $e \in \{x^*\}\backslash\{x\}$, $x + \mathcal{X}_e$ is a feasible solution. By Lemma 2 we can see there must exist $\lambda(e) \in [d]$ satisfying $f(\mathcal{X}_e \mid x) < \frac{c_{\lambda(e)}(e)v}{1+d}$ for any $e \in \{x^*\} \backslash \{x\}$. Define $E_j^* = \{e | e \in \{x^*\} \backslash \{x\} \wedge \lambda(e) = j\}$ and $x_j^* = \sum_{e \in E_j^*} \mathcal{X}_e$. We have

$$
\begin{aligned}
f(x + x_j^*) - f(x) &\leq \sum_{e \in \{x_j^*\}} f(\mathcal{X}_e \mid x) \leq \sum_{e \in \{x_j^*\}} \frac{c_j(e)v}{1+d} \\
&= \frac{v}{1+d} c_j^T x_j^* \leq \frac{v}{1+d},
\end{aligned}
$$

where the last inequality follows from the fact that $K_j = 1$ for $j \in [d]$.

Then, by Lemma 1 and the fact that $\{x^*\} \setminus \{x\} = \cup_{j=1}^{d}\{x_j^*\}$, i.e. $(x^* - x) \vee 0 = \sum_{j=1}^{d} x_j^*$, we have

$$f(x + x^*) - f(x) \leq \sum_{j=1}^{d}(f(x + x_j^*) - f(x)) \leq d \cdot \frac{v}{1 + d}. \tag{3}$$

Due to the monotonicity of f, we have $f(x + x^*) \geq f(x^*) = OPT$. By rearranging inequality (3) and utilizing the assumption that $v \leq OPT$, we have $f(x) \geq \frac{1}{1+d}OPT$.

From the above two cases, we have

$$f(x) \geq \min\{\frac{\lambda(1 - \alpha)}{1 + d}, \frac{1}{1 + d}\}OPT = \frac{\lambda(1 - \alpha)}{1 + d}OPT.$$

Lemma 4. *When $\alpha > 0.5$, we have $f(\hat{x}) \geq \frac{\lambda}{2(1+d)}OPT$, where $\hat{x}$ is the final output returned by Algorithm 1 and α is the upper bound for the cost of each item in the stream.*

Proof. According to the relation between γ and 0.5, we divide the proof into two cases.

Case 1. When $\gamma \geq 0.5$, it is obvious that $\gamma \geq (1 - \alpha)$. The problem is transformed into Case 1 in Lemma 3. Thus we have

$$f(x) \geq \frac{\gamma v}{1 + d}OPT \geq \frac{\lambda}{2(1 + d)}OPT.$$

Case 2. When $\gamma < 0.5$, the following proof proceeds in two cases depending on the reason why $e \in \{x^*\} \setminus \{x\}$ is not added to x. One is there exist some $j \in [d]$ such that e does not pass the threshold condition in Algorithm 1, the other is that it does not satisfy the d-knapsack constraints when it is added to x.

1. When $\gamma < (1 - \alpha)$, for $e \in \{x^*\} \setminus \{x\}$, $x + \mathcal{X}_e$ satisfies the d-knapsack constraints. Thus there must exist some $j \in [d]$ such that $f(\mathcal{X}_e|x_e) < \frac{c_j(e)v}{1+d}$, which is similar to the Case 2 in Lemma 3. Then we have

$$f(x) \geq \frac{1}{1 + d}OPT.$$

2. $e \in \{x^*\} \setminus \{x\}$ satisfies the threshold condition in Algorithm 1, but adding it breaks the d-knapsack constraints. Then for some $j \in [d]$ we have

$$f(l_e\mathcal{X}_e \mid x_e) \geq \frac{c_j^T(l_e\mathcal{X}_e)v}{1 + d}, \tag{4}$$

and

$$c_j^T(x_e + l_e\mathcal{X}_e) > 1. \tag{5}$$

By rearranging inequality (4), we have

$$f(\boldsymbol{x}_e + l_e \mathcal{X}_e) \geq f(\boldsymbol{x}_e) + f(l_e \mathcal{X}_e \mid \boldsymbol{x}_e)$$
$$\geq \frac{v}{1+d}(\boldsymbol{c}_j^T \boldsymbol{x}_e + \boldsymbol{c}_j^T(l_e \mathcal{X}_e))$$
$$\geq \frac{v}{1+d},$$

where the second inequality holds due to the definition of $\boldsymbol{x}_e$ and inequality (4), the last inequality holds due to the inequality (5).

Then using the monotonicity and DR-submodularity of f, we have

$$\frac{v}{1+d} \leq f(\boldsymbol{x}_e + l_e \mathcal{X}_e) \leq f(\boldsymbol{x}_e) + f(l_e \mathcal{X}_e) \leq f(\boldsymbol{x}) + f(\boldsymbol{b}(e_{max})\mathcal{X}_{emax}).$$

Thus we have $f(\boldsymbol{x}) \geq \frac{v}{2(1+d)}$ or $f(\boldsymbol{b}(e_{max})\mathcal{X}_{emax}) \geq \frac{v}{2(1+d)}$.

Consequently, from the above two cases we have $f(\hat{\boldsymbol{x}}) = \max\{f(\boldsymbol{x}), f(\boldsymbol{b}(e_{max}) \mathcal{X}_{emax})\} \geq \frac{v}{2(1+d)} \geq \frac{\lambda}{2(1+d)}OPT$.

By applying Lemma 3 and Lemma 4, we obtain the following result.

Theorem 1. *Denote $\hat{\boldsymbol{x}}$ be the final solution returned by Algorithm 1, for any $\lambda \in (0, 1]$ we have*

$$f(\hat{\boldsymbol{x}}) \geq \begin{cases} \dfrac{\lambda(1-\alpha)}{1+d}OPT, & \alpha \leq 0.5, \\[2ex] \dfrac{\lambda}{2(1+d)}OPT, & \alpha > 0.5, \end{cases} \tag{6}$$

where OPT is the optimal solution of Problem (1) and α is the upper bound for the cost of each item in the stream.

Theorem 2. *Algorithm 1 requires only one pass, at most β^{-1} space and $O(\log \|\boldsymbol{b}\|_\infty)$ update time per element, where β is the lower bound for the cost of each item in the stream.*

3.2   The Threshold-Streaming Algorithm

Algorithm 1 works on the assumption that OPT is known, which is unrealistic. To solve this problem, we utilize a traditional method to estimate OPT. When item e_i and its copies arriving, we use sequence $V_i = \{(1+\epsilon)^s \mid s \in \mathbf{Z}, m \leq (1+\epsilon)^s \leq M(1+d)\}$ to estimate OPT. For each $v \in V_i$, Algorithm 3 runs in parallel and finally outputs the best solution.

Lemma 5. *There must exist a guess $v \in V$ such that $(1-\epsilon)OPT \leq v \leq OPT$.*

Obviously, Lemmas 3 and 4 also work for Algorithm 3. Thus, by Lemmas 3, 4 and 5, we get the following results.

Algorithm 3. Threshold-Streaming/DR-Submodular

Require: function f, cost c_j, ground set E, $\boldsymbol{b}$ and $\epsilon \in (0,1)$
1: $V = \{(1+\epsilon)^s | s \in \mathbf{Z}\}$
2: for each $v \in V$, set $\boldsymbol{x}^v \leftarrow \boldsymbol{0}$
3: Initialize $m, M \leftarrow 0$ and $e_{max} \leftarrow \emptyset, \boldsymbol{b}(e_{max}) \leftarrow 0$
4: **for** $i = 1, 2, ..., n$ **do**
5:      **if** $f(\boldsymbol{b}(e_i)\mathcal{X}_{e_i}) > f(\boldsymbol{b}(e_{max})\mathcal{X}_{e_{max}})$ **then**
6:          $e_{max} \leftarrow e_i, \boldsymbol{b}(e_{max}) \leftarrow \boldsymbol{b}(e_i)$
7:      **end if**
8:      $\alpha_i = \max_{\forall j \in [d]} c_{ij}, \beta_i = \min_{\forall j \in [d]} c_{ij}$
9:      **if** $\frac{f(\mathcal{X}_{e_i})}{\beta_i} > M$ **then**
10:          $M \leftarrow \frac{f(\mathcal{X}_{e_i})}{\beta_i}$ and $m \leftarrow f(\mathcal{X}_{e_i})$
11:          $V_i = \{(1+\epsilon)^s | s \in \mathbf{Z}, m \le (1+\epsilon)^s \le M(1+d)\}$
12:          Delete $\boldsymbol{x}^v$ if $v \notin V_i$
13:      **end if**
14:      **for all** $v \in V_i$ **do**
15:          $l_i \leftarrow$ Binary Search $(f, \boldsymbol{x}^v, \boldsymbol{b}, e_i, c_j, d, \frac{v}{1+d})$
16:          **if** $c_j(\boldsymbol{x}^v + l_i\mathcal{X}_{e_i}) \le 1, j \in [d]$ **then**
17:             $\boldsymbol{x}^v \leftarrow \boldsymbol{x}^v + l_i\mathcal{X}_{e_i}$
18:          **end if**
19:      **end for**
20: **end for**
21: $\boldsymbol{x} \leftarrow \arg\max_{v \in V_n} f(\boldsymbol{x}^v)$
22: **return** $\tilde{\boldsymbol{x}} \leftarrow \arg\max(f(\boldsymbol{b}(e_{max})\mathcal{X}_{e_{max}}), f(\boldsymbol{x}))$

Theorem 3. *Denote $\tilde{\boldsymbol{x}}$ be the final solution returned by Algorithm 3, then we have $f(\tilde{\boldsymbol{x}}) \ge \frac{1-\theta}{1+d}OPT$, where $\theta = \min(\alpha + \epsilon, 0.5 + \epsilon)$ and α is the upper bound for the cost of each item in the stream.*

Theorem 4. *Algorithm 3 requires only one pass, at most $O(\frac{\log(d\beta^{-1})}{\beta\epsilon})$ space and $O(\frac{\log(d\beta^{-1})}{\epsilon} \log \|\boldsymbol{b}\|_\infty)$ update time per element, where β is the lower bound for the cost of each item in the stream.*

Proof. For each item e_i, the number of parameters v in V_i is $\lceil \log_{(1+\epsilon)} \beta^{-1}(1+d) \rceil$. In the final solution, at most β^{-1} items can be selected, thus the memory of Algorithm 3 is $O(\frac{\log(d\beta^{-1})}{\beta\epsilon})$. Further, for each item e, the time complexity of Algorithm 2 is $O(\log \|\boldsymbol{b}\|_\infty)$. Thus, the query complexity of Algorithm 3 is $O(\frac{\log(d\beta^{-1})}{\epsilon} \log \|\boldsymbol{b}\|_\infty)$ update time per element.

Algorithm 4. Accelerated-Streaming/DR-Submodular

Require: function f, cost c_j, ground set E, b and $\epsilon \in (0,1)$
1: $V = \{(1+\epsilon)^s | s \in \mathbf{Z}\}$
2: for each $v \in V$, set $\boldsymbol{x}^v \leftarrow \mathbf{0}$
3: Initialize $m, M, \varphi, \omega \leftarrow 0$ and $e_{max} \leftarrow \emptyset, \boldsymbol{b}(e_{max}) \leftarrow 0$
4: **for** $i = 1, 2, ..., n$ **do**
5:     **if** $f(\boldsymbol{b}(e_i)\mathcal{X}_{e_i}) > f(\boldsymbol{b}(e_{max})\mathcal{X}_{e_{max}})$ **then**
6:         $e_{max} \leftarrow e_i, \boldsymbol{b}(e_{max}) \leftarrow \boldsymbol{b}(e_i)$
7:     **end if**
8:     $\alpha_i = \max_{\forall j \in [d]} c_{ij}, \beta_i = \min_{\forall j \in [d]} c_{ij}$
9:     **if** $\frac{f(\mathcal{X}_{e_i})}{\beta_i} > M$ **then**
10:         $M \leftarrow \frac{f(\mathcal{X}_{e_i})}{\beta_i}, m \leftarrow f(\mathcal{X}_{e_i})$ and $\varphi \leftarrow \max\{m, \omega\}$
11:         $V_i = \{(1+\epsilon)^s | s \in \mathbf{Z}, \varphi \leq (1+\epsilon)^s \leq M(1+d)\}$
12:         Delete $\boldsymbol{x}^v$ if $v \notin V_i$
13:     **end if**
14:     **for all** $v \in V_i$ **do**
15:         $l_i \leftarrow$ Binary Search $(f, \boldsymbol{x}^v, \boldsymbol{b}, e_i, c_j, d, \frac{v}{1+d})$
16:         **if** $c_j(\boldsymbol{x}^v + l_i\mathcal{X}_{e_i}) \leq 1, j \in [d]$ **then**
17:             $\boldsymbol{x}^v \leftarrow \boldsymbol{x}^v + l_i\mathcal{X}_{e_i}$
18:         **end if**
19:     **end for**
20:     $\omega \leftarrow \max_{v \in V_i}\{\omega, f(\boldsymbol{x}^v)\}$
21: **end for**
22: $\boldsymbol{x} \leftarrow \arg\max_{v \in V_n} f(\boldsymbol{x}^v)$
23: **return** $\tilde{\boldsymbol{x}} \leftarrow \arg\max(f(\boldsymbol{b}(e_{max})\mathcal{X}_{e_{max}}), f(\boldsymbol{x}))$

---

4   Accelerated-Streaming Algorithm for DR-Submodular Maximization

In this section, we propose a new algorithm by modifying Algorithm 3 to improve the memory complexity. In each iteration, we reduce the sequence of OPT estimates, namely V_i, so as to reduce the memory. The specific process of streaming algorithm is as follows.

Theorem 5. *For any $\epsilon \in (0,1)$, Algorithm 4 is a one pass $\frac{1-\theta}{1+d}$-approximation streaming algorithm with an improved memory $O(\frac{d}{\beta\epsilon})$ and $O(\frac{\log(d\beta^{-1})}{\epsilon} \log \|\boldsymbol{b}\|_\infty)$ update time per element, where $\theta = \min(\alpha + \epsilon, 0.5 + \epsilon)$ and α, β are the upper and lower bounds for the cost of each item in the stream, respectively.*

References

1. Badanidiyuru, A., Mirzasoleiman, B., Karbasi, A., Krause, A.: Streaming submodular maximization: massive data summarization on the fly. In: 20th ACM SIGKDD International Conference on Knowledge Discovery and Data Mining, pp. 671–680. Association for Computing Machinery, New York (2014)

2. Buchbinder, N., Feldman, M., Schwartz, R.: Online submodular maximization with preemption. In: 26th ACM-SIAM Symposium on Discrete Algorithms, pp. 1202–1216. Society for Industrial and Applied Mathematics, Cambridge (2014)
3. Kempe, D., Kleinberg, J., Tardos, É.: Maximizing the spread of influence through a social network. In: 9th ACM SIGKDD International Conference on Knowledge Discovery and Data Mining, pp. 137–146. Association for Computing Machinery, Washington (2003)
4. Guo, J., Wu, W.: Adaptive influence maximization: if influential node unwilling to be the seed. ACM Trans. Knowl. Discov. Data 15(5), 1–23 (2021)
5. Guo, J., Wu, W.: Continuous profit maximization: a study of unconstrained Dr-submodular maximization. IEEE Trans. Comput. Soc. Syst. 8(3), 768–779 (2021)
6. Huang, C., Kakimura, N.: Improved streaming algorithms for maximising monotone submodular functions under a knapsack constraint. Algorithmica 83(3), 879–902 (2021). https://doi.org/10.1007/s00453-020-00786-4
7. Kapralov, M., Post, I., Vondrák, J.: Online submodular welfare maximization: greedy is optimal. In: 24th ACM-SIAM Symposium on Discrete Algorithms, pp. 1216–1225. Society for Industrial and Applied Mathematics, New Orleans (2013)
8. Kazemi, E., Mitrovic, M., Zadimoghaddam, M., Lattanzi, S., Karbasi, A.: Submodular streaming in all its glory: Tight approximation, minimum memory and low adaptive complexity. In: 36th International Conference on Machine Learning, pp. 3311–3320. International Machine Learning Society, Long Beach (2019)
9. Krause, A., Leskovec, J., Guestrin, C., VanBriesen, J., Faloutsos, C.: Efficient sensor placement optimization for securing large water distribution networks. J. Water Resour. Plan. Manag. 134(6), 516–526 (2008)
10. Norouzi-Fard, A., Tarnawski, J., Mitrovic, S., Zandieh, A., Mousavifar, A., Svensson, O.: Beyond 1/2-approximation for submodular maximization on massive data streams. In: 35th International Conference on Machine Learning, pp. 3829–3838. International Machine Learning Society, Stockholm (2018)
11. Soma, T., Kakimura, N., Inaba, K., Kawarabayashi, K.-I.: Optimal budget allocation: theoretical guarantee and efficient algorithm. In: 31th International Conference on Machine Learning, pp. 351–359. International Machine Learning Society, Beijing (2014)
12. Soma, T., Yoshida, Y.: Maximizing monotone submodular functions over the integer lattice. Math. Program. 172(1), 539–563 (2018). https://doi.org/10.1007/s10107-018-1324-y
13. Wang, Y., Li, Y., Tan, K.L.: Efficient representative subset selection over sliding windows. IEEE Trans. Knowl. Data Eng. 31(7), 1327–1340 (2018)
14. Wolsey, L.: Maximising real-valued submodular set function: primal and dual heuristics for location problems. Math. Oper. Res. 7(3), 410–425 (1982)
15. Yu, Q., Xu, L., Cui, S.: Streaming algorithms for news and scientific literature recommendation: monotone submodular maximization with a d-Knapsack constraint. IEEE Access 6, 53736–53747 (2018)

Stochastic Submodular Probing
with State-Dependent Costs

Shaojie Tang[✉] [ID]

Naveen Jindal School of Management, University of Texas at Dallas, Richardson, USA
shaojie.tang@utdallas.edu

Abstract. In this paper, we study a new stochastic submodular maximization problem with state-dependent costs and rejections. The input of our problem is a budget constraint B, and a set of items whose states (i.e., the marginal contribution and the cost of an item) are drawn from a known probability distribution. The only way to know the realized state of an item is to probe that item. We allow rejections, i.e., after probing an item and knowing its actual state, we must decide immediately and irrevocably whether to add that item to our solution or not. Our objective is to sequentially probe/select a best group of items subject to a budget constraint on the total cost of the selected items. We present a constant approximate solution to this problem. We show that our solution can be extended to an online setting.

Keywords: Stochastic submodular maximization · State-dependent costs · Knapsack constraints

1 Introduction

In this paper, we study a new stochastic submodular maximization problem. We introduce the state-dependent item costs and rejections into the classic stochastic submodular maximization problem. The input of our problem is a budget constraint B, and a set of items whose states are drawn from a known probability distribution. The marginal contribution and the cost of an item is dependent on its actual state. We must probe an item in order to reveal its actual state. After probing an item and knowing its actual state, one must decide immediately and irrevocably whether to add that item to our solution or not. Our objective is to sequentially probe/select a best group of items subject to a budget constraint on the total cost of the selected items. We present a constant approximate solution to this problem. Perhaps surprisingly, our algorithm also applies to an online setting described as follows: suppose there is a sequence of items arriving in an adversarial order, on the arrival of an item, we must decide immediately and irrevocably whether to select it or not after seeing its realization. For this online decision problem, our algorithm achieves the same approximation ratio as obtained under the offline setting.

© Springer Nature Switzerland AG 2021
W. Wu and H. Du (Eds.): AAIM 2021, LNCS 13153, pp. 170–178, 2021.
https://doi.org/10.1007/978-3-030-93176-6_15

Related Works. Stochastic submodular maximization has been extensively studied recently [5,6,8]. However, most of existing works assume that the cost of an item is deterministic and pre-known. We relax this assumption by introducing the state-dependent item cost. In particular, we assume that the actual cost of an item is decided by its realized state. We must probe an item in order to know its state. When considering linear objective function, our problem reduces to the stochastic knapsack problem with rejections [9]. [9] gave a constant approximate algorithm for this problem. Recently, [7] studied the stochastic submodular maximization problem with performance-dependent costs, however, their model does not allow rejections. Therefore, our problem does not coincide with their work. Moreover, it is not immediately clear how to extend their algorithm to online setting. Our work is also closely related to submodular probing problem [1], however, they assume each item has only two states, i.e., active or inactive, we relax this assumption by allowing each item to have multiple states and the item cost is dependent on its state. Furthermore, their model does not allow rejections, i.e., one can not reject an active item after it has been probed.

2   Preliminaries and Problem Formulation

Lattice-Submodular Functions. Let $[I] = \{1, 2, \cdots, I\}$ be a set of items and $[S] = \{1, 2, \cdots, S\}$ be a set of states. Given two vectors $u, v \in [S]^{[I]}$, $u \leq v$ means that $u(i) \leq v(i)$ for all $i \in [I]$. Define $(u \vee v)(i) = \max\{u(i), v(i)\}$ and $(u \wedge v)(i) = \min\{u(i), v(i)\}$. For each $i \in [I]$, define $\mathbf{1}_i$ as the vector that has a 1 in the i-th coordinate and 0 in all other coordinates. A function $f : [S]^{[I]} \to \mathbb{R}_+$ is called *monotone* if $f(u) \leq f(v)$ holds for any $u, v \in [S]^{[I]}$ such that $u \leq v$, and f is called *lattice submodular* if $f(u \vee s\mathbf{1}_i) - f(u) \geq f(v \vee s\mathbf{1}_i) - f(v)$ holding for any $u, v \in [S]^{[I]}$, $s \in [S]$, $i \in [I]$.

Items and States. We let vector $\Phi \in [S]^{[I]}$ denote random states of all items. For each item $i \in [I]$, let $\Phi(i) \in [S]$ denote the random state of item i. Let $\phi(i)$ denote a *realization* of $\Phi(i)$. The state of each item is unknown initially, one must probe an item before observing its realization. We allow rejections, i.e., after probing an item and knowing its state, we must decide immediately and irrevocably whether to pick that item or not. We assume there is a known prior probability distribution $\mathcal{D}_i$ over realizations for each item $i \in [I]$, i.e., $\mathcal{D}_i = \{\Pr[\Phi(i) = s] : s \in [S]\}$. The states of all items are decided independently at random, i.e., ϕ is drawn randomly from the product distribution $\mathcal{D} = \prod_{i \in [I]} \mathcal{D}_i$. For each $(i, s) \in [I] \times [S]$, we use $c_i(s)$ to denote the cost of an item i when its state is s.

Assumption 1. *We assume that $c_i(s) \geq c_i(s')$ for any $i \in I$ and $s, s' \in [S]$ such that $s \geq s'$, i.e., the cost of an item is larger if it is in a "better" state.*

The above assumption can also be found in [7]. For each set of item-state pairs $U \subseteq [I] \times [S]$, we define a vector $u \in [S]^{[I]}$ such that $u(i) = 0$ if $(i, s) \notin U$, otherwise $u(i) = \max\{s \mid (i, s) \in U\}$. Now we are ready to introduce a set function h over a new ground set $[I] \times [S]$: consider an arbitrary set of item-state pairs $U \subseteq [I] \times [S]$, define $h(U) = f(u)$. It is easy to verify that if f is monotone and lattice-submodular, then h is monotone and submodular. Given an $I \times S$ matrix $\mathbf{x}$, we define the multilinear extension H of h as:

$$H(\mathbf{x}) = \sum_{U \subseteq [I] \times [S]} h(U) \prod_{(i,s) \in U} x_{is} \prod_{(i,s) \notin U} (1 - x_{is})$$

The value $H(\mathbf{x})$ is the expected value of $h(R)$ where R is a random set obtained by picking each element $(i, s) \in [I] \times [S]$ independently with probability x_{is}.

Adaptive Policy and Problem Formulation. We model the adaptive strategy of probing/picking items through a policy π. Formally, a policy π is a function that specifies which item to probe/pick next based on the observations made so far. Consider an arbitrary policy π, assume that conditioned on $\Phi = \phi$, π picks a set of items (and corresponding states) $G(\pi, \phi) \subseteq [I] \times [S]$[1]. The expected utility of π is $f(\pi) = \sum_{\phi} \Pr[\Phi = \phi] h(G(\pi, \phi))$. We say a policy π is *feasible* if for any ϕ such that $\Pr[\Phi = \phi] > 0$, $\sum_{(i,s) \in G(\pi, \phi)} c_i(s) \leq B$ where B a budget constraint. Our goal is to identify the best feasible policy that maximizes its expected utility:

$$\max_{\pi} f(\pi) \text{ subject to } \pi \text{ is feasible.}$$

3   Algorithm Design

We next describe our algorithm and analyze its performance. Our algorithm is based on the contention resolution scheme [4], which is proposed in the context of submodular maximization with deterministic item cost. We extend their design by considering state-dependent item cost and rejections. Our algorithm, called StoCan, is composed of two phases.

The first phase is done offline, we use the continuous greedy algorithm (Algorithm 1) to compute a fractional solution over a down monotone polytope. The framework of continuous greedy algorithm is first proposed by [2] in the context of submodular maximization subject to a matroid constraint. In particular, Algorithm 1 maintains an $I \times S$ matrix $\mathbf{y}(t)$, starting with $\mathbf{y}(0) = \mathbf{0}$. Let $R(t)$ contain each (i, s) independently with probability $y_{is}(t)$. For each $(i, s) \in [I] \times [S]$, estimate its weight ω_{is} as follows

$$\omega_{is} = \mathbb{E}[h(R(t) \cup \{(i, s)\})] - \mathbb{E}[h(R(t))]$$

For each pair of i and s, let $p_i(s)$ denote $\Pr[\Phi(i) = s]$ for short. Solve the following linear programming problem **LP** and obtain the optimal solution $\mathbf{x}^{LP}$, then update the fractional solution at round t as $\forall (i, s) \in [I] \times [S], y_{is}(t + \delta) = y_{is}(t) + x_{is}^{LP}$.

---

[1] For simplicity, we only consider deterministic policy. However, all results can be easily extended to random policies.

> **LP:** *Maximize* $\sum_{(i,s)\in[I]\times[S]} \omega_{is} x_{is}$
> **subject to:**
>
> $$\begin{cases} \forall (i,s) \in [I] \times [S] : x_{is} \leq p_i(s) \\ \sum_{(i,s)\in[I]\times[S]} x_{is} c_i(s) \leq B \end{cases}$$

After $1/\delta$ rounds, $\mathbf{y}(1/\delta)$ is returned as the final solution. In the rest of this paper, let $\mathbf{y}$ denote $\mathbf{y}(1/\delta)$ for short.

In the second phase, we implement a simple randomized policy based on $\mathbf{y}$. Our policy randomly picks a policy from π^{small} (Algorithm 2) and π^{large} (Algorithm 3) with equal probability to execute. If π^{small} is picked, we discard all *large* items whose cost is larger than $B/2$ (Line 5 in Algorithm 2), and add the rest of items according to the corresponding distribution in (scaled) $\mathbf{y}$ (Line 8 in Algorithm 2). If Algorithm π^{large} is picked, we discard all *small* items whose cost is no larger than $B/2$ (Line 5 in Algorithm 3), and add the rest of items according to the corresponding distribution in (scaled) $\mathbf{y}$ (Line 8 in Algorithm 3).

Algorithm 1. Continuous Greedy

1: Set $\delta = 1/(IS)^2, t = 0, f(\emptyset) = 0, \mathbf{y}(0) = \mathbf{0}$.
2: **while** $t < 1$ **do**
3:    Let $R(t)$ contain each $(i, s) \in [I] \times [S]$ independently with probability $y_{is}(t)$.
4:    For each $(i, s) \in [I] \times [S]$, estimate $\omega_{is} = \mathbb{E}[h(R(t) \cup \{(i, s)\})] - \mathbb{E}[h(R(t))]$;
5:    Solve the following linear programming problem and obtain the optimal solution $\mathbf{x}^{LP}$
6:
> **LP:** *Maximize* $\sum_{(i,s)\in[I]\times[S]} \omega_{is} x_{is}$
> **subject to:**
>
> $$\begin{cases} \forall (i,s) \in [I] \times [S] : x_{is} \leq p_i(s) \\ \sum_{(i,s)\in[I]\times[S]} x_{is} c_i(s) \leq B \end{cases}$$

7:    Let $y_{is}(t + \delta) = y_{is}(t) + x_{is}^{LP}$;
8:    Increment $t = t + \delta$;
9: **return** $\mathbf{y}(1/\delta)$;

We next provide the main theorem of this paper.

Theorem 1. *Let π^* denote the optimal policy, the expected utility achieved by* StoCan *is at lest $\frac{1-1/e}{16} f(\pi^*)$.*

Before presenting the proof of Theorem 1, we first introduce four preparation lemmas.

Lemma 1. *Let $\mathbf{y}$ denote the fractional solution returned from Algorithm 1, $H(\mathbf{y}) \geq (1 - 1/e)f(\pi^*)$.*

Algorithm 2. π^{small}

1: Set $G = \emptyset$, $i = 1$.
2: **while** $i \le n$ **do**
3:    probe item i and observe its state s
4:    **if** $c_i(s) > B/2$ **then**
5:       $i = i + 1$; {discard all large items}
6:    **else**
7:       **if** the remaining budget is no less than $c_i(s)$ **then**
8:          add (i, s) to G with probability $y_{is}/4p_i(s)$;
9:       $i = i + 1$;
10: **return** G;

Algorithm 3. π^{large}

1: Set $G = \emptyset$, $i = 1$.
2: **while** $i \le n$ **do**
3:    probe item i and observe its state s
4:    **if** $c_i(s) \le B/2$ **then**
5:       $i = i + 1$; {discard all small items}
6:    **else**
7:       **if** the remaining budget is no less than $c_i(s)$ **then**
8:          add (i, s) to G with probability $y_{is}/4p_i(s)$;
9:       $i = i + 1$;
10: **return** G;

Proof. Given π^*, for each item-state pair $(i, s) \in [I] \times [S]$, let y_{is}^* denote the probability that $\Phi(i) = s$ and i is picked by π^*. Clearly, $\forall (i, s) \in [I] \times [S] : y_{is}^* \le p_i(s)$. Moreover, consider a fixed realization ϕ, for each $(i, s) \in [I] \times [S]$, let $1_{i,s}$ be an indicator that $\phi(i) = s$ and i is picked by π^*, we have $\sum_{(i,s) \in [I] \times [S]} 1_{i,s} c_i(s) \le B$, Thus,

$$\mathbb{E}\Big[\sum_{(i,s) \in [I] \times [S]} 1_{i,s} c_i(s) \Big] = \sum_{(i,s) \in [I] \times [S]} \mathbb{E}[1_{i,s}] c_i(s) = \sum_{(i,s) \in [I] \times [S]} y_{is}^* c_i(s) \le B$$

where the expectation is taken over Φ with respect to $\mathcal{D}$. It follows that $\mathbf{y}^*$ is a feasible solution to **LP**. Define 1_{is} as the matrix that has a 1 in the (i, s)-th entry and 0 in all other entries. Let $h_V((i, s)) = h(V \cup \{(i, s)\}) - h(V)$ and $H_{\mathbf{y}(t)}((i, s)) = H(\mathbf{y}(t) \vee 1_{is}) - H(\mathbf{y}(t))$ denote the marginal utility of (i, s) with respect to V and $\mathbf{y}(t)$, respectively. We next bound the increment of $H(\mathbf{y}(t))$ during one step of Algorithm 1.

$$f(\pi^*) \le \min_{V \subseteq [I] \times [S]} \Big(h(V) + \sum_{(i,s) \in [I] \times [S]} y_{is}^* h_V((i, s)) \Big) \tag{1}$$

$$\le H(\mathbf{y}(t)) + \sum_{(i,s) \in [I] \times [S]} y_{is}^* H_{\mathbf{y}(t)}((i, s)) \tag{2}$$

$$\leq H(\mathbf{y}(t)) + \sum_{(i,s)\in[I]\times[S]} x_{is}^{LP} H_{\mathbf{y}(t)}((i,s)) \tag{3}$$

The first inequality is proved in [2]. The third inequality is due to $\mathbf{x}^{LP}$ is an optimal solution to **LP**. Then this lemma follows from the standard analysis on submodular maximization.

Given the fractional solution $\mathbf{y}$ returned from Algorithm 1, we next introduce two new fractional solutions $\overline{\mathbf{y}}$ and $\underline{\mathbf{y}}$. Define $\overline{y}_{is} = y_{is}$ if $c_i(s) \leq B/2$, otherwise, $\overline{y}_{is} = 0$. Define $\underline{y}_{is} = y_{is}$ if $c_i(s) > B/2$, otherwise, $\underline{y}_{is} = 0$. Due to the submodularity of h, we have the following lemma.

Lemma 2. $H(\underline{\mathbf{y}}) + H(\overline{\mathbf{y}}) \geq H(\mathbf{y})$

We next bound the expected utility achieved by π^{small}.

Lemma 3. $f(\pi^{small}) \geq H(\underline{\mathbf{y}})/8$

Proof. Consider a modified version of π^{small} by removing Line 7, that is, after probing an item i and observing its state s, if $c_i(s) \leq B/2$, we select i with probability $y_{is}/4p_i(s)$ regardless of the remaining budget, otherwise, we discard i. Denote by G' the returned solution from the modified π^{small}. It is easy to verify that for each $(i,s) \in [I] \times [S]$ with $c_i(s) \leq B/2$, the probability that (i,s) is included in G' is $p_i(s)y_{is}/4p_i(s) = y_{is}/4$. Notice that since each item i can only have one state, the event that (i,s) is included in G' is not independent from the event that (i,s') is included in G' where s' is a different state from s and $c_i(s') \leq B/2$. However, as shown in Lemma 3.7 in [2], this dependency does not degrade the expected utility, i.e., $\mathbb{E}[h(G')] \geq H(\overline{\mathbf{y}}/4)$. Due to H is concave along any nonnegative direction [2], we have $H(\overline{\mathbf{y}}/4) \geq H(\overline{\mathbf{y}})/4$. It follows that

$$\mathbb{E}[h(G')] \geq H(\overline{\mathbf{y}}/4) \geq H(\overline{\mathbf{y}})/4 \tag{4}$$

Next we focus on proving that

$$f(\pi^{small}) = \mathbb{E}[h(G)] \geq \mathbb{E}[h(G')]/2 \tag{5}$$

This lemma follows from (4) and (5).

Recall that if the remaining budget is no less than $c_i(s)$, π^{small} adds (i,s) to G. Because $\mathbf{y}$ is a feasible solution to **LP**, $\overline{\mathbf{y}}$ is also a feasible solution to **LP**, it implies that $\sum_{(i,s)\in[I]\times[S]} \overline{y}_{is} c_i(s)/4 \leq B/4$. According to Markov's inequality, the probability that the remaining budget is less than $B/2$ is at most $1/2$. Because we assume $c_i(s) \leq B/2$, the probability that the remaining budget is less than $c_i(s)$ is at most $1/2$. Thus, the probability that (i,s) is included in G is at least $y_{is}/8$.

Let $G[i]$ (resp. $G'[i]$) denote all item-state pairs in G (resp. G') that involve items in $[i]$, i.e., $G[i] = G \cap \{(j,s) \mid j \in [i], s \in [S]\}$ and $G'[i] = G' \cap \{(j,s) \mid j \in [i], s \in [S]\}$. We next prove that for any $i \in [I]$,

$$\mathbb{E}[h(G[i]) - h(G[i-1])] \geq \frac{1}{2}\mathbb{E}[h(G'[i]) - h(G'[i-1])] \tag{6}$$

Notice that (6) implies (5) due to $\mathbb{E}[h(G)] = h(\emptyset) + \sum_{i=1}^{n} \mathbb{E}[h(G[i]) - h(G[i-1])] \geq f(\emptyset) + \sum_{i=1}^{n} \frac{1}{2}\mathbb{E}[h(G'[i]) - h(G'[i-1])] = \mathbb{E}[h(G')]/2$.

We first give an lower bound on $\mathbb{E}[h(G[i]) - h(G[i-1])]$. For each $(i,s) \in [I] \times [S]$, let $\mathbf{1}_{(i,s) \in G}$ be the indicator that (i,s) is included in G.

$$\mathbb{E}[h(G[i]) - h(G[i-1])]$$
$$= \sum_{s \in [S]} \mathbb{E}\left[\mathbf{1}_{(i,s) \in G}\big(h(G[i-1] \cup (i,s)) - h(G[i-1])\big)\right]$$
$$\geq \sum_{s \in [S]} \mathbb{E}\left[\mathbf{1}_{(i,s) \in G}\big(h(G'[i-1] \cup (i,s)) - h(G'[i-1])\big)\right]$$
$$\geq \sum_{s \in [S]} \mathbb{E}\left[\mathbf{1}_{(i,s) \in G}\right] \mathbb{E}\left[h(G'[i-1] \cup (i,s)) - h(G'[i-1])\right]$$
$$\geq \sum_{s \in [S]} \frac{y_{is}}{8} \mathbb{E}\left[h(G'[i-1] \cup (i,s)) - h(G'[i-1])\right]$$

The first inequality is due the submodularity of f. The second inequality follows from the same proof (monotonicity part) of Theorem 5 in [7] and Assumption 1. Moreover,

$$\mathbb{E}[h(G'[i]) - h(G'[i-1])] = \sum_{s \in [S]} \frac{y_{is}}{4} \mathbb{E}\left[h(G'[i-1] \cup (i,s)) - h(G'[i-1])\right]$$

Based on the above discussions, we have $\mathbb{E}[h(G[i]) - h(G[i-1])] \geq \frac{1}{2}\mathbb{E}[h(G'[i]) - h(G'[i-1])]$. This finishes the proof of (6) and hence (5).

Now consider the second option π^{large}. In the following lemma, we prove that the expected utility achieved by π^{large} is at least $H(\bar{\mathbf{y}})/8$.

Lemma 4. $f(\pi^{large}) \geq H(\bar{\mathbf{y}})/8$.

Proof. Because $\mathbf{y}$ is a feasible solution to **LP**, $\bar{\mathbf{y}}$ is also a feasible solution to **LP**, it implies that $\sum_{(i,s) \in [I] \times [S]} \bar{y}_{is} c_i(s)/4 \leq B/4$. Since we only consider those (i,s) whose cost is larger than $B/2$, the probability that $G = \emptyset$ is at least $1/2$. Consider any $(i,s) \in [I] \times [S]$, conditioned on $G[i-1] = \emptyset$, the probability that (i,s) is included in G is at least $\bar{y}_{is}/4$. Thus, the probability that (i,s) is included in G is at least $\bar{y}_{is}/8$. Recall that π^{large} only picks large items, G contains at most one item (and its state) due to budget constraint. Thus, the expected utility of π^{large} is at least $f(\pi^{large}) \geq \sum_{(i,s) \in [I] \times [S]} \frac{\bar{y}_{is} h((i,s))}{8}$. Due to the submodularity of h and Lemma 3.7 in [2], we have $\sum_{(i,s) \in [I] \times [S]} \bar{y}_{is} h((i,s))/8 \geq H(\bar{\mathbf{y}}/8)$. Since H is concave along any nonnegative direction [2], we have $H(\bar{\mathbf{y}}/8) \geq H(\bar{\mathbf{y}})/8$, thus $f(\pi^{large}) \geq H(\bar{\mathbf{y}})/8$.

Proof of Theorem 1: Now we are ready to present the proof of Theorem 1. Based on Lemma 3 and 4, we have $f(\pi^{small}) + f(\pi^{large}) \geq \frac{H(\mathbf{y}) + H(\bar{\mathbf{y}})}{8}$. This, together with Lemma 2, implies that $f(\pi^{small}) + f(\pi^{large}) \geq \frac{H(\mathbf{y})}{8}$. Because

$H(\mathbf{y}) \geq (1 - 1/e)f(\pi^*)$ as proved in Lemma 1, we have $f(\pi^{small}) + f(\pi^{large}) \geq \frac{1-1/e}{8}f(\pi^*)$. Since StoCan randomly picks one policy from π^{small} and π^{large} to execute, the expected utility of StoCan is at least $\frac{1-1/e}{16}f(\pi^*)$.

4 Extension to Online Setting: a Variant of Submodular Prophet Inequalities

One nice feature about StoCan is that the implementation of π^{small} and π^{large} does not require any specific order of items. Therefore, StoCan can be implemented in an online setting described as follows: suppose there is a sequence of items arriving with different states, on the arrival of an item, we observe its state and decide immediately and irrevocably whether to select it or not subject to a budget constraint. In this sense, the online version of our problem can be viewed as a variant of the submodular prophet inequalities [3,10]. Our setting differs from theirs in two ways: 1) our utility function is modeled as a lattice submodular function over items and states, and 2) our model incorporates state-dependent cost. Similar to the offline setting, StoCan first computes $\mathbf{y}$ using Algorithm 1 in advance, then randomly picks one policy from π^{small} and π^{large} to execute. Notice that the online version of π^{small} and π^{large} probes the items in order of their arrival. It is easy to verify that this does not affect the performance analysis of StoCan, i.e., our analysis does not rely on any specific order of items, thus StoCan achieves the same approximation ratio as obtained under the offline setting.

5 Conclusion

In this paper, we study the stochastic submodular probing problem with state-dependent costs and rejections. We present a constant approximate solution to this problem. We show that our solution can be implemented in an online fashion.

References

1. Adamczyk, M., Sviridenko, M., Ward, J.: Submodular stochastic probing on matroids. Math. Oper. Res. 41(3), 1022–1038 (2016)
2. Calinescu, G., Chekuri, C., Pál, M., Vondrák, J.: Maximizing a monotone submodular function subject to a matroid constraint. SIAM J. Comput. 40(6), 1740–1766 (2011)
3. Chekuri, C., Livanos, V.: On submodular prophet inequalities and correlation gap. arXiv preprint arXiv:2107.03662 (2021)
4. Chekuri, C., Vondrák, J., Zenklusen, R.: Submodular function maximization via the multilinear relaxation and contention resolution schemes. SIAM J. Comput. 43(6), 1831–1879 (2014)
5. Chen, Y., Krause, A.: Near-optimal batch mode active learning and adaptive submodular optimization. In: ICML, no. 1, vol. 28, pp. 160–168, 8–1 (2013)

6. Fujii, K., Kashima, H.: Budgeted stream-based active learning via adaptive submodular maximization. In: Advances in Neural Information Processing Systems, pp. 514–522 (2016)
7. Fukunaga, T., Konishi, T., Fujita, S., Kawarabayashi, K.I.: Stochastic submodular maximization with performance-dependent item costs. In: Proceedings of the AAAI Conference on Artificial Intelligence, vol. 33, pp. 1485–1494 (2019)
8. Golovin, D., Krause, A.: Adaptive submodularity: theory and applications in active learning and stochastic optimization. J. Artif. Intell. Res. **42**, 427–486 (2011)
9. Gupta, A., Krishnaswamy, R., Molinaro, M., Ravi, R.: Approximation algorithms for correlated knapsacks and non-martingale bandits. In: 2011 IEEE 52nd Annual Symposium on Foundations of Computer Science, pp. 827–836. IEEE (2011)
10. Rubinstein, A., Singla, S.: Combinatorial prophet inequalities. In: Proceedings of the Twenty-Eighth Annual ACM-SIAM Symposium on Discrete Algorithms, pp. 1671–1687. SIAM (2017)

Bi-criteria Adaptive Algorithms
for Minimizing Supermodular Functions
with Cardinality Constraint

Xiaojuan Zhang, Qian Liu, Min Li, and Yang Zhou$^{(\boxtimes)}$

School of Mathematics and Statistics, Shandong Normal University,
Jinan 250014, People's Republic of China
{liminEmily,zhouyang}@sdnu.edu.cn

Abstract. In this paper, we study the adaptability of minimizing a non-increasing supermodular function f with cardinality constraint k. We first propose an algorithm with $\mathcal{O}\left(\log^2 n \cdot \log \frac{f(S_0)}{\epsilon \cdot \mathsf{OPT}}\right)$ adaptive rounds which can return a solution set S with $|S| = |S_0| + O\left(k \log \frac{f(S_0)}{\epsilon \cdot \mathsf{OPT}}\right)$ satisfying $f(S) \leq (1+\epsilon)\mathsf{OPT}$, where S_0 is the initial solution set and OPT is the optimal value. The adaptivity is then improved to $\mathcal{O}\left(\log n \cdot \log \frac{f(S_0)}{\epsilon \cdot \mathsf{OPT}}\right)$ by the second algorithm. The application of the new algorithms to the fuzzy C-means problem is also discussed.

Keywords: Supermodular minimization · Bi-criteria analysis · Adaptive algorithm · Fuzzy C-means

1 Introduction

Supermodular or submodular optimization problem has a wide range of applications in machine learning and data mining. In this paper, we consider the problem of minimizing a non-negative and non-increasing supermodular function under a cardinality constraint, which has been widely applied to many academic areas, such as data clustering [2], economics [9,10] and many others.

Supermodular minimization problem can also be seen as a submodular maximization problem. Variety of algorithms on submodular maximization problem have been raised in recent decades. However, it is impossible to obtain similar approximation results for these algorithms to solve supermodular minimization problem, even under similar additional conditions such as monotonicity, normalization and non-negativity. When considering maximizing a non-negative, monotone and normalized submodular function f under a cardinality constraint, the classical greedy algorithm guarantees an approximate ratio of $(1 - 1/e)$ [3], and it can be proved that this result is tight for polynomial-time algorithms [4].

Supported by National Natural Science Foundation of China (No. 12001335), Shandong Provincial Natural Science Foundation, China (Nos. ZR2020MA029, ZR2019PA004).

W. Wu and H. Du (Eds.): AAIM 2021, LNCS 13153, pp. 179–189, 2021.
https://doi.org/10.1007/978-3-030-93176-6_16

In [2], Liberty and Sviridenko present a generic greedy algorithm for supermodular minimization. It obtains a $(1+\epsilon)$-approximation ratio in the bi-criteria sense, in which the cardinality of solution set exceeds the constraint. When the total curvature (denoted as c) of the function is considered, Sviridenko, Vondrák and Ward [6] present an approximation factor of $1 + \frac{c}{1-c}e^{-1} + \frac{1}{1-c}O(\epsilon)$ for minimizing a non-increasing supermodular function with matroid constraint.

Due to the increasing scales of problems in real applications, adaptive algorithms has been paid more and more attention to in recent decades. For an algorithm, its *adaptive complexity* (or *adaptivity*) denotes the minimum number of sequential rounds required to achieve constant factor approximation, assuming that in each round polynomial-time queries can be executed in parallel. Currently, adaptive algorithms for supermodular minimization is rare, whereas many adaptive algorithms have been proposed for submodular maximization, which could provide greatly inspiration to us. It has been shown that there is no algorithm that can be faster than $\mathcal{O}(\log n)$ adaptivity with a constant approximate ratio for submodular maximization [5]. For the cardinality constraint, Balkanski et al. [1] give an adaptive algorithm whose approximation is arbitrarily close to the optimal $1 - 1/e$ guarantee in $\mathcal{O}(\log n)$ adaptive rounds. Ene et al. [7] give a near-optimal algorithm that reaches $1 - 1/e - \epsilon$ approximation in $O(\log n/\epsilon^2)$ rounds of adaptivity. For the packing constraint, Chekuri et al. [8] propose adaptive algorithms which achieves a near-optimal $(1 - 1/e - \epsilon)$-approximation in $O(\log^2 m \log n/\epsilon^4)$ rounds, where n is the cardinality of the ground set and m is the number of packing constraints.

1.1 Main Contribution and Structure

In this paper we aim to explore an adaptive algorithm to solve the supermodular minimization problem with a cardinality constraint. According to the idea in [1], we add enough many elements in each round such that the algorithm can be executed within a logarithmic number of rounds. The key difference is that, it is impossible to obtain a constant approximation solution for the supermodular minimization problem. To deal with this difficulty, we allow that the constraint can be violated to some degree. Explicitly, for minimizing the non-increasing non-negative supermodular function f with cardinality constraint k, let S^* be its optimal solution and OPT be its optimal value. We first propose an algorithm that can return a solution set S with $|S| = |S_0| + O\left(k \log \frac{f(S_0)}{\epsilon \cdot \mathrm{OPT}}\right)$ satisfying $f(S) \leq (1+\epsilon)\mathrm{OPT}$ in $O\left(\epsilon^{-3} \log^2 n \cdot \log \frac{f(S_0)}{\epsilon \cdot \mathrm{OPT}}\right)$ adaptive rounds, where S_0 is an arbitrary initial solution set. Then the adaptivity is improved to $O\left(\epsilon^{-2} \log n \cdot \log \frac{f(S_0)}{\epsilon \cdot \mathrm{OPT}}\right)$ by the second algorithm.

We apply the novel algorithms to the fuzzy C-means problem, which is a well known soft clustering model with wide applications [11,12]. In the fuzzy C-means problem, the concept of membership degree is introduced to describe a fuzzy belonging of data point to every cluster. The algorithms could get a solution with $(4 + \epsilon)$-approximation guarantee with $O\left(\log \frac{k^{m-1}}{\epsilon}\right)$ cardinality violation, where m is the fuzzifier parameter.

The main structure of the paper is as follows. In Sect. 2, we give a brief notation introduction and propose a nonparallel threshold extension algorithm for further analysis.

In Sect. 3, we give a parallel algorithm named Iteration Threshold Extension (ITE) with violation of the cardinality constraint. In the outer loop of ITE we uniformly select k/ω elements from a candidate set X which is generated by a threshold subroutine each time, and continuously update the current solution S until enough iterations are completed. While for the inner loop of threshold subroutine, we discard elements that do not meet the threshold condition by judging the marginal contribution value of the set R on the current solution S.

In Sect. 4, we put forward the concept of "Phase" and obtain the so-called Phase Threshold Extension (PTE) Algorithm. In PTE we also add and remove points to make the remaining points X_τ has a upper bound of marginal contribution. Our result is that we can prove PTE gives a solution S such that $f(S) \leq (1 + \epsilon)\,\mathsf{OPT}$ for any given $\epsilon > 0$, and improves the adaptivity to $O\left(\epsilon^{-2} \log n \cdot \log \frac{f(S_0)}{\epsilon \cdot \mathsf{OPT}}\right)$.

In Sect. 5, we creatively proposed the relationship of the optimal solution between the constrained and unconstrained fuzzy C-means and give their related theorems based on the relationship.

2 Preliminaries

2.1 Notations

Given a ground set N with $|N| = n$, an integer k, a supermodular function $f : 2^N \to \mathbb{R}_+$ is defined as follows.

Definition 1. *A function $f : 2^N \to \mathbb{R}_+$ is supermodular if and only if for any $A, B \subseteq N$, there holds $f(A) + f(B) \leq f(A \cup B) + f(A \cap B)$.*

In this paper we also require that the supermodular function f is *non-negative* and *non-increasing* (or *non-increasing*), which means that $f(A) \geq 0$ and $f(A) \geq f(B)$ for all $A \subseteq B \subset N$. The *marginal contribution of a set $B \subseteq N$* to another set $A \subseteq N$ is denoted as $f_A(B) = f(A \cup B) - f(A)$. Also, we denote *marginal contribution of an element $a \in N$ to a set A* as $f_A(a)$ is defined as $f(A \cup \{a\}) - f(A)$. For simplicity, for a function f and an element $x \in N$, we use $f(x)$ instead of $f(\{x\})$. Then the supermodular minimization problem subject to a cardinality constraint we consider in this paper can be written as

$$(SupMMin) \quad \min_{S:|S| \leq k} f(S).$$

In following sections we also denote S^* and OPT as the optimal solution and the optimal value, respectively.

2.2 Threshold Extension Algorithm

For the readability of the paper, we propose a simple non-parallel algorithm as Algorithm 1 in the following, which can be seen as a threshold version of the greedy extension algorithm in [2]. In this algorithm we choose $\beta := \beta(S_0, k)$ elements, in which $\beta(S_0, k)$ is usually greater than k, and S_0 is some initial solution set (maybe $S_0 = \emptyset$) in 2^N.

Algorithm 1: Threshold-Extension (TE)

Input: OPT, k, S_0
Output: S_β

1 $S = S_0$,
2 **for** $i = 0, 1, \ldots, \beta$ **do**
3 $\quad$ Find an element x such that $f(S_i) - f(S_i + x) \geq \frac{1}{k}[f(S_i) - \text{OPT}]$;
4 $\quad$ $S_{i+1} \leftarrow S_i \cup \{x\}$

It can be shown that Algorithm 1 could achieve a $(1 + \epsilon)$-approximation guarantee when properly more elements are selected into the solution set. At a high level, the supermodularity of function f guarantees that the value $f(S) -$ OPT shrinks by a factor of $(1 - 1/k)$ at each iteration. The final approximation guarantee could then be derived iteratively.

Lemma 1. *The output of Algorithm 1 satisfies* $f(S_\beta) \leq (1 - e^{-\beta})\text{OPT} + e^{-\beta}f(S_0)$.

Then we can get the following theorem.

Theorem 1. *The output of the Algorithm 1 satisfies* $f(S_\beta) \leq (1 + \epsilon)\text{OPT}$ *when*

$$\beta = \lceil k \log \frac{f(S_0)}{\epsilon \cdot \text{OPT}} \rceil.$$

See Appendix A for the proofs of Lemma 1 and Theorem 1.

3 Iterative Threshold Extension Algorithm

Now we consider the parallelization of Algorithm 1. Inspired by the idea in [1] which solves the cardinality constrained monotone submodular maximization problem in adaptive rounds, we propose the following algorithms for the supermodular minimization problem. In each loop of Algorithm 2 we add k/ω elements chosen uniformly at random in a set X. The set X is generated by Algorithm 3, at each iteration of which we filter out elements N that cannot provide enough marginal contribution. When the outer loop ends it can be easily obtained that the cardinality of the candidate set S is at least $|S_0| + \frac{k}{1-\epsilon} \log \frac{f(S_0)}{\epsilon \cdot \text{OPT}}$, thus the constraint may be violated. The symbol $R \sim \mathcal{U}(X)$ means that we choose the set R uniformly at random in set X with $|R| \equiv k/\omega$.

Algorithm 2: Iterative-Threshold-Extension (ITE)

Input: OPT, k, ω, S_0
Output: S
1 **for** $\lceil \frac{\omega}{1-\epsilon} \log \frac{f(S_0)}{\epsilon \cdot \text{OPT}} \rceil$ *iterations* **do**
2 $\quad$ $X \longleftarrow$ Threshold(N, S_i, ω)
3 $\quad$ $S_{i+1} \longleftarrow S_i \cup R$, where $R \sim \mathcal{U}(X)$

Since greedy algorithm guarantees that for any given set S there exists an element whose marginal contribution to S is at least a $1/k$ fraction of the remaining optimal value $\text{OPT} - f(S)$. In each round of the subroutine Threshold, as long as the expected value of the marginal contribution of a random subset R to the current solution S is large enough, the algorithm will discard all elements x whose expected value of the marginal contribution of $S \cup (R \backslash \{x\})$ will be higher than the fixed threshold value from set X.

Algorithm 3: Threshold(X, S, ω)

Input: Remaining X, current solution S, OPT, ω
Output: X
1 **while** $\mathbb{E}_{R \sim \mathcal{U}(X)}[f_S(R)] > (1 - \epsilon)(\text{OPT} - f(S))/\omega$ **do**
2 $\quad$ $X \leftarrow X \backslash \{x : \mathbb{E}_{R \sim \mathcal{U}(X)}[f_{S \cup (R \backslash \{x\})}(x)] > (1 + \epsilon/2)(1 - \epsilon)(\text{OPT} - f(S))/k\}$

What should be noticed is that we cannot accurately estimate the expected value of OPT and the expected value involved in the algorithms. In fact, the expected value can be approximated by proper many times of parallelized random sampling, and OPT can be guessed to have its error at most ϵ time in $O(\log \epsilon)$ complexity.

3.1 Theoretical Analysis of ITE

In this subsection we would like to analyze the approximation guarantee and the adaptivity of Algorithm ITE. Different with TE, in ITE we choose k/ω elements instead of one. Similar with the key idea in the proof of TE, the approximation guarantee can also be proved by firstly showing that at each outer loop the value $f(S) - \text{OPT}$ shrinks in a constant factor dependent on k/ω. Inspired by this idea, we give an upper bound of the expected marginal contribution of S^* on $S \cup (\cup_{i=1}^{\tau} R_i)$ with a form relative to the $\text{OPT} - f(S)$.

Lemma 2. *Let $R_i \sim \mathcal{U}(X)$ be the random set at iteration i of Algorithm 3. For all $S \subseteq N$ and $\omega, \tau > 0$, if Threshold has not terminated after τ iterations, then we have*

$$\mathbb{E}_{R_1, \ldots, R_\tau} \left[f_{S \cup (\cup_{i=1}^{\tau} R_i)}(S^*) \right] \leq \left(1 - \frac{\tau}{\omega} \right) \cdot (\text{OPT} - f(S)) \leq 0.$$

From the monotonicity, we can prove that the contribution of the "good" element set X returned by Threshold to the current solution S is arbitrarily close to $|(\mathsf{OPT} - f(S))/\omega|$.

Lemma 3. *For all $S \subseteq N$ and $\epsilon > 0$, if $\omega \geq 20\tau\epsilon^{-1}$, the set of elements X_τ that still survive after τ iterations of* Threshold *satisfies*

$$f_S(X_\tau) \leq \frac{1}{\omega}(1 - \epsilon)(\mathsf{OPT} - f(S)).$$

Here we give a general summary of the proof. Refer to Appendix B for details. We first define a subset P of the optimal solution of OPT. Then we prove that the elements in P still exist after the τ round iteration of Threshold(X, S, ω). Moreover, we observe that the upper bound of the marginal contribution of P in S. Finally, from the monotonicity and $P \subseteq X_\tau$, we can get the set X_τ consisting of the remaining elements after deleting a part of the "bad" points and satisfying $f_S(X_\tau) \leq \frac{1}{\omega}(1 - \epsilon) \cdot (\mathsf{OPT} - f(S))$.

To obtain the adaptivity of ITE, we first limit the number of adaptive rounds of the Threshold subroutine. Lemma 4 proves that elements will be discarded in a certain proportion in each round.

Lemma 4. *Let X_i and X_{i+1} be the remaining elements at the beginning and end of the ith iteration of the Algorithm* Threshold(N, S, ω) *respectively. For all $S \subseteq N$ and $\omega, i, \epsilon > 0$, if* Threshold(X, S, ω) *does not terminate in the ith iteration, then*

$$\frac{|X_i|}{|X_{i+1}|} > 1 + \frac{\epsilon}{2}.$$

A proof sketch is as follows and for details refer to Appendix B. First, we estimate the value of $f(R_i \cap X_{i+1})$ of the surviving elements X_{i+1} in a random set $R_i \sim \mathcal{D}_{X_i}$. we can get $\mathbb{E}\left[f_S(R_i \cap X_{i+1})\right] \leq |X_{i+1}| \cdot \frac{1}{\tau|X_i|} \cdot (1+\epsilon/2)(1-\epsilon) \cdot (\mathsf{OPT} - f(S))$. Next, since elements are discarded, the algorithm shows that the value of a random set must higher, i.e. $\mathbb{E}\left[f_S(R_i)\right] > \frac{1}{\omega}(1 - \epsilon)(\mathsf{OPT} - f(S))$. Then by monotonicity, we obtain $\mathbb{E}\left[f_S(R_i)\right] \leq \mathbb{E}\left[f_S(R_i \cap X_{i+1})\right]$. Finally, by combining the above inequalities and simplification, we get $|X_i| / |X_{i+1}| > (1 + \epsilon/2)$.

Combined with Lemma 3, we can get Lemma 5 that the Threshold subroutine has $O(\log n)$ adaptivity when $\omega \geq 40\epsilon^{-2} \log n$ for all $\epsilon > 0$.

Lemma 5. *For all $S \subseteq N$,* Threshold(N, S, ω) *is $O(\log n)$-adaptive with parameter $\omega \geq 40\epsilon^{-2} \log n$ for all $\epsilon > 0$.*

Now we can prove the main conclusion for ITE. Considering the total number of iterations of the algorithm, we fix parameter $\omega = 40\epsilon^{-2} \log n$. ITE runs up to at most $\lceil \omega \log (f(S_0)/(\epsilon \cdot \mathsf{OPT})) \rceil$ rounds. According to Lemma 3 and Lemma 4 that the inner loop of Threshold has $O(\log n)$ adaptive rounds, the whole algorithm is $\mathcal{O}\left(\log^2 n \cdot \log \frac{f(S_0)}{\epsilon \cdot \mathsf{OPT}}\right)$-adaptive.

Theorem 2. *For any constant $\epsilon > 0$, ITE returns a solution S_η with*

$$|S_\eta| \leq |S_0| + \left\lfloor \frac{k}{1-\epsilon} \log \frac{f(S_0)}{\epsilon \cdot \mathrm{OPT}} \right\rfloor + \frac{k}{\omega},$$

satisfying $f(S_\eta) \leq (1+\epsilon)\,\mathrm{OPT}$. The adaptivity is

$$O\left(\epsilon^{-3} \log^2 n \cdot \log \frac{f(S_0)}{\epsilon \cdot \mathrm{OPT}} \right),$$

where $\omega = 40\epsilon^{-2} \log n$.

See Appendix B for the proofs of Lemma and Theorem in Sect. 3.

4 Phased Threshold Extension Algorithm

In this section, we would like to improve the $\mathcal{O}\left(\log^2 n \cdot \log \frac{f(S_0)}{\epsilon \cdot \mathrm{OPT}} \right)$-adaptivity of ITE to $\mathcal{O}\left(\log n \cdot \log \frac{f(S_0)}{\epsilon \cdot \mathrm{OPT}} \right)$-adaptivity by proposing the Phased Threshold Extension Algorithm (PTE), as is shown in Algorithm 4. The main idea is that we perform fewer element removal processes, while still maintaining the approximate quality as that in the previous section. A concept named "phase" is applied in PTE, in each of whom the value of the solution S is decreased by at least $|(\mathrm{OPT} - f(S))/20|$. PTE will also invoke the subroutine Threshold to add a block of k/ω elements randomly to the current solution S. What is different from ITE is that the second input of Threshold is set to be N at the beginning of each phase but modified as $S \cup T$ otherwise.

Algorithm 4: Phased-Threshold-Extension (PTE)

Input: OPT, k, ω
Output: S

1 $S \leftarrow \emptyset$
2 **for** $\lceil \frac{20}{\epsilon(1-\epsilon)} \log \frac{f(S_0)}{\epsilon \cdot \mathrm{OPT}} \rceil$ phases **do**
3       $X \leftarrow N, T \leftarrow \emptyset$
4   **while** $f_S(T) > (\epsilon/20)(\mathrm{OPT} - f(S))$ and $|S \cup T| < \lceil \frac{k}{1-\epsilon} \log \frac{f(S_0)}{\epsilon \cdot \mathrm{OPT}} \rceil$ **do**
5           $X \leftarrow \mathrm{Threshold}(X, S \cup T, \omega)$
6           $T \leftarrow T \cup R$, where $R \sim \mathcal{U}(X)$
7       $S \leftarrow S \cup T$

4.1 Analysis of the PTE

Like ITE, PTE constantly updates the candidate set X by observing the contribution of point $x \in X$. But it should be noted that, unlike ITE, in any phase of PTE, when we call Threshold(X, S, ω), we need to look at the expected value of the marginal contribution of the random set R on $S \cup T$ rather than S, and remove some elements whose contribution is high enough on $S \cup T \cup (R \backslash \{x\})$ rather than $S \cup (R \backslash \{x\})$. In addition, every time Threshold(X, S, ω) is called, our set $S \cup T$ is always changing, which should be paid more attention to. A phase is composed of multiple iterations of PTE, each of which consists of multiple iterations of Threshold(X, S, ω). Then in a phase, the Threshold(X, S, ω) will end until the gain of the set T on the solution S is not higher than $(\epsilon/20)(\mathtt{OPT} - f(S))$ or the number of $S \cup T$ reaches our upper bound. The number of our phases is at most $20\epsilon^{-1} \log{(f(S_0)/(\epsilon \cdot \mathtt{OPT}))}$, which is a constant with respect to n and k.

We first need to introduce some new notation. We say that An element *survives τ iterations of phase* at phase j if it has not been discard at iteration i of Threshold with $i \in [\tau]$. Let S_j denote the solution S at phase $j \in \left[20\epsilon^{-1} \log{(f(S_0)/(\epsilon \cdot \mathtt{OPT}))} \right]$, S_j^+ denote $S_j \cup T$ during the last iteration of PTE during phase j, i.e., the last T such that $f_S(T) < (\epsilon/20)(\mathtt{OPT} - f(S_j))$, then $S_j \subseteq S_j^+ \subseteq S_{j+1}$. Let $S_{j,i}$ denote $S_j \cup T$ at the i-th iteration of thresholding during phase j, so for all $i_1 < i_2$, we have $S_{j,i_1} \subseteq S_{j,i_2} \subseteq S_j^+$. Thus, we can get

$$f\left(S_j^+\right) - f(S_j) > (\epsilon/20)(\mathtt{OPT} - f(S_j))$$

by the algorithm.

For any given phase, similar to the previous section, we want to analyze the marginal contribution of S^* to $S_j^+ \cup (\cup_{i=1}^{\tau} R_i)$, which can be arbitrarily close to the desired value $|\mathtt{OPT} - f(S)|$, where the random sets $\{R_i\}_{i=1}^{\tau}$ is selected from corresponding current remaining elements X when there are τ filtering iterations during the phase. However, different from the analysis process in the previous section, in each phase, since the elements in set T are constantly updated, attention should be paid more to the changes in set $S \cup T$ when judging whether to discard points and how to remove points.

Lemma 6. *For any phase j and $\epsilon > 0$, for all $\omega, \tau > 0$, if phase j has not ended after τ iterations of* Threshold, *then we have*

$$\mathop{\mathbb{E}}_{R_1, \ldots, R_\tau} \left[f_{S_j^+ \cup (\cup_{i=1}^{\tau} R_i)}(S^*) \right] \leq \left(1 - \frac{\tau}{\omega} - \frac{\epsilon}{20} \right) \cdot (\mathtt{OPT} - f(S_j)),$$

where $R_i \sim \mathcal{U}(X)$ is the random set at filtering iteration during phase j with $i \in [\tau]$.

Next, during a phase, similar to Lemma 3, we bound on the contribution value of the remaining elements set X_τ to the current solution S_j^+, where we suppose X_τ survives τ iterations of filtering during an phase.

Lemma 7. *For any phase j and $\epsilon > 0$, if $\omega \geq 20\tau\epsilon^{-1}$, then the elements X_τ that survive τ iterations of filtering at phase j satisfies*

$$f_{S_j^+}(X_\tau) \leq (\epsilon/4)(1 - \epsilon)(\mathtt{OPT} - f(S_j)).$$

Since the loop condition of $f_S(T)$ and $|S \cup T|$ will always invoke Threshold, the number of invocations to Threshold in a phase is the number of invocations to PTE. By Lemma 4, after multiple rounds there will be at most k/ω elements left in the set X during every phase. According to Lemma 7, we can prove that every phase will terminate and in each phase, the total number of iterations of filtering is $O(\log n)$ rounds.

Lemma 8. *In any phase of PTE and for any $0 < \epsilon < 1/2$, if $\omega \geq 40\epsilon^{-2}\log n$, then there are at most $2\epsilon^{-1}\log n$ iterations of filtering during the phase.*

Now we can prove the main theorem of PTE. There are two situations when the algorithm stops. One is that the algorithm ends after $\lceil \omega \log(f(S_0)/(\epsilon \cdot \mathsf{OPT})) \rceil$ rounds of iterative filtering with $|S \cup T| = \lceil k \log(f(S_0)/(\epsilon \cdot \mathsf{OPT})) \rceil$, and the other is that the algorithm has completed $\lceil (20/(\epsilon(1-\epsilon))) \log(f(S_0)/(\epsilon \cdot \mathsf{OPT})) \rceil$ phases. When $\omega = O(\log n)$, by the algorithm we know that the number of rounds for adding elements to T is at most $O(\log n)$. Moreover, there are at most $O(1)$ phases with $O(\log n)$ Threshold iterations per phase. Therefore, the total number of adaptive rounds is $O(\log n)$.

Theorem 3. *For any constant $\epsilon > 0$, PTE returns a solution S_η satisfying*

$$f(S_\eta) \leq (1 + \epsilon)\,\mathsf{OPT},$$

with complexity as

$$O\left(\epsilon^{-2}\log n \cdot \log \frac{f(S_0)}{\epsilon \cdot \mathsf{OPT}}\right).$$

See Appendix C for the proofs of Lemma and Theorem in Sect. 4.

5 Applications on Fuzzy C-means

In this section we show that the new adaptive algorithms could be applied to the well known fuzzy C-means problem, so as to the classical k-means problem. Given a set of nodes $N = \{x_1, x_2, \ldots, x_n\} \subseteq \mathbb{R}^n$, the fuzzy C-means problem is to find an k-configuration point set $S = \{s_1, s_2, \ldots, s_k\} \subseteq \mathbb{R}^n$ and a matrix of membership degree $\boldsymbol{\mu} = (\mu_{i,j}) \in [0, 1]^{n \times k}$ to minimize the potential function

$$\phi(S, \boldsymbol{\mu}) = \sum_{i=1}^{n}\sum_{j=1}^{k}\mu_{ij}^{m}\|x_i - s_j\|^2,$$

with the constraint

$$\boldsymbol{\mu} \in [0, 1]^{n \times k} \quad \text{and} \quad \boldsymbol{\mu}e = e,$$

where $m \geq 1$ is the fuzzifier parameter and e is an all-one vector with dimension k.

If $m = 1$, it is easy to see that fuzzy C-means problem is reduced to k-means problem. A well known result is that when $m > 1$, for a given S, through the Lagrangian multiplier method we can obtain the optimal solution of μ to minimize $\phi(S, \mu)$ as

$$\mu_{ij} = \frac{d_{ij}^{-\frac{2}{m-1}}}{\sum_{l=1}^{k} d_{il}^{-\frac{2}{m-1}}}, \quad \forall\, i = 1, 2, \ldots, n; \; j = 1, 2, \ldots, k,$$

where $d_{ij} = \|x_i - s_j\|$. By this fact we can rewrite the potential function as

$$\phi(S) := \phi(S, \mu) = \sum_{i=1}^{n} \left(\sum_{j=1}^{k} d_{ij}^{-\frac{2}{m-1}} \right)^{1-m}. \tag{1}$$

Especially when $m = 2$, we have

$$\phi(S) = \sum_{i=1}^{n} \left(\sum_{j=1}^{k} d_{ij}^{-2} \right)^{-1}. \tag{2}$$

Additionally, we can define

$$\phi(\emptyset) = \max \left\{ \phi(s_u) + \phi(s_v) - \phi(\{s_u, s_v\}) : u \neq v \right\}. \tag{3}$$

Theorem 4. *The set function $\phi(S)$ defined by (1) and (3) is supermodular.*

The fuzzy C-means problem could not be solved by the algorithms proposed in this paper but its constrained version could be. The constrained fuzzy C-means problem is nothing different from the fuzzy C-means problem but forcing the point set S to be contained in N. That is, for a given set $N = \{x_1, x_2, \ldots, x_n\}$, the constrained fuzzy C-means problem is to find a k-configuration point set $S \subseteq N$ and a matrix of membership degree $\mu = (\mu_{i,j}) \in [0,1]^{n \times k}$ to minimize the potential function $\phi(S, \mu)$ with the same constraint on μ. The following lemma shows that relationship of the two problems.

Lemma 9. *The optimal solution of a constrained fuzzy C-means problem is a 4-approximation solution of its unconstrained version.*

Lemma 10. *For the constrained fuzzy C-means problem, if there exists an algorithm to obtain a feasible solution S_0 satisfying $\phi(S_0) \leq \alpha \phi(S^*)$. Then one can find a set S of size $O(k \log \frac{\alpha}{\epsilon})$ satisfying $\phi(S) \leq (1+\epsilon)\phi(S^*)$. The computational time from S_0 to S is*

$$O\left(\frac{nkd}{\epsilon^2} \log n \left(\log \frac{\alpha}{\epsilon} \right)^2 \right),$$

where S^* is the optimal solution.

See Appendix D for the proofs of lemmas and theorem in Sect. 5.

The following corollary then could improve the computational complexity of the algorithm proposed in [2], which is $O\left(n^2 kd \log \frac{1}{\epsilon}\right)$ for the k-means clustering problem.

Corollary 1. *For the fuzzy C-means problem, there is an algorithm that with a high probability it can find a set S of size*

$$|S| = O\left(k\left(\log\frac{1}{\epsilon} + (m-1)\log k\right)\right) = O\left(k\log\frac{k^{m-1}}{\epsilon}\right)$$

such that $\phi(S) \leq (4+\epsilon)\phi(S^*)$, in time

$$O\left(\frac{nkd}{\epsilon^2}\log n \left(\log\frac{k^{m-1}}{\epsilon}\right)^2\right).$$

References

1. Balkanski, E., Rubinstein, A., Singer, Y.: An exponential speedup in parallel running time for submodular maximization without loss in approximation. In: Proceedings of the Thirtieth Annual ACM-SIAM Symposium on Discrete Algorithms, pp. 283–302. Society for Industrial and Applied Mathematics (2019). https://doi.org/10.5555/3310435.3310454
2. Liberty, E., Sviridenko, M.: Greedy minimization of weakly supermodular set functions. In: Approximation, Randomization, and Combinatorial Optimization. Algorithms and Techniques (APPROX/RANDOM 2017), vol. 81, pp. 19:1–19:11. Schloss Dagstuhl-Leibniz-Zentrum fuer Informatik (2017)
3. Nemhauser, G.L., Wolsey, L.A., Fisher, M.L.: An analysis of approximations for maximizing submodular set function—I. Math. Program. **14**(1), 265–294 (1978). https://doi.org/10.1007/BF01588971
4. Nemhauser, G.L., Wolsey, L.A.: Best algorithms for approximating the maximum of a submodular set function. Math. Oper. Res. **3**(3), 177–188 (1978)
5. Balkanski, E., Singer, Y.: The adaptive complexity of maximizing a submodular function. In: Proceedings of the Annual ACM SIGACT Symposium on Theory of Computing, pp. 1138–1151. Association for Computing Machinery, Los Angeles (2018). https://doi.org/10.1145/3188745.3188752
6. Sviridenko, M., Vondrák, J., Ward, J.: Optimal approximation for submodular and supermodular optimization with bounded curvature. Math. Oper. Res. **42**(4), 1197–1218 (2017)
7. Ene, A., Nguyễn, H.L.: Submodular maximization with nearly-optimal approximation and adaptivity in nearly-linear time. In: Proceedings of the Annual ACM-SIAM Symposium on Discrete Algorithms, pp. 274–282. Society for Industrial and Applied Mathematics (2019). https://doi.org/10.1137/1.9781611975482
8. Chekuri, C., Quanrud, K.: Submodular function maximization in parallel via the multilinear relaxation. In: Proceedings of the Annual ACM-SIAM Symposium on Discrete Algorithms, pp. 303–322. Society for Industrial and Applied Mathematics (2019). https://doi.org/10.1137/1.9781611975482.20
9. Chen, X., Long, D.Z., Qi, J.: Preservation of supermodularity in parametric optimization: necessary and sufficient conditions on constraint structures. Oper. Res. **69**(1), 1–12 (2021)
10. Liu, B., Zhang, Q., Yuan, Z.: Two-stage distributionally robust optimization for maritime inventory routing. Comput. Chem. Eng. **149**, 107307 (2021)
11. Dunn, J.: A fuzzy relative of the ISODATA process and its use in detecting compact well-separated clusters. J. Cybern. **3**(3), 32–57 (1973)
12. Zass, R., Shashua, A.: A unifying approach to hard and probabilistic clustering. In: Proceedings of ICCV, vol. 1, pp. 294–301 (2005)

Improved Algorithms
for Non-submodular Function
Maximization Problem

Zhicheng Liu[1], Hong Chang[2(✉)], Donglei Du[3], and Xiaoyan Zhang[2]

[1] College of Taizhou, Nanjing Normal University,
Taizhou 225300, People's Republic of China
[2] School of Mathematical Science and Institute of Mathematics,
Nanjing Normal University, Nanjing 210023, People's Republic of China
{changh,zhangxiaoyan}@njnu.edu.cn
[3] Faculty of Management, University of New Brunswick,
Fredericton, NB E3B 5A3, Canada
ddu@unb.ca

Abstract. We consider a set function maximization problem where the objective function is the sum of a monotone γ-weakly submodular function f and a supermodular function g. This problem can be seen as the generalization of maximization of the BP function (when $\gamma = 1$) and γ-weakly submodular function. We give offline and streaming algorithms for this generalized problem respectively and our algorithms can improve several previous results.

Keywords: Non-submodular maximization · Uniform matroid · Streaming model

1 Introduction

We consider a set function maximization problem where the objective function is the sum of a monotone γ-weakly submodular function f and a supermodular function g. Given a ground set $V = \{1, \ldots, n\}$, the objective is to select a set $S \subseteq V$ of cardinality no more than a given parameter k to maximize the following objective function

$$F(S) = f(S) + g(S),$$

where $f : 2^V \to \mathbb{R}_+$ is a non-negative monotone γ-submodular function and $g : 2^V \to \mathbb{R}_+$ is a non-negative monotone supermodular function. A set function $f : 2^V \to \mathbb{R}_+$ is monotone if $f(S) \leq f(T), \forall S \subseteq T \subseteq V$. It is submodular if $f(S) + f(T) \geq f(S \cap T) + f(S \cup T), \forall S, T \subseteq V$. An equivalent definition is that $\sum_{e \in B \setminus A} f(e|A) \geq (f(A \cup B) - f(A))$ for all $A \subseteq B$ where $f(e|A) = f(e \cup A) - f(A)$. It is supermodular if its negative is submodular. It is modular if it is both submodular and supermodular. A monotone set function f is γ-weakly submodular for $\gamma \in (0, 1]$ if $\sum_{e \in B \setminus A} f(e|A) \geq \gamma(f(A \cup B) - f(A))$ holds for all $A \subseteq B$. Note that f is submodular if and only if it is γ-weakly submodular with $\gamma = 1$.

This problem can be seen as the generalization of several problems which have been studied widely. When $\gamma = 1$, the problem can be called as BP maximization problem. When $g(S) = 0$, the problem can be called as γ-weakly submodular maximization problem.

For this problem, we design offline and streaming algorithms respective whose approximation ratios depend on the curvatures of the functions involved. For a given non-negative function $f : 2^V \to \mathbb{R}_+$, its curvature [3] is the smallest scalar α

$$f(i|S \setminus \{i\} \cup \Omega) \geq (1 - \alpha)f(i|S \setminus \{i\}), \forall \, \Omega, \, S \subseteq V, \, i \in S \setminus \Omega. \tag{1.1}$$

When f is a submodular function, α is equal to the submodular curvature k_f [4] which has been defined as follows

$$k_f = 1 - \min_{v \in V} \frac{f(V) - f(V \setminus \{v\})}{f(v)}. \tag{1.2}$$

For a given supermodular function $g : 2^V \to \mathbb{R}_+$, its curvature [2] is defined as follows

$$k^g = 1 - \min_{v \in V} \frac{g(v)}{g(V) - g(V \setminus \{v\})}. \tag{1.3}$$

The supermodular curvature is a natural dual to the submodular curvature, and both are computationally feasible to compute, requiring only linear time in the oracle model.

For the first offline model, we can require access to the complete dataset repeatedly. In this case, we construct a new modular function and in every step, we use a new distorted greedy to choose the maximum marginal value. When $\gamma = 1$, our first algorithm yields guarantee $\min\{1 - k_f e^{-1}, 1 - (k^g)^2\}$ for maximizing the sum of a suBmodular and suPermodular (BP) function under a cardinality constraint, which improves the approximate ratio $k_f^{-1}\left(1 - e^{-k_f(1-k^g)}\right)$ in [2]. When $g(S) = 0$, the approximation guarantee is $1 - (1 - \gamma + \gamma\alpha)e^{-\gamma}$ for cardinality constrained maximization of non-submodular nondecreasing set functions, which improves the approximate ratio $\alpha^{-1}\left(1 - e^{-\alpha\gamma}\right)$ in [3]. The explanation for improvement is shown in Sect. 3.

For the second streaming model, unlike the first problem require access to the complete dataset repeatedly, we assume elements arrive over time at a fast pace. And, the memory is restricted to limited space which is sublinear with respect to the input size, we pick elements whose marginal gain is above a suitable threshold. It is worth mentioning that if $g(S) = 0$, then the approximation guarantee in our second algorithm is $\frac{\gamma}{\gamma+1}$ for maximizing a normalized monotone non-submodular set function subject to a cardinality constraint, which improves the approximate ratio $1 - \frac{1}{2^7}$ in [9]. The explanation for improvement is shown in Sect. 4.

The rest of the paper is organized as follows. In Sect. 2, we review relevant literature. In Sects. 3, 4, we present algorithms and their approximation ratio analysis for the problems, respectively. Finally, the concluding remarks are in Sect. 5.

2   Related Work

Submodular curvature and supermodular curvature are two important parameters in this work since they can improve approximation performance. For example, [4] designs a $\frac{1}{k_f}(1 - e^{-k_f})$ guarantee for cardinality constraints and a $\frac{1}{p+k_f}$ guarantee for p matroid constraints. Authors in [8] improve the previous results and give a $(1 - k_f e^{-1})$-approximation algorithm under the down-closed constraint. For supermodular curvature, authors in [2] consider the problem of maximizing the sum of submodular and supermodular functions. In their paper, they define supermodular curvature as k^g and give algorithms for cardinality constraint and matroid constraint, respectively. Recently, authors in [6] propose a new curvature called generic submodularity ratio.

In the offline case, submodular maximization has been widely studied in the literature and the readers are referred to relevant literature such as [7]. Now we move to the submodular optimization on streaming model. Authors in [1] propose the first efficient method for the submodular maximization problem with cardinality constraint about massive data summarization on the fly, without any assumption of the stream order. Authors in [10] study the budgeted influence maximization problem under credit distribution models which can be formulated as a submodular maximization problem under cardinality constraint, and give a streaming algorithm of which threshold is different from the algorithm in [1] while having the same approximate ratio. Besides the above results on submodular optimization, many non-submodular optimization problems arise in machine learning. Authors in [5] consider the weakly submodular functions and propose the first streaming algorithm for weakly submodular case. Authors in [9] utilize the concept of diminishing-return ratio to design and analyze four diminishing-return sieve-streaming (DRSS) algorithms.

3   An Offline Algorithm for a Cardinality Constraint

At the beginning of this section, we introduce two equivalent definitions and properties of submodular function, which are useful for our proof.

Definition 1 *(Submodular function: Definition 1). A function $f: 2^V \to \mathbb{R}_+$ is submodular if for every $X, Y \subseteq V$,*

$$f(X) + f(Y) \geq f(X \cup Y) + f(X \cap Y).$$

Definition 2 *(Submodular function: Definition 2). A function $f: 2^V \to \mathbb{R}_+$ is submodular if for every $X \subseteq Y \subseteq V, a \in V \backslash Y$,*

$$f(X \cup \{a\}) - f(X) \geq f(Y \cup \{a\}) - f(Y).$$

Property 1. *For any submodular function $f: 2^V \to \mathbb{R}_+$ and $X, Y \subseteq V$, we have*

$$\sum_{u \in X} [f(Y \cup \{u\}) - f(Y)] \geq f(X \cup Y) - f(Y).$$

In this section, we consider the problem under the offline case. We first consider the problem of maximizing the sum of a γ-weakly submodular, a supermodular and a modular function. Construct the following functions

$$\ell_1(S) = (1 - \alpha) \sum_{j \in S} f(j),$$

$$\ell_2(S) = \sum_{j \in S} g(j),$$

$$\ell(S) = \ell_1(S) + \ell_2(S),$$

$$f_1(S) = f(S) - \ell_1(S),$$

$$g_1(S) = g(S) - \ell_2(S).$$

Note that $f_1(S)$ is a monotone nonnegative γ-weakly submodular function, $g_1(S)$ is a monotone nonnegative supermodular function, $\ell(S)$ is a modular function, and $f(S) + g(S) = f_1(S) + g_1(S) + \ell(S)$.

Next, our new results for the sum of a submodular function and a supermodular function make use of an algorithm for the following problem: given a ground set $V = \{1, \ldots, n\}$, the problem is to select a set $S \subseteq V$ of cardinality no more than a given parameter k to maximize the following objective function

$$f_1(S) + g_1(S) + \ell(S).$$

3.1   Algorithm

The algorithm works in k rounds, where each round chooses an element e_i to maximize the increment based on the distorted objective. It starts with an empty set $S_0 = \varnothing$.

Let

$$\Delta_i(x, A) = \left(1 - \frac{\gamma}{k}\right)^{k - (i+1)} f_1(x|A) + (1 - k^g) g_1(x|V \backslash x) + \ell(x).$$

Our algorithm is presented below.

Algorithm 1. Distorted Greedy
___

1: $S_0 \leftarrow \varnothing$
2: **for** $0 \leq i \leq k - 1$ **do**
3:     $e_i \leftarrow \arg\max_{e \in V} \Delta_i(x, A)$
4:     $S_{i+1} \leftarrow S_i \cup \{e_i\}$
5: **end for**
6: Return S_k

3.2   Analysis

Our analysis relies on the following distorted objective function Φ. Let k denote the cardinality constraint. For any $i = 0, 1, \cdots, k$ and any set S, we define

$$\Phi_i(S) = \left(1 - \frac{\gamma}{k}\right)^{k-i} f_1(S) + (1 - k^g) \sum_{e \in S} g_1(e|V \backslash e) + \ell(S).$$

Lemma 1. *In each iteration of Algorithm 1,*

$$\Phi_{i+1}(S_{i+1}) - \Phi_i(S_i) = \Delta_i(e_i, S_i) + \frac{\gamma}{k}\left(1 - \frac{\gamma}{k}\right)^{k-(i+1)} f_1(S_i).$$

Proof.

$\Phi_{i+1}(S_{i+1}) - \Phi_i(S_i)$

$= \left(1 - \frac{\gamma}{k}\right)^{k-(i+1)} f_1(S_{i+1}) + (1 - k^g) \sum_{e \in S_{i+1}} g_1(e|V \backslash e) + \ell(S_{i+1})$

$\quad - \left(1 - \frac{\gamma}{k}\right)^{k-i} f_1(S_i) - (1 - k^g) \sum_{e \in S_i} g_1(e|V \backslash e) - \ell(S_i)$

$= \left(1 - \frac{\gamma}{k}\right)^{k-(i+1)}(f_1(S_{i+1}) - f_1(S_i)) + (1 - k^g)g_1(e_i|V \backslash e_i) + \ell(e_i) + \frac{\gamma}{k}\left(1 - \frac{\gamma}{k}\right)^{k-(i+1)} f_1(S_i)$

$= \Delta_i(e_i, S_i) + \frac{\gamma}{k}\left(1 - \frac{\gamma}{k}\right)^{k-(i+1)} f_1(S_i).$

∎

Lemma 2. *In each iteration of Algorithm 1, we have*

$$\Delta_i(e_i, S_i) \geq \frac{\gamma}{k}\left(1 - \frac{\gamma}{k}\right)^{k-(i+1)}(f_1(OPT) - f_1(S_i)) + \frac{1 - k^g}{k}g_1(OPT) + \frac{1}{k}\ell(OPT).$$

Proof.

$\Delta_i(e_i, S_i) \geq \frac{1}{k} \sum_{e \in OPT} \Delta_i(e, S_i)$

$= \frac{1}{k} \sum_{e \in OPT} \left(\left(1 - \frac{\gamma}{k}\right)^{k-(i+1)} f_1(e|S_i) + (1 - k^g)g_1(e|V \backslash e) + \ell(e)\right)$

$\geq \frac{1}{k}\left(\left(1 - \frac{\gamma}{k}\right)^{k-(i+1)} \gamma\,(f_1(OPT) - f_1(S_i)) + (1 - k^g)g_1(OPT) + \ell(OPT)\right)$

$= \frac{\gamma}{k}\left(1 - \frac{\gamma}{k}\right)^{k-(i+1)}(f_1(OPT) - f_1(S_i)) + \frac{1 - k^g}{k}g_1(OPT) + \frac{1}{k}\ell(OPT),$

where the second inequality follows from the definition of f and the supermodularity of g. ∎

Finally, the main result is summarized as follows.

Theorem 1. *Algorithm 2 returns a set S_k of size k such that*

$$f_1(S_k) + g_1(S_k) + \ell(S_k) \geq \left(1 - e^{-\gamma}\right) f_1(OPT) + \left(1 - k^g\right) g_1(OPT) + \ell(OPT).$$

Proof. According to the definition of Φ, we have

$$\Phi_0(S_0) = \left(1 - \frac{\gamma}{k}\right)^{k-0} f_1(S_0) + (1 - k^g)g_1(S_0) + \ell(S_0) = 0,$$

$$\Phi_k(S_k) = \left(1 - \frac{\gamma}{k}\right)^{k-k} f_1(S_k) + (1 - k^g) \sum_{e \in S_k} g_1(e|V \backslash e) + \ell(S_k)$$

$$\leq f_1(S_k) + g_1(S_k) + \ell(S_k).$$

Applying Lemma 1, we have

$$\Phi_{i+1}(S_{i+1}) - \Phi_i(S_i) = \Delta_i(e_i, S_i) + \frac{\gamma}{k}\left(1 - \frac{\gamma}{k}\right)^{k-(i+1)} f_1(S_i)$$

$$\geq \frac{\gamma}{k}\left(1 - \frac{\gamma}{k}\right)^{k-(i+1)} f_1(OPT) + \frac{1-k^g}{k} g_1(OPT) + \frac{1}{k}\ell(OPT).$$

Finally,

$$f_1(S_k) + g_1(S_k) + \ell(S_k)$$

$$\geq \sum_{i=0}^{k-1} (\Phi_{i+1}(S_{i+1}) - \Phi_i(S_i))$$

$$\geq \sum_{i=0}^{k-1} \left(\frac{\gamma}{k}\left(1 - \frac{\gamma}{k}\right)^{k-(i+1)} f_1(OPT) + \frac{1-k^g}{k} g_1(OPT) + \frac{1}{k}\ell(OPT) \right)$$

$$\geq \left(1 - e^{-\gamma}\right) f_1(OPT) + (1 - k^g)g_1(OPT) + \ell(OPT).$$

∎

We are now ready to give an approximation ratio of this problem.

Theorem 2. *There exists an algorithm returning a set S_k of size k such that*

$$f(S_k) + g(S_k) \geq \min\{1 - (1 - \gamma + \gamma\alpha)e^{-\gamma}, 1 - (k^g)^2\}(f(OPT) + g(OPT)).$$

Proof. According to the submodular and supermodularity curvatures, we have,

$$\ell_1(S) = (1 - \alpha) \sum_{j \in S} f(j) \geq (1 - \alpha)\gamma f(S),$$

$$\ell_2(S) = \sum_{j \in S} g(j) \geq (1 - k^g)g(S).$$

So,

$$f(S) + g(S) = f_1(S) + g_1(S) + \ell(S)$$
$$\geq (1 - e^{-\gamma})f_1(OPT) + (1 - k^g)g_1(OPT) + \ell(OPT)$$
$$= (1 - e^{-\gamma})(f(OPT) - \ell_1(OPT)) + (1 - k^g)(g(OPT) - \ell_2(OPT))$$
$$+ \ell_1(OPT) + \ell_2(OPT)$$
$$= (1 - e^{-\gamma})f(OPT) + (1 - k^g)g(OPT) + e^{-\gamma}\ell_1(OPT) + k^g\ell_2(OPT)$$
$$\geq (1 - e^{-\gamma})f(OPT) + (1 - k^g)g(OPT) + e^{-\gamma}(1 - \alpha)\gamma f(OPT)$$
$$+ k^g(1 - k^g)g(OPT)$$
$$= \left(1 - (1 - \gamma + \gamma\alpha)e^{-\gamma}\right)f(OPT) + \left(1 - (k^g)^2\right)g(OPT)$$
$$\geq \min\{1 - (1 - \gamma + \gamma\alpha)e^{-\gamma}, 1 - (k^g)^2\}(f(OPT) + g(OPT)).$$

∎

Theorem 2 gives a lower bound of distorted greedy in terms of the curvature α and the supermodular curvature k^g. We notice that this bound improves known results.

(1) If $\gamma = 1$, then f is a submodular function. We use k_f to replace α and the approximation guarantee is $\min\{1 - k_f e^{-1}, 1 - (k^g)^2\}$, which improves the approximate ratio $k_f^{-1}\left(1 - e^{-k_f(1-k^g)}\right)$ in [2]. It can be shown that

$$1 - k_f e^{-1} \geq 1 - k_f e^{-k_f} = \max_{0 \leq k^g \leq 1} 1 - k_f e^{-k_f(1-k^g)} \geq 1 - k_f e^{-k_f(1-k^g)},$$

$$1 - (k^g)^2 \geq 1 - k^g = \max_{0 \leq k_f \leq 1} 1 - k_f e^{-k_f(1-k^g)} \geq 1 - k_f e^{-k_f(1-k^g)}.$$

(2) If $g(S) = 0$, then the approximation guarantee is $1 - (1 - \gamma + \gamma\alpha)e^{-\gamma}$, which improves the approximate ratio $\alpha^{-1}\left(1 - e^{-\alpha\gamma}\right)$ in [3]. Let $F(\alpha, \gamma) = \alpha\left(1 - (1 - \gamma + \gamma\alpha)e^{-\gamma}\right) - (1 - e^{-\alpha\gamma})$. Notice that $F(\alpha, \gamma)$ has no extreme point for $0 < \alpha < 1$ and $0 < \gamma < 1$. It follows that $F(\alpha, \gamma)$ maximizes or minimizes at four boundary points $(0, 0)$, $(0, 1)$, $(1, 0)$, $(1, 1)$. Thus, it is computed that $F(\alpha, \gamma) \geq 0$, which implies that $1 - (1 - \gamma + \gamma\alpha)e^{-\gamma} \geq \alpha^{-1}\left(1 - e^{-\alpha\gamma}\right)$.

(3) If $f(S) = 0$, then we get $1 - (k^g)^2$, which is a new curvature-based bound for monotone supermodular maximization subject to a cardinality constraint.

4   A Streaming Algorithm for a Cardinality Constraint

In this section, we consider the problem $\max_{|S| \leq k} f(S) + g(S)$ in the streaming model. We consider the scaled objective $f(S) + \sum_{e \in S} g(e)$. Now, instead of picking elements whose (scaled) marginal gain is positive, we pick elements whose (scaled) marginal gain is above a suitable threshold.

4.1 Algorithm

Let

$$h(S) = f(S) + \sum_{e \in S} g(e).$$

Our algorithm is presented below.

Algorithm 2. Scaled single-threshold Greedy

1: $S \leftarrow \varnothing$
2: while stream not empty
3: $e \leftarrow$ next stream element
4: if $h(e|S) \geq \tau$ and $|S| < k$
5: $S \leftarrow S \cup e$
6: Return S

4.2 Analysis

We are now ready to give an approximation ratio of the streaming algorithm.

Theorem 3. *When run with threshold* $\tau = \frac{1}{(\gamma+1)k}(\gamma f(OPT) + (1-k^g)g(OPT))$, *Algorithm 2 returns a solution* S *satisfying* $f(S) + g(S) \geq \frac{\gamma}{\gamma+1}f(OPT) + \frac{1-k^g}{\gamma+1}g(OPT)$.

Proof. Let us now consider the following cases:

Case 1: $|S| = k$. We have

$$f(S) + g(S)$$
$$\geq f(S) + \sum_{e \in S} g(e)$$
$$\geq \tau k$$
$$\geq \frac{\gamma}{\gamma+1}f(OPT) + \frac{1-k^g}{\gamma+1}g(OPT),$$

where the first inequality follows from the supermodularity of g.

Case 2: $|S| < k$. For every item $e \in OPT \backslash S$, we have

$$\tau > f(e|S) + g(e).$$

Therefore we have

$$\tau k \geq \tau |OPT|$$
$$\geq \sum_{e \in OPT} f(e|S) + \sum_{e \in OPT} g(e)$$
$$\geq \gamma f(OPT \cup S) - \gamma f(S) + (1-k^g)g(OPT)$$
$$\geq \gamma f(OPT) - \gamma f(S) + (1-k^g)g(OPT),$$

where the third inequality follows from the definition of f and the supermodularity of g.

Rearrange the inequality, we have

$$f(S) + g(S) \geq f(S)$$
$$\geq f(OPT) + \frac{1 - k^g}{\gamma} g(OPT) - \frac{\tau k}{\gamma}$$
$$= \frac{\gamma}{\gamma + 1} f(OPT) + \frac{1 - k^g}{\gamma + 1} g(OPT).$$

∎

Theorem 3 gives a lower bound of scaled single-threshold greedy in terms of the supermodular curvature k^g. We notice that this bound improves known results.

(1) If $g(S) = 0$, then the approximation guarantee is $\frac{\gamma}{\gamma+1}$, which improves the approximate ratio $1 - \frac{1}{2^\gamma}$ in [9]. It is easy to obtain that $\frac{\gamma}{\gamma+1} \geq 1 - \frac{1}{2^\gamma}$ for $\gamma \in [0, 1]$.
(2) If $f(S) = 0$, then we get $\frac{1-k^g}{\gamma+1}$, which is a new curvature-based bound for monotone supermodular maximization subject to a cardinality constraint under streaming algorithm.

5   Conclusion

In this paper, we consider offline and streaming model for our generalized non-submodular maximization problems. We design some non-obvious greedy algorithms, which may be of independent interest because of its potential to be applicable to other problems.

Acknowledgements. The research is supported by NSFC (No. 11871280) and Qinglan Project, Natural Science Foundation of Jiangsu Province (No. BK20200723), and Natural Science Foundation for institutions of Higher Learning of Jiangsu Province (No. 20KJB110022).

Data Availability Statement. The data used to support the findings of this study are available from the corresponding author upon request.

References

1. Badanidiyuru, A., Mirzasoleiman, B., Karbasi, A., Krause, A.: Streaming submodular maximization: massive data summarization on the fly. In: Proceedings of SIGKDD, pp. 671–680 (2014)

2. Bai, W., Bilmes, J.A.: Greed is still good: maximizing monotone submodular+supermodular (BP) functions. In: ICML, pp. 304–313 (2018)

3. Bian, A., Buhmann, J., Krause, A., Tschiatschek, S.: Guarantees for greedy maximization of non-submodular functions with applications. In: ICML 2017, pp. 498–507 (2017)

4. Conforti, M., Cornuéjols, G.: Submodular set functions, matroids and the greedy algorithm: tight worst-case bounds and some generalizations of the Rado-Edmonds theorem. Discrete Appl. Math. **7**(3), 251–274 (1984)

5. Elenberg, E., Dimakis, A.G., Feldman, M., Karbasi, A.: Streaming weak submodularity: interpreting neural networks on the fly. In: Proceedings of NIPS, pp. 4044–4054 (2017)

6. Gong, S., Nong, Q., Liu, W., Fang, Q.: Parametric monotone function maximization with matroid constraints. J. Glob. Optim. **75**(3), 833–849 (2019). https://doi.org/10.1007/s10898-019-00800-2

7. Krause, A., Golovin, D.: Submodular function maximization. Tractability **3**, 71–104 (2014). http://www.cs.cmu.edu/dgolovin/papers/submodularsurvey12.pdf

8. Sviridenko, M., Vondrák, J., Ward, J.: Optimal approximation for submodular and supermodular optimization with bounded curvature. Math. Oper. Res. **42**(4), 1197–1218 (2017)

9. Wang, Y., Xu, D., Wang, Y., Zhang, D.: Non-submodular maximization on massive data streams. J. Glob. Optim. **76**(4), 729–743 (2019). https://doi.org/10.1007/s10898-019-00840-8

10. Yu, Q., Li, H., Liao, Y., Cui, S.: Fast budgeted influence maximization over multi-action event logs. IEEE Access **6**, 14367–14378 (2018)

Fixed Observation Time-Step: Adaptive Influence Maximization

Yapu Zhang[1], Shengminjie Chen[2], Wenqing Xu[3]([✉]), and Zhenning Zhang[1]

[1] Department of Operations Research and Information Engineering,
Beijing University of Technology, Beijing 100124, People's Republic of China
{zhangyapu,zhangzhenning}@bjut.edu.cn
[2] School of Mathematical Sciences, University of Chinese Academy of Sciences,
Beijing 100049, People's Republic of China
chenshengminjie19@mails.ucas.ac.cn
[3] Beijing Institute for Scientific and Engineering Computing, Beijing University
of Technology, Beijing 100124, People's Republic of China
xwq@bjut.edu.cn

Abstract. The vanilla influence maximization problem requires some kind of seeds before the diffusion process so as to maximize the expected influence spread in a social network. This problem has been extensively studied due to its applications in viral marketing. However, most studies require selecting all seeds at once, which wastes part of the budget due to not utilizing the observation results. This paper considers adaptive influence maximization and adaptive stochastic influence maximization problems under a general feedback model, where seeds can be selected after a fixed number of observation time-steps. Generally, the objective function lacks the adaptive submodularity property, making it difficult to construct effective approximate solutions. We introduce a comparative factor and present a theoretical analysis of the solution using an adaptive greedy framework to solve them. In addition, a feasible approximation algorithm based on the reverse sampling technique is used to solve the adaptive stochastic influence maximization problem.

Keywords: Social network · Adaptive influence maximization · Approximation algorithm

1 Introduction

Online social network platforms such as WeChat and Facebook have flourished in the past decades. More and more people are willing to share and discuss their ideas on these platforms. Ideas can spread in the networks through word-of-mouth effects. In order to utilize this effect to promote products, opinions, and innovations, many companies will provide free or discounted samples to

This work is supported by National Natural Science Foundation of China under grants No. 12131003 and No. 12001025.

W. Wu and H. Du (Eds.): AAIM 2021, LNCS 13153, pp. 200–211, 2021.
https://doi.org/10.1007/978-3-030-93176-6_18

influential users in exchange for their help in publicity, making a wide range of influence in the social network [5,15]. Kempe, Kleinberg, and Tardos [9] first proposed the influence maximization problem from the perspective of discrete optimization. The goal is to find a small set of users (i.e., seed nodes), hoping the expected number of users influenced by the seeds can be maximized. They proved that the problem is NP-hard under two models, namely the independent cascade model and the linear threshold model. Due to the submodularity of the objective function, there is a $(1 - 1/e)$-approximate solution using the greedy algorithm [12].

Due to essential applications in marketing [10,11], rumor control [13,23,25] and other fields, it has become a focus of concern in both academia and industry. The majority of studies focus on non-adaptive strategies, which require all seed nodes to be selected at once without observing the node status and diffusion process. However, during the diffusion process, some seed nodes selected by the non-adaptive strategy may influence other seed nodes, causing a waste of the budget. Compared with the non-adaptive strategy, adaptive strategies allow the selection of seed nodes after observing specific propagation results, making better use of the budget.

As a variant of the influence maximization problem, the adaptive influence maximization has attracted attention recently. However, most works are based on the full feedback model [6]. That is, we select the subsequent seeds when the current diffusion process completes. In reality, it is impractical to wait for the end of the diffusion process before the next seed selection. Therefore, we consider a general feedback model, which allows a fixed observation time-step and then executes the selection of the next seeds. This feedback model includes the full feedback model, and it is hard to figure out due to the lack of adaptive submodularity. To solve it, we define a comparative factor and present an analysis of the approximation ratio using an adaptive greedy algorithm. Furthermore, we study the adaptive stochastic influence maximization when considering the randomness of the policy during algorithm execution. Combined with the reverse sampling technique [1,19,21,22], we design a feasible algorithm that can return an approximate solution to the problem.

2   Related Works

Golovin and Krause [6] first discussed adaptive influence maximization. They proposed two special feedback models, namely the full feedback model and the myopic feedback model. Before choosing the next seed node, the full feedback model always waits for the termination of the diffusion process, while the myopic feedback model waits for a round of propagation.

Full Feedback Model. Golovin and Krause [6] defined adaptive submodularity and adaptive monotonicity and proved that a simple adaptive greedy algorithm could guarantee a $(1-1/e)$-approximate solution if the objective function satisfies these two properties. Fortunately, the adaptive influence maximization problem under the full feedback model satisfies these two properties and thus has a $(1 -$

$1/e$)-approximate solution. However, the theoretical analysis of the algorithm needs to know the exact influence spread of each node. Since the computation of the influence spread is usually #P-difficult [3,4], it is difficult to guarantee the approximation ratio in practical implementation. Based on the framework of the full feedback model, researchers have also studied other related adaptive optimization problems [2,7,8,17,18,20,24].

Myopic Feedback Model. The objective function of the adaptive influence maximization problem lacks adaptive submodularity under the myopic feedback model, which presents challenges to the design of an approximation algorithm. Facing this difficulty, Salha et al. [16] modified the independent cascade model to make the objective function possess adaptive submodularity and obtained the $(1-1/e)$-approximate solution. Golovin and Krause [6] once guessed that there is an adaptive algorithm with a constant approximation ratio for the problem under the myopic feedback model. Peng and Chen [14] gave a deterministic answer to this conjecture. They proposed the concept of the adaptive gap, which is the ratio between the optimal solution to the adaptive problem and the optimal solution to the non-adaptive problem. They proved that the range of the adaptive gap under the myopic feedback model is from 4 to $e/(e-1)$. Then, both adaptive and non-adaptive greedy algorithms can get $0.25(1-1/e)$-approximate solution.

In this paper, we study the adaptive influence maximization and adaptive stochastic influence maximization problems under a general feedback model. In fact, the general feedback model is equivalent to the full feedback model when the observation time-step is equal to the longest distance between two nodes in the social network. And the general feedback model is the myopic feedback model when the observation time-step is equal to 1. Therefore, the general feedback model is a generalization of both the full feedback model and the myopic feedback model.

3   Problem Formulation

In this section, we first introduce some notations and definitions used in this paper. Then, we present the adaptive influence maximization and adaptive stochastic influence maximization, respectively.

3.1   Notations

Let $G = (V, E)$ be a social network, in which V is the set of n nodes, E is the set of m edges, and each directed edge $(u, v) \in E$ is associated with an influence probability $p(u, v) \in [0, 1]$. We consider the influence diffusion under the independent cascade (IC) model and the definition of the IC model is as follows.

Definition 1 (IC Model). *Given a graph $G = (V, E)$ and a seed set S, the discrete-time diffusion process under the IC model is as follows. Initially, all seeds in S are activated. Then, at each subsequent time step, each newly-activated* *node u tries to activate its out-neighbor node v with only one chance and success probability $p(u, v)$. This process terminates when no nodes can be activated.*

In diffusion process, we say that an edge (u, v) is live if node u activates its neighbor v successfully, and the edge (u, v) is dead otherwise.

Definition 2 (Realization). *Define a realization as $\phi = (L(\phi), D(\phi))$, where $L(\phi) \subseteq E$ contains all live edges, $D(\phi) \subseteq E$ contains all dead edges, and $L(\phi) \cap D(\phi) = \emptyset$. We say that realization ϕ is full realization if $L(\phi) \cup D(\phi) = E$ and ϕ is partial realization otherwise.*

In [9], it is shown that the following two diffusion processes are equivalent:

- Execute a stochastic diffusion process on graph $G = (V, E)$;
- Randomly generate a full realization ϕ of G and execute the deterministic diffusion process on graph $G' = (V, L(\phi))$.

Then, the influence spread of nodes set S, i.e., the expected number of activated nodes of S can be written as

$$I(S) = \mathbb{E}_\phi[I_\phi(S)] = \sum_\phi \Pr[\phi] \cdot I_\phi(S), \tag{1}$$

where ϕ is the full realization, $I_\phi(S)$ is the number of active nodes influenced by set S under the realization ϕ and $\Pr[\phi] = \prod_{(u,v) \in L(\phi)} p(u, v) \cdot \prod_{(u,v) \in D(\phi)} (1 - p(u, v))$.

For ease of reference, ϕ and φ represent the full realization and partial realization in the following, respectively. We define $\phi \sim \varphi$ as φ is consistent with ϕ if $L(\varphi) \subseteq L(\phi)$ and $D(\varphi) \subseteq D(\phi)$. In addition, $\varphi_1 \prec \varphi_2$ represents that φ_1 is a subrealization of φ_2 if $L(\varphi_1) \subseteq L(\varphi_2)$ and $D(\varphi_1) \subseteq D(\varphi_2)$.

3.2 Problem Definition

Different from the traditional influence maximization, the adaptive influence maximization has a feedback model.

Definition 3 (General Feedback Model). *Let ϕ and φ be the known full realization and partial realization with $\phi \sim \varphi$, respectively. Given a nodes set S and a time-step T, we define by $I_\phi^T(S)$ the set of nodes which are activated by S within time $T - 1$ under the realization ϕ. Then, a general feedback model is a function $F_{\phi,\varphi}^T$ that maps a set S to a partial realization ψ such that $L(\psi) = L(\varphi) \cup \{(u, v) \in L(\phi) : u \in I_\phi^T(S)\}$ and $D(\psi) = D(\varphi) \cup \{(u, v) \in D(\phi) : u \in I_\phi^T(S)\}$.*

Notice that the feedback model is equivalent to the myopic feedback model when time-step $T = 1$ and the full feedback model when time-step T is no less than the longest distance between any two nodes in graph G, respectively. For convenience, we denote D as the longest distance between any two nodes in graph G.

In this paper, we assume that the number of seeding processes is k, and B nodes can be selected at each seeding process.

Definition 4 (Adaptive Policy). *An adaptive policy π is a function that maps a partial realization φ and a budget B to a nodes set, determining which B seeds to pick next based on φ returned by the feedback model.*

Definition 5 (Policy Truncation). *Given a policy π, we denote π_i as the policy truncation, illustrating π_i only executes the first i seeding processes, and these processes are totally the same as π. Notice that $\pi_0 = \emptyset$ for any policy π.*

Definition 6 (Policy Concatenation). *For any two policies π and π', we denote $\pi \oplus \pi'$ as the policy concatenation, illustrating π runs first, and then runs π' without considering the results observed from π.*

Given a time-step T for the feedback model, let $\zeta^T(\pi, \phi)$ be the set of nodes selected by π under a full realization ϕ. More specifically, $\zeta^T(\pi, \phi)$ can be obtained as follows:

- Initially $S = \emptyset$, $\varphi = (\emptyset, \emptyset)$, and $I_\phi^T(S) = \emptyset$;
- For each seeding process, compute $S = S \cup \pi(\varphi, B)$, $\varphi = F_{\phi,\varphi}^T(S)$ and $I_\phi^T(S)$;
- The process terminates after k seeding processes and return $\zeta^T(\pi, \phi) = S$.

Based on the above definitions, we can present our problems formally.

Definition 7 (Adaptive Influence Maximization). *Suppose that we execute k seeding processes and the size of selected seeds is B at each process. Given a graph G and a time-step T, let $f^T(\pi) = \mathbb{E}_\phi[I_\phi(\zeta^T(\pi, \phi))] = \sum_\phi \Pr[\phi] \cdot I_\phi(\zeta^T(\pi, \phi))$ and Π be all possible adaptive policies. The adaptive influence maximization problem aims at finding a policy π such that*

$$\max_{\pi \in \Pi} f^T(\pi) \tag{2}$$

$$\text{s.t. } |\zeta^T(\pi, \phi)| \leq k \cdot B \text{ for all } \phi \tag{3}$$

Furthermore, we consider the randomness of policy during algorithm execution and then study the adaptive stochastic influence maximization.

Definition 8 (Adaptive Stochastic Influence Maximization). *Suppose that we execute k seeding processes and the size of selected seeds is B at each process. Let $\pi(\omega)$ be a random adaptive policy corresponding to variable ω, where ω represents the random factors that affect the policy. The adaptive stochastic influence maximization problem aims at finding a policy $\pi(\omega)$ such that*

$$\max_{\pi \in \Pi} \mathbb{E}_\omega[f^T(\pi(\omega))] \tag{4}$$

$$\text{s.t. } |\zeta^T(\pi(\omega), \phi)| \leq k \cdot B \text{ for all } \phi \tag{5}$$

4 Algorithm Design

In this section, we aim to design approximation algorithms for our problems. For ease of reference, let π^* be the optimal policy, $\phi^* = (E, \emptyset)$. Given any policy π, we denote by $\hat{\pi}_{i,j} = \pi_i \oplus \pi_j^*$ for each $i, j \geq 0$. In addition, for any $i, j \geq 0$, let $\varphi(\pi_i, \phi)$(resp. $\varphi(\hat{\pi}_{i,j}, \phi)$) be the partial realization being observed after adopting policy π_i(resp. $\hat{\pi}_{i,j}$) under the realization ϕ within time-step $i \cdot T$.

First, we define the comparative factor. For convenience, given a set S, a policy π and a realization φ, let $\Delta_S(\pi|\varphi) = \mathbb{E}_\phi[I_\phi(\zeta^T(\pi, \phi) \cup S) - I_\phi(\zeta^T(\pi, \phi))|\phi \sim \varphi]$. Similarly, given policies π and π', we also denote $\Delta_{\pi'}(\pi|\varphi) = \mathbb{E}_\phi[I_\phi(\zeta^T(\pi', \phi)) - I_\phi(\zeta^T(\pi, \phi))|\phi \sim \varphi]$.

Definition 9 (Comparative Factor). *A parameter β_i is the comparative factor if β_i is the minimum value satisfying*

$$\max_{S \subseteq V, |S|=B} \Delta_S(\hat{\pi}_{i,j-1}|\varphi(\hat{\pi}_{i,j-1}, \phi')) \leq \beta_i \cdot \max_{S \subseteq V, |S|=B} \Delta_S(\hat{\pi}_{i,j-1}|\varphi(\pi_i, \phi')), \quad (6)$$

for each full realization ϕ' and $i, j \geq 0$.

If $T \geq D$, then $\beta_i = 1$ due to the adaptive submodularity of the objective function [6]. If $T < D$, $\beta_i \leq \max_{v \in V} I_{\phi^*}(\{v\})$ since $\max_{S \subseteq V, |S|=B} \Delta_S(\hat{\pi}_{i,j-1}|\varphi(\hat{\pi}_{i,j-1}, \phi')) \leq B \cdot \max_{v \in V} I_{\phi^*}(\{v\})$ and $\max_{S \subseteq V, |S|=B} \Delta_S(\hat{\pi}_{i,j-1}|\varphi(\pi_i, \phi')) \geq B$. Notice that for each realization ϕ, we suppose that the number of nodes that can be activated by set $\zeta^T(\hat{\pi}_{i,j-1}, \phi)$ is at least B here.

Lemma 1. *Given any adaptive policy π, if it satisfies*

$$\alpha_i \cdot \max_{S \subseteq V, |S|=B} \Delta_S(\pi_i|\varphi(\pi_i, \phi')) \leq \Delta_{\pi_{i+1}}(\pi_i|\varphi(\pi_i, \phi')) \quad (7)$$

for any realization ϕ' and $i \geq 0$, then $f^T(\pi^*) - f^T(\pi_i) \leq \frac{k\beta_i}{\alpha_i} \cdot (f^T(\pi_{i+1}) - f^T(\pi_i))$, where $\alpha_i \in [0, 1]$ and β_i is the comparative factor.

Proof. According to the monotonicity of the objective function, we have

$$f^T(\pi^*) - f^T(\pi_i) \leq f^T(\hat{\pi}_{i,k}) - f^T(\pi_i)$$
$$= \sum_{j=1}^{k} (f^T(\hat{\pi}_{i,j}) - f^T(\hat{\pi}_{i,j-1})).$$

Next, for any $j = 1, \ldots, k$, we have

$$f^T(\hat{\pi}_{i,j}) - f^T(\hat{\pi}_{i,j-1}) = \mathbb{E}_\phi[I_\phi(\zeta^T(\hat{\pi}_{i,j}, \phi)) - I_\phi(\zeta^T(\hat{\pi}_{i,j-1}, \phi))]$$
$$= \mathbb{E}_{\phi'}[\mathbb{E}_\phi[I_\phi(\zeta^T(\hat{\pi}_{i,j}, \phi)) - I_\phi(\zeta^T(\hat{\pi}_{i,j-1}, \phi))|\phi \sim \varphi(\hat{\pi}_{i,j-1}, \phi')]]$$
$$= \mathbb{E}_{\phi'}[\Delta_{\hat{\pi}_{i,j}}(\hat{\pi}_{i,j-1})|\varphi(\hat{\pi}_{i,j-1}, \phi')]$$
$$\leq \mathbb{E}_{\phi'}[\max_{S \subseteq V, |S|=B} \Delta_S(\hat{\pi}_{i,j-1})|\varphi(\hat{\pi}_{i,j-1}, \phi')].$$

According to the definition of comparative factor, it holds that

$$\max_{S \subseteq V, |S|=B} \Delta_S(\hat{\pi}_{i,j-1}|\varphi(\hat{\pi}_{i,j-1}, \phi')) \leq \beta_i \cdot \max_{S \subseteq V, |S|=B} \Delta_S(\hat{\pi}_{i,j-1}|\varphi(\pi_i, \phi')).$$

Then,

$$
\begin{aligned}
f^T(\hat{\pi}_{i,j}) - f^T(\hat{\pi}_{i,j-1}) &\le \beta_i \cdot \mathbb{E}_{\phi'}[\max_{S \subseteq V, |S|=B} \Delta_S(\hat{\pi}_{i,j-1}|\varphi(\pi_i, \phi'))] \\
&\le \beta_i \cdot \mathbb{E}_{\phi'}[\max_{S \subseteq V, |S|=B} \Delta_S(\pi_i|\varphi(\pi_i, \phi'))] \\
&\le \frac{\beta_i}{\alpha_i} \cdot \mathbb{E}_{\phi'}[\Delta_{\pi_{i+1}}(\pi_i|\varphi(\pi_i, \phi'))] \\
&\le \frac{\beta_i}{\alpha_i} \cdot \mathbb{E}_{\phi'}\left[\mathbb{E}_{\phi}[I_\phi(\zeta^T(\pi_{i+1}, \phi)) - I_\phi(\zeta^T(\pi_i, \phi))|\phi \sim \varphi(\pi_i, \phi')]\right] \\
&= \frac{\beta_i}{\alpha_i} \cdot \left(f^T(\pi_{i+1}) - f^T(\pi_i)\right).
\end{aligned}
$$

Based on above results, the lemma follows.

Algorithm 1 adaptively selects k seed sets $S_0, S_1, \ldots, S_{k-1}$. Initially, let $G_0 = G$ and $X_0 = \emptyset$. At each seeding process i, this algorithm needs to identify an α_i-approximate solution S_i for problem $\max_{S \subseteq V, |S|=B} I(S \cup X_i)$ corresponding to G_i. After the seed selection, we observe the influence of S_i within time-step T. On the one hand, the algorithm removes all nodes activated before time T and creates a new graph G_{i+1}. On the other hand, it generates a set X_i which contains all nodes activated exactly in time T.

Algorithm 1. Adaptive Greedy

Input: graph $G = (V, E)$, the number of seeding processes k, the size of selected nodes B in each seeding process

Output: adaptively return seed sets $S_0, S_1, \ldots, S_{k-1}$

1: Initialize $G_0 = G$ and $X_0 = \emptyset$
2: **for** $i = 0$ *to* $k - 1$ **do**
3:     Identify a size-B set S_i such that S_i is an α_i-approximate solution for the problem $\max_{S \subseteq V, |S|=B} I(S \cup X_i)$ corresponding to G_i
4:     Observe the influence diffusion of S_i within time-step T
5:     Remove all nodes activated before time T and generate a new graph G_{i+1}
6:     Identify a set X_{i+1} which contains all nodes activated exactly at time T
7: **end for**
8: **return** $S_0, \ldots, S_{k-1}$

Theorem 1. *The adaptive policy π returned by Algorithm 1 satisfies $f^T(\pi) \ge (1 - e^{-\sum_{i=0}^{k-1} \frac{\alpha_i}{k\beta_i}}) \cdot f^T(\pi^*)$, where $\alpha_i \in [0,1]$ and β_i is the comparative factor.*

Proof. Let $a_i = f^T(\pi^*) - f^T(\pi_i)$. For each $i \ge 0$, Lemma 1 gives

$$
a_i \le \frac{k\beta_i}{\alpha_i} \cdot (a_i - a_{i+1}), \tag{8}
$$

and

$$
a_{i+1} \le (1 - \frac{\alpha_i}{k\beta_i}) \cdot a_i. \tag{9}
$$

Notice that $1 - x \leq e^{-x}$ for all $x \in \mathbb{R}$, we have

$$a_k \leq \prod_{i=0}^{k-1} (1 - \frac{\alpha_i}{k\beta_i}) \cdot a_0 \leq e^{-\sum_{i=0}^{k-1} \frac{\alpha_i}{k\beta_i}} \cdot a_0. \qquad (10)$$

Rearranging Eq. 10, $f^T(\pi^*) - f^T(\pi_k) \leq e^{-\sum_{i=0}^{k-1} \frac{\alpha_i}{k\beta_i}} \cdot f^T(\pi^*)$ holds and Theorem 1 follows.

Notice that Algorithm 1 needs to identify an α_i-approximate solution for the problem $\max_{S \subseteq V, |S|=B} I(S \cup X_i)$ for each seeding process i. Using the greedy algorithm, we can obtain a $(1 - 1/e)$-approximate solution since the objective function is non-negative monotone non-decreasing and submodular [12]. However, the computation of the objective function is #P-hard [3] and the approximation ratio is $1 - 1/e - \varepsilon$ using the greedy algorithm based on Monte-Carlo simulation, where $\varepsilon \in (0, 1)$. Here, it is difficult to give the exact value of ε. Based on the reverse influence sampling strategy, we can obtain a $(1 - 1/e - \varepsilon)$-approximate solution with at least probability $1 - \delta$, where $\varepsilon, \delta \in (0, 1)$ and they are the inputs of the algorithm. Due to the exponential number $O(2^m)$ of realizations, the failure probability is at most $O(2^m \cdot \delta)$. Then, it is hard to ensure an approximate solution for the adaptive influence maximization using this method. Next, we utilize this idea to solve the adaptive stochastic influence maximization problem.

Definition 10 (RR set). *Given a graph $G = (V, E)$, we can generate a subgraph $g = (V, E')$, where $E' \subseteq E$ and each edge $(u, v) \in E$ is in g with probability $p(u, v)$. A random reverse reachable (RR) set is a set of nodes that can reach v in g with node $v \in V$ uniformly at random.*

Given a random RR set R, we can conclude that $I(S) = n \cdot \Pr[R \cap S \neq \emptyset]$ for any sets $S \subseteq V$. Let $\mathcal{R}$ be a collection of RR sets and $\Lambda_{\mathcal{R}}(S)$ be the fraction of the sets in $\mathcal{R}$ covered by S. We can use $|\mathcal{R}| \cdot \Lambda_{\mathcal{R}}(S)$ to estimate $I(S)$ when the size of $\mathcal{R}$ is large enough. Given set X, we suppose that S^* is the optimal solution for $\max_{S \subseteq V, |S|=B} I(S \cup X)$. Given sets S and $\mathcal{R}$, an upper bound of $\Lambda_{\mathcal{R}}(S^* \cup X)$ can be written as follows:

$$\Lambda_{\mathcal{R}}^u(S^* \cup X) = \min_{0 \leq i \leq B} \left(\Lambda_{\mathcal{R}}(S_i \cup X) + \sum_{v \in \mathrm{maxMC}(S_i, B)} (\Lambda_{\mathcal{R}}(S_i \cup X \cup \{v\}) - \Lambda_{\mathcal{R}}(S_i \cup X)) \right),$$

$$(11)$$

where S_i is the first i nodes in S and $\mathrm{maxMC}(S, l)$ is the first l nodes with the largest marginal coverage in $\mathcal{R}$ corresponding to S.

Inspired by the OPIM-C method for non-adaptive influence maximization [19], we use Algorithm 2 to solve $\max_{S \subseteq V, |S|=B} I(S \cup X)$, where X is given. First, we generate two collections of RR sets $\mathcal{R}_1$ and $\mathcal{R}_2$. At each iteration, we obtain a solution based on $\mathcal{R}_1$ and compute its approximation ratio α. If the approximation ratio is no less than $1 - 1/e - \varepsilon$, then the algorithm returns this solution. Otherwise, we generate new RR sets and insert them into $\mathcal{R}_1$ and

---

Algorithm 2. Seeding Process

Input: graph $G = (V, E)$, budget B, set X, error threshold ε, failure probability threshold δ

Output: set S

1: $\theta_{max} \leftarrow \dfrac{2n\left((1-1/e)\sqrt{\ln\frac{6}{\delta}} + \sqrt{(1-1/e)(\ln\frac{n}{B} + \ln\frac{6}{\delta})}\right)^2}{\varepsilon^2(B+|X|)}$

2: $\theta_0 \leftarrow \dfrac{\theta_{max} \cdot \varepsilon^2(B+|X|)}{n}$

3: Generate two collections of RR sets $\mathcal{R}_1$ and $\mathcal{R}_2$, with $|\mathcal{R}_1| = |\mathcal{R}_2| = \theta_0$

4: $i_{max} \leftarrow \lceil \log_2 \frac{\theta_{max}}{\theta_0} \rceil$

5: **for** $i = 1 \, to \, i_{max}$ **do**

6:      $S \leftarrow$ a size-B set based on $\mathcal{R}_1$ using Algorithm 3

7:      $I^u(S^* \cup X) \leftarrow \left(\sqrt{\Lambda_{\mathcal{R}_1}^u(S^* \cup X) + \dfrac{\ln(1/\delta_1)}{2}} + \sqrt{\dfrac{\ln(1/\delta_1)}{2}}\right)^2 \cdot \dfrac{n}{|\mathcal{R}_1|}$, where $\delta_1 = \delta/(3i_{max})$

8:      $I^l(S \cup X) \leftarrow \left(\left(\sqrt{\Lambda_{\mathcal{R}_2}(S \cup X) + \dfrac{2\ln(1/\delta_2)}{9}} - \sqrt{\dfrac{\ln(1/\delta_2)}{2}}\right)^2 - \dfrac{\ln(1/\delta_2)}{18}\right) \cdot \dfrac{n}{|\mathcal{R}_2|}$,

     where $\delta_2 = \delta/(3i_{max})$

9:      $\alpha \leftarrow I^l(S \cup X)/I^u(S^* \cup X)$

10:      **if** $\alpha \geq 1 - 1/e - \varepsilon$ **then**

11:         **return** S

12:      **end if**

13:      Double the sizes of $\mathcal{R}_1$ and $\mathcal{R}_2$ with new RR sets, respectively

14: **end for**

15: **return** S

$\mathcal{R}_2$, respectively. Algorithm 2 completes when the number of iterations is equal to i_{max}. Here, the solution corresponding to $\mathcal{R}_1$ is obtained from Algorithm 3. Algorithm 3 can return a set so that its union with a given set X can cover the most sets in $\mathcal{R}_1$.

Lemma 2. *Given a set X, an error threshold $\varepsilon \in (0,1)$ and a failure probability threshold $\delta \in (0,1)$, Algorithm 2 returns a $(1-1/e-\varepsilon)$-approximate solution for $\max_{S \subseteq V, |S|=B} I(S \cup X)$ with at least probability $1 - \delta$.*

Proof. When $1 \leq i \leq i_{max} - 1$, the approximation ratio obtained from $I^l(S \cup X)/I^u(S^* \cup X)$ is incorrect with at least probability $1 - 2\delta/(3i_{max})$ at each iteration i. When $i = i_{max}$, according to the results in [21], Algorithm 2 achieves a $(1 - 1/e - \varepsilon)$-approximate solution with at least probability $1 - \delta/3$ when $|\mathcal{R}_1| = \theta_{max}$. Using the union bound, the lemma follows.

Theorem 2. *Combined with Algorithm 2, the adaptive policy π returned by Algorithm 1 satisfies $\mathbb{E}_\omega[f^T(\pi)] \geq (1 - e^{-\sum_{i=0}^{k-1} \frac{\alpha_i}{k\beta_i}}) \cdot \mathbb{E}_\omega[f^T(\pi^*)]$, where $\alpha_i = 1 - 1/e - \varepsilon_i \cdot (1 - \delta_i) - \delta_i$, $\varepsilon_i, \delta_i \in (0,1)$ and β_i is the comparative factor.*

Algorithm 3. Max-Coverage

Input: a collection of RR sets $\mathcal{R}$, budget B, set X
Output: set S
1: Initialize $S = \emptyset$
2: Remove from $\mathcal{R}$ all sets covered by X
3: **for** $i = 1\,to\,B$ **do**
4:    $v \leftarrow$ the node that covers the most sets in $\mathcal{R}$
5:    $S \leftarrow S \cup \{v\}$
6:    Remove from $\mathcal{R}$ all sets in which v appears
7: **end for**

Proof. At each seeding process i, Algorithm 2 returns a $(1 - 1/e - \varepsilon_i(\omega))$-approximate solution for $\max_{S \subseteq V, |S| = B} I(S \cup X_i)$, where $X_i \subseteq V$. According to Lemma 2, $\varepsilon_i(\omega) \leq \varepsilon_i$ with at least probability $1 - \delta_i$. Then, we have

$$\mathbb{E}_\omega[\varepsilon_i(\omega)] \leq \varepsilon_i \cdot (1 - \delta_i) + \delta_i. \tag{12}$$

Combined with Theorem 1, this theorem holds.

Theorem 3. *When $\delta \leq 1/2$, Algorithm 1 runs in $O((kB \ln n + k \ln(1/\delta))(n + m)\varepsilon^{-2}) + kBdT)$ expected time, where d is the average degree of the network.*

Proof. At each iteration i, we first identify a size-B set S_i using Algorithm 2 and Algorithm 2 runs in $O((k \ln n + \ln(1/\delta)(n + m)\varepsilon^{-2}))$ expected time if $\delta \leq 1/2$ [19]. Then, we observe the influence spread of S_i within time-step T. Based on this observation, we generate a new graph G_{i+1} and a set X_{i+1}. These processes run in $O(BdT)$ expected time. The number of iterations of the Algorithm 1 is k and the theorem follows.

5   Conclusions

Different from the classic influence maximization problem, the selection of seed nodes are not completed at one time for the adaptive influence maximization. We can select subsequent seed nodes according to the observation of the diffusion results. Therefore, the adaptive strategy can make better use of the budget. Notice that there is a feedback model for adaptive influence maximization, and different feedback models will lead to different strategies. In this paper, we consider a general feedback model, including the full feedback model and myopic feedback model. More specifically, the general feedback model allows a fixed observation time-step and the next seeds selection after this observation results. Generally, the objective function lacks adaptive submodularity, and it is hard to derive an approximate solution. To solve it, we propose a comparative factor and provide a theoretical analysis of the solution using an adaptive greedy algorithm. In addition, considering the randomness of the policy during algorithm execution, we discuss the adaptive stochastic influence maximization and design a feasible approximation algorithm.

References

1. Borgs, C., Brautbar, M., Chayes, J., Lucier, B.: Maximizing social influence in nearly optimal time. In: Proceedings of the Twenty-Fifth Annual ACM-SIAM Symposium on Discrete Algorithms, pp. 946–957. SIAM (2014)
2. Chen, W., Peng, B., Schoenebeck, G., Tao, B.: Adaptive greedy versus non-adaptive greedy for influence maximization. In: Proceedings of the AAAI Conference on Artificial Intelligence (2020)
3. Chen, W., Wang, C., Wang, Y.: Scalable influence maximization for prevalent viral marketing in large-scale social networks. In: Proceedings of the 16th ACM SIGKDD International Conference on Knowledge Discovery and Data Mining, pp. 1029–1038 (2010)
4. Chen, W., Yuan, Y., Zhang, L.: Scalable influence maximization in social networks under the linear threshold model. In: 2010 IEEE International Conference on Data Mining, pp. 88–97. IEEE (2010)
5. Domingos, P., Richardson, M.: Mining the network value of customers. In: Proceedings of the Seventh ACM SIGKDD International Conference on Knowledge Discovery and Data Mining, pp. 57–66. ACM (2001)
6. Golovin, D., Krause, A.: Adaptive submodularity: theory and applications in active learning and stochastic optimization. J. Artif. Intell. Res. **42**(1), 427–486 (2012)
7. Guo, J., Wu, W.: Adaptive influence maximization: if influential node unwilling to be the seed. ACM Trans. Knowl. Discov. Data (TKDD) **15**(5), 1–23 (2021)
8. Huang, K., et al.: Efficient approximation algorithms for adaptive influence maximization. VLDB J. **29**(6), 1385–1406 (2020). https://doi.org/10.1007/s00778-020-00615-8
9. Kempe, D., Kleinberg, J., Tardos, É.: Maximizing the spread of influence through a social network. In: Proceedings of the Ninth ACM SIGKDD International Conference on Knowledge Discovery and Data Mining, pp. 137–146. ACM (2003)
10. Lin, Y., Lui, J.: Analyzing competitive influence maximization problems with partial information: an approximation algorithmic framework. Perform. Eval. **91**, 187–204 (2015)
11. Lu, W., Chen, W., Lakshmanan, L.V.: From competition to complementarity: comparative influence diffusion and maximization. Proc. VLDB Endow. **9**(2), 60–71 (2015)
12. Nemhauser, G.L., Wolsey, L.A., Fisher, M.L.: An analysis of approximations for maximizing submodular set functions. Math. Program. **14**(1), 265–294 (1978)
13. Nguyen, N.P., Yan, G., Thai, M.T., Eidenbenz, S.: Containment of misinformation spread in online social networks. In: ACM Web Science Conference (2012)
14. Peng, B., Chen, W.: Adaptive influence maximization with myopic feedback. Adv. Neural. Inf. Process. Syst. **32**, 5574–5583 (2019)
15. Richardson, M., Domingos, P.: Mining knowledge-sharing sites for viral marketing. In: Proceedings of the Eighth ACM SIGKDD International Conference on Knowledge Discovery and Data Mining, pp. 61–70. ACM (2002)
16. Salha, G., Tziortziotis, N., Vazirgiannis, M.: Adaptive submodular influence maximization with myopic feedback (2017)
17. Sun, L., Huang, W., Yu, P.S., Wei, C.: Multi-round influence maximization. In: the 24th ACM SIGKDD International Conference (2018)
18. Tang, J., et al.: Efficient approximation algorithms for adaptive seed minimization. In: Proceedings of the 2019 International Conference on Management of Data, SIGMOD 2019, New York, NY, USA, pp. 1096–1113. Association for Computing Machinery (2019). https://doi.org/10.1145/3299869.3319881

19. Tang, J., Tang, X., Xiao, X., Yuan, J.: Online processing algorithms for influence maximization. In: Proceedings of the 2018 International Conference on Management of Data, pp. 991–1005 (2018)
20. Tang, S., Tong, G., Wu, W., Du, D.Z.: Adaptive influence maximization in dynamic social networks. In: IEEE/ACM Transactions on Networking: A Joint Publication of the IEEE Communications Society, the IEEE Computer Society, and the ACM with Its Special Interest Group on Data Communication (2017)
21. Tang, Y., Shi, Y., Xiao, X.: Influence maximization in near-linear time: a martingale approach. In: Proceedings of the 2015 ACM SIGMOD International Conference on Management of Data, pp. 1539–1554. ACM (2015)
22. Tang, Y., Xiao, X., Shi, Y.: Influence maximization: near-optimal time complexity meets practical efficiency. In: Proceedings of the 2014 ACM SIGMOD International Conference on Management of Data, pp. 75–86. ACM (2014)
23. Tong, A., Du, D.Z., Wu, W.: On misinformation containment in online social networks. In: Advances in Neural Information Processing Systems, pp. 341–351 (2018)
24. Vaswani, S., Lakshmanan, L.V.: Adaptive influence maximization in social networks: why commit when you can adapt? arXiv preprint arXiv:1604.08171 (2016)
25. Zhang, Y., Yang, W., Du, D.Z.: Rumor correction maximization problem in social networks. Theoret. Comput. Sci. **861**(1), 102–116 (2021)

Measured Continuous Greedy with Differential Privacy

Xin Sun, Gaidi Li, Yapu Zhang$^{(\boxtimes)}$, and Zhenning Zhang

Department of Operations Research and Information Engineering,
Beijing University of Technology, Beijing 100124, People's Republic of China
athossun@emails.bjut.edu.cn, {ligd,zhangyapu,zhangzhenning}@bjut.edu.cn

Abstract. In the paper, we design a privacy algorithm for maximizing a general submodular set function over a down-monotone family of subsets, which includes some typical and important constraints such as matroid and knapsack constraints. The technique is inspired by the measured continuous greedy (MCG) which compensates for the difference between the residual increase of elements at a given point and the gradient of it by distorting the original direction with a multiplicative factor. It directly makes the continuous greedy approach fit to the problem of maximizing a non-monotone submodular function. We generate the MCG algorithm in the framework of differential privacy. It is accepted as a robust mathematical guarantee and can provide the protection to sensitive and personal data. We propose a $1/e$-approximation algorithm for the general submodular function. Moreover, for monotone submodular objective functions, our algorithm achieves an approximation ratio that depends on the density of the polytope defined by the problem at hand, which is always at least as good as the previously known best approximation ratio of $1 - 1/e$.

Keywords: Approximation algorithm · Submodular maximization · Differential privacy · Down-monotone family

1 Introduction

The theory of *submodular maximization* provides a general and unified framework for various combinatorial optimization problems including Maximum Coverage, Maximum Cut and Facility Location [13]. Furthermore, it also appears in a wide variety of applications such as viral marketing [19], information gathering [21], feature selection for classification [20], influence maximization in social networks [19], document summarization [23], speeding up satisfiability solvers [32], computer vision [18], social welfare [34] and data privacy [31]. As a consequence of its importance in these applications, a wide range of efficient approximation algorithms have been developed for maximizing submodular functions subject to different constraints.

In this paper, given a non-negative submodular function $f : 2^{\mathcal{X}} \to \mathbb{R}_{\geq 0}$ over a ground set $\mathcal{X}$, we focus on the basic problem of seeking a subset of $2^{\mathcal{X}}$ maximizing

© Springer Nature Switzerland AG 2021
W. Wu and H. Du (Eds.): AAIM 2021, LNCS 13153, pp. 212–226, 2021.
https://doi.org/10.1007/978-3-030-93176-6_19

f and satisfying a down-monotone family of subsets $\mathcal{I} \subseteq 2^{\mathcal{X}}$ in the framework of *differential privacy*. A family of subsets $\mathcal{I} \subseteq 2^{\mathcal{X}}$ is *down-monotone* if $B \in \mathcal{I}$ and $A \subseteq B$ imply $A \in \mathcal{I}$. Note that many natural families of subsets $\mathcal{I}$ are down-monotone, e.g., families introduced by matroid and knapsack constraints.

1.1 The Continuous Greedy Approach

For now, the *continuous greedy* is the most popular approach for solving various constrained submodular maximization problems. This approach has been used to obtain improved approximations to various problems [6,8–10,22,24]. Most notable of these results is an asymptotically tight approximation for maximizing a monotone submodular function given a single matroid constraint [6,28,33]. The *matroid constraint* is one of the important instances for the problem we consider, since it gives a pithy unifying abstract treatment of dependence in linear algebra and graph theory. Besides, the theory of matroid has its the most successful applications in the areas of combinatorial optimization and network theory. For maximizing a monotone submodular function over a general matroid constraint, a renowned 1/2-approximation greedy algorithm is given by Nemhauser et al. [29]. Then, Nemhauser and Wolsey [28] points out that there exists no polynomial-time algorithm with an approximation guarantee better than $(1 - 1/e)$ even for the special case of cardinality constraint. In 2011, Calinescu et al. [6] proposed an optimal randomized $(1-1/e)$-approximation algorithm based on the *continuous greedy* plus *pipage rounding* [2], which can be seen as a milestone that firstly achieves the bound of approximation guarantee. This continuous approach formulates the objective to a continuous function known as *multilinear extension*. A function $F : [0,1]^{\mathcal{X}} \rightarrow \mathbb{R}_{\geq 0}$ is the multilinear extension of f if

$$F(\mathbf{y}) = \mathbb{E}_{R \sim \mathbf{y}}[f(R)] = \sum_{R \subseteq \mathcal{S}} f(R) \prod_{i \in R} y_i \prod_{j \notin R}(1 - y_j),$$

where the random set R contains element i independently with probability y_i. Commonly, it selects a feasible point $\mathbf{x} \in \mathcal{P}$ greedily by solving $x = \arg\max\{w(\mathbf{y}) \cdot \mathbf{x} | \mathbf{x} \in \mathcal{P}\}$ where the feasibility polytope is down monotone and solvable and the weight vector $w(\mathbf{y}) \in \mathbb{R}^{\mathcal{X}}$ is $w(\mathbf{y})_e = F(\mathbf{y} \vee \mathbf{1}_e) - F(\mathbf{y})$, for every $e \in \mathcal{X}$. Thus, $\mathbf{x}$ is chosen according to the *residual increase* of each element e, i.e., $F(\mathbf{y} \vee \mathbf{1}_e) - F(\mathbf{y})$. Then, the algorithm returns a fractional solution with $1 - 1/e$ approximation guarantee when f is monotone and submodular and the polytope $\mathcal{P}$ is solvable. As for the rounding techniques, the swap rounding [8] or pipage rounding can efficiently meet our needs without any loss of approximation, since the set function is submodular. However, this method is only known to work for the multilinear extensions of monotone submodular functions.

In contrast, the general problem with a *non-monotone* objective has proved to be considerably more challenging. It is well-studied [5,7,12–15,17,24,26], particularly under a cardinality constraint [15,17,24,26] and matroid constraints [13] with guarantee $1/e - o(1)$ [5]. A refined approach for continuous greedy is the *measured continuous greedy* given by [14] in 2011. It compensates for the difference between the residual increase of elements at point $\mathbf{y}$, and $\nabla F(\mathbf{y})$, by distorting the direction x as to mimic the value of $\nabla F(\mathbf{y})$. This is done by decreasing $\mathbf{x}_e$, for every $e \in \mathcal{X}$, by a multiplicative factor of $1 - \mathbf{y}_e$. Moreover, Buchbinder and Feldman [4] give the currently best algorithm for general matroid constraints with 0.385 ratio.

1.2 Differential Privacy and CPP Problem

The need for efficient optimization methods that guarantee the privacy of individuals is wide-spread across many applications concerning sensitive data about individuals, e.g., medical data, web search query data, salary data, social networks. Differential privacy [11] is a kind of privacy concept that gives a standard paradigm for confidential data analysis. Informally, differential privacy guarantees that the distribution of outcomes of the computation does not change significantly when one individual changes its input data. We say two datasets are *neighboring* if they differ in a single record and we denote this relationship by $D \sim D'$. Formally, we give the definition of differential privacy here, which is induced by Dwork et al. [11] in 2006.

Definition 1.1. *For $\epsilon, \delta \in \mathbb{R}_{\geq 0}$, we say that a randomized computation M is (ϵ, δ)-differentially private if for any neighboring datasets $D \sim D'$, and for any set of outcomes $S \subseteq range(M)$,*

$$\Pr[M(D) \in S] \leq \exp(\epsilon) \Pr[M(D') \in S] + \delta.$$

When $\delta = 0$, we say M is ϵ-differentially private.

Gupta et al. [16] considered an important case of the problem we consider called the *Combinatorial Public Projects* (CPP problem). The CPP problem was introduced by Papadimitriou et al. [30] and is as follows. For a dataset $D = (x_1, \ldots, x_n)$, each individual x_i submits a *private* non-decreasing and submodular valuation function $G_{x_i} : 2^{\mathcal{X}} \to [0, 1]$. The goal is to select a subset $S \subseteq \mathcal{X}$ to maximize function G_D that takes the particular form $G_D(S) = \frac{1}{n} \sum_{i=1}^{n} G_{x_i}(S)$.

Gupta et al. [16] gave an (ϵ, δ)-differentially private algorithm with $(1 - \frac{1}{e})OPT - O(\frac{k \ln(e/\delta) \ln |\mathcal{X}|}{\epsilon})$ approximation ratio under cardinality constraint. Mitrovic et al. [27] gave differentially private algorithms for monotone submodular maximization under different constraints. It proposed an $1 - 1/e$ approximation private algorithm for cardinality and $1/2$ for matroid, and p-extendible system constraints. Recently, Rafiey and Yoshida [31] presented a $(1 - \frac{1}{e})$-approximation algorithm based on the continuous greedy approach for monotone submodular maximization subject to matroid constraints in the framework of differential privacy. Moreover, differential privacy has great applications in deep learning, called private learning [1], to prevent that the models expose private information in the datasets. In 2021, Bu et al. [3] generates the techniques to many useful optimizers like SGD and Adam.

In our case, a dataset D consists of private submodular functions $f_1, \ldots, f_n :$ $2^{\mathcal{X}} \to [0, 1]$. Recall that two datasets D and D' are neighboring if all but one submodular function in those datasets are equal. The submodular function f_D depends on the dataset D in different ways, for example $f_D(S) = \sum_{i=1}^n f_i(S)/n$ (CPP problem), or much more complicated ways than averaging functions associated to each individual. Differentially private algorithms must be calibrated to the sensitivity of the function of interest with respect to small changes in the input dataset, defined formally as follows.

Definition 1.2. *The sensitivity of a function $f_D : X \to Y$, parameterized by a dataset D, is defined as $\max_{D':D' \sim D} \max_{x \in X} |f_D(x) - f_{D'}(x)|$. A function with sensitivity Δ is called Δ-sensitive.*

Note that in this setting, the sensitivity can be always bounded from above by $\frac{1}{n}$.

1.3 Our Contribution

We clearly present a private version of the powerful technique measured continuous greedy, which is one of the best approaches for solving the problem of maximizing the general submodular set function subject to a down-monotone family of subsets, which includes some typical and important constraints such as matroid and knapsack constraints. This $1/e$-approximation algorithm inherits from the well-known continuous greedy for monotone submodular maximization. It compensates for the difference between the residual increase of elements at a given point and the gradient of it by distorting the original direction with a multiplicative factor. From the perspective of sensitive data privacy, we generates the MCG in the framework of differential privacy, which guarantees the distribution of outcomes of the computation does not change significantly when one individual changes its input data. Following the exponential mechanism, we present a brand new analysis process for this algorithm in Sect. 3.1 for all theoretical results, although the structure is inherited from [14]. Besides, we give a privacy analysis (stated as Theorem 3.1) for this algorithm. The whole analysis finally leads to a $(Te^{-T} - o(1))$ approximation guarantee with $O(\frac{\Delta}{n^4})$ privacy loss, where T is the stopping time and Δ denotes the sensitivity of the set function. It is obvious that we can nearly get $1/e$ when we set $T = 1$. Moreover, our private algorithm is also suitable for the monotone case with the help of the density of the polytope with respect to the constraint. The approximation ratio is at least as good as the previously known best guarantee $1 - 1/e$.

1.4 Organization

We first cover preliminary definitions and basic properties in Sect. 2. Then, the private algorithm and its analysis are proposed in Sect. 3. We give three main results for different settings respectively. Moreover, the privacy analysis is also showed in Sect. 3, which measures how safe of our data can be protected by our algorithm. Finally, we conclude the paper in Sect. 4.

2   Preliminaries

In the section we provide the necessary notations and terms appeared in the rest sections. Given a *ground set* $\mathcal{X}$ including n elements, a set function $f : 2^{\mathcal{X}} \to \mathbb{R}$ is *submodular* if and only if $f(A)+f(B) \geq f(A \cup B)+f(A \cap B)$ for any $A, B \subseteq \mathcal{X}$. Besides, we say f is *monotone* if $f(A) \leq f(B)$ for all $A \subseteq B \subseteq \mathcal{X}$. Also, f is *non-negative* and *normalized* if $f(A) \geq 0$ for $A \in \mathcal{X}$ and $f(\emptyset) = 0$, respectively. For briefness, we use the shorthands $S + u$ and $S - u$ for $S \cup \{u\}$ and $S \backslash \{u\}$, respectively.

Given a non-negative submodular function $f : 2^{\mathcal{X}} \to \mathbb{R}_+$, our goal is to find a subset belonging to a down-monotone family of $\mathcal{I} \subseteq 2^{\mathcal{X}}$ in the framework of differential privacy. A family of subsets $\mathcal{I} \subseteq 2^{\mathcal{X}}$ is *down-monotone* if $B \in \mathcal{I}$ and $A \subseteq B$ imply $A \in \mathcal{I}$. Formally, the problem can be simply formulated as:

$$\max_{S \in \mathcal{I}} f(S)$$

where $\mathcal{I} \subseteq 2^{\mathcal{X}}$ can be viewed as the constraint family. Also, we assume that we have a *value oracle* which returns the value of $f(S)$ immediately and answers whether $S \in \mathcal{I}$.

2.1   Multilinear Relaxation

The *multilinear extension* of a set function f is defined as $F : [0,1]^n \to \mathbb{R}_+$, which maps a point $\mathbf{x} \in [0,1]^{\mathcal{X}}$ to the expected value of a random set $R(\mathbf{x}) \subseteq \mathcal{X}$ containing each element $e \in \mathcal{X}$ with probability x_e independently, i.e. $F(\mathbf{x}) := \mathbb{E}[f(R(\mathbf{x}))] = \sum_{S \subseteq \mathcal{X}} f(S) \prod_{e \in S} \mathbf{x}_e \prod_{e \notin S} (1 - \mathbf{x}_e)$. For vectors $\mathbf{x}, \mathbf{y} \in [0,1]^n$, we denote $(\mathbf{x} \vee \mathbf{y})_e = \max\{\mathbf{x}_e, \mathbf{y}_e\}$ and $(\mathbf{x} \wedge \mathbf{y})_e = \min\{\mathbf{x}_e, \mathbf{y}_e\}$ as the coordinate-wise maximum and minimum of these two vectors, respectively.

The first order and second order properties of multilinear extension are well-known as follows:

Lemma 2.1 ([6]). *Let $F : [0,1]^n \to \mathbb{R}$ be the multilinear extension of a monotone submodular function $f : 2^{\mathcal{X}} \to \mathbb{R}$. Then*

(1) $\partial_e F(\mathbf{x}) := \frac{\partial F(\mathbf{x})}{\partial \mathbf{x}_e} = \frac{F(\mathbf{x} \vee 1_e) - F(\mathbf{x})}{1 - \mathbf{x}_e} = \frac{F(\mathbf{x}) - F(\mathbf{x} \wedge 1_{\bar{e}})}{\mathbf{x}_e} = F(\mathbf{x} \vee 1_e) - F(\mathbf{x} \wedge 1_{\bar{e}})$;
(2) F *is monotone, meaning* $\frac{\partial F}{\partial \mathbf{x}_e} \geq 0$. *Hence,* $\nabla F(\mathbf{x}) = (\frac{\partial F}{\partial \mathbf{x}_1}, \dots, \frac{\partial F}{\partial \mathbf{x}_n})$ *is a non-negative vector;*
(3) F *is concave along any direction* $\mathbf{d} \geq 0$.

Also, we present the local nearly linearity of multilinear extension as a technical lemma for the analysis.

Lemma 2.2 ([13]). *Consider two vectors $\mathbf{x}, \mathbf{x}' \in [0,1]^n$ such that $|\mathbf{x}_e - \mathbf{x}'_e| \leq \delta$ for every $e \in \mathcal{X}$, and let F be the multilinear extension of a non-negative submodular function f. Then,*

$$F(\mathbf{x}') - F(\mathbf{x}) \geq (\mathbf{x}' - \mathbf{x}) \cdot \partial F(\mathbf{x}) - O(n^3 \delta^2) \cdot \max_{e \in \mathcal{X}} f(e).$$

Even though the submodular function defined by the multilinear extension is neither convex nor concave, it is still possible to efficiently compute an approximate feasible fractional solution for the relaxation, assuming its feasibility polytope $\mathcal{P}$ is down-monotone and solvable. A polytope $\mathcal{P} \subseteq [0,1]^n$ is *donw-monotone* if $x \in \mathcal{P}$ and $0 \leq \mathbf{y} \leq \mathbf{x}$ imply $\mathbf{y} \in \mathcal{P}$. A polytope $\mathcal{P}$ is *solvable* if linear functions can be maximized over it in polynomial time. For the problem we consider, the underlying polytope $\mathcal{P}$ can be defined by m inequality constraints, i.e., $\sum_{e \in \mathcal{X}} a_{i,e} \mathbf{x}_e \leq b_i$ for all $1 \leq i \leq m$. Also, we denote the *density* of $\mathcal{P}$ by $d(\mathcal{P}) = \min_{1 \leq i \leq m} \frac{b_i}{\sum_{e \in \mathcal{X}} a_{i,e}}$. It is straightforward to assume that all parameter $a_{i,e}$ and b_i are non-negative, and $0 < d(\mathcal{P}) \leq 1$. Also, we denote the ρ-covering of the polytope $\mathcal{P}$ as the neighboring polytope inspired by [31].

Definition 2.1. *Let $K \subseteq \mathbb{R}^{\mathcal{X}}$ be a set. For $\rho > 0$, a set $C \subseteq K$ of points is called a ρ-covering of K if for any $\mathbf{x} \in K$, there exists $\mathbf{y} \in C$ such that $\|\mathbf{x} - \mathbf{y}\| \leq \rho$.*

And we can construct it in $O(n^{1/\epsilon^2})$ time due to [36].

2.2 Lovász Extension

In addition to the multilinear extension, we make use of the Lovász extension. Given a vector $\mathbf{x} \in [0,1]^n$ and a scalar $\lambda \in [0,1]^n$, let $T_\lambda(\mathbf{x})$ be the set of elements in $\mathcal{X}$ whose coordinate in $\mathbf{x}$ is at least λ, i.e., $T_\lambda(\mathbf{x}) = \{e \in \mathcal{X} : \mathbf{x}_e \geq \lambda\}$. The Lovász extension of a set function $f : 2^{\mathcal{X}} \to \mathbb{R}$ is defined as

$$\hat{f}(x) = \int_0^1 f(T_\lambda(\mathbf{x})) d\lambda = \mathbb{E}_{\lambda \sim \text{Unif}[0,1]}[f(T_\lambda(\mathbf{x}))].$$

Beside its use in relaxations for minimization problems, the Lovász extension can also be used to lower bound the multilinear extension via the following lemma.

Lemma 2.3 ([35]). *Let $F(\mathbf{x})$ denote the multilinear extension and $\hat{f}(\mathbf{x})$ denote the Lovász extension of submodular function $f : 2^{\mathcal{X}} \to \mathbb{R}$. Then $F(\mathbf{x}) \geq \hat{f}(\mathbf{x})$.*

2.3 Exponential Mechanism

One particularly general tool that we will use is the *exponential mechanism* introduced by McSherry and Talwar [25]. The exponential mechanism is defined in terms of a *quality function* $q_D : \mathcal{R} \to \mathbb{R}$, which is parameterized by a dataset D and maps a candidate result $R \in \mathcal{R}$ to a real-valued score.

Definition 2.2. *Let $\epsilon, \Delta > 0$ and let $q_D : \mathcal{R} \to \mathbb{R}$ be a quality score. Then, the exponential mechanism $EM(\epsilon, \Delta, q_D)$ outputs $R \in \mathcal{R}$ with probability proportional to $\exp(\frac{\epsilon}{2\Delta} \cdot q_D(R))$.*

The following result shows the chance of obtaining a good enough output by the instructions of the exponential mechanism.

Theorem 2.1 ([25]). *Suppose that the quality score $q_D : \mathcal{R} \to \mathbb{R}$ is Δ-sensitive. Then, $EM(\epsilon, \Delta, q_D)$ is ϵ-differentially private, and for every $\beta \in (0,1)$ outputs $R \in \mathcal{R}$ with*

$$\Pr\left[q_D(R) \geq \max_{R' \in \mathcal{R}} q_D(R') - \frac{2\Delta}{\epsilon} \ln\left(\frac{|\mathcal{R}|}{\beta}\right)\right] \geq 1 - \beta.$$

3   The Privacy Algorithm

This section starts from the differential private algorithm, which is a modification of the *measured continuous greedy* algorithm [14] and showed in Algorithm 1. Instead of finding the optimal solution for the linear programming $\max_{\mathbf{x} \in \mathcal{P}} \mathbf{x} \cdot \nabla f_D(\mathbf{y}(t))$, we sample a vector $I(t)$ in the feasible polytope with probability proportional to $\exp\left(\epsilon'\langle I(t), w(t)\rangle\right)$, where $\langle \cdot \rangle$ denotes the inner product of two vectors. Apparently, we will obtain better solutions with higher probability by following this rule.

Algorithm 1. Differentially Private Measured Continuous Greedy

---

Input: Submodular function $f_D : 2^{\mathcal{X}} \to [0,1]$, dataset D, polytope $\mathcal{P}$, ρ-covering $C_\rho^\mathcal{P}$
   of $\mathcal{P}$ $\rho \geq 0$ and $\epsilon > 0$.
Output: $\mathbf{y}(t)$
1: Initialization: $\delta \leftarrow T(\lceil n^5 T \rceil)^{-1}$, $t \leftarrow 0$, $\mathbf{y}(0) \leftarrow \mathbf{1}_\emptyset$, $\epsilon' \leftarrow \frac{\epsilon}{2\Delta}$
2: **while** $t < T$ **do**
3:    **for** $e \in \mathcal{X}$ **do**
4:       $w_e(t) \leftarrow F(\mathbf{y} \vee \mathbf{1}_e) - F(\mathbf{y}(t))$
5:    **end for**
6:    Sample $I(t) \in C_\rho^\mathcal{P}$ with probability proportional to $\exp\left(\epsilon'\langle I(t), w(t)\rangle\right)$
7:    $\mathbf{y}(t + \delta) \leftarrow \mathbf{y}(t) + \delta I(t) \cdot (1 - \mathbf{y}(t))$
8:    $t \leftarrow t + \delta$
9: **end while**

3.1   Analysis

In this subsection, we analyse Algorithm 1 for general non-negative submodular functions and give the results when f is non-monotone and monotone, respectively. Firstly, we present the result of privacy analysis, which describes the privacy level of Algorithm 1.

Theorem 3.1. *Algorithm 1 preserves $O(\epsilon \cdot d_\mathcal{P}^2)$-differential privacy.*

   Then, we show one of the most important results in our analysis, which is given in the next lemma. It helps us to give the lower bound on the improvement achieved by the algorithm in each iteration.

Lemma 3.1. *For every time* $0 \leq t \leq T$, $I(t)$ *is the sampled vector by Algorithm 1 and* $I'(t)$ *is the solution of the linear programming in* measured continuous greedy *algorithm. Then,*

$$I(t)w(t)(1 - \mathbf{y}(t)) \geq F(\mathbf{y}(t) \vee \mathbf{1}_{OPT}) - F(\mathbf{y}(t)) - O\left(\sqrt{\epsilon} + \frac{2\Delta \ln n}{\epsilon^3}\right).$$

Together with multilinearity of the multilinear extension (Lemma 2.2), we get a lower bound on the improvement achieved by the algorithm in each iteration.

Lemma 3.2. *For every time* $0 \leq t < T$,

$$F(\mathbf{y}(t + \delta)) - F(\mathbf{y}(t))$$
$$\geq \delta \cdot \left[F(\mathbf{y}(t) \vee \mathbf{1}_{OPT}) - F(\mathbf{y}(t)) - O\left(\sqrt{\epsilon} + \frac{2\Delta \ln n}{\epsilon^3}\right)\right] - O(n^3\delta^2) \cdot f(OPT).$$

Then, we take advantage of the relationship between the improvement of each iteration and the residual value to lower bound $F(\mathbf{y}(t) \vee \mathbf{1}_{OPT})$ with the help of Lovász extension (see Lemma 3.5 and Lemma 3.6 in [13]), which leads to the result below. It is a lower bound on the improvement achieved by the algorithm in each iteration.

Lemma 3.3. *For every time* $0 \leq t < T$,

$$F(\mathbf{y}(t) + \delta) - F(\mathbf{y}(t))$$
$$\geq \delta \cdot \left[(e^{-t} - O(\delta))f(OPT) - F(\mathbf{y}(t)) - O\left(\sqrt{\epsilon} + \frac{2\Delta \ln n}{\epsilon^3}\right)\right] - O(n^3\delta^2) \cdot f(OPT)$$
$$= \delta \cdot \left[e^{-t} \cdot f(OPT) - F(\mathbf{y}) - O\left(\sqrt{\epsilon} + \frac{2\Delta \ln n}{\epsilon^3}\right)\right] - O(n^3\delta^2) \cdot f(OPT).$$

In order to get the approximate ratio, we still need the help of two auxiliary functions. First we define $g(t) : g(0) = 0$ and $g(t + \delta) = (1 - \delta)g(t) + \delta e^{-t}f(OPT)$.

Lemma 3.4. *For every* $0 \leq t \leq T$,

$$g(t) \leq F(\mathbf{y}(t)) + O(n^3\delta) \cdot t f(OPT) + \delta O\left(\sqrt{\epsilon} + \frac{2\Delta \ln n}{\epsilon^3}\right).$$

Second we define $h(t) = te^{-t}f(OPT)$ and we use the former auxiliary function to bound it.

Lemma 3.5. *For every time* $0 \leq t \leq T$, $g(t) \geq h(t)$.

Our next corollary can help us easily proof the first theorem for the non-monotone case.

Corollary 3.1. $F(\mathbf{y}(t)) \geq \left[Te^{-T} - o(1)\right] \cdot f(OPT) - \delta O\left(\sqrt{\epsilon} + \frac{2\Delta \ln n}{\epsilon^3}\right).$

The theorem below shows the approximation guarantee achieved by our algorithm for the multilinear relaxation problem of maximizing non-monotone submodular set function while satisfying a down-monotone family of subsets.

Theorem 3.2. *Suppose f_D is Δ-sensitive and $\mathcal{P} \subseteq [0,1]^n$ is down-monotone solvable convex polytope and stopping time $T \in [0,1]$. Then, Algorithm 1, with high probability, finds $\mathbf{y} \in [0,1]^n$ such that $\mathbf{y}/T \in \mathcal{P}$ and*

$$F(\mathbf{y}) \geq [Te^{-T} - o(1)] \cdot f(OPT) - \delta \cdot O\left(\sqrt{\epsilon} + \frac{2\Delta \ln n}{\epsilon^3}\right).$$

For the existence of the output of Algorithm 1, we can easily get $\mathbf{y}(t) \in [0,1]^n$ for any $t < T$ and $\mathbf{y}/T \in \mathcal{P}$ by convexity and the down-monotonicity of $\mathcal{P}$. These two results make sure the fractional solution always satisfies the definition of multilinear extension and the solution scaled by the execution time stays in the constraint polytope. And we can get the discrete solution by using pipage rounding or swap rounding without any loss of approximation.

Moreover, we give the approximation guarantee of Algorithm 1 for monotone submodular set function with differential privacy. Combine the analysis process of the non-monotone case and section 3.2 in [13], we can easily get the coming theorem. Therefore, we only present it here without proof.

Theorem 3.3. *Suppose f_D is Δ-sensitive and $\mathcal{P} \subseteq [0,1]^n$ is down-monotone solvable convex polytope and stopping time $T \geq 0$. Then, Algorithm 1, with high probability, finds $\mathbf{x} \in [0,1]^n$ such that*

$$F(\mathbf{x}) \geq [1 - e^{-T} - o(1)] \cdot f(OPT) - \delta \cdot O\left(\sqrt{\epsilon} + \frac{2\Delta \ln n}{\epsilon^3}\right).$$

Additionally, we have

(1) $\mathbf{x}/T \in \mathcal{P}$.
(2) Let $T_\mathcal{P} = -\ln(1 - d(\mathcal{P}) + n\delta)/d(\mathcal{P})$. Then, $T \leq T_\mathcal{P}$ implies $\mathbf{x} \in \mathcal{P}$,

where, $d(\mathcal{P}) := \min_{1 \leq i \leq m} \frac{b_i}{\sum_{e \in \mathcal{X}} a_{i,e}}$ defines the density of the polytope $\mathcal{P}$.

At last, we present a further conclusion when the polytope $\mathcal{P}$ binary polytope, which can be viewed as a special case of our constraints.

Theorem 3.4. *Let $\mathbf{x}$ be the output of the measured continuous greedy, assuming $T = T_\mathcal{P}$. Then,*

$$F(\mathbf{y}) \geq \left[1 - (1 - d(\mathcal{P}))^{1/d(\mathcal{P})} - O(n^{-2})\right] \cdot f(OPT) - \delta O\left(\sqrt{\epsilon} + \frac{2\Delta \ln n}{\epsilon^3}\right)$$

$$= \left[1 - (1 - d(\mathcal{P}))^{1/d(\mathcal{P})} - o(1)\right] \cdot f(OPT) - \delta O\left(\sqrt{\epsilon} + \frac{2\Delta \ln n}{\epsilon^3}\right).$$

4   Conclusion

We proposed a differentially private algorithm for maximizing a non-negative submodular function under a down-monotone family of subsets. For general non-monotone submodular objective functions, our algorithm achieves an $1/e$

approximation ratio. For monotone submodular objective functions, our algorithm achieves an approximation ratio that depends on the density of the polytope defined by the problem at hand, which is always at least as good as the previously known best approximation ratio of $1 - 1/e$. Moreover, we present a more elegant approximation guarantee for the special case of our constraints, when the polytope $\mathcal{P}$ is binary. Finally, we can easily get the discrete solution of all these results without any approximation loss by commonly rounding techniques, such as pipage rounding or swap rounding.

Acknowledgements. The first author is supported by Beijing Natural Science Foundation Project No. Z200002 and National Natural Science Foundation of China (No. 12131003). The fourth author is supported by National Natural Science Foundation of China (No. 12001025) and Science and Technology Program of Beijing Education Commission (No. KM201810005006).

Appendix: Missing Proofs

Theorem 3.1. *Algorithm 1 preserves $O(\epsilon \cdot d_{\mathcal{P}}^2)$-differential privacy.*

Proof. Let D and D' be two neighboring datasets and f_D, $f_{D'}$ be their associated functions. For a fixed $\mathbf{y}_t \in C_\rho$, we consider the relative probability of Algorithm 1 (denoted by M) choosing $\mathbf{y}_t$ at time step t given multilinear extensions of f_D and $f_{D'}$. Let $M_t(f_D|\mathbf{x}_t)$ denote the output of M at time step t given dataset D and point $\mathbf{x}_t$. Similarly, $M_t(f_{D'}|\mathbf{x}_t)$ denotes the output of M at time step t given dataset D' and point $\mathbf{x}_t$. Further, write $d_{\mathbf{y}} = \langle \mathbf{y}, \nabla f_D(\mathbf{x}_t) \rangle$ and $d'_{\mathbf{y}} = \langle \mathbf{y}, \nabla f_D(\mathbf{x}_t) \rangle$. We have

$$\frac{\Pr[M_t(f_D|\mathbf{x}_t) = \mathbf{y}_t]}{\Pr[M_t(f_{D'}|\mathbf{x}_t) = \mathbf{y}_t]} = \frac{\exp(\epsilon' \cdot d_{\mathbf{y}_t})}{\exp(\epsilon' \cdot d'_{\mathbf{y}_t})} \cdot \frac{\sum_{\mathbf{y} \in C_\rho} \exp(\epsilon' \cdot d'_{\mathbf{y}})}{\sum_{\mathbf{y} \in C_\rho} \exp(\epsilon' \cdot d_{\mathbf{y}})}.$$

For the first factor, we have

$$\frac{\exp(\epsilon' \cdot d_{\mathbf{y}_t})}{\exp(\epsilon' \cdot d'_{\mathbf{y}_t})}$$
$$= \exp\left(\epsilon'(d_{\mathbf{y}_t} - d'_{\mathbf{y}_t})\right)$$
$$= \exp\left(\epsilon'(\langle \mathbf{y}_t, \nabla f_D(\mathbf{x}_t) - \nabla f_{D'}(\mathbf{x}_t)\rangle)\right)$$
$$\leq \exp\left(\epsilon' \|\mathbf{y}_t\|_1 \|\nabla f_D(\mathbf{x}_t) - \nabla f_{D'}(\mathbf{x}_t)\|_\infty\right)$$
$$= \exp\left(\epsilon' \sum_{e \in \mathcal{X}} \mathbf{y}_t(e) \cdot \left(\max_{e \in \mathcal{X}} \mathbb{E}_{R \sim \mathbf{x}_t}\left[f_D(R \cup \{e\}) - f_D(R) - f_{D'}(R \cup \{e\}) + f_{D'}(R)\right]\right)\right)$$
$$\leq \exp(O(\epsilon' \cdot m d_{\mathcal{P}} \cdot 2\Delta)) = \exp(O(\epsilon \cdot d_{\mathcal{P}})).$$

Note that the last inequality holds since $\mathbf{y}_t$ is a member of the polytope $\mathcal{P}$ and by definition we have $\sum_{e \in \mathcal{X}} a_{i,e} \mathbf{y}_t(e) \leq b_i$ and $d_{\mathcal{P}} = \min_{1 \leq i \leq m} \frac{b_i}{\sum_{e \in \mathcal{X}} a_{i,e}}$. Moreover, recall that f_D is Δ-sensitive.

For the second factor, let us write $\beta_{\mathbf{y}} = d'_{\mathbf{y}} - d_{\mathbf{y}}$ to be the deficit of the probabilities of choosing direction $\mathbf{y}$ in instances $f_{D'}$ and f_D. Then, we have

$$\frac{\sum_{\mathbf{y} \in C_\rho} \exp(\epsilon' \cdot d'_{\mathbf{y}})}{\sum_{\mathbf{y} \in C_\rho} \exp(\epsilon' \cdot d_{\mathbf{y}})} = \frac{\sum_{\mathbf{y} \in C_\rho} \exp(\epsilon' \cdot \beta_{\mathbf{y}}) \exp(\epsilon' \cdot d_{\mathbf{y}})}{\sum_{\mathbf{y} \in C_\rho} \exp(\epsilon' \cdot d_{\mathbf{y}})}$$

$$= \mathbb{E}_{\mathbf{y}}[\exp(\epsilon' \cdot \beta_{\mathbf{y}})] \leq \exp(O(\epsilon' \cdot m d_{\mathcal{P}} \cdot 2\Delta))$$

$$= \exp\left(O(\epsilon \cdot d_{\mathcal{P}})\right).$$

$\square$

Lemma 3.1. *For every time* $0 \leq t \leq T$, $I(t)$ *is the sampled vector by Algorithm 1 and* $I'(t)$ *is the solution of the linear programming in measured continuous greedy algorithm. Then,*

$$I(t)w(t)(1 - \mathbf{y}(t)) \geq F(\mathbf{y}(t) \vee 1_{OPT}) - F(\mathbf{y}(t)) - O\left(\sqrt{\epsilon} + \frac{2\Delta \ln n}{\epsilon^3}\right).$$

Proof.

$$w(t) \cdot 1_{OPT} = \sum_{e \in OPT} w_e(t) = \sum_{e \in OPT} [F(\mathbf{y}(t) \vee 1_e) - F(\mathbf{y}(t))]$$

$$= \mathbb{E}\left[\sum_{e \in OPT} f(R(\mathbf{y}(t)) + e) - f(R(\mathbf{y}(t)))\right]$$

$$\geq \mathbb{E}\left[f(R(\mathbf{y}(t)) \cup OPT) - f(R(\mathbf{y}(t)))\right] = F(\mathbf{y}(t) \vee 1_{OPT}) - F(\mathbf{y}(t)),$$

where the inequality is followed by submodularity.

Since $1_{OPT} \in \mathcal{P}$, we get from Algorithm 1

$$I(t)' \cdot w(t) \geq F(\mathbf{y}(t) \vee 1_{OPT}) - F(\mathbf{y}(t)).$$

Hence,

$$\sum_{e \in \mathcal{X}} I'_e(t) \cdot (1 - \mathbf{y}_e(t)) \cdot \partial_e F(\mathbf{y}(t))$$

$$= \sum_{e \in \mathcal{X}} (1 - \mathbf{y}_e(t)) \cdot I'_e(t) \cdot [F(\mathbf{y}(t) \vee 1_e) - F(\mathbf{y}(t) \wedge 1_{\hat{e}})]$$

$$= \sum_{e \in \mathcal{X}} I'_e(t) \cdot [F(\mathbf{y}(t) \vee 1_e) - F(\mathbf{y}(t))] = I'(t) \cdot w(t)$$

$$\geq F(\mathbf{y}(t) \vee 1_{OPT}) - F(\mathbf{y}(t)).$$

Recall we define a neighboring feasible field, i.e., the ρ-*covering* of $\mathcal{P}$. And we get the followings by the Theorem 2.2 of exponential mechanism:

$$I(t)w(t)(1 - \mathbf{y}(t)) \geq \sum_{e \in \mathcal{X}} I'_e(t) \cdot (1 - \mathbf{y}_e(t)) \cdot \partial_e F(\mathbf{y}(t)) - O\left(\sqrt{\epsilon} + \frac{2\Delta \ln n}{\epsilon^3}\right)$$

$$\geq F(\mathbf{y}(t) \vee 1_{OPT}) - F(\mathbf{y}(t)) - O\left(\sqrt{\epsilon} + \frac{2\Delta \ln n}{\epsilon^3}\right).$$

$\square$

Lemma 3.2. *For every time* $0 \leq t < T$,

$$F(\mathbf{y}(t+\delta)) - F(\mathbf{y}(t))$$
$$\geq \delta \cdot \left[F(\mathbf{y}(t) \vee \mathbf{1}_{OPT}) - F(\mathbf{y}(t)) - O\left(\sqrt{\epsilon} + \frac{2\Delta \ln n}{\epsilon^3}\right) \right] - O(n^3\delta^2) \cdot f(OPT).$$

Proof.

$$F(\mathbf{y}(t+\delta)) - F(\mathbf{y}(t))$$
$$\geq (\mathbf{y}(t+\delta) - \mathbf{y}(t)) \cdot \partial F(\mathbf{y}(t)) - O(n^3\delta^2) \cdot f(OPT)$$
$$= \delta \cdot I(t)(1 - \mathbf{y}(t))w(t) - O(n^3\delta^2) \cdot f(OPT)$$
$$\geq \delta \cdot \left[F(\mathbf{y}(t) \vee \mathbf{1}_{OPT}) - F(\mathbf{y}(t)) - O\left(\sqrt{\epsilon} + \frac{2\Delta \ln n}{\epsilon^3}\right) \right] - O(n^3\delta^2) \cdot f(OPT),$$

where the first and last inequalities are given by Lemma 2.2 and Lemma 3.1. And the algorithm makes the equality hold. $\qquad \square$

Lemma 3.3. *For every* $0 \leq t \leq T$,

$$g(t) \leq F(\mathbf{y}(t)) + O(n^3\delta) \cdot tf(OPT) + \delta O\left(\sqrt{\epsilon} + \frac{2\Delta \ln n}{\epsilon^3}\right).$$

Proof. Assume the big O notation in Lemma 3.3 to be $cn^3\delta^2$. Prove by induction on t that $g(t) \leq F(\mathbf{y}(t)) + cn^3\delta tf(OPT)$. For $t = 0$, $g(0) = 0 \leq F(\mathbf{y}(0))$. Assume that the claim holds for some t. Then

$$g(t+\delta) = (1-\delta)g(t) + \delta f(OPT)$$
$$\leq (1-\delta)\left[F(\mathbf{y}(t) + cn^3\delta tf(OPT))\right] + \delta e^{-t}f(OPT)$$
$$= F(\mathbf{y}(t)) + \delta[e^{-t}f(OPT) - F(\mathbf{y}(t))] + c(1-\delta)n^3\delta tf(OPT)$$
$$\leq F(\mathbf{y}(t+\delta)) + cn^3\delta^2 f(OPT) + c(1-\delta)n^3\delta tf(OPT) + \delta O\left(\sqrt{\epsilon} + \frac{2\Delta \ln n}{\epsilon^3}\right)$$
$$\leq F(\mathbf{y}(t+\delta)) + cn^3\delta(t+\delta)f(OPT) + \delta O\left(\sqrt{\epsilon} + \frac{2\Delta \ln n}{\epsilon^3}\right),$$

where the inductive assumption and Lemma 3.3 give the first two inequalities and the last one is hold by $\delta \in [0, 1]$. $\qquad \square$

Lemma 3.4. *For every time* $0 \leq t \leq T$, $g(t) \geq h(t)$.

Proof. The proof is by induction on t. For $t = 0$, $g(0) = 0 = h(0)$. Assume that the lemma holds for some t. Then, we can easily get

$$h(t+\delta) = h(t) + \int_t^{t+\delta} h'(l)dl = h(t) + f(OPT) \cdot \int_t^{t+\delta} e^{-l}(1-l)dl$$
$$\leq h(t) + f(OPT) \cdot \delta e^{-t}(1-t) = (1-\delta)h(t) + \delta e^{-t} \cdot f(OPT)$$
$$\leq (1-\delta)g(t) + \delta e^{-t} \cdot f(OPT) = g(t+\delta).$$

$\qquad \square$

Corollary 3.1. $F(\mathbf{y}(t)) \geq \left[Te^{-T} - o(1)\right] \cdot f(OPT) - \delta O\left(\sqrt{\epsilon} + \frac{2\Delta \ln n}{\epsilon^3}\right).$

Proof. By Lemma 3.3 and Lemma 3.4,

$$F(\mathbf{y}(T)) \geq g(T) - O(n^3\delta) \cdot T \cdot f(OPT) - \delta O\left(\sqrt{\epsilon} + \frac{2\Delta \ln n}{\epsilon^3}\right)$$

$$\geq h(T) - O(n^3\delta) \cdot f(OPT) - \delta O\left(\sqrt{\epsilon} + \frac{2\Delta \ln n}{\epsilon^3}\right)$$

$$= \left[Te^{-T} - O(n^3\delta)\right] \cdot f(OPT) - \delta O\left(\sqrt{\epsilon} + \frac{2\Delta \ln n}{\epsilon^3}\right).$$

Recall that $\delta \leq n^{-5}$, hence, $O(n^3\delta) = o(1)$ and the proof is complete. □

References

1. Abadi, M.: Deep learning with differential privacy. In: Proceedings of the 23rd ACM SIGSAC Conference on Computer and Communications Security, pp. 308–318 (2016)
2. Alon, N., Spencer, J.H.: The Probabilistic Method, vol. 3, pp. 307–314. Wiley, New York (2004)
3. Bu, Z.Q., Gopi, S., Kulkarni, J., Lee, Y.T., Shen, J.H., Tantipongpipat, U.: Fast and memory efficient differentially private-SGD via JL projections. arXiv: 2102.03013 (2021)
4. Buchbinder, N., Feldman, M.: Constrained submodular maximization via a non-symmetric technique. Math. Oper. Res. **44**(3), 988–1005 (2019)
5. Buchbinder, N., Feldman, M., Naor, J.S., Schwart, R.: Submodular maximization with cardinality constraints. In: Proceedings of the 25th Annual ACM-SIAM Symposium on Discrete Algorithms, pp. 1433–1452 (2014)
6. Călinescu, G., Chekuri, C., Pál, M., Vondrák, J.: Maximizing a monotone submodular function subject to a matroid constraint. SIAM J. Comput. **40**(6), 1740–1766 (2011)
7. Chekuri, C., Jayram, T.S., Vondrák, J.: On multiplicative weight updates for concave and submodular function maximization. In: Proceedings of the 6th Innovations in Theoretical Computer Science, pp. 201–210 (2015)
8. Chekuri, C., Vondrák, J., Zenklusen, R.: Dependent randomized rounding for matroid polytopes and applications. In: Proceedings of the 51st Annual Symposium on Foundations of Computer Science, pp. 575–584 (2010)
9. Chekuri, C., Vondrák, J., Zenklusen, R.: Multi-budgeted matchings and matroid intersection via dependent rounding. In: Proceedings of the 22nd Annual ACM-SIAM Symposium on Discrete Algorithms, pp. 1080–1097 (2011)
10. Chekuri, C., Vondrák, J., Zenklusen, R.: Submodular function maximization via the multilinear relaxation and contention resolution schemes. SIAM J. Comput. **43**(6), 1831–1879 (2014)
11. Dwork, C., Kenthapadi, K., McSherry, F., Mironov, I., Naor, M.: Our data, ourselves: privacy via distributed noise generation. In: Vaudenay, S. (ed.) EUROCRYPT 2006. LNCS, vol. 4004, pp. 486–503. Springer, Heidelberg (2006). https://doi.org/10.1007/11761679_29

12. Ene, A., Nguyen, H.L.: Constrained submodular maximization: beyond $1/e$. In: Proceedings of the 57th Annual Symposium on Foundations of Computer Science, pp. 248–257 (2016)
13. Feige, U., Mirrokni, V.S., Vondrák, J.: Maximizing non-monotone submodular functions. SIAM J. Comput. **40**(4), 1133–1153 (2011)
14. Feldman, M., Naor, J., Schwartz, R.: A unified continuous greedy algorithm for submodular maximization. In: Proceedings of the 52nd Annual Symposium on Foundations of Computer Science, pp. 570–579 (2011)
15. Gharan, S.O., Vondrák, J.: Submodular maximization by simulated annealing. In: Proceedings of the 22 Annual ACM-SIAM Symposium on Discrete Algorithms, pp. 1098–1116 (2011)
16. Gupta, A., Ligett, K., McSherry, F., Roth, A., Talwar, K.: Differentially private combinatorial optimization. In: Proceedings of the 31st Annual ACM-SIAM Symposium on Discrete Algorithms, pp. 1106–1125 (2010)
17. Gupta, A., Roth, A., Schoenebeck, G., Talwar, K.: Constrained non-monotone submodular maximization: offline and secretary algorithms. In: Saberi, A. (ed.) WINE 2010. LNCS, vol. 6484, pp. 246–257. Springer, Heidelberg (2010). https:// doi.org/10.1007/978-3-642-17572-5_20
18. Hochbaum, D.S.: An efficient algorithm for image segmentation, Markov random fields and related problems. J. ACM **48**(4), 686–701 (2001)
19. Kempe, D., Kleinberg, J.M., Tardos, E.: Maximizing the spread of influence through a social network. In: Proceedings of the 9th ACM SIGKDD International Conference on Knowledge Discovery and Data Mining, pp. 137–146 (2003)
20. Krause, A., Guestrin, C.: Near-optimal nonmyopic value of information in graphical models. In: Proceedings of the 21st Conference in Uncertainty in Artificial Intelligence, pp. 324–331 (2005)
21. Krause, A., Guestrin, C.: Near-optimal observation selection using submodular functions. In: Proceedings of the 32nd AAAI Conference on Artificial Intelligence, pp. 1650–1654 (2007)
22. Kulik, A., Shachnai, H., Tamir, T.: Maximizing submodular set functions subject to multiple linear constraints. In: Proceedings of the 20th Annual ACM SIAM Symposium on Discrete Algorithms, pp. 545–554 (2009)
23. Lin, H., Bilmes, J.A.: A class of submodular functions for document summarization. In: Proceedings of the 49th Annual Meeting of the Association for Computational Linguistics: Human Language Technologies, pp. 510–520 (2011)
24. Lee, J., Mirrokni, V.S., Nagarajan, V., Sviridenko, M.: Non-monotone submodular maximization under matroid and knapsack constraints. In: Proceedings of the 41th Annual ACM Symposium on Theory of Computing, pp. 323–332 (2009)
25. McSherry, F., Talwar, K.: Mechanism design via differential privacy. In: Proceedings of the 48th Annual IEEE Symposium on Foundations of Computer Science, pp. 94–103 (2007)
26. Mirzasoleiman, B., Badanidiyuru, A., Karbasi, A.: Fast constrained submodular maximization: personalized data summarization. In: Proceedings of the 33rd International Conference on Machine Learning, pp. 1358–1367 (2016)
27. Mitrovic, M., Bun, M., Krause, A., Karbasi, A.: Differentially private submodular maximization: data summarization in disguise. In: Proceedings of the 34th International Conference on Machine Learning, pp. 2478–2487 (2017)
28. Nemhauser, G.L., Wolsey, L.A.: Best algorithms for approximating the maximum of a submodular set function. Math. Oper. Res. **3**(3), 177–188 (1978)
29. Nemhauser, G.L., Wolsey, L.A., Fisher, M.L.: An analysis of approximations for maximizing submodular set functions-I. Math. Program. **14**(1), 265–294 (1978)

30. Papadimitriou, C.H., Schapira, M., Singer, Y.: On the hardness of being truthful. In: Proceedings of the 49th Annual Symposium on Foundations of Computer Science, pp. 250–259 (2008)
31. Rafiey, A., Yoshida, Y.: Fast and private submodular and k-submodular functions maximization with matroid constraints. In: Proceedings of the 37th International Conference on Machine Learning, pp. 7887–7897 (2020)
32. Streeter, M.J., Golovin, D.: An online algorithm for maximizing submodular functions. In: Proceedings of the 22nd International Conference on Advances in Neural Information Processing Systems, pp. 1577–1584 (2008)
33. Sviridenko, M.: A note on maximizing a submodular set function subject to knapsack constraint. Oper. Res. Lett. **32**(1), 41–43 (2004)
34. Vondrák, J.: Optimal approximation for the submodular welfare problem in the value oracle model. In: Proceedings of the 40th Annual ACM Symposium on Theory of Computing, pp. 67–74 (2008)
35. Vondrák, J.: Symmetry and approximability of submodular maximization problems. SIAM J. Comput. **42**(1), 265–304 (2013)
36. Yoshida, Y.: Cheeger inequalities for submodular transformations. In: Proceedings of the 30th Annual ACM Symposium on Discrete Algorithms, pp. 2582–2601 (2019)

Network Problems

Robust *t*-Path Topology Control Algorithm in Wireless Ad Hoc Networks

Xiujuan Zhang[1,2], Yongcai Wang[1], Deying Li[1(✉)], Wenping Chen[1], and Xingjian Ding[3]

[1] School of Information, Renmin University of China, Beijing 100872, China
{ycw,deyingli,chenwenping}@ruc.edu.cn
[2] School of Computer Science, Qufu Normal University, Rizhao 276826, China
xiujuanzhang@qfnu.edu.cn
[3] School of Software Engineering, Beijing University of Technology,
Beijing 100124, China
dxj@bjut.edu.cn

Abstract. Topology control protocol aims at efficiently adjusting the network topology to improve the performance and scalability for networks, for example, spanner topology characteristics can decrease communication links and ensure that the distance of any pair of communication nodes is only within a little increase from that in the original communication graph. This is especially essential to large-scale wireless ad hoc networks, which can be the front-end of ubiquitous Internet of Things (IoT). Since nodes failure is common because of their limited resource, a large number of previous works focused on fault tolerance topologies or fault tolerance spanner topologies. But these topologies are no longer sparse. In this paper, we propose an intuitive solution to construct sparse and robust spanner topology, which builds a sparse spanner topology quickly and reverts to the required spanner topology rapidly without any outside intervention when any node fails. Extensive simulations confirm the effectiveness and efficiency of our solution.

Keywords: Ad hoc network · Spanner · Robust · Distributed algorithm

1 Introduction

The front-end of the Internet of Things (IoT) is often formed by large-scale wireless ad hoc networks used to collect information from a wide area or harsh environment [1]. Such an infrastructure-free wireless ad hoc network may consist of a few hundreds or even thousands of autonomous nodes with low power equipped [2]. Topology control (TC) tries to maintain a subgraph structure that can be used for efficient routing or improving the overall networking performance

This work was supported in part by National Natural Science Foundation of China under Grant No. (12071478, 61972404) and Natural Science Foundation of Shandong Province under Grant No. (ZR2019ZD10, ZR2020MF149).

W. Wu and H. Du (Eds.): AAIM 2021, LNCS 13153, pp. 229–239, 2021.
https://doi.org/10.1007/978-3-030-93176-6_20

[3]. In traditional ad hoc networks, all the nodes are equal, i.e. they take part in the routing and forwarding packets equally. If only a small number of links are been maintained while the network is still connected and the paths that connect any pair of nodes are linearly bounded compared with the original network, the result topology will reduce energy consumption and still have efficient routing [4,5]. Such topology is called a spanner and has t-path property (t is a small positive real number). Therefore our topology considers not only the basic criteria for TC, which are connectivity and sparseness (i.e., the number of links is linear with the number of nodes) but also t-path property.

TC should be robust, i.e., the topology should survive a certain level of node/link failures and maintain topological characteristics [6]. The reason is that node failure is common due to energy consumption, severe or unattended environments, intentional damages, etc. and that it will result in serious problems such as data lost, link broken and even network fragmented. A $(k+1)$-connected topology remains connected if at most k nodes are removed. Hence, fault tolerance topology approaches generally adopt $(k + 1)$-connected technique. So a mass of previous works focused on $(k + 1)$-connected topologies, such as [7,8], and some previous works focused on $(k + 1)$-connected spanner topologies, such as [9]. However, these topologies with n nodes are not sparse since a $(k + 1)$-connected graph has at least $n(k+1)/2$ edges. So how to find sparse and robust spanner topology protocol becomes our research goal.

Our solution is distributed. Since the algorithms have to work without global information and coordinated central control, distributed algorithms are inherent requirements for wireless ad hoc networks. Moreover, TC aims to efficiently adjust the network topology in a self-adaptive fashion to improve the performance and reliability of networks [10]. These are especially essential to large-scale wireless ad hoc networks because of the nodes' limited resources. Actually, our algorithms are distributed and only use 1-hop neighbor information. Thus, the topology can be robust and self-adaptive.

The rest of this paper is organized as follows. Section 2 reviews most of the related work. Section 3 defines some graph notations, the network model, and problem definitions. A distributed directed t-spanner algorithm is presented in Sect. 4. Section 5 gives a robust spanner algorithm. Simulation results are presented in Sect. 6. Section 7 concludes the paper.

2   Related Work

Generally, $(k + 1)$-connected topology schemes are used to improve TC reliability, which is named as k-fault-tolerant topology. Bahramgiri et al. [7] proposed a variation of the Cone Based Topology Control (CBTC) algorithm to preserve the k-connectivity. Li and Hou [8] presented a fault-tolerant Local Spanning Subgraph (FLSSk) algorithm to construct the $(k + 1)$-connected topology. [11] constructed $(k + 1)$-connected topology based on FLSSk in cooperative communication networks. Additionally, there is some literature about fault-tolerant spanner in wireless ad hoc networks, such as [9,12].

Li's [13] k-fault-tolerant spanner topology control protocol in wireless ad hoc networks is based on the modified Yao structure, called as $Yao_{c,k+1}$. In each cone, node u chooses the $k+1$ closest nodes in that cone. If there is any, add directed links from u to these nodes. This structure can sustain k nodes faults if the original communication graph is $(k+1)$-connected. The structure is still a spanner even with k nodes faults. But the probability that the underlying communication graph is $(k+1)$-connected is also extremely small when the topology is required to be sparse. Moreover, the value of k is quite small compared to n, i.e., the number of fault nodes must be too small. So we propose a robust sparse spanner topology problem for wireless ad hoc networks and try to solve it using Yao_c structure, not $Yao_{c,k+1}$.

3   Preliminaries

In this section, we give some graph notations, the network model, and problem definitions used in the paper.

First, we give some graph notations. Let $G = (V, E)$ be a directed graph with V a set of n nodes and E a set of directed edges where a directed edge $<u, v> \in E$ leads from node u to node v. For two nodes $u, v \in V$, let $|uv|$ be the Euclidean distance between u and v. The length of a directed edge $<u, v> \in E$ is $|uv|$. The length of the shortest path from u to v in G which is defined as the sum of the lengths of its edges is denoted by $d_G(u, v)$. If $S \subseteq V$, let $V \setminus S$ be the nodes set deleting all nodes in S from V. And $G \setminus S$ is the subgraph of G induced by $V \setminus S$, i.e., $G \setminus S$ is a graph only deleting all nodes in S and all the edges adjacent to these nodes in S from G. And K_n denotes the complete graph on V.

We consider a network with n nodes deployed arbitrarily in a 2-dimensional geographic plane, modeled as a graph $G = (V, E)$. Autonomous and decentralized nodes will need to communicate with one another to build networks and maintain them. Assume that all nodes are time-synchronized and that the system time is subdivided into slots. Within each slot, a node v can transmit and listen. For simplicity, we assume that all nodes have the same transmission power which get the same transmission range r. Each node knows nothing about the network except its ID and coordinates. Each node can receive all messages from its neighbors since we do not consider interference caused by other nodes simultaneously transmitting in this paper.

Communication Graph. The communication graph $C(V, E)$ of a given network consists of all network nodes and edges $<u, v> \in E$ and $<v, u> \in E$ such that $|uv| \leq r$. Any edge in the communication graph is called a communication edge.

t-**Spanner and** t-**Path.** Let $t > 1$ be a real number. A spanning subgraph $H(V, E_H)$ of $G(V, E)$ is said to be a t-spanner of G, if for any two nodes u and v in V, the length of the shortest path between u and v in H is at most t times that of the shortest path in G, i.e.,

$$d_H(u, v) \leq t \cdot d_G(u, v)(G). \tag{1}$$

The constant t is called the *stretch factor* of H (w.r.t. G). Any path satisfying the above condition is called t-path. Note that t-spanner and t-path may be directed. Obliviously, if $G(V, E)$ is connected, $H(V, E_H)$ is connected.

Note that G can be the complete graph K_n or a communication graph $C(V, E)$ that is a spanning subgraph of K_n.

Yao Graph (Yao). The directed Yao graph over a (directed) graph G with an integer parameter c, denoted by $\overrightarrow{Yao}_c(G)$, is defined as follows. At each node u, any c equal-separated rays originated at u define c equal cones, and the shortest directed edge $<u, v>$ in G is added to $\overrightarrow{Yao}_c(G)$ in each cone. Ties are broken arbitrarily. Any two orientation of the cone partition for each node is the same.

$C_{u,w}$ is the cone of the node u containing w. An undirected Yao graph, in which the edge directions in $\overrightarrow{Yao}_c(G)$ are ignored, is denoted by $Yao_c(G)$.

Now we give the main problem of this paper.

Robust dIrected t-Spanner for Autonomous Nodes (RISA). Given a graph $H(V, E_H)$ which is the directed t-spanner of the communication graph $C(V, E)$ where all n autonomous nodes is denoted by the set V, an integer $0 \leq k < n$, any k fault nodes set S, we aim to find a directed t-spanner H' of $C \setminus S$ by adding up to μ communication edges to $H \setminus S$, such that the value of μ is minimized.

In order to distributedly implement the RISA problem, we should implement the following subproblem first.

Distributed Directed t-Spanner (DDS Problem). Given a set V of n autonomous nodes deployed randomly in a 2-dimensional geographic plane, a real number $t \geq 1$, the goal is to design a distributed algorithm to find a directed t-spanner of the corresponding communication graph $C(V, E)$, such that the number of messages exchanged is minimized.

4   Distributed Directed t-Spanner Algorithm (DDspanner)

Our first distributed algorithm is a directed t-spanner algorithm called DDspanner for the DDS problem. Its basic idea is that each node locally broadcasts its own message to its neighbor nodes and then computes the nearest neighbor in each cone based on the received messages. This algorithm can gain the required directed t-spanner and solve the DDS problem which will be proved. DDspanner is described in detail as follows.

Given the real number t, each node u computes the number of cones c according to the equation $t = \frac{1}{1 - 2\sin(\pi/c)}$. Each node has its outcoming neighbors' list $O[c]$. Each node independently sends an INIT message containing its ID and coordinates. When u receives an INIT message from some node v, u computes the index i of the cone that v belongs to the node u. Then $O[i] = u$ if in the ith sector a neighbor node did not be found before or v is the nearest neighbor node. Finally, each $O[c]$ stores all outcoming neighbors for each node. Thus $\overrightarrow{Yao}_c(C)$ is formed.

The messages are used in this algorithm, called INIT containing each node's ID and coordinate. The pseudo-code for node u is given in Algorithm 1.

Algorithm 1. DDspanner(u)

1: Initializes the stretch factor t;
2: Initializes the number of cones $c = \lceil \frac{\pi}{\arcsin((t-1)/(2t))} \rceil$;
3: Initializes the outcoming neighbors list $O[i] = -1$ for $i = 1, 2, \cdots, c$;
4: Sends an INIT message containing its ID and coordinates;
5: **if** Receives an INIT message from some node v **then**
6:     Computes the index i of the cone that v belongs to the node u;
7:     **if** $O[i] == -1$ or the distance $|O[i], u| > |vu|$ **then**
8:         $O[i] = v$
9:     **end if**
10: **end if**

Before we prove that the DDspanner algorithm can solve the DDS problem, we first give the following lemma.

Lemma 1. *Let* $t = \frac{1}{1-2\sin(\pi/c)}$ *for* $c > 6$, *the node* $v_i \in V$, *and* $j(\leq n-1)$ *is an integer. For* $\forall u, w \in V$, *there exists a path* $u = v_0, v_1, \cdots, v_i, \cdots, v_j = w$ *from* u *to* w *in* $\overrightarrow{Yao}_c(C)$, *whose length is at most* t *times the length of the short path from* u *to* w *in the corresponding communication graph* C. *The edge* $<v_i, v_{i+1}>$ *in this* t-path *of* $\overrightarrow{Yao}_c(C)$, *where* $0 \leq i \leq j - 1$, *is the edge in the cone* $C_{v_i, w}$.

Due to the lack of space we omit the proof of Lemma 1.

Theorem 1. *Let* $t(> 1)$ *be a real number. The* $\overrightarrow{Yao}_c(C)$ *obtained after DDspanner is the required* t-Spanner, *i.e., it can solve the DDS problem. Moreover, the message complexity of DDspanner is* $O(n)$ *where* n *is the number of nodes.*

Proof. First, after the DDspanner algorithm, each node's outcoming neighbors are stored in each node's $O[c]$, the $\overrightarrow{Yao}_c(C)$ forms. From Lemma 1, the result $\overrightarrow{Yao}_c(C)$ is the required directed t-spanner. Next, each node sends an INIT message once and only once, so the total number of messages is n. Thus, the number of messages exchanged is minimized. □

5 Robust Topology

In this section, we propose the RobustSpanner algorithm based on the DDspanner algorithm to build robust spanner topology and to solve special instance of the RISA problem. After the DDspanner algorithm, the RobustSpanner algorithm runs in background and never stop. If no node fails, no update will be performed. If some node fails, the t-spanner topology will update locally to keep t-spanner property. So the RobustSpanner algorithm is a self-stabilizing algorithm.

Our detailed idea is the following. Our algorithm is divided into three actions. In Action 1, if the node u is going to fail, u locally broadcasts a FAIL message to its neighbors. In Action 2, when the node u receives a FAIL message from its failed neighbor node v and v is in u's outcoming neighbors list $O[c]$, delete v, then u which have received a FAIL message locally broadcasts a Rebuild message containing its ID u and coordinate. In Action 3, when u receives a Rebuild message from its neighbor node v, u computes the index i of the cone that v belongs to the node u, and then if v becomes the nearest node in the cone, v is added to u's outcoming neighbor list. Finally, the $\overrightarrow{Yao_c}(C \setminus S)$ forms, where S is the set of the k failed nodes.

The pseudo-code for node u is given in Algorithm 2.

Algorithm 2. RobustSpanner(u)

Action 1:
1: **if** u is going to fail **then**
2:     u sends a FAIL message;
3: **end if**

Action 2:
1: **if** Receives a FAIL message from some node v **then**
2:     **if** v is in u's outcoming neighbors list $O[c]$ **then**
3:         Deletes v from $O[c]$;
4:     **end if**
5:     Sends a Rebuild message containing its ID u and coordinates;
6: **end if**

Action 3:
1: **if** Receives a Rebuild message from v **then**
2:     Computes the index i of the cone that v belongs to the node u;
3:     **if** $O[i]$ is NULL or $|uv|$ is smaller than the distance from u to any node in $O[i]$ **then**
4:         $O[i] = v$;
5:     **end if**
6: **end if**

Before we prove that the RobustSpanner algorithm can solve the SISA problem, we first give the following lemma.

Lemma 2. *If node w is in the cone $C_{u,v}$ with $|uv| < |uw|$ and $c > 6$ as Fig. 1, then $|vw| < |uw|$.*

Proof. For $\triangle uvw$ shown in Fig. 1, since $c > 6$, $\angle wuv < \pi/3$. Since $|uv| < |uw|$, $\angle uwv < \angle uvw$ using the law of sines. So $\angle uvw > \pi/3 > \angle wuv$. Finally applying the law of sines again, $|vw| < |uw|$.   □

We now prove that the effectiveness and efficiency of the RobustSpanner algorithm in the following theorem.

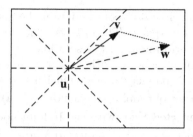

Fig. 1. When v is going to fail

Theorem 2. *Let* $t(> 1)$ *be a real number,* n *be the number of nodes, and* $0 \leq k < n$ *be the number of failed nodes. If* k *nodes in the set* S *fail, the* t-*spanner topology w.r.t the communication graph* C *obtained by the DDspanner algorithm will rebuild locally into the* t-*spanner topology w.r.t* $C \setminus S$ *after the RobustSpanner algorithm runs* 2 *timeslots in a self-stabilizing style. The number of the reconnected edges is up to* $k(n - k)$.

Proof. Given t, suppose c is the number of cones that construct the corresponding t-spanner topology obtained by the DDspanner algorithm. Assume that some node v is failed, node w is in the cone $C_{u,v}$, and w is going to add to u's outcoming neighbors list $O[c]$ as Fig. 1. Since $|uv| < |uw| < r$, $|vw| < |uw| < r$ using Lemma 2. Hence, u and w can receive v's Fail message, and u can receive w's Rebuild message. So u can delete v from u's outcoming neighbors list $O[c]$ and add w to $O[c]$. Thus, $\overrightarrow{Yao_c}(C \setminus S)$ forms. $\overrightarrow{Yao_c}(C \setminus S)$ is the required t-spanner topology w.r.t $C \setminus S$ applying Lemma 1.

In the first timeslot, the failed nodes send Fail messages, the corresponding nodes receive the messages and process accordingly. In the second timeslot, if the node has received a Fail message, it sends a Rebuild message containing its ID and coordinates and then handles Rebuild messages. Hence, the t-spanner topology w.r.t the communication graph C obtained by DDspanner will rebuild into the t-spanner topology w.r.t $C \setminus S$ after the RobustSpanner algorithm runs 2 timeslots in a self-stabilizing style. Moreover, the construction presented here only uses 1-hop neighbor information.

Each failed node can delete at most one piece of outcoming neighbor information from one normal node and then try to reconnect. Hence, each failed node causes **at most the number of its incoming neighbor nodes**, which is up to $n - k$, to reconnect links. So k failed nodes can delete at most $k(n - k)$ pieces of outcoming neighbor information and then try to reconnect. □

6 Simulation

In this section, we generate n ad hoc nodes randomly and uniformly distributed in a unit-area square to confirm the accuracy and efficiency of our topology protocols. First, we use a number of intuitive legends to verify that our solution is able to efficiently and self-stabilizedly maintain the given t even if the communication graphs are very sparse. Then, we conduct extensive simulations to investigate the average performance of our solution.

Firstly, we use a sparse instance with $r = 0.15$ and $n = 100$ to show our protocol can reconstruct t-spanner topology efficiently even with a quite sparse C in Fig. 2. It displays a set of topologies, including C (a), $\overrightarrow{Yao}_8$ (b), $\overrightarrow{Yao}_8$ with 10 fault nodes (c), reconnected $\overrightarrow{Yao}_8$ with the 10 fault nodes (d) and $\overrightarrow{Yao}_{17}$ (e), $\overrightarrow{Yao}_{17}$ with 10 fault nodes (f), and reconnected $\overrightarrow{Yao}_{17}$ with the 10 fault nodes (g). For the same instance, the indegrees, the outdegrees, the average degrees, and the stretch factors w.r.t. C are reported in Table 1. From Fig. 2 and Table 1, compared with $\overrightarrow{Yao}_8$ with the 10 fault nodes, the reconnected one reconnected 15 links, and the stretch factor w.r.t. C changed from infinity to 1.4147, which satisfied the given t, meanwhile, compared with $\overrightarrow{Yao}_{17}$ with the 10 fault nodes, the reconnected one reconnected 7 links which marked in red in Fig. 2 (g), and the stretch factor w.r.t. C changed from infinity to 1.2361, which also satisfied the given t.

Table 1. The performance of original and reconnected t-spanner topologies with $\overrightarrow{Yao}_c$.

	n	r	Indegree	Outdegree	Average degree	Stretch factor w.r.t. C
C	100	0.15	12	12	5.7	1
$\overrightarrow{Yao}_8$	100	0.15	11	8	4.22	1.3948
$\overrightarrow{Yao}_8$ with the 10 fault nodes	100	0.15	10	8	3.54	inf
Reconnected $\overrightarrow{Yao}_8$ with the 10 fault nodes	100	0.15	10	8	3.69	1.4147
$\overrightarrow{Yao}_{17}$	100	0.15	11	10	4.96	1.2361
$\overrightarrow{Yao}_{17}$ with the 10 fault nodes	100	0.15	10	10	4.2	inf
Reconnected $\overrightarrow{Yao}_{17}$ with the 10 fault nodes	100	0.15	10	10	4.27	1.2361

Next, we conduct extensive simulations to investigate the average performance of our solution on various node numbers and different transmission ranges. For each case, we generate 500 random node sets and set up 200 sets of fault nodes at random for each. The results are shown in Fig. 3 and 4 for the parameter settings given by the caption of each subfigure. Note that we have tested other parameter settings and obtained very similar results.

Figure 3 depicts that the average number of reconnected edges increases with the increase of the number of fault nodes. As the number of fault nodes remains unchanged, the average number of reconnected edges increases when n increases in Fig. 3(a) and when r increases in Fig. 3(b). However, the number of reconnected edges is far below $k(n - k)$, and it is not only related to n and k but also related to r. We analyze that the average degree is an important factor affecting

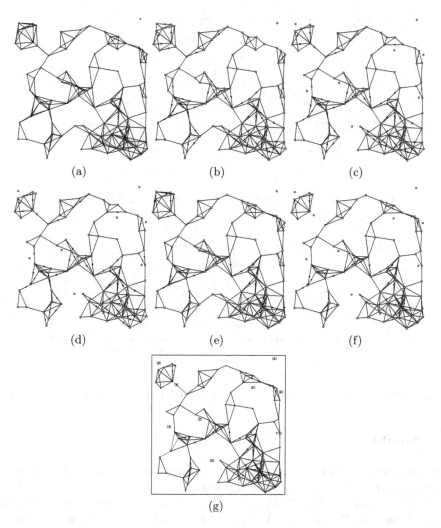

Fig. 2. The comparison of original and reconnected t-spanner topologies with $\overrightarrow{Yao_c}$.

the number of reconnected edges. Both n and r affect the average degree, so they make an impact on the number of reconnected edges. When n is large enough or r is large enough, the communication graph C is close to the complete graph, the number of reconnected edges does not change much as n and r vary, such as $n = 120, 140$, and $r = 0.4, 0.5$.

Figure 4 demonstrates that the stretch factors increase as n decreases in (a) and as r increases in (b) when the number of fault nodes doesn't change. So the average degree is the main factor affecting the stretch factors. Whereas the stretch factors have only a small increase with the increase of the number of fault nodes. And all the stretch factors can greatly meet the requirements of given t.

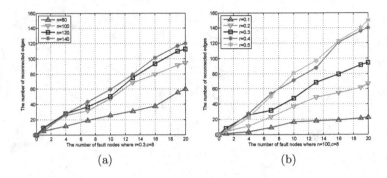

Fig. 3. Influences of n and r on the number of reconnected edges.

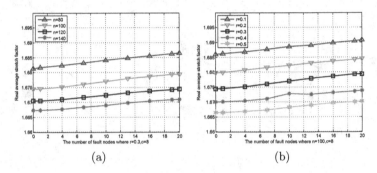

Fig. 4. Influences of n and r on the real average stretch factor of reconnected topologies.

7   Summary

In this paper, we present a self-stabilizing solution to construct sparse spanner topology, which has the following characteristics: (1) giving a detailed definition of the corresponding RISA problem, (2) being distributed and having no central daemon, (3) applying Yao graph idea, and (4) theory and simulation guaranteed. In future research, we will consider constructing robust and sparse spanner topology in 3D, and other approaches for spanner construction except Yao graphs are also worthy of investigating.

References

1. Cheng, M.X., Sun, J., Min, M., Li, Y., Wu, W.: Energy-efficient broadcast and multicast routing in multihop ad hoc wireless networks. Wirel. Commun. Mob. Comput. **6**(2), 213–223 (2006)
2. Luo, C., Hong, Y., Li, D., Wang, Y., Chen, W., Hu, Q.: Maximizing network lifetime using coverage sets scheduling in wireless sensor networks. Ad Hoc Netw. **98**, 102037 (2020)
3. Park, M.A., Willson, J., Wang, C., Thai, M., Wu, W., Farago, A.: A dominating and absorbent set in a wireless ad-hoc network with different transmission ranges.

In: Proceedings of the 8th ACM International Symposium on Mobile Ad Hoc Networking and Computing (MobiHoc), Montréal, Canada, pp. 22–31. ACM (2007)

4. Zhang, X., Yu, J., Li, W., Cheng, X., Yu, D., Zhao, F.: Localized algorithms for Yao-graph based spanner construction in wireless networks under SINR. IEEE/ACM Trans. Netw. **25**(4), 2459–2472 (2017)

5. Yu, D., Ning, L., Zou, Y., Yu, J., Cheng, X., Lau, F.C.: Distributed spanner construction with physical interference: constant stretch and linear sparseness. IEEE/ACM Trans. Netw. **25**(4), 2138–2151 (2017)

6. Wang, W., et al.: On construction of quality fault-tolerant virtual backbone in wireless networks. IEEE/ACM Trans. Netw. **21**(5), 1499–1510 (2013)

7. Bahrmgiri, M., Hajiaghayi, M.T., Andmirrokni, V.S.: Fault tolerant and three-dimensional distributed topology control algorithms in wireless multi-hop networks. Wirel. Netw. **12**(2), 179–188 (2006)

8. Li, N., Hou, J.: Localized fault-tolerant topology control in wireless ad hoc networks. IEEE Trans. Parallel Distrib. Syst. **17**(4), 307–320 (2006)

9. Sun, R., Wang, Y., Yuan, J., Shan, X., Ren, Y.: Localized fault-tolerant spanner based topology control in wireless networks. In: Proceedings of Cross Strait Quad-Regional Radio Science and Wireless Technology Conference, Harbin, China, pp. 752–756. IEEE (2011)

10. Kadivar, M.: An adaptive Yao-based topology control algorithm for wireless ad-hoc networks. In: 10th International Conference on Computer and Knowledge Engineering (ICCKE), Ferdowsi University of Mashhad, Iran, pp. 457–462. IEEE (2020)

11. Guo, J., Liu, X., Jiang, C., et al.: Distributed fault-tolerant topology control in cooperative wireless ad hoc networks. IEEE Trans. Parallel Distrib. Syst. **26**(10), 2699–2710 (2015)

12. Abam, M.A., Baharifard, F., Borouny, M.S., Zarrabi-Zadeh, H.: Fault-tolerant spanners in networks with symmetric directional antennas. Theoret. Comput. Sci. **704**(12), 18–27 (2017)

13. Li, X., Wan, P., Wang, Y., Yi, C.: Fault tolerant deployment and topology control in wireless ad hoc networks. Wirel. Commun. Mob. Comput. **4**(1), 109–125 (2004)

Multi-attribute Based Influence Maximization in Social Networks

Qiufen Ni[1], Jianxiong Guo[2,3]([✉]), and Hongmin W. Du[4]

[1] School of Computers, Guangdong University of Technology,
Guangzhou 510006, China
niqiufen@gdut.edu.cn
[2] BNU-UIC Institute of Artificial Intelligence and Future Networks,
Beijing Normal University at Zhuhai, Zhuhai 519087, Guangdong, China
jianxiongguo@bnu.edu.cn
[3] Guangdong Key Lab of AI and Multi-Modal Data Processing, BNU-HKBU
United International College, Zhuhai 519087, Guangdong, China
[4] Accounting and Information Systems Department, Rutgers University,
Piscataway, NJ 08854, USA
hd255@scarletmail.rutgers.edu

Abstract. Viral marketing on social networks is an important application and hot research problem. Most of the related work on viral marketing focuses on the spread of single information, while a product may associate with multi-attribute in real life. Information on multiple attributes of a product propagates in the social networks simultaneously and independently. The attribute information that a user receives will determine whether he would purchase the product or not. We extend the traditional single information influence maximization problem to the Multi-attribute based Influence Maximization Problem (MIMP). We present the Multi-dimensional IC model (MIC model) for the proposed problem. The objective function for MIMP is proved to be non-submodular, then we solve the problem with the Sandwich Algorithm, which can get a $\max\left\{\frac{f(S_U)}{\bar{f}(S_U)}, \frac{f(S_L^*)}{f(S_o^*)}\right\}(1 - 1/e)$ approximation ratio to the optimal solution. Experiments are conducted in two real world datasets to verify the correctness and effectiveness of the proposed algorithm.

Keywords: Social network · Influence maximization · Multi-attribute information · Approximation algorithm

1 Introduction

Online social networks are an important class of graph data. Data by generated by users have been growing rapidly through various online social networks, such as Facebook, LinkedIn, ResearchGate, and messengers like Skype and WeChat [1]. The widespread use of these social platforms leads to an increasing interest in mining important and useful but implicit patterns. Efficient techniques for extracting information from graph data are crucial to applications

© Springer Nature Switzerland AG 2021
W. Wu and H. Du (Eds.): AAIM 2021, LNCS 13153, pp. 240–251, 2021.
https://doi.org/10.1007/978-3-030-93176-6_21

across many domains, including public safety, environment management, election, and especially viral marketing [2].

Viral marketing promotes products through giving away free products to a small customer set to spread the product information, which makes good use of the word-of-mouth effect of the social networks to propagate the influence of a product. Kempe *et al.* in [3] formulated the spread process of information in social networks as the influence maximization (IM) problem which aims at maximizing the number of people who adopt the product by choosing a small number of users as a seed set. Then they propose two classic information spread models, IC (Independent Cascade) model and LT (Linear Threshold) model. In [3], the influence maximization problem is formulated as a monotone and submodular objective function, and a greedy algorithm is presented to solve it, the returned solution can get an $1 - 1/e$ approximation guarantee. The submodular property is very important in optimization theory. A function f over all subsets of a finite set V is said to be submodular if for any two subsets A and B, $f(A) + f(B) \geq f(A \cup B) + f(A \cap B)$ satisfies. When the inequality is reversed, function f is supermodular. Moreover, f is monotone non-decreasing if for $A \subset B$, $f(A) \leq f(B)$.

There exists various situations about the product and the company in viral marketing, which lead to different problem formulations. Most of the existing research focuses on the single diffusion, which has only one piece of product information spreading on the social network. Usually, a product may have multiple attributes considered by a customer to make a decision "buy or not" since different persons have various emphasis points. We summarize the main contributions in this paper as follows:

- We formulate the Multi-attribute based Influence Maximization Problem (MIMP) under the MIC model in social networks. The objective function of the proposed MIMP is proved to be non-submodular.
- We consider the MIC model as a multi-dimensional IC model to analyze the MIMP. A Sandwich Approximation Algorithm for MIMP is proposed, which can get a $\max\left\{ \frac{f(S_U)}{\bar{f}(S_U)}, \frac{f(S_L^*)}{f(S_o^*)} \right\} (1 - 1/e)$-approximation solution.

The rest of this paper can be arranged as follows. The related work is reviewed in Sect. 2. The MIC model is introduced and the Multi-attribute based Influence Maximization Problem (MIMP) is formulated in Sect. 3. In Sect. 4, we prove the properties of the objective function for MIMP and present Sandwich Approximation Algorithm. The experimental results are described in Sect. 5. We conclude our work at last in Sect. 6.

2   Related Work

Influence maximization is a fundamental problem in the study of social networks, which attracts a lot of researchers studying on it. It can be described as: given a social network G with a diffusion model m, and a positive integer b, find at most

b seeds to maximize the influence spread, i.e., the expected number of influenced nodes. We summarize the common studies on influence maximization as follows.

Most of the existing research consider single diffusion in the social network. Different from the classical influence maximization problem, some research focus on the influence probability maximization of a target user set [4,5]. In [4], they propose an Acceptance Probability Maximization (APM) problem which chooses b seeds to maximize the acceptance probability of a target set T, they solve this problem in the context of social networks with community structures, and transform this problem into a Maximum Weight Hitting Set problem. Proposed algorithm can get an $1 - 1/e$ approximation ratio.

Some related work studies the discount allocation problem [6,7], which is a transformation of the influence maximization. In [6], they adopt the non-adaptive and adaptive setting to allocate discount to users, respectively, which aim at maximizing the expected number of users who purchase the product. The proposed greedy algorithm with the adaptive setting can obtain a bounded approximation guarantee, while the non-adaptive setting can get an $1 - 1/e$ approximation ratio.

There are few studies extend the influence maximization of single product to the situation of multiple products [8,9]. S. Bharathi et $al.$ [8] introduce the competitive influence maximization problem that several companies market their products on the social network competitively. They solve the problem with game theory, the returned solution achieves an $1 - 1/e$ approximation.

Some existing work studies the group influence maximization [10,11]. J. Zhu et $al.$ [10] consider the group influence problem and model it with a hypergraph $G = (V, E, P)$, which aims at maximizing the activated users. They prove that the objective function is non-submodular and non-supermodular. Then a sandwich algorithm is proposed to solve the problem which has a bounded approximation ratio. A D-SSA algorithm is also presented and obtains an $(1 - 1/e - \epsilon)$ approximation ratio.

There are few studies related to the multi-attribute information diffusion problem. J. Guo [12] consider the multi-feature based rumor blocking problem. They devise a multi-feature diffusion model (MF-model), and propose a novel Multi-Sampling method to estimate the influence spread function $f(S_p)$. A Revised-IMM algorithm is presented, which can achieve a good approximate performance guarantee.

3 Network Model and Problem Formulation

3.1 The Network Model

A social network is modeled as a directed graph $G = (V, E)$, where each vertex $v \in V$ represents a user, and each edge $(u, v) \in E$ is the relationship between user u and v. In the social network, the incoming neighbor set and the outgoing neighbor set of a node v is denoted as $N^-(v)$ and $N^+(v)$, respectively. In

[3], Kemp *et al.* present the IC model and LT model to formulate the propagation process of information in social networks. We extend the IC model for our problem as follows.

Multi-attribute Independent Cascade (MIC) Model: Given a product with r attributes, then we list important facts about this model consisting of discrete steps:

- Each node represents a customer. An active node represents a customer who has purchased the product; vice versa.
- Each edge $(u, v) \in E$ is associated with r probability $p_{uv}^1, p_{uv}^2, \cdots, p_{uv}^r$, where $p_{uv}^i \in [0, 1]$. These probabilities represent the expertise of u from the view point of v. When u just becomes active, v will be influenced by u and accepts the attribute ϕ with probability p_{uv}^ϕ.
- When v receives influence from more than one active inneighbors, v treats them as independent events.
- Each customer v has a threshold θ_v and a weight w_v^ϕ for each attribute ϕ, where $\sum_{\phi=1}^r w_v^\phi = 1$. Customer v decides to purchase the product if and only if the total weight of accepted attributes reaches at least θ_v.
- Initially, a set of seeds are selected and activated. At each subsequent step, every node checks if the activation condition is satisfied. Each newly influenced user v in step t only has a single chance to influence his uninfluenced outgoing neighbors in step $t + 1$. After time step $t + 1$, v could not activate any of its outgoing neighbors. This process ends if no node becomes active at current step.

3.2   Problem Formulation

In this section, we formulate our problem based on the MIC model described above.

Give a social network $G = (V, E)$ and a product with r attributes. r kinds of attribute information propagate in the social networks on the MIC model at the same time. The influence maximization problem for the multi-attribute information aims at choosing k initial users as seeds to maximize the influence of the product. This problem can be formulated as follows:

Problem 1 *(Multi-attribute based Influence Maximization Problem (MIMP)). Given a graph $G = (V, E)$, r kinds of attribute information of a product, the MIC model. Our target is to select an optimal node set S such that the number of expected customers who purchased the given product is maximized, i.e., $S = \arg\max_{S \subseteq V, |S| \leq k} f(S)$.*

In [3], they proved that the influence maximization problem is NP-hard. The multi-attribute based influence maximization problem is a generalization of the influence maximization problem, therefore, the MIMP is also a NP-hard problem.

4   Solution for MIMP

In this section, we solve the MIMP. Let us introduce an important definition "realization" firstly.

Definition 1 (Realization). *Given a graph $G = (V, E)$, a realization $g = (V, E(g))$ is an instance of influence sampled from a given model. In the MIC model, we assume there are r edges between two nodes u and v because of r attributes. For each edge $(u, v) \in E$, it can be decomposed into the r edges as $(u_1, v_1), \cdots, (u_i, v_i), \cdots, (u_r, v_r)$. Based on the graph G and MIC model, we can create a constructed graph $G' = (V, E')$ such that $E' = \{\cup_{i=1}^{r}(u_i, v_i) : (u, v) \in E\}$. For each virtual edge $(u_i, v_i) \in E'$, there is a probability p_{uv}^i associated with it. We sample the r edges between any two nodes by removing each edge $(u_i, v_i) \in E$ with probability p_{uv}^i, the remaining subgraph g is a realization. So a realization $g = (V, E(g))$ is a subgraph of $G' = (V, E')$, where all the edges $e \in E(g)$ are live edges.*

Let g^* be the set of all realizations generated from G, clearly, there are $2^{E(G) \cdot r}$ possible realizations. Based on the definition of realization, we can give the probability of a realization g which is generated from graph G in MIC model:

$$p_g = \prod_{e \in E(g)} p_e \prod_{e \in E' \setminus E(g)} (1 - p_e).$$

For each node v, if $\sum_{(u_i, v_i) \in E(g)} w_v^i \geq \theta_v$, the node v can be activated by node u. Thus, given a realization g and a seed set S, the number of nodes that can be activated by S is deterministic, denoted by $\sigma_g(S)$. Let $\sigma(S)$ be the expected number of nodes which is activated by seed set S, then, in the MIC model, $\sigma(S)$ can be expressed as:

$$\sigma(S) = \sum_{g \in g^*} p_g \cdot \sigma_g(S),$$

where $\sigma_g(S)$ is the number of nodes that can be activated by S in the realization g. Then, we analyze the properties of the objective function for MIMP.

4.1   Properties of Influence Propagation Function f

The influence propagation function f for MIMP is clearly monotone since more seeds would boost the overall spread of influence.

Next, we explore the submodularity of the objective function f in Lemma 1 as follows.

Lemma 1. *The influence propagation function f for the MIMP under the MIC model is non-submodular.*

Proof. We omit the proof as the limitation of pages. ∎

4.2   Sandwich Approximation Algorithm

For a non-submodular function, we do not have a traditional approach to solve it. We can use the sandwich approximation method which is proposed by Lu *et al.* [13]. An approximate solution can be obtained for the objective function by using an upper bound function and a lower bound function to define it. Next, we look for the upper bound function and lower bound function for the objective function f, respectively.

4.2.1   Lower Bound Function $\underline{f}$ of f

Firstly, we consider the MIC model as a multi-dimensional IC model for understanding the problem more easily. In the multi-dimensional IC model, each attribute propagates independently in the social network. When some of r attributes of the product are accepted by a user v, no matter v is active or not, it can spread those accepted attributes to influence his outgoing neighbor u. We define the Multi-dimensional IC model as follows.

Definition 2 (Multi-dimensional IC model). *Given a social network $G = (V, E)$ and a product with r attributes, the multi-dimensional IC model is a r dimensional graph $G^* = (V^*, E^*) = G^1 \cup G^2 \cup \cdots \cup G^r$, where $G^i = (V^i, E^i)$ and graph G^i is a copy of graph G. The i represents the i-th attribute. In graph G^i, the corresponding node set and edge set are denoted by V^i and E^i, respectively. For the edge $(u_i, v_i) \in E^i$, where $i = 1, \cdots, r$, the influence probability on (u_i, v_i) is p_{uv}^i, which is the probability that user u tries to activate user v to accept attribute i.*

We find that there are some special constraints on weights and threshold that each customer chooses, which can make the influence spread submodular. Let us explain this conclusion as follows.

If a customer decides to purchase the product if and only if all attributes are accepted by him. Clearly, the influence spread in the model with the constraint on thresholds and weights is a lower bound for the influence spread in the general case of the MIC model. In this situation, we treat the MIC model as the Multi-dimensional IC model. So each of the r kinds of attribute information diffuses in its own dimension with the IC model and only consults with other dimensions when try to obtain the ability of influence others. According to a method given in [3], the IC model can be transformed into an equivalent general threshold model with monotone submodular threshold function at each node. Next, we explain the general threshold model. For the one-attribute influence, the general threshold model is a generalization of the LT model. In the definition of LT model, each node is assigned a monotone set function f_v which is a threshold function on the set of active inneighbors. f_v is used for determining whether v is activated or not. At each step, every node evaluates the value of f_v. If the value of f_v reaches at least the threshold θ_v, then v is activated; otherwise, v keeps inactive. Kemp, Kleinburg and Tados [3] conjecture that if for every node, the

---

Algorithm 1. Greedy Algorithm

Input: Graph $G^* = (V^*, E^*)$, r, and k.
Output: S'.
1: Initialize $S' \leftarrow 0$;
2: **for** 1 to k **do**
3:     $v = argmax_{v \in V}(h_{G^*}(S' \cup \{v\}) - h_{G^*}(S'))$.
4:      $S' = S' \cup \{v\}$.
5: **end for**
6: **return** S'.

---

threshold function is monotone and submodular, then the influence spread is a monotone and submodular function. This conjecture is proved by Mossel and Roch in [14]. We are able to extend this proof to the multi-dimensional general threshold model. In the multi-dimensional general threshold model, each node v is assigned with r threshold function f_v^ϕ for $\phi = \{1, \cdots, r\}$. At beginning, every node v chooses r thresholds θ_v^ϕ randomly and uniformly from $[0, 1]$. At each step, every node v checks the value of threshold function f_v^ϕ for all r kinds of attributes, if for all ϕ, the value of f_v^ϕ on the set of active inneighbors is at least θ_v^ϕ, then v is activated; otherwise, v keeps inactive. We are able to show that if at every node, all threshold functions are monotone and submodular, then the influence spread is a monotone and submodular function.

4.2.2   Upper Bound Function $\overline{f}$ of f

In this section, we construct a monotone submodular upper bound for the objective function f.

Consider the Multi-dimensional IC model defined above. Let each node v select a positive weight w_v^ϕ for each attribute ϕ and a threshold $\theta_v = min_\phi w_v^\phi$. This selection will make each node to be activated if and only if at least one attribute is accepted. Clearly, the influence spread in the Multi-dimensional IC diffusion model with this constraint is an upper bound for the influence spread in the general case. Moreover, the influence spread in this special case is also a monotone and submodular function with respect to the seed set. To this end, we show that the multi-dimensional IC diffusion model with this constraint can be transformed into an equivalent IC model. In fact, this can be done by, on each edge (u, v), replacing r probabilities $p_{uv}^1, \cdots, p_{uv}^r$ by a probability $p_{uv} = 1 - (1 - p_{uv}^1) \cdots (1 - p_{uv}^r)$, where p_{uv}^i is the probability that v accepts at least one attribute's influence from u.

4.2.3   Greedy Algorithm for Bound Function

For the monotone and submodular lower bound function $\underline{f}$ and upper bound function $\overline{f}$, we can use greedy algorithm to solve both of them, and the solutions

---

Algorithm 2. Sandwich Approximation Algorithm

Input: Graph $G = (V, E)$, $G^* = (V^*, E^*)$, r, and k.
Output: S.
1: Initialize $S \leftarrow 0$, $S_o \leftarrow 0$, $S_U \leftarrow 0$, $S_L \leftarrow 0$;
2: Let S_L be $(1 - 1/e)$-approximation by greedy algorithm to $\underline{f}$.
3: Let S_U be $(1 - 1/e)$-approximation by greedy algorithm to $\overline{f}$.
4: Let S_o be a solution by greedy to f.
5: $S = \arg\max_{S_A \in \{S_U, S_L, S_o\}} f(S_A)$.
6: **return** S.

---

returned by the greedy algorithm for each of the two functions can obtain an $1 - 1/e$ approximation ratio. The greedy algorithm shows in Algorithm 1.

In Algorithm 1, we use $h_{G^*}(\cdot)$ to represent the influence propagation function $\underline{f}$ or $\overline{f}$. S' represents the selected seeds for the upper bound problem or the lower bound problem.

4.2.4 Sandwich Approximation Framework

Motivated from the monotone and submodular lower bound and upper bound of the influence spread, it is intuitive to employ the sandwich method for solving the MIMP.

Before employing the sandwich approximation framework, we need to get a solution for the original objective function f. We use greedy algorithm directly to choose top k users as seed nodes, the solution is denoted as S_o. Let S_L, S_U be the solutions which are obtained by greedy algorithms for lower bound function $\underline{f}$ and the upper bound function $\overline{f}$, respectively. Finally, The sandwich algorithm returns the final optimal approximation solution S which can maximize the influence spread among S_o, S_L, S_U. The sandwich approximation algorithm shows as follows.

Although the influence maximization function f for MIMP is non-submodular, we construct a submodular lower bound function $\underline{f}$ and a submodular upper bound function $\overline{f}$, which satisfies $\underline{f} \leq f \leq \overline{f}$. Then we can get a data-dependent approximation ratio for the solution, which is returned by the sandwich approximation algorithm in Algorithm 2. The conclusion shows in Theorem 1 as follows.

Theorem 1. *Let S_o^*, S_L^* and S_U^* be the optimal solution for the original objective function f, the lower bound function $\underline{f}$, and the upper bound function $\overline{f}$, respectively. Algorithm 2 derives a $\max\{\frac{f(S_U)}{\overline{f}(S_U)}, \frac{f(S_L^*)}{\overline{f}(S_o^*)}\}(1 - 1/e)$-approximate solution.*

Proof. As S_o^*, S_L^* and S_U^* are the optimal solutions to maximize f, $\underline{f}$, $\overline{f}$, respectively. We can get that

$$f(S_U) = \frac{f(S_U)}{\overline{f}(S_U)}\overline{f}(S_U) \geq \frac{f(S_U)}{\overline{f}(S_U)}(1 - 1/e)\overline{f}(S_U^*)$$

$$\geq \frac{f(S_U)}{\overline{f}(S_U)}(1 - 1/e)\overline{f}(S_o^*)$$

$$\geq \frac{f(S_U)}{\overline{f}(S_U)}(1 - 1/e)f(S_o^*).$$

For the lower bound, we have

$$f(S_L) \geq \underline{f}(S_L) \geq (1 - 1/e)\underline{f}(S_L^*) = \frac{\underline{f}(S_L^*)}{f(S_o^*)}(1 - 1/e)f(S_o^*).$$

Then, let $S = \arg max_{S^* \in \{S_U, S_L, S_o\}} f(S_A)$, we can get

$$f(S) = max\{f(S_U), f(S_L), f(S_o)\}$$

$$\geq max\{f(S_U), f(S_o)\}$$

$$= max\{\frac{f(S_U)}{\overline{f}(S_U)}, \frac{\underline{f}(S_L^*)}{f(S_o^*)}\}(1 - 1/e)f(S_o^*).$$

■

5   Experiments

In this section, we compare the efficiency and effectiveness of the proposed sandwich approximation algorithm with other algorithms. All the experiments are done on two different datasets.

5.1   Experimental Setup

Datasets: Dataset 1 is a NetScience dataset [15], which is a co-authorship network on scientists to publish papers in the field of network science. Dataset 2 is a Wikivote dataset [15], which is from a Wikipedia voting set and represents the relationship of who votes to whom. The statistics of the two datasets are summarized in Table 1.

Table 1. Statistics of two datasets.

Dataset	Nodes	Edges	Type	Average degree
Dataset 1	400	1010	Directed	5.0
Dataset 2	914	2914	Directed	6.2

Influence Model: In this paper, we use IC model as the basic influence propagation model, the simulation is based on the MIC model. The spread probability on each edge (u_i, v_i) is set as $p_{uv}^i = 1/|N^-(v_i)|$, which is widely used in previous work [16]. The activation threshold θ_v for each node v is randomly generated from $[0, 1]$. The number of Monte Carlo simulations which is used to estimate the maximum marginal gain in one iteration is set as 200.

Baseline Methods: Max Degree. It firstly ranks the nodes based on the out-degrees and then selects k nodes which have the highest out-degree as seeds. This is a classical baseline algorithm.

Random. It is also a classical baseline algorithm which randomly selects nodes as seeds.

Given a product with r attributes, let $w = (w^1, w^2, \cdots, w^r)$ be the weight vector for r attributes. Then, we evaluate the performance of the proposed algorithms from two subcases: (1) $r = 2$, $w = (0.3, 0.7)$; (2) $r = 3$, $w = (0.2, 0.3, 0.5)$. The number of nodes and edges on graph G^* vary as the number of attributes. When $r = 2$, they will be doubled; when $r = 3$, they will be tripled.

5.2   Experimental Results

The performance of the proposed algorithms will be evaluated in this section. The results in Fig. 1 (a) and (b) are collected from dataset 1 when the budget k increases from 0 to 50 for $r = 2$ and $r = 3$, respectively. From the two sub-figures (a) and (b), we can see that the performance of our proposed sandwich method is between the performance of the upper bound and lower bound no matter $r = 2$ or $r = 3$, which verifies the correctness of our sandwich method. We can also observe that the influence propagation of the sandwich method is better than that of the max degree algorithm and random algorithm for both $r = 2$ and $r = 3$. Random algorithm has the worst performance among all the algorithms since it selects seeds randomly. Note that the advantage of the sandwich algorithm over the max degree algorithm becomes more and more obvious as the budget k increases. For all the algorithms, the influence propagation increases as the

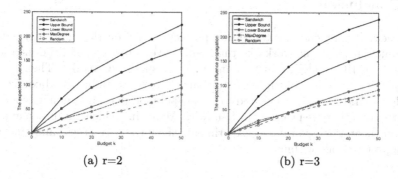

(a) r=2   (b) r=3

Fig. 1. Performance comparison achieved by different algorithms with the changing of budget on dataset 1

increasing of budget k, which is because that more seeds can spread information more widely. Comparing the subfigures (a) and (b) of Fig. 1, we can find that the increase of the number of attribute r increases the difference between upper and lower bounds. Comparing subfigure (a) and (b) in Fig. 1, we can find that for the same budget, the influence propagation for $r = 3$ is different from that for $r = 2$, which shows that increasing the number of r will change the user group who buy the product.

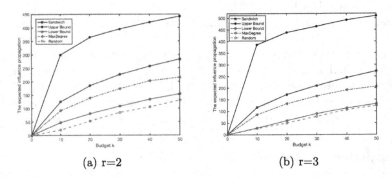

(a) r=2   (b) r=3

Fig. 2. Performance comparison achieved by different algorithms with the changing of budget on dataset 2

Figure 2 shows the influence propagations of all the compared algorithms on dataset 2 when the budget k increases from 0 to 50 for $r = 2$ and $r = 3$, respectively. We can find that the results we conclude from Fig. 1 are also can be observed from Fig. 2, which further verifies the correctness and effectiveness of the proposed sandwich method. We compare the results in Fig. 2 with that in Fig. 1, we can find that when the budget k is fixed, the larger dataset can produce greater influence propagation no matter $r = 2$ or $r = 3$.

6   Conclusion

We study the influence maximization problem based on the multi-attribute of a product in social networks in this paper. We propose a Multi-dimensional IC model (MIC model), which is an extension of the IC model. Then we formulate the proposed problem as a Multi-attribute based Influence Maximization Problem (MIMP), which is proved to be non-submodular. For a non-submodular objective function, we propose the Sandwich Algorithm which obtains an approximation theoretical guarantee. Experiments show the correctness and efficiency of the proposed algorithm.

References

1. Ni, Q., Guo, J., Huang, C., Weili, W.: Community-based rumor blocking maximization in social networks: algorithms and analysis. Theoret. Comput. Sci. **840**, 257–269 (2020)
2. Ni, Q., Guo, J., Huang, C., Weili, W.: Information coverage maximization for multiple products in social networks. Theoret. Comput. Sci. **828**, 32–41 (2020)
3. Kempe, D., Kleinberg, J., Tardos, É.: Maximizing the spread of influence through a social network. In: Proceedings of the Ninth ACM SIGKDD International Conference on Knowledge Discovery and Data Mining, pp. 137–146. ACM (2003)
4. Yan, R., Zhu, Y., Li, D., Wang, Y.: Community based acceptance probability maximization for target users on social networks: algorithms and analysis. Theoret. Comput. Sci. **803**, 116–129 (2020)
5. Guo, J., Li, Y., Weili, W.: Targeted protection maximization in social networks. IEEE Trans. Netw. Sci. Eng. **7**(3), 1645–1655 (2019)
6. Yuan, J., Tang, S.-J.: Adaptive discount allocation in social networks. In: Proceedings of the 18th ACM International Symposium on Mobile Ad Hoc Networking and Computing, pp. 1–10 (2017)
7. Yang, Yu., Mao, X., Pei, J., He, X.: Continuous influence maximization. ACM Trans. Knowl. Discov. Data (TKDD) **14**(3), 1–38 (2020)
8. Bharathi, S., Kempe, D., Salek, M.: Competitive influence maximization in social networks. In: Deng, X., Graham, F.C. (eds.) WINE 2007. LNCS, vol. 4858, pp. 306–311. Springer, Heidelberg (2007). https://doi.org/10.1007/978-3-540-77105-0_31
9. Singh, S.S., Singh, K., Kumar, A., Biswas, B.: MIM2: multiple influence maximization across multiple social networks. Phys. A Stat. Mech. Appl. **526**, 120902 (2019)
10. Zhu, J., Zhu, J., Ghosh, S., Weili, W., Yuan, J.: Social influence maximization in hypergraph in social networks. IEEE Trans. Netw. Sci. Eng. **6**(4), 801–811 (2018)
11. Zhu, J., Ghosh, S., Weili, W.: Group influence maximization problem in social networks. IEEE Trans. Comput. Soc. Syst. **6**(6), 1156–1164 (2019)
12. Guo, J., Chen, T., Weili, W.: A multi-feature diffusion model: rumor blocking in social networks. IEEE/ACM Trans. Netw. **29**(1), 386–397 (2020)
13. Lu, W., Chen, W., Lakshmanan, L.V.S.: From competition to complementarity: comparative influence diffusion and maximization. Proc. VLDB Endow. **9**(2), 60–71 (2015)
14. Mossel, E., Roch, S.: On the submodularity of influence in social networks. In: Proceedings of the Thirty-Ninth Annual ACM Symposium on Theory of Computing, pp. 128–134 (2007)
15. Rossi, R., Ahmed, N.: The network data repository with interactive graph analytics and visualization. In: Twenty-Ninth AAAI Conference on Artificial Intelligence, pp. 4292–4293 (2015)
16. Tang, Y., Xiao, X., Shi, Y.: Influence maximization: near-optimal time complexity meets practical efficiency. In: Proceedings of ACM SIGMOD International Conference on Management of Data, pp. 75–86 (2014)

A Parallel Algorithm for Constructing Multiple Independent Spanning Trees in Bubble-Sort Networks

Shih-Shun Kao[1,2(✉)], Ralf Klasing[1], Ling-Ju Hung[3], and Sun-Yuan Hsieh[2,4]

[1] CNRS, LaBRI, Université de Bordeaux, 351 Cours de la Libération, 33405 Talence, France
{shih-shun.kao,ralf.klasing}@labri.fr
[2] Department of Computer Science and Information Engineering, National Cheng Kung University, No. 1, University Road, Tainan, Taiwan
[3] Department of Creative Technologies and Product Design, National Taipei University of Business, No. 100, Sec. 1, Fulong Road, Taoyuan, Taiwan
ljhung@ntub.edu.tw
[4] Institute of Medical Informatics, National Cheng Kung University, No. 1, University Road, Tainan, Taiwan
hsiehsy@mail.ncku.edu.tw

Abstract. The use of multiple independent spanning trees (ISTs) for data broadcasting in networks provides a number of advantages, including the increase of fault-tolerance and secure message distribution. Thus, the designs of multiple ISTs on several classes of networks have been widely investigated. Kao *et al.* [*Journal of Combinatorial Optimization* 38 (2019) 972–986] proposed an algorithm to construct independent spanning trees in bubble-sort networks. The algorithm is executed in a recursive function and thus is hard to parallelize. In this paper, we focus on the problem of constructing ISTs in bubble-sort networks B_n and present a non-recursive algorithm. Our approach can be fully parallelized, i.e., every vertex can determine its parent in each spanning tree in constant time. This solves the open problem from the paper by Kao *et al.* Furthermore, we show that the total time complexity $\mathcal{O}(n \cdot n!)$ of our algorithm is asymptotically optimal, where n is the dimension of B_n and $n!$ is the number of vertices of the network.

Keywords: Independent spanning trees · Bubble-sort networks · Interconnection networks

1 Introduction

The design of modern interconnected networks faces several critical demands, such as how to perform fault-tolerant transmission and secure message distribution in

This research was supported by the LaBRI under the "Projets émergents" program. This study has been carried out in the frame of the "Investments for the future" Programme IdEx Bordeaux - SysNum (ANR-10-IDEX-03-02).

© Springer Nature Switzerland AG 2021
W. Wu and H. Du (Eds.): AAIM 2021, LNCS 13153, pp. 252–264, 2021.
https://doi.org/10.1007/978-3-030-93176-6_22

a reliable communication network. The practical solution to meet the above requirements is to design a multi-path routing mechanism, which requires the network to provide disjoint paths between each pair of vertices. Therefore, if the transmission fails due to a disconnection in the current transmission path, we can resume the data transmission via another disjoint backup path. This dramatically increases the performance of fault-tolerant communication [2,14]. In addition, disjoint paths could be used in secure message distribution over a fault-free network in the following way [2,30]. A message can be divided into several packets where the source node sends each packet to its destination via different paths. Thus, each node in the network receives at most one of the packets except for the destination node that receives all the packets.

Usually, an interconnection network is modeled by a simple undirected graph $G = (V, E)$, where the vertex set $V(G)$ and the edge set $E(G)$ represent the set of processors and the set of communication links between the processors, respectively. A *spanning tree* T in G is a connected acyclic subgraph of G such that $V(T) = V(G)$. Two spanning trees rooted at a specific vertex, say r, are called *independent spanning trees* (ISTs for short) if, for any vertex $v \in V(G) \setminus \{r\}$, the two paths from v to r in any two trees share no common edge and no common vertex except for v and r. Accordingly, the provision of multiple ISTs suffices to meet the requirement of reliable communication in a network.

Research on ISTs has been conducted for nearly three decades. In 1989, Zehavi and Itai [40] conjectured that there exist k ISTs rooted at an arbitrary vertex in a k-connected graph. From then on, this conjecture has been confirmed only for k-connected graphs with $k \leq 4$ (see [9,10,14]). Since this conjecture is still unsolved for general k-connected graphs for $k \geq 5$, the follow-up research mainly focused on the study of constructing ISTs on specific interconnection networks, e.g., the construction of ISTs on some variations of hypercubes [3, 20,29,30,37], torus networks [28], recursive circulant graphs [34,35], and special subclasses of Cayley networks [7,8,12,13,15,19,39]. In particular, special topics related to ISTs include the research on reducing the height of the ISTs [31,33,36] and parallel construction of ISTs [4–6,32,37,38].

Note that there is a similar problem called the construction of *completely independent spanning trees* (CISTs for short) in a network. A set of k unrooted spanning trees are called CISTs if they are pairwise edge-disjoint and inner-node-disjoint (i.e., for each pair of vertices u and v in any two spanning trees, there exist no common edge and vertex in the paths between u and v except for the two end vertices). In particular, if $k = 2$, the two CISTs are called a *dual-CIST*. Hasunuma [11] showed that the problem of determining whether there exists a dual-CIST in a graph is NP-complete. He also conjectured that there exist k CISTs in a $2k$-connected graph. Currently, this conjecture has been proved to fail by counterexamples [21,26]. For recent research results on CISTs and their applications, the reader is referred to [22–25] and references quoted therein. Here, we explicitly point out that the construction of multiple ISTs and CIST are two different problems.

For the construction of ISTs on bubble-sort networks, Kao *et al.* [15] proposed an algorithm to construct $n - 1$ ISTs of B_n and showed that the algorithm has optimal amortized efficiency for multiple trees construction. In particular, every vertex can determine its parent in each spanning tree in constant amortized time. The algorithm is executed in a recursive function and thus is hard to parallelize. In this paper, we present a parallel algorithm to construct $n-1$ ISTs in bubble-sort networks B_n. Our approach can be fully parallelized, i.e., every vertex can determine its parent in each spanning tree in constant time. This solves the open problem from [15]. Furthermore, we show that the total time complexity $\mathcal{O}(n \cdot n!)$ of our algorithm is asymptotically optimal, where n is the dimension of B_n and $n!$ is the number of vertices of the network.

The rest of this paper is organized as follows. In Sect. 2, we introduce the bubble-sort graphs and some notations. In Sect. 3, we introduce the algorithm for constructing independent spanning trees of B_n. In Sect. 4, we show the correctness of our algorithm and give the complexity analysis. Finally, conclusions and future works are given in Sect. 5.

2   Preliminaries

Let Σ_n be the set of all permutations on $\{1, 2, \ldots, n\}$. For a permutation $p \in \Sigma_n$ and an integer $i \in \{1, 2, \ldots, n\}$, we use the following notations. The symbol at the ith position of p is denoted by p_i, and the position where the symbol i appears in p is denoted by $p^{-1}(i)$. A symbol i is said to be at the *right position* of p if $p_i = i$, and for $p \neq 12 \cdots n$ the position of the first symbol i from the right which is not in the right position is denoted by $r(p)$. For $i \in \{1, \cdots, n-1\}$, let $p\langle i \rangle = p_1 p_2 \cdots p_{i-1} p_{i+1} p_i p_{i+2} \cdots p_n$ be the permutation of Σ_n obtained from p by swapping two consecutive symbols at positions i and $i+1$. The *bubble-sort network*, denoted by B_n, is an undirected graph consisting of the vertex set $V(B_n) = \Sigma_n$ and the edge set $E(B_n) = \{(\mathbf{x}, \mathbf{x}\langle i \rangle): \mathbf{x} \in \Sigma_n, 1 \leqslant i \leqslant n-1\}$, where the edge $(\mathbf{x}, \mathbf{x}\langle i \rangle)$ is called an *i-edge* of B_n. Thus, B_n is a Cayley graph generated by the transposition set $\{(i, i+1): 1 \leqslant i \leqslant n-1\}$, which is specified by an n-path $P_n = (1, 2, \ldots, n)$ as its transposition graph [1,16]. For example, Fig. 1 depicts B_4. Clearly, for B_n, the transposition graph P_n contains only two subgraphs isomorphic to an $(n-1)$-path: one is $(1, 2, \ldots, n-2)$ and the other is $(2, 3, \ldots, n-1)$. Thus, for $n \geqslant 3$, there are exactly two ways to decompose B_n into n disjoint subgraphs that are isomorphic to B_{n-1}. Let B_n^i denote the graph obtained from B_n by removing the set of all i-edges. Then, both B_n^1 and B_n^{n-1} consist of n disjoint subgraphs isomorphic to B_{n-1}.

3   Constructing ISTs on B_n

In this section, we present an algorithm for constructing $n-1$ ISTs of B_n. Since B_n is vertex-transitive, without loss of generality, we may choose the identity $\mathbf{1}_n = 12 \cdots n$ as the common root of all ISTs. Also, since B_n has connectivity

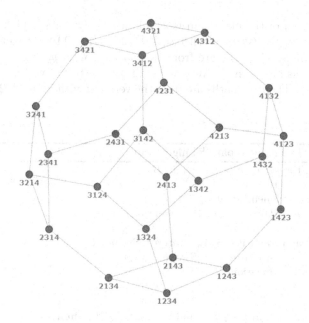

Fig. 1. The bubble-sort network B_4

$n-1$, the root in every spanning tree has a unique child. For $1 \leqslant t \leqslant n-1$, if the root of a spanning tree takes $\mathbf{1}_n\langle t \rangle = 12 \cdots (t-1)(t+1)t(t+2) \cdots n$ as its unique child, then the spanning tree of B_n is denoted by T_t^n. To describe such a spanning tree, for each vertex $v = v_1 \cdots v_n \in V(B_n)$ except the root $\mathbf{1}_n$, we denote by $\mathrm{Parent}(v, t, n)$ the parent of v in T_t^n.

The Case $n = 3$. Since B_3 is isomorphic to a 6-cycle, we have

$$\mathrm{Parent}(v,1,3) = \begin{cases} 123 & \text{if } v = 213; \\ 213 & \text{if } v = 231; \\ 231 & \text{if } v = 321; \\ 321 & \text{if } v = 312; \\ 312 & \text{if } v = 132; \end{cases} \quad \text{and} \quad \mathrm{Parent}(v,2,3) = \begin{cases} 231 & \text{if } v = 213; \\ 321 & \text{if } v = 231; \\ 312 & \text{if } v = 321; \\ 132 & \text{if } v = 312; \\ 123 & \text{if } v = 132. \end{cases}$$

That is, the two paths $T_1^3 = (132, 312, 321, 231, 213, 123)$ and $T_2^3 = (213, 231, 321, 312, 132, 123)$ are ISTs of B_3 that take $\mathbf{1}_3 = 123$ as the common root.

The Case $n \geqslant 4$. In general, for B_n with $n \geqslant 4$, the construction of the ISTs of B_n can be accomplished by Algorithm 1 to determine the parent of each vertex (except the root) in every spanning tree.

The main idea of the algorithm is as follows. In T_t^n for $t \in \{1, 2, \ldots, n-2\}$ all paths are from the vertex x with $x_n \in \{1, 2, \ldots, n-1\} \setminus \{t\}$ to the vertex y with $y_n = t$. Then, all paths are from the vertex y with $y_n = t$ to the root r. In T_{n-1}^n all paths are from the vertex v with $v_n = n$ to the vertex u with $u_n \in \{1, 2, \ldots, n-1\}$. Then, all paths are from the vertex u with $u_n \in \{1, 2, \ldots, n-1\}$ to the root r.

Algorithm 1: The new parallel algorithm

    Input   : v: the vertex $v = v_1 \cdots v_n$ in B_n
                t: the t-th tree T_t^n in IST
                n: the dimension of B_n
    Output: p: $p = \mathrm{Parent}(v, t, n)$ the parent of v in T_t^n

1 **if** $v_n = n$ **then**
2   **if** $t = 2$ and $\mathrm{Swap}(v, t) = \mathbf{1}_n$ **then** $p = \mathrm{Swap}(v, t-1)$
3   **else if** $t = n-1$ **then** $p = \mathrm{Swap}(v, v_{n-1})$
4   **else** $p = \mathrm{FindPosition}(v)$
5 **end**
6 **else**
7   **if** $v_n = n-1$ and $v_{n-1} = n$ and $\mathrm{Swap}(v, n) \neq \mathbf{1}_n$ **then**
8   **if** $t = 1$ **then** $p = \mathrm{Swap}(v, n)$
9   **else** $p = \mathrm{Swap}(v, t-1)$
10   **end**
11   **else**
12   **if** $v_n = t$ **then** $p = \mathrm{Swap}(v, n)$
13   **else** $p = \mathrm{Swap}(v, t)$
14   **end**
15 **end**
16 **return** p

Function $\mathrm{FindPosition}(v)$

    Input   : v: the vertex $v = v_1 \cdots v_n$ in B_n
    Output: p: $p = \mathrm{Parent}(v, t, n)$ the parent of v in T_t^n

1 **if** $v_{n-1} \in \{t, n-1\}$ **then** $j = r(v), p = \mathrm{Swap}(v, v_j)$
2 **else** $p = \mathrm{Swap}(v, t)$
3 **return** p

Function $\mathrm{Swap}(v, x)$

    Input   : v: the vertex $v = v_1 \cdots v_n$ in B_n
                x: the symbol in the vertex $v_1 \cdots v_n$
    Output: p: $p = \mathrm{Parent}(v, t, n)$ the parent of v in T_t^n

1 $i = v^{-1}(x), p = v\langle i \rangle$
2 **return** p

Table 1. The parent of every vertex $v \in V(B_4) \setminus \{\mathbf{1}_4\}$ in T_t^4 for $t \in \{1, 2, 3\}$ calculated by Algorithm 1

v	t	v_4	p	v	t	v_4	p
1234	-	-	-	3124	1	4	3214
					2		1324
					3		3142
1243	1	3	2143	3142	1	2	3412
	2		1423		2		3124
	3		1234		3		1342
1324	1	4	3124	3214	1	4	2314
	2		1234		2		3124
	3		1342		3		3241
1342	1	2	3142	3241	1	1	3214
	2		1324		2		3421
	3		1432		3		2341
1423	1	3	4123	3412	1	2	3421
	2		1432		2		3142
	3		1243		3		4312
1432	1	2	4132	3421	1	1	3241
	2		1342		2		3412
	3		1423		3		4321
2134	1	4	1234	4123	1	3	4213
	2		2314		2		4132
	3		2143		3		1423
2143	1	4	2134	4132	1	2	4312
	2		2413		2		1432
	3		1243		3		4123
2314	1	4	2134	4213	1	3	4231
	2		3214		2		4123
	3		2341		3		2413
2341	1	1	2314	4231	1	1	2431
	2		3241		2		4321
	3		2431		3		4213
2413	1	3	2431	4312	1	2	4321
	2		4213		2		3412
	3		2143		3		4132
2431	1	1	2341	4321	1	1	3421
	2		4231		2		4312
	3		2413		3		4231

Note that in a pre-processing stage, each node $v = v_1v_2 \cdots v_n$ $(v \neq 1_n)$ computes its inverse permutation, i.e., $v^{-1}(1)v^{-1}(2) \cdots v^{-1}(n)$, and the position of the first symbol i from the right which is not in the right position, i.e., $r(v)$. This can be done efficiently in $\mathcal{O}(n)$ time for each vertex. Algorithm 1 uses two functions FindPosition(v) and Swap(v, x). The function FindPosition(v) finds the rightmost symbol x in v which is not in the right position, and then calls the Swap(v, x) function. The function Swap(v, x) swaps the symbol x in v in its position i with the symbol in position $i+1$. Since we have the pre-processing stage, the two functions FindPosition(v) and Swap(v, x) can be calculated in constant time.

Table 1 shows the parent of every vertex $v \in V(B_4) \setminus \{1_4\}$ in T_t^4 for $t \in \{1, 2, 3\}$ calculated by Algorithm 1. For example, we consider $v = 3214$ and $t = 3$. Since $v_4 = 4$, $p = $ Swap(v, v_{4-1}) $= 3241$. Also, we consider $v = 4321$ and $t = 1$. Since $v_4 = 1$, $p = $ Swap($v, 4$) $= 3421$. The corresponding three ISTs rooted at vertex 1_4 for B_4 are shown in Fig. 2.

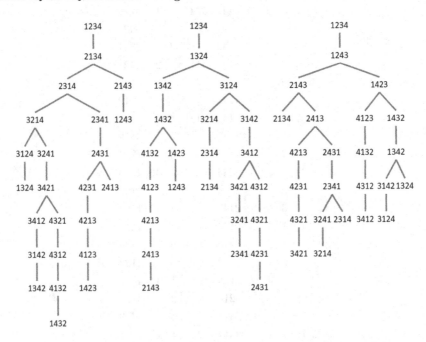

Fig. 2. The three ISTs of B_4 calculated by Algorithm 1

4   Correctness and Complexity Analysis

In this section, we first show the correctness of Algorithm 1. Let T be a tree and $u, v \in V(T)$, we use $T(u, v)$ to denote the unique path joining u and v in T. For two spanning trees T_t^n and $T_{t'}^n$ for $t, t' \in \{1, 2, \ldots, n-1\}$ with $t \neq t'$, we denote by $T_t^n(v, r)$ and $T_{t'}^n(v, r)$ the two paths from v to the common r.

Theorem 1. For $n \geqslant 4$, $T_1^n, T_2^n, \ldots, T_{n-1}^n$ are $n - 1$ ISTs of B_n.

Proof. Suppose that $n \geqslant 4$, let $r = 1_n (= 12 \cdots n)$, the proof is by showing that for any vertex $v \in V(B_n) \setminus \{r\}$, the two paths from v to r in any two trees of $T_1^n, T_2^n, \ldots, T_{n-1}^n$ share no common edge and no common vertex except for v and r, and thereby proving the independence. Consider the following three cases:

CASE 1: $v_n = n$.

Each vertex of the two paths $T_t^n(v, r)$ and $T_{t'}^n(v, r)$ (apart from $T_{n-1}^n(v, r)$) swaps symbol t (resp., t') to the position v_{n-1} for $t, t' \in \{1, 2, \ldots, n-2\}$. Then, the rightmost symbol i which is not in the right position swaps to the right position. Therefore, $T_t^n(v, r)$ and $T_{t'}^n(v, r)$ are vertex-disjoint. Now consider $T_{n-1}^n(v, r)$, each vertex of the path swaps the position v_{n-1} to v_n. Then, the vertex v with $v_n = n$ swaps the symbol $n - 1$ to the position v_n. Hence, $T_t^n(v, r)$, $T_t^n(v, r)$ and $T_{n-1}^n(v, r)$ are vertex-disjoint. See Fig. 3, the paths from the vertex v with $v_n = n$ to r are marked in red, in $T_{n-1}^n(v, r)$ each vertex of the path has symbol $n - 1$ in v_n. The other trees $T_t^n(v, r)$ have symbol t in position v_n for $t \in \{1, 2, \ldots, n-2\}$.

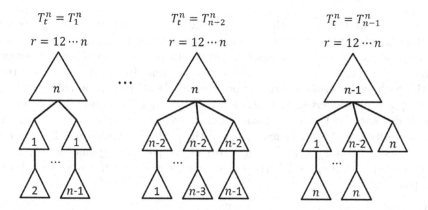

Fig. 3. An illustration of the paths described in the proof of Case 1 of Theorem 1

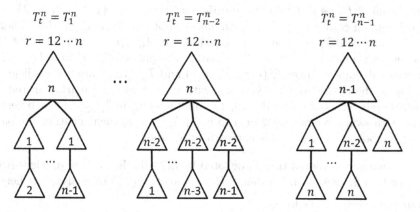

Fig. 4. An illustration of the paths described in the proof of Case 2 of Theorem 1

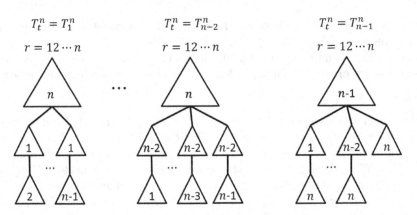

Fig. 5. An illustration of the paths described in the proof of Case 3 of Theorem 1

CASE 2: $v_n = n - 1$.

Each vertex of the two paths $T_t^n(v, r)$ and $T_{t'}^n(v, r)$ (apart from $T_{n-1}^n(v, r)$) swaps symbol t (resp., t') to the position v_n for $t, t' \in \{1, 2, \ldots, n - 2\}$. On the other hand each vertex of the path has symbol t (resp., t') in different position. Therefore, $T_t^n(v, r)$ and $T_{t'}^n(v, r)$ are vertex-disjoint. In $T_{n-1}^n(v, r)$ each vertex of the path swaps symbol n to the position v_n. By CASE 1, the paths $T_1^n(v, r)$ and $T_{n-2}^n(v, r)$ are vertex-disjoint. Hence, $T_t^n(v, r)$, $T_{t'}^n(v, r)$ and $T_{n-1}^n(v, r)$ are vertex-disjoint. See Fig. 4, the paths from the vertex v with $v_n = n - 1$ to r are marked in red, in $T_1^n(v, r)$ each vertex of the path has symbol $n - 1$, 1 or n in the position v_n, in $T_{n-2}^n(v, r)$ each vertex of the path has symbol $n - 1$, $n - 2$ or n in the position v_n, in $T_{n-1}^n(v, r)$ each vertex of the path swaps symbol n to the position v_n.

CASE 3: $v_n = j$ for $j \in \{1, 2, \ldots, n - 2\}$.

Each vertex of the two paths $T_t^n(v, r)$ and $T_{t'}^n(v, r)$ (apart from $T_{n-1}^n(v, r)$) swaps symbol t (resp., t') to the position v_n for $t, t' \in \{1, 2, \ldots, n - 2\}$. On the other hand each vertex of the path has symbol t in different position. Therefore, $T_t^n(v, r)$ and $T_{t'}^n(v, r)$ are vertex-disjoint. In $T_{n-1}^n(v, r)$ each vertex of the path swaps symbol $n - 1$ to v_n. By CASE 2, the paths $T_1^n(v, r)$ and $T_{n-2}^n(v, r)$ are vertex-disjoint. Hence, $T_t^n(v, r)$, $T_{t'}^n(v, r)$ and $T_{n-1}^n(v, r)$ are vertex-disjoint. See Fig. 5, the paths from the vertex v with $v_n = 1$ to r are marked in red, in $T_1^n(v, r)$ each vertex of the path swaps symbol n to v_n, in $T_{n-2}^n(v, r)$ each vertex of the path swaps symbol $n - 2$ or n to v_n, in $T_{n-1}^n(v, r)$ each vertex of the path swaps symbol $n - 1$ to v_n. This completes the proof. □

The *height* of a rooted tree T, denoted by $h(T)$, is the number of edges from the root to a farthest leaf. We define $H_n = \max_{1 \leqslant t \leqslant n-1} h(T_t^n)$ to analyze the height of our constructed ISTs for B_n.

Theorem 2. *For the bubble-sort graph* B_n, *Algorithm 1 correctly constructs* $n-1$ *ISTs of* B_n *with height at most* $n(n+1)/2 - 1$. *In particular, every vertex can determine its parent in each spanning tree in constant time.*

Proof. From Algorithm 1, the path from the vertex v with $v_n = 2$ to the vertex u with $u_n = 1$ has at most $n-1$ edges, and the path from the vertex u with $u_n = 1$ to the vertex x with $x_n = n$ has at most $n-1$ edges. Moreover, the path from the vertex w with $w_n = n$ to the vertex x with $x_n = n$ and $x_{n-1} = t$ has at most $n-1$ edges, and the path from the vertex x with $x_n = n$ and $x_{n-1} = t$ to the vertex y with $y_n = n$ and $y_{n-1} = n-1$ has at most $n-2$ edges, and the path from the vertex y with $y_n = n$ and $y_{n-1} = n-1$ to the vertex z with $z_n = n$, $z_{n-1} = n-1$ and $z_{n-2} = n-2$ has at most $n-3$ edges. Since $(n-2) + (n-3) + \cdots + 1 = (n-1)(n-2)/2$, the path from the vertex x with $x_n = n$ and $x_{n-1} = t$ to the root r has at most $(n-1)(n-2)/2$ edges. The path from the vertex v with $v_n = 2$ to the root r has at most $(n-1)(n-2)/2 + (n-1) + (n-1) = (n^2 - 3n + 2 + 4n - 4)/2 = (n^2 + n - 2)/2 = n(n+1)/2 - 1$ edges. Hence, $H_n \leq n(n+1)/2 - 1$. Obviously, each vertex in Algorithm 1 can determine its parent in each spanning tree in constant time. This completes the proof. □

Corollary 1. *The total time complexity* $\mathcal{O}(n \cdot n!)$ *of Algorithm 1 is asymptotically optimal.*

Proof. There are $n-1$ ISTs, each IST contains $n!$ vertices, hence the lower bound $\Omega(n \cdot n!)$ is obvious. Since each vertex in Algorithm 1 can determine its parent in each spanning tree in constant time, the total time complexity of the proposed Algorithm 1 is $\mathcal{O}(n \cdot n!)$. Hence, the total time complexity $\mathcal{O}(n \cdot n!)$ of Algorithm 1 is asymptotically optimal. This completes the proof. □

5   Conclusion

In this paper, we have proposed an algorithm for constructing $n-1$ ISTs rooted at an arbitrary vertex of the bubble-sort network B_n. Our approach can be fully parallelized, i.e., every vertex can determine its parent in each spanning tree in constant time. Furthermore, we show that the total time complexity $\mathcal{O}(n \cdot n!)$ of our algorithm is asymptotically optimal, where n is the dimension of B_n and $n!$ is the number of vertices of the network.

Since B_n is a regular graph with connectivity $n-1$, the number of constructed ISTs is the maximum possible. For future work, a problem remaining open from our work is whether our algorithm can be extended to the (n, k)-bubble-sort graph [27,41,42] which is a generalization of bubble-sort networks. Moreover, the butterfly graph [17,18] has good structural symmetries, is regular of degree 4, and the recursive construction properties are similar to bubble-sort networks. Thus, it is of interest to study the construction of ISTs on butterfly graphs.

References

1. Akers, S.B., Krishnamurty, B.: A group theoretic model for symmetric interconnection networks. IEEE Trans. Comput. **38**(4), 555–566 (1989)
2. Bao, F., Funyu, Y., Hamada, Y., Igarashi, Y.: Reliable broadcasting and secure distributing in channel networks. In: Proceedings of 3rd International Symposium on Parallel Architectures, Algorithms and Networks, ISPAN 1997, Taipei, December, pp. 472–478 (1997)
3. Chang, J.-M., Wang, J.-D., Yang, J.-S., Pai, K.-J.: A comment on independent spanning trees in crossed cubes. Inf. Process. Lett. **114**(12), 734–739 (2014)
4. Chang, J.-M., Yang, T.-J., Yang, J.-S.: A parallel algorithm for constructing independent spanning trees in twisted cubes. Discret. Appl. Math. **219**, 74–82 (2017)
5. Chang, Y.-H., Yang, J.-S., Chang, J.-M., Wang, Y.-L.: A fast parallel algorithm for constructing independent spanning trees on parity cubes. Appl. Math. Comput. **268**, 489–495 (2015)
6. Chang, Y.-H., Yang, J.-S., Hsieh, S.-Y., Chang, J.-M., Wang, Y.-L.: Construction independent spanning trees on locally twisted cubes in parallel. J. Comb. Optim. **33**(3), 956–967 (2016). https://doi.org/10.1007/s10878-016-0018-8
7. Cheng, D.-W., Chan, C.-T., Hsieh, S.-Y.: Constructing independent spanning trees on pancake networks. IEEE Access **8**, 3427–3433 (2020)
8. Cheng, D.-W., Yao, K.-H., Hsieh, S.-Y.: Constructing independent spanning trees on generalized recursive circulant graphs. IEEE Access **9**, 74028–74037 (2021)
9. Cheriyan, J., Maheshwari, S.N.: Finding nonseparating induced cycles and independent spanning trees in 3-connected graphs. J. Algorithms **9**(4), 507–537 (1988)
10. Curran, S., Lee, O., Yu, X.: Finding four independent trees. SIAM J. Comput. **35**(5), 1023–1058 (2006)
11. Hasunuma, T.: Completely independent spanning trees in maximal planar graphs. In: Goos, G., Hartmanis, J., van Leeuwen, J., Kučera, L. (eds.) WG 2002. LNCS, vol. 2573, pp. 235–245. Springer, Heidelberg (2002). https://doi.org/10.1007/3-540-36379-3_21
12. Huang, J.-F., Cheng, E., Hsieh, S.-Y.: Two algorithms for constructing independent spanning trees in (n, k)-star graphs. IEEE Access **8**, 175932–175947 (2020)
13. Huang, J.-F., Kao, S.-S., Hsieh, S.-Y., Klasing, R.: Top-down construction of independent spanning trees in alternating group networks. IEEE Access **8**, 112333–112347 (2020)
14. Itai, A., Rodeh, M.: The multi-tree approach to reliability in distributed networks. Inf. Comput. **79**(1), 43–59 (1988)
15. Kao, S.-S., Pai, K.-J., Hsieh, S.-Y., Wu, R.-Y., Chang, J.-M.: Amortized efficiency of constructing multiple independent spanning trees on bubble-sort networks. J. Comb. Optim. **38**(3), 972–986 (2019). https://doi.org/10.1007/s10878-019-00430-0
16. Lakshmivarahan, S., Jwo, J., Dhall, S.K.: Symmetry in interconnection networks based on Cayley graphs of permutation groups: a survey. Parallel Comput. **19**(4), 361–407 (1993)
17. Leighton, F.T.: Introduction to Parallel Algorithms and Architectures: Arrays, Trees, Hypercubes. Morgan Kaufmann Publishers, Burlington (1992)
18. Liu, L.-H., Chen, J.-E., Chen, S.-Q., Jia, W.-J.: An new representation for interconnection network structures. J. Cent. South Univ. Technol. **9**(1), 47–53 (2002)
19. Lin, C.-F., Huang, J.-F., Hsieh, S.-Y.: Constructing independent spanning trees on transposition networks. IEEE Access **8**, 147122–147132 (2020)

20. Lin, J.-C., Yang, J.-S., Hsu, C.-C., Chang, J.-M.: Independent spanning trees vs. edge-disjoint spanning trees in locally twisted cubes. Inf. Process. Lett. **110**(10), 414–419 (2010)

21. Pai, K.-J., Yang, J.-S., Yao, S.-C., Tang, S.-M., Chang, J.-M.: Completely independent spanning trees on some interconnection networks. IEICE Trans. Inf. Syst. **E97–D**(9), 2514–2517 (2014)

22. Pai, K.-J., Chang, J.-M.: Dual-CISTs: configuring a protection routing on some Cayley networks. IEEE/ACM Trans. Netw. **27**(3), 1112–1123 (2019)

23. Pai, K.-J., Chang, R.-S., Wu, R.-Y., Chang, J.-M.: A two-stages tree-searching algorithm for finding three completely independent spanning trees. Theoret. Comput. Sci. **784**, 65–74 (2019)

24. Pai, K.-J., Chang, R.-S., Wu, R.-Y., Chang, J.-M.: Three completely independent spanning trees of crossed cubes with application to secure-protection routing. Inf. Sci. **541**, 516–530 (2020)

25. Pai, K.-J., Chang, R.-S., Chang, J.-M.: Constructing dual-CISTs of pancake graphs and performance assessment of protection routings on some Cayley networks. J. Supercomput. https://doi.org/10.1007/s11227-020-03297-9

26. Péterfalvi, F.: Two counterexamples on completely independent spanning trees. Discret. Math. **312**(4), 808–810 (2012)

27. Shawash, N.: Relationships among popular interconnection networks and their common generalization. Ph.D. thesis, Oakland University (2008)

28. Tang, S.-M., Yang, J.-S., Wang, Y.-L., Chang, J.-M.: Independent spanning trees on multidimensional torus networks. IEEE Trans. Comput. **59**(1), 93–102 (2010)

29. Wang, Y., Fan, J., Zhou, G., Jia, X.: Independent spanning trees on twisted cubes. J. Parallel Distrib. Comput. **72**(1), 58–69 (2012)

30. Yang, J.-S., Chan, H.-C., Chang, J.-M.: Broadcasting secure messages via optimal independent spanning trees in folded hypercubes. Discret. Appl. Math. **159**(12), 1254–1263 (2011)

31. Yang, J.-S., Chang, J.-M.: Optimal independent spanning trees on Cartesian product of hybrid graphs. Comput. J. **57**(1), 93–99 (2014)

32. Yang, J.-S., Chang, J.-M., Pai, K.-J., Chan, H.-C.: Parallel construction of independent spanning trees on enhanced hypercubes. IEEE Trans. Parallel Distrib. Syst. **26**(11), 3090–3098 (2015)

33. Yang, J.-S., Chang, J.-M., Tang, S.-M., Wang, Y.-L.: Reducing the height of independent spanning trees in chordal rings. IEEE Trans. Parallel Distrib. Syst. **18**(5), 644–657 (2007)

34. Yang, J.-S., Chang, J.-M., Tang, S.-M., Wang, Y.-L.: On the independent spanning trees of recursive circulant graphs $G(cd^m, d)$ with $d > 2$. Theoret. Comput. Sci. **410**(21–23), 2001–2010 (2009)

35. Yang, J.-S., Chang, J.-M., Tang, S.-M., Wang, Y.-L.: Constructing multiple independent spanning trees on recursive circulant graphs $G(2^m, 2)$. Int. J. Found. Comput. Sci. **21**(1), 73–90 (2010)

36. Yang, J.-S., Luo, S.-S., Chang, J.-M.: Pruning longer branches of independent spanning trees on folded hyper-stars. Comput. J. **58**(11), 2972–2981 (2015)

37. Yang, J.-S., Tang, S.-M., Chang, J.-M., Wang, Y.-L.: Parallel construction of optimal independent spanning trees on hypercubes. Parallel Comput. **33**(1), 73–79 (2007)

38. Yang, J.-S., Wu, M.-R., Chang, J.-M., Chang, Y.-H.: A fully parallelized scheme of constructing independent spanning trees on Möbius cubes. J. Supercomput. **71**(3), 952–965 (2014). https://doi.org/10.1007/s11227-014-1346-z

39. Yang, Y.-C., Kao, S.-S., Klasing, R., Hsieh, S.-Y., Chou, H.-H., Chang, J.-M.: The construction of multiple independent spanning trees on burnt pancake networks. IEEE Access **9**, 16679–16691 (2021)
40. Zehavi, A., Itai, A.: Three tree-paths. J. Graph Theory **13**(2), 175–188 (1989)
41. Zhao, S.-L., Hao, R.-X.: The generalized connectivity of (n, k)-bubble-sort graphs. Comput. J. **62**(9), 1277–1283 (2019)
42. Zhao, S.-L., Hao, R.-X.: The fault tolerance of (n, k)-bubble-sort networks. Discret. Appl. Math. **285**, 204–211 (2020)

A Fast FPTAS for Two Dimensional Barrier Coverage Using Sink-Based Mobile Sensors with MinSum Movement

Wenjie Zou[1], Longkun Guo[1], Chunlin Hao[2(✉)], and Lei Liu[2]

[1] College of Mathematics and Statistics, Fuzhou University,
Fuzhou 350116, People's Republic of China
wenjie@foxmail.com, lkguo@fzu.edu.cn
[2] Department of Operations Research and Information Engineering,
Beijing University of Technology, Beijing 100124, People's Republic of China
{haochl,liuliu_leilei}@bjut.edu.cn

Abstract. Energy efficiency is a critical issue that attracts numerous interest of many researchers in wireless mobile sensor networks. Emerging IoT applications have brought the MinSum Sink-based Linear Barrier Coverage (MinSum SLBC) problem which aims to use sink-based mobile sensors (such as drones) to cover a line barrier (such as borders possibly for monitoring illegal intrusion). In the scenario, all the sensors are initially located at k sink stations, while the aim is to find the final positions of the sensors on the line barrier, such that the line barrier is completely covered and the total movement of the sensors is minimized. In this paper, we first study geometry properties of an optimal solution of MinSum SLBC, and reveal that an optimal solution of MinSum SLBC actually consist of intersecting segments of tangent sensors. Then, we devise a segmentation algorithm for computing a near-optimal position of each segment that is possibly part of the optimum. Lastly, by selecting segments consisting of tangent sensors via transforming to the shortest path problem, we eventually derive a factor-$(1 + \varepsilon)$ approximation algorithm with a time complexity $O(k^2(\log \frac{2r}{\varepsilon} + \log k))$, where $\varepsilon > 0$ is any given positive real number.

Keywords: Mobile sensor network · Barrier coverage · MinSum movement · Approximation algorithm · Sink station

1 Introduction

Along with the development of sensor technology, new coverage problems with sink-based mobile sensors [1–3] have appeared in emerging IoT applications. In such problems, the mobile sensors are originated at sink stations that are distributed on the plane. Assume that we are given a barrier of length L located on the X-axis from point $(0,0)$ to $(L,0)$ on the plane, and a set of sink stations $\Psi = \{S_1, S_2, \cdots, S_k\}$ in which each sink station can emit an infinite number of

© Springer Nature Switzerland AG 2021
W. Wu and H. Du (Eds.): AAIM 2021, LNCS 13153, pp. 265–276, 2021.
https://doi.org/10.1007/978-3-030-93176-6_23

mobile sensors with identical sensing radii. The goal of the MinSum Sink-based Linear Barrier Coverage problem (MinSum SLBC) is to move a set of mobile sensors from the sinks to incorporate a complete coverage over the barrier, such that the total movement of the sensors for the coverage is minimized. The formal definition of MinSum SLBC is as follows:

Definition 1. *Let Ψ be a set of sink stations distributed on the plane, where each sink station $S_i \in \Psi$ is with position (x_i, y_i) and can emit mobile sensor with identical sensing radius r for covering the line barrier. Given a line barrier of length L on the X-axis, the problem aims to find new positions of a set of emitted sensors on the line containing the barrier, such that each point of the barrier is within the sensing area of at least one sensor while the total movement of the sensors is minimized.*

In the above definition, the movement of a sensor is measured by the Euclidean distance between the position of its originated sink station and its final position on the line accommodating the barrier. Formally, for points p and q with position coordinates (x_p, y_p) and (x_q, y_q), the Euclidean distance between them is denoted as $d(p, q) = \sqrt{(x_p - x_q)^2 + (y_p - y_q)^2}$. Besides, for briefness we say $p \prec q$ if and only if $x_p < x_q$. W.l.o.g. we assume $x_i \neq x_j$ holds for every two sink stations S_i and S_j, since otherwise we could only use sensors from the sink station with smaller perpendicular distance for coverage. Again for briefness, we assume the sink stations satisfy the ordering: $S_1 \prec S_2 \prec \cdots \prec S_k$, which means, $x_1 < x_2 < \cdots < x_k$.

1.1 Related Works

In recent years, several papers in the field of sensor networks considered the new coverage problems with sink-based mobile sensors, see for example [1,3]. Gao *et al.* [1] first studied the new coverage problem of sink-based sensors with identical sensing radii covering targets distributed on the plane, which is defined as k-sink minimum movement target coverage (kMMTC) problem, where k is the number of sink stations and all sensors in the problem have identical sensing radii. The purpose of the research is to minimize the total movement of the sensors covering the set of targets distributed on the plane. They proved that kMMTC is NP-hardness and proposed a polynomial-time $1 + \varepsilon$ approximation scheme with a runtime of $n^{O(\frac{1}{\varepsilon^2})}$. Our previous work studied the barrier coverage problem of sink-based sensors, and proved that the MinSum SLBC problem is NP-complete, where the radius of the sensor is not necessarily the same. We also studied the MinMax SLBC problem [3] in which the radius of the sensors are the same and the aim is to minimize the maximum movement of the sensors, and proposed an optimal algorithm with linear time.

Besides using sensors originated at sink station, there exist many research works in literature considering the coverage problem with sensors distributed on the plane. When the radius of the sensor is identical, for the 2D MinSum problem where the sensors are distributed on the plane, Cherry *et al.* [4] presented a factor-$\sqrt{2}$ approximation algorithm with the time complexity $O(n^4)$ for MinSum problem of covering a single line barrier. Interestingly, the algorithm by Cherry *et al.* also works for covering k parallel barriers with an approximation ratio $\sqrt{2}$ and a runtime of $O(kn^{2k+2})$. Later, Erzin and Lagutkina [5] improved the runtime to $O(n^2)$. More recently, the same authors eventually devised a fully polynomial time approximation scheme (*FPTAS*) for the problem with a runtime of $O(\frac{n^3}{\epsilon^2})$ [6]. When the barrier is a circular boundary, and the sensors are initially distributed in the interior of the boundary, Bhattacharya *et al.* [7] designed two approximation algorithms, where the first is with a runtime of $O(n^2)$ and a ratio $(1 + \pi)$ that was later improved to 3 by Chen *et al.* [8]; where the second is a fully polynomial time approximation scheme (*FPTAS*) with a ratio $(1 + \varepsilon)$ and a runtime $O(\frac{1}{\varepsilon}n^4)$ for a given constant $\varepsilon > 0$, which was later improved to $O(\frac{1}{\varepsilon^{O(1)}}n^{2+\varepsilon'})$ by Carmi *et al.* [9] for $\varepsilon > 0$ being any constant and $\varepsilon' = O(\varepsilon)$. The 1D MinSum problem in which the sensors are distributed on the line containing the barrier was studied much earlier by Czyzowicz *et al.* [10], who proposed an exact algorithm with the time complexity $O(n^2)$. Later, Andrews and Wang [11] improved the time complexity to $O(n \log n)$.

When the barrier is a circular boundary and the sensors are initially distributed on the boundary, Tan and Wu [12] proposed an exact algorithm with $O(n^4)$ time by moving n sensors to form a regular n-gon on the circular boundary. When the radii of the sensors are not necessarily the same, Czyzowicz *et al.* [10] proved that the MinSum barrier coverage problem is $\mathcal{NP}$-hard even if the initial positions of sensors are on the targeted line. Benkoczi *et al.* [13] gave an *FPTAS* for the special case where the sensing range of the sensor is not allowed to overlap. For 2D MinSum problem, where the sensors can perform only perpendicular movement against the barrier and given k parallel barriers need to be covered, Dobrev *et al.* [14] designed an algorithm with a runtime of $O(kn^{k+1})$.

1.2 Our Contribution and Organization

As the main contribution of this paper, we design a factor-$(1 + \varepsilon)$ approximation algorithm with a time complexity $O(k^2(\log \frac{2r}{\varepsilon} + \log k)$ for MinSum SLBC. The algorithm is derived through analyzing the geometric properties of an optimal solution of MinSum SLBC, particularly analyzing the overlap of the sensor coverage area in the optimal solution. The runtime of our algorithm compares favorably to the existing algorithms in literature. We achieve a significantly lower time complexity exponentially decreasing the factor of polynomial $\frac{1}{\epsilon}$ to the logarithm of $\log(\frac{1}{\epsilon})$, improving the previous FPTAS with runtime $O\left(\frac{n^3}{\epsilon^2}\right)$ for the sensor-based version of MinSum SLBC [6], and the additive FPTAS with runtime $O\left(k^2 \left(\frac{L}{\epsilon}\right)^2\right)$ for MinSum SLBC with sensors of non-uniform radii [15].

The remainder of this paper is organized as follows: Sect. 2 describes the formal definition of the problem, as well as some necessary definitions and notations; Sect. 3 gives important properties on an optimal solution of MinSum SLBC, based on which Sect. 4 proposes the near-optimal algorithm; Sect. 5 concludes the paper.

2 Preliminaries

In this section, we shall give some notations and definitions that will be used when introducing our algorithm and related proofs.

Definition 2. *The point u_{ij} on the barrier is a movement parity point for S_i and S_j, if and only if the distance between u_{ij} and S_i equals that between u_{ij} and S_j, that is, $d(S_i, u_{ij}) = d(u_{ij}, S_j)$.*

Proposition 3. *For S_i, S_j with $S_i \prec S_j$ and u_{ij} being the movement parity point between them, for the point q on the line segment containing the barrier, $q \preceq u_{ij}$ is true iff $d(q, S_j) \geq d(q, S_i)$ is true.*

From the above proposition, it can be seen from left to right that the order of sensors covering the barrier is the same as the order of the sink stations. Because if there are two sensors on the barrier, the order is opposite to the order of the sink stations emitting them, we can change the order of the sink stations emitting the two sensors without changing other sensors to reduce the total movement of the sensors. Of course, some sink stations that are too far away from the barrier to be used are not considered. Therefore, before giving our algorithm, we must first remove those sink stations that cannot be used because the perpendicular distance from the barrier is too far.

3 Properties on Optimum Solution of MinSum SLBC

In this section, we will give the relevant geometric properties by observing the structure of an optimal solution of MinSum SLBC, which inspires us to derive the near-optimal algorithm. We first denote an optimal solution of MinSum SLBC as OPT, and then give the geometric properties of OPT.

Lemma 4. *In OPT, a left-moving sensor does not overlap with a sensor on its left side; the right-moving sensor not overlap with the sensor on its right side.*

Proof. We will prove it by contradiction. Assume an left-moving sensor have overlap with sensor on its left side. Let the length of the overlap be η. There are two cases: (i) The left-moving sensor and the sensor on its left side are both emitted by the same sink station S_i; (ii) The left-moving sensor and the sensor on its left side are emitted by different sink stations S_i and S_j.

For case (i), assume that the two sensors are respectively with final positions $(a, 0)$ and $(b, 0)$, and sink station S_i is with position (x_i, y_i). Obviously $x_i \lneq a \lneq b$ holds. The total movement of the two sensors is $d = d(S_i, (a, 0)) + d(S_i, (b, 0))$. It's easy to find that increasing a can reduce the movement of the left-moving sensor, and until the overlap η reduce to 0 or the left-moving sensor becomes perpendicular-moving sensor. The above process will reduce the total movement d but will not change the coverage. This contradicts the minimality of OPT.

For case (ii), assume that the two sensors are respectively with final positions $(a, 0)$ and $(b, 0)$, and the different sink stations S_i and S_j are respectively with positions (x_i, y_i) and (x_j, y_j). Obviously $b < a < x_i$ holds. The total movement of the two sensors is $d = d(S_i, (a, 0)) + d(S_j, (b, 0))$. It's easy to find that increasing a can reduce the overlap η, and until the overlap η reduce to 0 or the left-moving sensor becomes perpendicular-moving sensor. The above process will reduce the total movement d but will not change the coverage. This contradicts the minimality of OPT. Therefore, the left-moving sensors have no overlap with sensor on its left side. This completes the proof since the cases of right-moving sensors are symmetric. □

Definition 5. *For sink station S_i with position (x_i, y_i), the point p_i on the barrier closest to sink station S_i is at $(x_i, 0)$, we call this point the projection point of sink station S_i, denoted as p_i. The area on the barrier covered by the sensor with the final position $(x_i, 0)$ is called the projection area of sink station S_i, denoted as $Area_i$.*

According to Lemma 4 and Definition 5, we find that if there is an overlap between the sensors' coverage areas in OPT, the overlap will definitely appear in the area covered by the sensors moving vertically to the barrier.

Lemma 6. *If there is an overlap in OPT, the overlap must appear in the projection areas of the sink stations emitting sensors in OPT.*

Proof. In OPT, according to Lemma 4, the left-moving sensor will not overlap with the sensor on its left side; the right-moving sensor will not overlap with the sensor on its right side. Therefore, if there is an overlap between the sensors' coverage areas in OPT, there are three cases in which the overlap appear: (i) The vertically-moving sensor overlaps with the sensor on its left (or right) side; (ii) The left-moving sensor overlaps with the sensor on its right side; (iii) The right-moving sensor overlaps with the sensor on its left side. Obviously, for case (i), according to the Definition 5, the overlap must appear in projection area. For case (ii), there are only two possibilities of the right sensor on the left-moving sensor: vertical-moving and right-moving. Vertical movement is equivalent to case (i), and the overlap between the left-moving sensor and the right-moving sensor on its right must be in the projection area. Similar to case (ii), we immediately have the correctness of case (iii). □

Through Lemma 4 and Lemma 6, we can know that in OPT, the overlap of the coverage area of the sensors definitely appear in the projection area of the sink stations. Then we can find the segment formed by the coverage areas of the maximal set of tangent sensors existing between the two projection areas, which is called a tangent segment. Obviously, there is no overlap between the sensors' coverage areas that form the tangent segment.

Definition 7. *For the projection areas $Area_i$ and $Area_j$, the line segment between them covered by the maximal set of sensors is called a tangent segment, where the sensing areas of adjacent sensors are tangent, that is, there is no overlap. We denote this tangent segment as T_{ij}.*

Because the coverage area of the sensor located at the projection point of each sink station may individually intersect the tangent segments on both sides of it, we regard the projection area of the sink station as a special tangent segment. In summary, we know that OPT can be seen as composed of intersection of multiple tangent segments. Then we need to consider how to find these tangent segments and the best position of each tangent segment. And each tangent segment is composed of the coverage area of the maximal set of tangent sensors existing between the two projection areas, so finding the tangent segment is to find the corresponding maximal set, and the best position of tangent segment is the best position of the corresponding maximal set.

Lemma 8. *In OPT, for the tangent segment T_{ij}, there are two possible numbers of sensors in the corresponding maximal set, $\left\lceil \frac{x_j - x_i}{2r} \right\rceil - 1$ and $\left\lfloor \frac{x_j - x_i}{2r} \right\rfloor + 1$.*

Proof. Apparently, the tangent segment T_{ij} is between the projection areas $Area_i$ and $Area_j$. So the leftmost and the rightmost points of T_{ij} are in $Area_i$ and $Area_j$, respectively. That is, the two points are in the range of $[p_i - r, p_i + r]$ and $[p_j - r, p_j + r]$, respectively. Therefore, the number of sensors is at most $\left\lfloor \frac{(x_j + r) - (x_i - r)}{2r} \right\rfloor = \left\lfloor \frac{x_j - x_i}{2r} \right\rfloor + 1$, and at least $\left\lceil \frac{(x_j - r) - (x_i + r)}{2r} \right\rceil = \left\lceil \frac{x_j - x_i}{2r} \right\rceil - 1$. □

When tangent segment T_{ij} is completely covered by the sensors and the total movement of sensors is minimized, we say that T_{ij} is in its optimal position.

Theorem 9. *In OPT, tangent segment T_{ij} is either in its optimal position or tangent to its left or right sensor.*

Proof. Suppose tangent segment T_{ij} in OPT is not in its optimal position, and is neither tangent to its left nor right sensor. Then the leftmost sensor and the rightmost sensor of T_{ij} overlap with the left and right sensors of T_{ij}, respectively. At this time, according to the meaning of T_{ij} is not in its optimal position, it can be seen that T_{ij} can move to the left or right to reduce the total movement of the sensors without affecting the coverage, which means that OPT is not a optimal solution. Contradicts the hypothesis. This completes the proof. □

4 The Near-Optimal Algorithm

In this section, we will first give a high-level overview of our near-optimal algorithm for MinSum SLBC, and then describe its details in three phases.

4.1 Overview of the Algorithm

Based on the key observation of Theorem 9, a natural idea is first to enumerate all the tangent segments together their optimal positions, and then to select the tangent segments to eventually compose a near-optimal solution. However, there are two difficulties: (1) how to find all possible tangent segments and the best position of all tangent segments, provided that the best position involves the positions of the sink stations, the radii of the sensors, and the number of sensors; (2) how to select the segments for optimally cover the barrier with minimum total movement. Regarding the difficulties, our algorithm can be divided into three phases in a high level manner. The first phase is to enumerates all possible tangent segments with their rightmost positions. Then the second is to compute the near-optimal position of each possible tangent segment, based on the observation that the function, which captures the movement change of a segment when moving leftwards from its rightmost position, is concave. The third phase is to construct an auxiliary graph in which selection among all the possible tangent segments is transformed to finding a shortest path therein.

4.2   Enumeration of Possible Tangent Segments

We shall give the algorithm for enumerating all possible tangent segments based on the observation of Lemma 8. The key idea of the algorithm is to first consider the leftmost and rightmost point of the barrier as two point projection areas, and then calculate the tangent segment between any two projection areas from left to right. First, according to Lemma 8, calculate the number of sensors contained in the tangent segment between any two projection areas, and then find the rightmost position of each tangent segment as the initial position and write down the initial position coordinate. The details of the algorithm are as in Algorithm 1. According to the algorithm, we can find the rightmost positions of all possible tangent segments, and then we consider moving the tangent segments to the left to find the best position of each tangent segment.

4.3   Computation of Near-Optimal Positions for Tangent Segments

We shall show how to compute near-optimal positions for the sensors of each segment. Let T_{ij} be a tangent segment between the projection areas $Area_i$ and $Area_j$, whose coverage sensors are emitted by the sink stations S_i to S_j. Assume the position of the leftmost sensor in T_{ij} is with a position $(x_0, 0)$, and the number of sensors emitted by sink station S_t with position (x_t, y_t) is denoted as n_t. Then the total movement of the sensors of T_{ij} is:

$$d_0 = \sum_{t=i}^{j} \sum_{m=1}^{n_t} d\left(S_t, (x_0 + 2r(m-1), 0)\right)$$

Considering to move the segment leftwards for distance ξ, then the total movement of the sensors of T_{ij} is:

$$d_1 = \sum_{t=i}^{j} \sum_{m=1}^{n_t'} d\left(S_t, (x_0 - \xi + 2r(m-1), 0)\right).$$

Algorithm 1. Calculation of the possible tangent segments

Input: The set of sink stations $\Psi = \{S_1, S_2, \cdots, S_k\}$ in which of sink station S_i
        is with position (x_i, y_i), the barrier $[0, L]$ and the radius r of sensor;

Output: The set C of all tangent segments.

1: Set $C = \emptyset$;
2: **For** $i = 1$ to k **do**
3:     Set $C = C \cup \{[x_i - r, x_i + r]\}$, $a = \max\{x_i, 0\}$;
4:     **If** $0 < x_i < L$ **then**
5:        Calculate the tangent segment T_{0i} between $(0,0)$ and $Area_i$, which
          contains $\left\lceil \frac{x_i - r}{2r} \right\rceil$ tangent sensors; /*$l_{0i} = \left\lceil \frac{x_i - r}{2r} \right\rceil \cdot 2r - (x_i - r)$*./
6:        Calculate T_{ik+1} between $Area_i$ and $(L,0)$, which contains $\left\lceil \frac{L-(x_i+r)}{2r} \right\rceil$
          tangent sensors; /*$l_{ik+1} = x_i + r + \left\lceil \frac{L-(x_i+r)}{2r} \right\rceil \cdot 2r - L$.*/
7:        Set $C = C \cup \{T_{0i}, T_{ik+1}\}$;
8:     **EndIf**
9:     **For** $j = i + 1$ to k **do**
10:      **If** $0 < x_j < L$ **then**
11:        Set $b = x_j$, $C = C \cup \{[x_j - r, x_j + r]\}$;
12:        Calculate T_{ij} between $Area_i$ and $Area_j$, where $T_{ij}^{(1)}$ contains $\left\lceil \frac{b-a}{2r} \right\rceil - 1$
          tangent sensors, and $T_{ij}^{(2)}$ contains $\left\lfloor \frac{b-a}{2r} \right\rfloor + 1$ tangent sensors;
          /*$l_{ij}^{(1)} = a - b + \left\lceil \frac{b-a}{2r} \right\rceil \cdot 2r$, $l_{ij}^{(2)} = b - a - \left\lfloor \frac{b-a}{2r} \right\rfloor \cdot 2r$.*/
13:        Set $C = C \cup \{T_{ij}^{(1)}, T_{ij}^{(2)}\}$;
14:      **EndIf**
15:     **EndFor**
16: **EndFor**
17: Return the set C.

Then we could define the function to capture the movement change when moving the segment T_{ij} leftwards for distances ξ:

$$\mathcal{F}_d(\xi) = d_0 - d_1$$

$$= \sum_{t=i}^{j} \left(\sum_{m=1}^{n_t} d\left(S_t, (x_0 + 2r(m-1), 0)\right) - \sum_{m=1}^{n_t'} d\left(S_t, (x_0 - \xi + 2r(m-1), 0)\right) \right)$$

Lemma 10. *The function $\mathcal{F}_d(\xi)$ is concave.*

Proof. We need only to show the second derivative of $\mathcal{F}_d(\xi)$ with respect to ξ is not greater than zero:

$$\mathcal{F}_d''(\xi) = \sum_{t=i}^{j} \sum_{m=1}^{n_t'} \frac{-y_t^2}{\sqrt[3]{[x_0 - \xi + 2r(m-1) - x_t]^2 + y_t^2}} \leq 0.$$

From Lemma 10, we can conclude that in the process of moving a tangent segment from the rightmost initial position to the left, there exists a position where the total movement of the sensors in the tangent segment attains minimum. Next, we will give the second algorithm to calculate the near-optimal position.

The key idea of the second algorithm is to first calculate the decrement function $\mathcal{F}_d$ corresponding to the input tangent segment, and then determine whether the tangent segment moves to the left and the distance of the left-moving according to the first derivative of $\mathcal{F}_d(\xi)$ where ξ is the distance of the tangent segment moved to the left. In this process, the divide and conquer method is used to determine the maximum value of ξ.

Theorem 11. *A placement of a tangent segment between two projection areas, with at most ε more movement than an optimal placement, can be computed in $O(\log \frac{2r}{\varepsilon})$ time.*

The proof is omitted due to length limitation.

4.4   Transformation to the Shortest Path Problem

The key idea of the transformation is essentially to construct an auxiliary graph. For the vertices of the graph, we add a vertex to correspond each tangent segment, and additionally add a source vertex s and a destination vertex t. For the edges between the vertices, we add an edge between two vertices if the two corresponding tangent segments overlap. Lastly, we use the total movement of a segment as the cost of the edge entering the corresponding vertex. Figure 1 gives an example of executing the whole near-optimal algorithm for MinSum SLBC, including the process of calculating all tangent segments, finding the near-optimal positions of tangent segments and constructing the auxiliary graph.

For the correctness of the transformation, we have the following theorem:

Theorem 12. *There is an st-path in the auxiliary graph iff there is a complete coverage for the instance of MinSum SLBC.*

Proof. Assume that P is a path in the auxiliary graph. The proof will be given by induction on the length of P, $L(P)$. If $L(P) = 1$, then we have $P = (v_1 \rightarrow v_2)$. Note that the condition of an edge between the vertex v_1 and v_2 is that their corresponding binary combinations $(c_{1,1}, c_{1,2})$ and $(c_{2,1}, c_{2,2})$ should satisfy that $c_{1,1} < c_{2,1} \leq c_{1,2} < c_{2,2}$. Then, using tangent sensors to cover the segment $[c_{1,1}, c_{1,2}]$ and $[c_{2,1}, c_{2,2}]$, respectively. That is, the above description is a complete coverage for the segment between $c_{1,1}$ and $c_{2,2}$. By hypothesis, using sensors to individually cover the segment $[c_{i,1}, c_{i,2}]$, $i \in [h]$, is a coverage for the segment between $c_{1,1}$ and $c_{h,2}$. Then only need to explain that there is a complete coverage for the segment between $c_{1,1}$ and $c_{h+1,2}$ in the situation of $P = (v_1 \rightarrow v_2 \rightarrow \ldots \rightarrow v_h \rightarrow v_{h+1})$. According to the auxiliary graph construction, $c_{h,1} < c_{h+1,1} \leq c_{h,2} < c_{h+1,2}$, that is, $c_{h,2}$ is in the segment $[c_{h+1,1}, c_{h+1,2}]$. So a path from v_1 to v_{h+1} is corresponding to a coverage for the segment between $c_{1,1}$ and $c_{h+1,2}$. Then when P is an st-path,

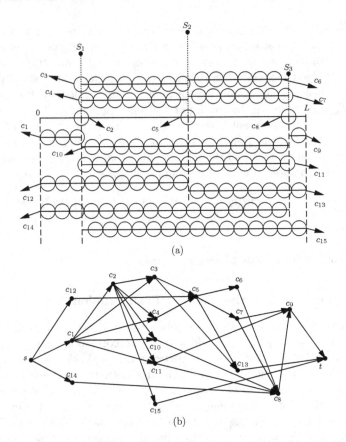

Fig. 1. Execution of the algorithm: (a) The barrier, sink stations, and the tangent segments; (b) The auxiliary graph corresponding to the near-optimally placed segments.

$P = (s \rightarrow v_1 \rightarrow v_2 \rightarrow \ldots \rightarrow v_h \rightarrow v_{h+1} \rightarrow t)$. Therefore, the binary combinations of the source node s and vertex v_1, and the binary combinations of the vertex v_{h+1} and destination node t should satisfy $c_{1,1} \leq 0$ and $c_{h+1,2} \geq L$. Then an st-path corresponds to a complete coverage for the barrier. Similarly, we have the proof of the "if" direction. This completes the proof. □

Next, the size of the auxiliary graph G can be stated below:

Theorem 13. *The auxiliary graph G has $O(k^2)$ vertices.*

Proof. According to the construction method of the auxiliary graph G, all tangent segments and the projection areas of all sink stations corresponds to vertices in $G \setminus \{s, t\}$ and vice versa. According to the calculation method of the tangent segment in Algorithm 1, there is only one tangent segment between the leftmost or rightmost point of the barrier and any sink station, and the barrier has two endpoints, then there are $2k$ tangent segments related to the barrier endpoint; There are two tangent segments between any two sink stations, so there are $2\sum_{i=1}^{k-1} i = k(k-1)$. Therefore, the number of tangent segments is $2k + k(k-1) = O(k^2)$. There are m sink stations and each sink station has a projection area, so there are k projection areas. So the number of vertices in the auxiliary graph is $O(k^2 + k) = O(k^2)$. □

Theorem 14. *Our near-optimal algorithm produces an approximate solution to MinSum SLBC with movement bounded by $d(OPT) + \varepsilon$.*

Proof. In the algorithm that calculate the near-optimal positions for tangent segments, the accuracy error is at most ε' for each tangent segment. So the accumulated accuracy error is at most $k^2\varepsilon'$, since there are at most k^2 tangent segments. Then our algorithm produces an approximate solution to MinSum SLBC with movement bounded by $d(OPT) + \varepsilon$ for $\varepsilon = k^2\varepsilon'$. □

Moreover, combining the above theorem with Theorem 11, we have the time complexity of the whole algorithm:

Corollary 1. *Our algorithm runs in $O(k^2 \log \frac{2r}{\varepsilon} + k^2 \log k)$ time, and computes a near-optimal solution with movement bounded by $1 + \varepsilon$ times of optimum, where $\varepsilon > 0$ is any given positive real number.*

Note that $O(k^2 \log \frac{2r}{\varepsilon})$ is the time needed to find all the near-optimal positions for all the segments, while $O(k^2 \log k)$ is the time needed to find the shortest path in the auxiliary graph which is acyclic admits algorithms with runtime linearly depend on the number of its edges.

5   Conclusion

In this paper, we study the MinSum SLBC problem where all sensors are emitted by the sink stations have identical sensing radii. We first observe the geometric properties of an optimal solution of MinSum SLBC, and then design a factor-$(1 + \varepsilon)$ approximation algorithm with a time complexity $O(k^2(\log \frac{2r}{\varepsilon} + \log k))$ based on the geometric properties. The algorithm consists of three phases: The first is to find all possible segments formed by the tangent sensor, the second to calculate the near-optimal position of each segment, and the third to select the segments via transforming to the shortest path problem.

Acknowledgment. This research work is supported by Natural Science Foundation of China (No. 61772005) and Natural Science Foundation of Fujian Province (No 2017J01753).

References

1. Gao, X., Chen, Z., Fan, W., Chen, G.: Energy efficient algorithms for k-sink minimum movement target coverage problem in mobile sensor network. IEEE/ACM Trans. Netw. **25**(6), 3616–3627 (2017)
2. Boubrima, A., Bechkit, W., Rivano, H.: On the deployment of wireless sensor networks for air quality mapping: optimization models and algorithms. IEEE/ACM Trans. Netw. **27**(4), 1629–1642 (2019)
3. Zou, W., Guo, L., Huang, P.: Min-max movement of sink-based mobile sensors in the plane for barrier coverage. In: 2019 20th International Conference on Parallel and Distributed Computing, Applications and Technologies (PDCAT), pp. 1–6. IEEE (2019)
4. Cherry, A., Gudmundsson, J., Mestre, J.: Barrier coverage with uniform radii in 2D. In: Fernández Anta, A., Jurdzinski, T., Mosteiro, M.A., Zhang, Y. (eds.) ALGO-SENSORS 2017. LNCS, vol. 10718, pp. 57–69. Springer, Cham (2017). https://doi.org/10.1007/978-3-319-72751-6_5
5. Erzin, A., Lagutkina, N.: Barrier coverage problem in 2D. In: Gilbert, S., Hughes, D., Krishnamachari, B. (eds.) ALGOSENSORS 2018. LNCS, vol. 11410, pp. 118–130. Springer, Cham (2019). https://doi.org/10.1007/978-3-030-14094-6_8
6. Erzin, A., Lagutkina, N.: FPTAS for barrier covering problem with equal touching circles in 2D. Optim. Lett. **15**(4), 1397–1406 (2021)
7. Bhattacharya, B., Burmester, M., Yuzhuang, H., Kranakis, E., Shi, Q., Wiese, A.: Optimal movement of mobile sensors for barrier coverage of a planar region. Theor. Comput. Sci. **410**(52), 5515–5528 (2009)
8. Chen, D.Z., Tan, X., Wang, H., Wu, G.: Optimal point movement for covering circular regions. Algorithmica **72**(2), 379–399 (2015)
9. Carmi, P., Katz, M.J., Saban, R., Stein, Y.: Improved PTASs for convex barrier coverage. In: Solis-Oba, R., Fleischer, R. (eds.) WAOA 2017. LNCS, vol. 10787, pp. 26–40. Springer, Cham (2018). https://doi.org/10.1007/978-3-319-89441-6_3
10. Czyzowicz, J., et al.: On minimizing the sum of sensor movements for barrier coverage of a line segment. In: Nikolaidis, I., Wu, K. (eds.) ADHOC-NOW 2010. LNCS, vol. 6288, pp. 29–42. Springer, Heidelberg (2010). https://doi.org/10.1007/978-3-642-14785-2_3
11. Andrews, A.M., Wang, H.: Minimizing the aggregate movements for interval coverage. Algorithmica **78**(1), 47–85 (2017)
12. Tan, X., Wu, G.: New algorithms for barrier coverage with mobile sensors. In: Lee, D.-T., Chen, D.Z., Ying, S. (eds.) FAW 2010. LNCS, vol. 6213, pp. 327–338. Springer, Heidelberg (2010). https://doi.org/10.1007/978-3-642-14553-7_31
13. Benkoczi, R., Friggstad, Z., Gaur, D., Thom, M.: Minimizing total sensor movement for barrier coverage by non-uniform sensors on a line. In: Bose, P., Gąsieniec, L.A., Römer, K., Wattenhofer, R. (eds.) ALGOSENSORS 2015. LNCS, vol. 9536, pp. 98–111. Springer, Cham (2015). https://doi.org/10.1007/978-3-319-28472-9_8
14. Dobrev, S., et al.: Complexity of barrier coverage with relocatable sensors in the plane. Theor. Comput. Sci. **579**, 64–73 (2015)
15. Guo, L., Zou, W., Wu, C., Xu, D., Du, D.-Z.: Minsum movement of barrier and target coverage using sink-based mobile sensors on the plane. In: 41th IEEE International Conference on Distributed Computing Systems, ICDCS 2021, Washington DC, USA, 7–10 July 2021, pp. 696–706. IEEE (2021)

Time Sensitive Sweep Coverage with Multiple UAVs

Huizhen Wang[✉]

Department of Computer Science and Technology, Key Laboratory of Internet Information Collaboration, Harbin Institute of Technology (Shenzhen), Shenzhen, China
20S151167@stu.hit.edu.cn

Abstract. The sweep coverage of UAVs can provide emergency communication support and conduct disaster surveys in emergency disaster relief. These rescue operations are extremely time sensitive. If the UAVs cannot arrive on time, the best rescue opportunity may be missed. When the number and energy of UAVs are limited, it is likely that some Points of Interest (POI) cannot be covered. Thus it's crucial to design a collaborative sweep coverage scheme for multiple UAVs to achieve the maximum effective coverage rate. In this paper, we first propose a novel problem named MEC-TS. Under the constraints of time sensitivity and the number of UAVs, the objective of the MES-TS problem is to maximize the effective coverage rate in sweep coverage. We prove that the MEC-TS problem is NP-hard. Accordingly, we propose the GCS algorithm for the MEC-TS problem. Finally, we conduct extensive simulations and the experimental results demonstrate GCS has improved the performance by 30% compared with existing algorithms.

Keywords: Sweep coverage · UAVs · Time sensitivity

1 Introduction

In recent years, with the development of unmanned aerial vehicle (UAV) technology, UAVs have attractive application prospects in many fields, such as smart logistics, agricultural planting, infrastructure inspection, public safety, and aerial media. In this paper, we study the time sensitive sweep coverage with multiple UAVs in the context of emergency and disaster relief.

The concept of sweep coverage originates from Wireless Sensor Networks (WSNs). In some monitoring tasks, it is not necessary to continuously monitor the Point Of Interest (POI) with static sensors, but only need to patrol the POI with mobile sensors periodically. In this way, a small number of mobile sensors can be used to cover more POIs, and this coverage mode is called sweep coverage [1,2].

In emergency and disaster relief, UAVs are often required to quickly cover designated locations within a specified time to provide emergency communications support, material supply, disaster surveys, and other special tasks. These

© Springer Nature Switzerland AG 2021
W. Wu and H. Du (Eds.): AAIM 2021, LNCS 13153, pp. 277–288, 2021.
https://doi.org/10.1007/978-3-030-93176-6_24

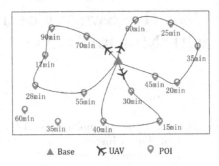

Fig. 1. Time sensitive of POIs

rescue operations are extremely time sensitive. If the UAV cannot arrive on time, the best rescue opportunity may be missed. And each coverage target has different time sensitivity. As shown in Fig. 1, if the time sensitivity of a POI is 60 min, it wants an UAV to cover it in 60 min. For example, UAV equipped with aerial base stations flying over the disaster area can bring communication signals to the disaster area. When communication is interrupted in the disaster area, the time sensitivity is higher in densely populated areas. It is expected that UAVs equipped with aerial base stations can cover the area in a shorter time, so that more people in the disaster area can communicate with the outside world in a timely manner, reducing the casualty rate. In uninhabited mountainous areas, time sensitivity is low, giving UAVs a longer time limit to reach the area. Therefore, how to plan the UAVs' sweep coverage path to meet the time sensitivity of different POIs is the focus of this paper. In addition, when the number and power of UAVs are limited, it is likely that some POIs cannot be covered. At this time, it is necessary to increase the effective coverage rate of the UAVs as much as possible, and those POIs that are covered within an acceptable time can be regarded as effectively covered. How to improve the effective coverage is another research focus of this paper.

Since the distance of the POI is not related to its time sensitivity, it is difficult to coordinate well whether UAVs should prioritize access to POI that are close in distance or POI that are time-critical. If the relationship between the two cannot be coordinated well, it will seriously affect the effective coverage rate. Therefore, how to coordinate the relationship between distance and time sensitivity and make every step of the decision for UAV path planning is a challenge to our algorithm design. The details of our contributions can be summarized as follows:

1) This paper consider a new and practical problem named Maximum Effective Coverage Rate in Time Sensitive Sweep Coverage with Multiple UAVs (MEC-TS). Given a limited number of UAVs and the time sensitivity of each POI, considering the performance constraints of UAVs, the objective of MEC-TS problem is to design a collaborative sweep coverage scheme for multiple UAVs to maximize the effective coverage. We prove that MEC-TS problem is NP-hard.

2) A concise and fast algorithm called GCS is proposed to solve the MEC-TS problem, with time complexity $O(mn^2)$.
3) We evaluate the performance of the proposed algorithm, the experimental results demonstrate the performance of the GCS algorithm is up to 30% higher than that of existing algorithms.

The rest of the paper is organized as follows. Section 2 describes the related work. Section 3 formulates the MEC-TS problem. Section 4 presents the GCS algorithm and the corresponding complexity analysis. Simulation results are displayed in Sect. 5. Finally, Sect. 6 concludes this paper.

2 Related Work

2.1 Traditional Sweep Coverage

Research on sweep coverage is mostly concentrated in traditional mobile sensor networks. Liu et al. [3] studied the problem of sweep coverage with return time constraint, which requires that the data collected from different POIs should be delivered to the base station within different preset time windows. They proposed heuristic algorithms G-MSCR and MinD-Expand to solve this problem. In addition, Liu also proposed group sweep coverage [5] and t,k sweep coverage [6] to further extend the sweep coverage problem. In [7], Nie et al. proposed an approximate better approximation algorithm IERSC for the energy limitation problem in sweep coverage, and for the first time studied the different energy consumption under different road sections. Chen et al. considered the sensing range for the first time to shorten the scanning path [8–10]. The author gives three algorithms for three situations: single sensing point, general scene and extended scene respectively to coordinate the sensors to complete the scanning requirements. Zhang et al. [11] investigated how to use the minimum number of mobile nodes to cover all POIs under the sensing and transmission delay constraints. But they simplified the problem and set the sensing delay of POIs to be the same. In our research, we schedule the sweep coverage path for UAVs based on the condition that different POIs have different time sensitivity.

2.2 Sweep Coverage of UAVs

At present, there is not much research literature on the sweep coverage of UAVs. The author of [12] proposed a new coverage path planning method for the energy limitation of multiple UAVs. They proposed a path-based optimization model that tracks the energy required for different mission phases, effective for path planning for both single and multiple UAVs. In [13], Luo et al. conducted fine-grained trajectory planning for the data collection of multiple UAVs in wireless sensor networks, and used approximate algorithms to minimize the maximum flight time of UAVs, but did not consider the UAV's maximum flight time. Turning angle and turning time. Sun et al. [14, 15] proposed the problem of target detection in UAV-based wireless sensor network (UWSN), while considering static and dynamic targets to plan the path of UAVs, and generate the best of multiple UAVs. Mobile plan. Parikshit et al. [16] considered polygonal obstacles and turning Angle constraints in uav path planning, and found feasible paths for UAV by improving Dijkstra algorithm and viewable reverse search method. Li et al. [4] proposed the min-time max-coverage problem. They uses objective function for path planning, and achieves a good result. However, they do not consider the return time of the UAVs. Therefore, we will make improvement to this problem to ensure that UAVs can return to the base station before the end of the endurance time.

3   Problem Formulation

In this section, we first introduce some assumptions and definitions. With these definitions, the MEC-TS problem is formally modeled.

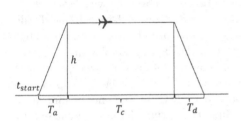

Fig. 2. Endurance of an UAV

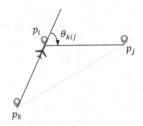

Fig. 3. Turning angle of an UAV

Assume that n POIs, denoted by $P = \{p_1, p_2, ..., p_n\}$ are randomly distributed in the target area. All of them are static with known position. And m UAVs $U = \{u_1, u_2, ..., u_m\}$ are responsible for covering these POIs. Each UAV starts from the base station B_0, performs the sweep coverage mission, and finally returns to the B_0.

The endurance of an UAV can be divided into three phases: ascent phase, cruise phase and descent phase. As shown in Fig. 2, the UAV takes off at time t_{start}. T_a represents the time of ascent phase, in which the UAV climbs from the base to a specified altitude. T_c represents the time of cruise phase, during which UAV performs the sweep coverage mission. T_d stands for the time of descent phase and the UAV returns to the base in this phase. In this paper, we require the UAVs to complete the sweep coverage mission before the descent phase, and the timing of the mission starts at the time of takeoff, so the maximum mission time is set to T_{max}:

$$T_{max} = T_a + T_c \tag{1}$$

In cruise phase, the UAVs fly in fixed altitude h and fixed speed v to sweep covering the POIs, so that the three-dimensional path planning problem for UAVs can be reduced to two dimensions. Thus we simplify that all the POIs are in a Euclidean plane, and the distance between any two POIs p_i and p_j are their Euclidean distance d_{ij}.

The m UAVs perform sweep coverage mission simultaneously. A POI is said to be covered when an UAV fly to its position. In a sweep coverage mission, each POI needs to be covered by an UAV only once. As mentioned in [4], it's necessary to consider the turning time in the flight of UAV, because it takes a certain amount of time for the UAV to turn, which may affect the path planning of the UAV for time-sensitive tasks. Figure 3 shows the turning angle of UAV.

The UAV flies from POI p_k to p_i, and then from p_i to p_j. θ_{kij}, the turning angle at p_i, can be calculated by the cosine theorem:

$$\theta_{kij} = \pi - \arccos \frac{d_{ij}^2 + d_{ki}^2 - d_{kj}^2}{2 \cdot d_{ij} \cdot d_{ki}} \tag{2}$$

The time for an UAV flies from POI p_i to p_j is consist of two parts [4]:

$$t_j = \frac{\theta_{kij}}{\omega} + \frac{d_{ij}}{v} \tag{3}$$

The first part is turning time, and the second part is distance time. ω and v represent the angular speed and moving speed of UAV respectively.

If p_j is the last POI visited by the UAV, then the UAV will return to base B_0 from p_j. The total time to choose the p_j to cover and then return to base B_0 is defined as t_{jr}.

$$t_{jr} = t_j + \frac{\theta_{ijB_0}}{\omega} + \frac{d_{jB_0}}{v} \tag{4}$$

Due to different emergency situations in different regions, n POIs have their own time sensitivity $Ts = \{ts_1, ts_2, ..., ts_n\}$, ts_i represents POI p_i expected to be covered by an UAV within time ts_i. In order to be more realistic, we set up a tolerance coefficient e, $0 \le e \le 1$. $e \cdot ts_i$ represents the time that the POI p_i allows the UAVs to be late, because the time sensitivity is not so strict sometimes. Although p_i expects to be covered within ts_i time, it is acceptable for the UAVs to arrive at $(1+e) \cdot ts_i$ time. When the UAVs take off at time t_{start}, the timing starts.

Definition 1 (On time rate, R_o). *A POI p_i is said to be covered on time when an UAV visit it within its time sensitivity ts_i. The on time rate is defined as follows:*

$$R_o = \frac{\sum_{k=1}^m \sum_{i=1}^n x_{ik}}{n} \tag{5}$$

x_{ik} represents whether the i_{th} POI p_i can be covered by the k_{th} UAV u_k on time, which only has two values of 0 and 1. If p_i can be covered by u_k on time, $x_{ik} = 1$; otherwise, $x_{ik} = 0$. The on time rate R_o is the ratio of the number of POIs covered on time to the total number of POIs.

Definition 2 (Effective coverage rate, R_e). *A POI p_i is said to be effectively covered when an UAV visit it within its acceptable time $(1+e) \cdot ts_i$. Before time*

$(1+e) \cdot ts_i$, we call POI alive, after time $(1+e) \cdot ts_i$, POI is dead. The effective coverage rate is defined as follows:

$$R_e = \frac{\sum_{k=1}^{m} \sum_{i=1}^{n} c_{ik}}{n} \tag{6}$$

c_{ik} represents whether the i_{th} POI p_i can be effectively covered by the k_{th} UAV u_k, which only has two values of 0 and 1. If p_i can be covered by u_k effectively, $c_{ik} = 1$; otherwise, $c_{ik} = 0$. The effective coverage rate R_e is the ratio of the number of POIs covered effectively to the total number of POIs.

Formally, the problem of achieving the maximum effective coverage rate in time sensitive sweep coverage with multiple UAVs (MEC-TS) can be defined as follows:

Definition 3 (MEC-TS). *Given a set of POIs $P = \{p_1, p_2, ..., p_n\}$, n POIs have their own time sensitivity $Ts = \{ts_1, ts_2, ..., ts_n\}$. With m UAVs $U = \{u_1, u_2, ..., u_m\}$, the goal of MEC-TS is to maximize the effective coverage rate R_e in time sensitive sweep coverage while ensuring that m UAVs end their sweep coverage mission in maximum mission Time T_{max} and return to Base B_0.*

The specific mathematical description of the problem is as follows:

$$\max \quad R_e \tag{7}$$

subject to

$$T_k(i) \le (1+e) \cdot ts_i, \tag{8}$$
$$\forall k \in \{1, 2, ..., m\}, \forall i \in \{1, 2, ..., n\}, 0 \le e \le 1$$

$$T_k = T_a + \sum_{i=1}^{n} x_{ik} t_i \le T_{max} \tag{9}$$
$$\forall k \in \{1, 2, ..., m\}$$

$$\sum_{k=1}^{m} x_{ik} \le 1, \forall i \in \{1, 2, ..., n\} \tag{10}$$

where $T_k(i)$ represents the cumulative mission time of the k_{th} UAV when arrives at POI p_i and T_k represents the total time it takes for the k_{th} UAV to complete its mission and return to the base. The optimization goal is to maximize the effective coverage R_e of the entire sweep coverage mission. The first constraint in (8) represents satisfying the time sensitivity of each POI, that is, covering POIs within their acceptable time. The second constraint in (9) means that the mission time of each UAV does not exceed the maximum mission time T_{max}. The third constraint in (10) means that each POI can be covered at most once in a sweep coverage mission to prevent POI from being repeatedly covered.

Furthermore, the MEC-TS problem can be proved to be NP-hard. The details of the proof are as follows:

Theorem 1. *The MEC-TS problem is NP-hard.*

Proof. The MEC-TS's decision problem is that, in a sweep coverage mission, whether there exist such sweep paths for m UAVs that the effective coverage rate can achieve R_e and return to Base. If we consider a special case, *i.e.*, $ts_i \gg T_{max}, \forall i \in \{1, 2, ..., n\}$, which means the time sensitivity of POIs on our problem can be ignored. Further, if we set $m = 1$ and $R_e = 1$, then the decision problem becomes if one UAV can access all POIs while ensuring that its travel distance is no more than vT_{max}. Apparently, the MEC-TS's decision problem under the special case is equivalent to the TSP decision problem. Since TSP problem is NP-hard, the MEC-TS problem is also NP-hard.

4 GCS Algorithm

In this section, we present the basic idea and details of the Greedy cost selection algorithm(GCS). Meanwhile, the complexity analysis of the GCS algorithm is also included.

The basic idea of GCS is to generate the sweep paths for each of the UAVs successively in a sweep coverage mission, and the starting point and ending point of each sweep path are the base station B_0. During path planning, we designed a cost function to calculate the cost of accessing each POI. This cost function takes into account the time required to access the POI, the time sensitivity of the POI, and the sweep coverage progress of the current UAV. We adopt a greedy strategy and choose the POI with the least cost to cover every time, getting the optimum sweep path for the current UAV.

4.1 Details of GCS

Algorithm 1. GCS

Input: The POIs set $P = \{p_1, p_2, ..., p_n\}$, the time sensitivity set $Ts = \{ts_1, ts_2, ..., ts_n\}$, the UAVs set $U = \{u_1, u_2, ..., u_m\}$, the base station B_0, the maximum mission time T_{max}, the moving speed of UAVs v and angular speed ω.

Output: The sweep paths $O = \{O_1, O_2, ..., O_m\}$; the on time rate R_o, the effective coverage rate R_e.

1: Check Ts and make it reasonable.
2: **for** $k = 1 \rightarrow m$ **do**
3:     Set $O_k = \emptyset$
4:     Set the $T_k = T_a$
5:     Set xik,
6:     **while** $P \neq \emptyset$ *and* $B_0 \notin O_k$ **do**
7:         **for** $p_i \in P$ **do**
8:             calculate t_i and t_{ir}.
9:             $condition_1 = T_k + t_i \leq (1 + e) \cdot ts_i$
10:            $condition_2 = T_{max} - T_k - t_{ir} \geq 0$

11:            **if** condition1 and condition2 **then**
12:                $\varphi = T_k/T_{max}$
13:                $cost_i = t_i + (ts_i - T_k)^\varphi$
14:            **else**
15:                $cost_i = +\infty$
16:            **end if**
17:        **end for**
18:        select the $p_j \in P$ who has the smallest cost.
19:        **if** p_j exists **then**
20:            add p_j into O_k.
21:            remove p_j from P.
22:            $T_k = T_k + t_j$
23:            $c_{jk} = 1$
24:            **if** $T_k < ts_j$ **then**
25:                $x_{jk} = 1$
26:            **end if**
27:        **else**
28:            add B_0 to O_k.
29:        **end if**
30:    **end while**
31: **end for**
32: Calculate the effective coverage rate R_e and on time rate R_o.
33: **return** O, R_e, R_o

The details of GCS are presented in Algorithm 1. Firstly, check whether the Ts is reasonable before starting. If ts_i is smaller than the direct distance time d_{iB_0}/v, it is inaccessible, and ts_i needs to be reset. The GCS plans the sweep paths $O = \{O_1, O_2, \ldots, O_m\}$ for m UAVs one by one. The initial sweep path O_k for the k_{th} UAV is an empty set. The variable T_k is a timer, which records the time spent by the UAV u_k after t_{start}. The initial value of T_k is T_a, which is the time of ascent phase. For each p_i in P, we firstly calculate the t_i and t_{ir} by Eq. (3) and (4). Then judge whether p_i is accessible by two conditions. The first condition is, when the drone flies to p_i, p_i is still alive. That is, when arriving at p_i, the time is within its acceptable range $(1 + e) \cdot ts_i$. The second condition is, the remaining mission time is enough for the u_k to fly to p_i and then return to the base B_0. Only when these two conditions are met at the same time, we say that p_i is accessible. Then we design a cost function to evaluate the cost of visiting p_i.

$$\varphi = T_k/T_{max}$$
$$cost_i = t_i + (ts_i - T_k)^\varphi, \forall i \in \{1, 2, \ldots, n\} \tag{11}$$

The coefficient φ reflects the sweep coverage progress of the current UAV. $(ts_i - T_k)$ represents the remaining access time of p_i. The closer to the end of the sweep coverage mission, the more priority should be given to access the POI with the short remaining access time. In the cost function, the POI that is closest to the current location and has the shortest remaining access time will have the least cost and will be accessed first. If p_i is inaccessible, we set the $cost_i$ to infinity. After calculating the cost for all POIs in P, select the POI p_j who has the smallest cost to cover. And then perform line 20 to 26, add p_j into O_k, remove p_j from P, update the timer T_k, and judge the effective coverage status and on time coverage status of p_j. If p_j does not exist, that is, the cost of all the POIs in current P are infinity, then it means that there is no suitable POI to continue to cover. In this case, the UAV u_k should return to the base B_0. When u_k returns to base or POI set P becomes empty, the path planning of u_k is completed. After planning the sweep paths for m UAVs, calculate the effective coverage rate R_e and on time rate R_o according to Eq. (6) and (5). The algorithm ends.

4.2   Complexity Analysis

In GCS, the time complexity for calculating the cost of all POIs in P from line 7 to line 17 is $O(n)$, since there are at most n POIs in P. Similarly, adding POI to the sweep path of UAV u_k from line 6 to 30 requires time complexity $O(n)$, because a sweep path can add n POIs at most. Finally, planning sweep paths for m UAVs, the number of iterations from line 3 to line 30 is m, thus the complexity of the GCS algorithm is $O(mn^2)$.

5   Performance Evaluation

In this section, simulations are conducted to demonstrate the advantage of the proposed algorithm on solving the MEC-TS problem. In the simulation, several algorithms from previous literature, G-MSCR and WTSC, are also implemented for performance comparison [3,4]. G-MSCR solves the problem that POIs should be covered and the UAVs should return to the base station within a preset time window, which is similar to the MEC-TS. WTSC also considers the maximum coverage problem with weighted POIs in sweep coverage, we replace the weight of POIs with time sensitivity in simulation experiment.

5.1   Simulation Configuration

In the simulation, the target area is a square with the width of 50 km. The base station B_0 is at the bottom left corner of the square, which is more in line with emergency rescue situations. A number of POIs are randomly distributed in the area and the time sensitivity ts_i is a randomly generated value within a certain range. The maximum mission time is 180 min and the time of ascent phase for UAV is 10 min. The moving speed v of each UAV is set to 25 m/s while the angular speed ω is 0.1 rad/s with the minimum turning radius 100 m. In order to study the influence of different variables on the experiment, i.e., the number of UAVs m, the number of POIs n, the range of time sensitivity Ts, we carry out the experiment under three scenarios:

1) $n = 100$, range of $Ts = (50\,\text{min}, 140\,\text{min})$, m varies from 0 to 10.
2) $m = 5$, range of $Ts = (50\,\text{min}, 140\,\text{min})$, n varies from 1 to 400.
3) $m = 5$, $n = 100$, the range of Ts are $(30\,\text{min}, 120\,\text{min})$, $(50\,\text{min}, 140\,\text{min})$, $(70\,\text{min}, 160\,\text{min})$ respectively.

These simulations are repeated for 50 times and the average value is taken as the result.

5.2   Simulation Results

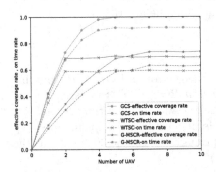

Fig. 4. $n = 100$, range of $Ts = (50\,\text{min}, 140\,\text{min})$

Fig. 5. $m = 5$, range of $Ts = (50\,\text{min}, 140\,\text{min})$

Scenario 1: Figure 4 illustrates the effective coverage rate R_e and on time rate R_o varying with the number of UAVs m. The solid line represents the effective coverage rate R_e, and the dotted line represents the on time rate R_o. The simulation results show that, regardless of the number of UAVs, the R_e and R_o of GCS are higher than those of WTSC and G-MSCR. When the number of UAV is small, the performance of GCS is similar to that of WTSC, and the performance of GCS is slightly higher than that of WTSC. But with the increase of the number of UAVs, the performance advantages of GCS are becoming more and more obvious, almost 30% higher than the other two algorithms. When the number of UAVs exceeds 5, the effective coverage of GCS can reach 100%. The performance of GCS is relatively good. Note that the effective coverage rate is almost 20% higher than the on time rate, this is because we set the tolerance coefficient e to 0.2 in the experiment. It can be seen that even if the tolerance coefficient e is not set, the on time rate of GCS is maintained at a high level.

Scenario 2: Figure 5 illustrates the effective coverage rate R_e and on time rate R_o varying with the number of POIs n. Similarily, regardless of the number of POIs, the R_e and R_o of GCS are much higher than those of WTSC and G-MSCR, about 30% higher than G-MSCR and 20% higher than WTSC. Note that as the number of POIs increases, the R_e and R_o of the three algorithms is declining. This is because the number and energy of UAVs are fixed, thus the

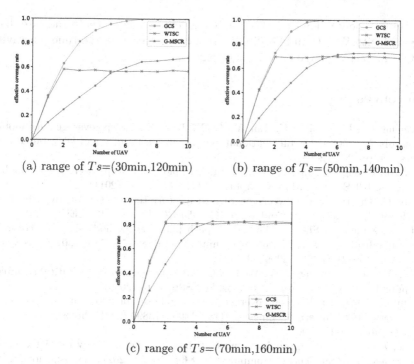

(a) range of $Ts=$(30min,120min)   (b) range of $Ts=$(50min,140min)

(c) range of $Ts=$(70min,160min)

Fig. 6. $m = 5$, $n = 100$, the range of Ts are in three different ranges.

number of POIs that can be covered is limited. The simulation results show that the proposed algorithm GCS has the best performance among these algorithms.

Scenario 3: Figure 6 illustrates the relationship between effective coverage rate R_e and the range of time sensitivity Ts. Range of $Ts = (30\,\mathrm{min},\ 120\,\mathrm{min})$ means that some POIs require UAVs to cover it within 30 min, which is relatively difficult. The shorter the time, the more difficult. However, in the three subfigures of Fig. 6, the effective coverage of GCS is still better than the other two algorithms. It shows that GCS is effective in different time sensitivity ranges. And under the more severe conditions, the advantages of GCS are more obvious.

To sum up, the simulation results indicate that the proposed algorithm GCS achieves better performance than the comparison algorithm in terms of effective coverage rate under all these scenarios.

6   Conclusions and Future Work

In this paper, we study the MEC-TS problem, the objective of which is to find such sweep paths that UAVs can achieve the maximum effective coverage rate. We prove the NP-hardness of the MEC-TS problem and devise an algorithm called GCS which delivers an effective solution in the MEC-TS problem. In the experiment, we compare the proposed algorithm with algorithms from previous literature, namely WTSC, and G-MSCR. The experimental results indicate that, compared with WTSC and G-MSCR, GCS can better meet the time sensitivity of POIs and achieve high effective coverage rate.

References

1. Cheng, W., Li, M., Liu, K., Liu, Y., Li, X., Liao, X.: Sweep coverage with mobile sensors. In: 2008 IEEE International Symposium on Parallel and Distributed Processing, pp. 1–9 (2008)
2. Li, M., Cheng, W., Liu, K., He, Y., Li, X., Liao, X.: Sweep coverage with mobile sensors. IEEE Trans. Mob. Comput. **10**(11), 1534–1545 (2011)
3. Liu, C., Du, H., Ye, Q.: Sweep coverage with return time constraint. In: 2016 IEEE Global Communications Conference (GLOBECOM), pp. 1–6. IEEE (2016)
4. Li, J., Xiong, Y., She, J., Wu, M.: A path planning method for sweep coverage with multiple UAVs. IEEE Internet Things J. **7**(9), 8967–8978 (2020). https://doi.org/10.1109/JIOT.2020.2999083
5. Cl, A., Hd, A., Qiang, Y.B., Wen, X.C.: Group sweep coverage with guaranteed approximation ratio. Theor. Comput. Sci. **836**, 1–15 (2020)
6. Liu, C., Du, H.: t, k-sweep coverage with mobile sensor nodes in wireless sensor networks. IEEE Internet Things J. **8**(18), 13888–13899 (2021). https://doi.org/10.1109/JIOT.2021.3070062
7. Nie, Z., Du, H.: An approximation algorithm for general energy restricted sweep coverage problem. Theor. Comput. Sci. **864**, 70–79 (2021). https://doi.org/10.1016/j.tcs.2021.02.028
8. Chen, Z., Wu, S., Zhu, X., Gao, X., Gu, J., Chen, G.: A route scheduling algorithm for the sweep coverage problem, pp. 750–751. IEEE (2015)
9. Feng, Y., Gao, X., Fan, W., Chen, G.: Shorten the trajectory of mobile sensors in sweep coverage problem. In: IEEE Global Communications Conference (2015)
10. Gao, X., Chen, Z., Pan, J., Wu, F., Chen, G.: Energy efficient scheduling algorithms for sweep coverage in mobile sensor networks. IEEE Trans. Mob. Comput. **19**(6), 1332–1345 (2020)
11. Zhang, D., Zhao, D., Ma, H.: On timely sweep coverage with multiple mobile nodes. In: 2019 IEEE Wireless Communications and Networking Conference (WCNC), pp. 1–6 (2019). https://doi.org/10.1109/WCNC.2019.8886067
12. Choi, Y., Choi, Y., Briceno, S., Mavris, D.N.: Energy-constrained multi-UAV coverage path planning for an aerial imagery mission using column generation. J. Intell. Rob. Syst. (2019)
13. Luo, C., Satpute, M.N., Li, D., Wang, Y., Chen, W., Wu, W.: Fine-grained trajectory optimization of multiple UAVs for efficient data gathering from WSNs. IEEE/ACM Trans. Netw. **29**(1), 162–175 (2021)
14. Sun, P., Boukerche, A.: Performance modeling and analysis of a UAV path planning and target detection in a UAV-based wireless sensor network. Comput. Netw. **146**(DEC.9), 217–231 (2018)
15. Peng, S., Boukerche, A., Tao, Y.: Theoretical analysis of the area coverage in a UAV-based wireless sensor network. In: The 13th International Conference on Distributed Computing in Sensor Systems, DCOSS 2017 (2018)
16. Maini, P., Sujit, P.B.: Path planning for a UAV with kinematic constraints in the presence of polygonal obstacles. In: International Conference on Unmanned Aircraft Systems (2016)

Recursive Merged Community Detection Algorithm Based on Node Cluster

Ailian Wang$^{(\boxtimes)}$, Liang Meng, and Lu Cui

Taiyuan University of Technology, Taiyuan 030024, Shanxi, China

Abstract. There exist problems in study of social network community detection, such as some algorithms detection result having high time complexity with comparatively satisfactory, existing fast algorithms in low quality because of stochastic iteration partition results for large scale network, and lacking of model and mechanism of individual and link attributes expressing and utilizing. To solve these problems, this paper proposes a recursive merged community detection model based on node cluster, which can express the tightness of relationship between individuals according to the closed preconditions. Based on this, an effective community detection algorithm is designed and implemented. The proposed recursive merging model has high generality and is applicable to both weighted and non-weighted networks. A series of experiments show that the proposed algorithm based on node cluster recursive model and following linked list is effective for community detection in social networks with relatively less time cost. The algorithm can also be applied to the need to fuse integrate individuals and links attributes of community detection algorithm with a comparatively fast speed and high quality partition.

Keywords: Social network · Recursion · Community detection · Linked list · Modularity

1 Introduction

Online social network has been booming as a new type of complex network in recent years. The issue of community detection in online social networks has drawn much attention. Community detection is an important means of researching the topology of social networks, discovering user aggregation patterns and promoting information dissemination. In essence, community detection in online social networks is the process of subdividing subgraph by network nodes according to the close connection of internal topological structure. Therefore, in the field of computer science, with the aid of mathematical tools such as graph theory, the collection discovery problem of the nodes connected tightly in the graph is described as a graph segmentation problem, and a representative algorithm based on greedy optimization strategy [1] and spectral bisector [2] emerges. The GN split algorithm proposed by Givanhe and Newman [3] is also the most used algorithm. In this algorithm, the concept of modularity is proposed as an important index to measure the discovery of online community for better detecting community. Since then, many scholars have taken modularity as an objective function and proposed

W. Wu and H. Du (Eds.): AAIM 2021, LNCS 13153, pp. 289–302, 2021.
https://doi.org/10.1007/978-3-030-93176-6_25

a community detection algorithm based on optimization theory. Some researchers proposed a community detection algorithm based on multi-objective optimization, by using genetic algorithms combined with multi-objective functions to characterize the community structure, the social communities was achieved [4–6]. Gergely Palla et al. used the concept of factions and groups to discuss the issue of community detection, focusing on the nodes of overlapping community and community boundaries in the network [7–11]. Infomap Algorithm [12–14] discovered communities by transforming the network topology into a coded construct with the aid of information-based data coding. There is also a community structure discovery algorithm based on probabilistic models [15, 16], which divides the community structure by analyzing the maximum likelihood probability. The time complexity of the above algorithm is relatively high, more emphasis on the accuracy of community structure, more suitable for social groups in smaller social groups detection. In the large-scale social network, because of the variety of node information, many scholars hope to develop a community detection algorithm with higher accuracy and lower time complexity. Therefore, from different perspectives, some new community detection algorithms are proposed.

In order to improve the efficiency of community detection, a typical dynamic community detection algorithm based on label propagation is proposed. The algorithm can discover the social structure of social networks with the given topology in a linear time and has low time complexity and is suitable for large-scale online social networks. In order to characterize the social structure of social networks from multiple perspectives, a cellular automata community detection algorithm is proposed, which combines the principle of cellular automata with the heuristic algorithm to optimize the community structure.

This paper will study the comparison of dynamic computing algorithm and static computing algorithm, and then proposes a recursive merging model based on node cluster and corresponding algorithms with high computing efficiency. Experiments show that the model and algorithm starting from node cluster are suitable for medium- and large-scale data set with good results, and it also has good applicability to all kinds of social networks with different degrees of completeness of information.

2   Background

2.1   Community Structure

Social networks have the phenomenon of non-uniform relations. Some individuals are densely connected with each other and some are sparsely connected with each other, forming social structure in social networks. Social networks can be regarded as a set of nodes with high cohesion characteristics in the network topology. Network nodes are the real people in the virtual network mapping, while the connected edges of the network represent the exchange and communication between network users. The community structure can be defined as several subsets of a complex network node set. The nodes in each subsection are relatively densely connected to each other, while the connecting edges between different subsets of nodes are relatively sparse. As shown in Fig. 1 below.

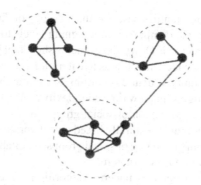

Fig. 1. Structure of SNS

For network nodes, many scholars have proposed a virtual community detection algorithm based on the similarity of nodes, such as the similarity-based aggregation algorithm [17–19], in which the EAGLE algorithm uses a clustering framework to find all the largest factions, and merge the largest sub-community; Also the vector similarity formula based on the cosine formula [20], each network node as a node with the community structure, through the local maximum edge concept to find the connection in the local area closer nodes to be merged, the combined community and then as a virtual node to participate in the subsequent merger process. The complexity of the algorithm is relatively high, and not very stable.

2.2 Label Propagation Algorithm

The label propagation algorithm (LPA) is a graph-based semi-supervised learning algorithm. Usha Nandini Raghavan et al. applied the label propagation algorithm to the network detection community [21]. The main steps of the algorithm are: setting different labels to the nodes in the network, the labels are iteratively propagated on the network according to the specific propagation rules, and the labels of all nodes are updated every time they are propagated. The criterion is to set the same label with the maximum number of neighbor nodes, when each node label propagation are stable, nodes with the same label will be divided into the same community.

Specific steps are as follows:

(1) First, all nodes are assigned a unique label.
(2) All nodes are scanned in a random order. The label of each node is replaced by the label with the largest number of labels assigned by the neighbors. If the maximum number of neighbors has multiple labels at the same time, one label can be selected. The choice at this time is random.
(3) When the label of each node is consistent with the label assigned by the maximum number of neighbors, proceed to step (4). If not, return to (2).
(4) Use nodes with the same tag in the network as a community.

The algorithm finds communities based on the topology of the network, has the time complexity close to linear, the complexity of assigning labels to nodes is $O(n)$, and the time complexity of one iteration is $O(m)$. Experiments show that [21] as many as 95% of the nodes can be correctly divided in as few as five iterations. This algorithm, which only needs to define the community structure according to intuitionistic heuristic rules, has the advantages of simple algorithm, low time complexity and fast convergence. The core idea is similar to that of the formation of social groups in online social networks. The algorithm is to find the community naturally based on network topology, without having to know in advance the number and size of social groups, is suitable for large-scale social network division of the community structure.

However, since this algorithm is an uncertain algorithm, the asynchronous tag update algorithm is used multiple times on the same online social network; the algorithm needs to adopt a new random order for each iteration, therefore, if the algorithm is run multiple times on the same data set, the nodes are updated in a different order, which may result in the eventual generation of more than one different community structure. This is an important reason for the unstable labeling algorithm.

2.3   Cellular Automata's Community Detection Algorithm

In order to show the various structural characteristics and attribute characteristics of social organizations in online social networks, many experts and scholars describe the characteristics of social organizations from multiple perspectives, and optimize the problem through heuristic algorithms such as cellular automata theory and genetic algorithm to realize virtual community structure of the division.

Yuxin Zhao proposed a CLA-net algorithm [22] based on cellular automata learning machine. With irregular cellular automaton machines, each network node is regarded as an automatic learning machine, and each auto learning machine characterizes the community structure from two aspects: the overall network structure of the network and the local community structure of the node, so as to realize the excavation of the structure of the virtual community. The irregular cellular automatic learning machine refers to the biological self-reproduction phenomenon and designs a local dynamic model which is discreet in time and space dimensions. The evolution of the model does not follow rigorous mathematical equations or functions. Instead, by defining a series of rules of cellular state changes, some seemingly simple rules are evolved through repeated computations to develop extremely complex dynamic models. The online social network is mapped to an irregular cellular learning machine. Through the customized evolution rules, the status of each node is dynamically adjusted so that the social structure in the entire network is gradually rationalized.

In each iteration t, the learning and updating process of the automatic learning machine L_i can be described as:

(1) Choose a behavior $\alpha_i(t)$ randomly according to the behavioral probability vector P_i.

(2) Interaction with the local environment (other neighboring nodes) and the global environment (the entire network) to obtain the feedback signal $\beta_i(t)$. If the satisfying node i belongs to the same community with most of its neighbor nodes, and the module degree $Q(t) \geq Q_b$ obtained in this iteration is obtained, then the obtained feedback signal $\beta_i(t) = 0$; otherwise, $\beta_i(t) = 1$.

(3) Assuming $\alpha_i(t) = \alpha_{iq}$, α_{iq} updates $W_{iq}(t)$ and $Z_{iq}(t)$ according to the feedback signal $\beta_i(t)$ for the elements in the behavior set α_i.

$$\begin{cases} W_{iq}(t) = W_{iq}(t-1) + (1 - \beta_i(t)) \\ Z_{iq}(t) = Z_{iq}(t-1) + 1 \end{cases} \tag{1}$$

(4) Update the optimal behavior of the current automatic learning machine L_i according to the following formula, and the behavior with the largest value of $D_{ij}(t)$ is the current optimal behavior of the automatic learning machine L_i:

$$D_{ij}(t) = \frac{W_y(t)}{Z_y(t)} \tag{2}$$

(5) Assuming that the current optimal behavior of the automatic learning machine L_i is α_{im}, the behavior probability vector P_i of the current automaton L_i is updated according to the following formula, where α is the reward coefficient and P_j is the jth component of the vector P_i:

$$P_i(t+1) = \begin{cases} P_i(t) + \alpha(1 - P_i(t)) & j = m \\ (1 - \alpha)P_i(t) & j \neq m \end{cases} \tag{3}$$

The specific process of the algorithm is as follows:

(1) Randomly initialize the automatic learning machine in each node in the network.
(2) All automatic learning machines in the network choose their own behavior according to their own behavioral probability vectors, and obtain the corresponding community structure after decoding.
(3) All automatic learning machines in the network interact with the local environment and the global environment to learn and update.
(4) Repeat step (2) until the community structure obtained no longer changes.

The objective function of the algorithm is the modularity function and $k_i(c_i(t)) \geq k_i(c')$, where $c_i(t)$ denotes the community number of the node after the t times iteration, $k_i(C) = \sum_{j \in C} A_{ij}$, A is the network adjacency matrix.

3 Recursive Merged Community Detection Algorithm Based on Node Cluster

There is an organizational characteristic in social networks, such as Renren's social network composed of employees of the same company can form a virtual social network according to different departments. Each department's virtual community can be divided into different sub-groups according to different projects community structure. From a physical point of view, these community structures represent a complex system or a collection of elements that have the same or similar functions in a complex network. These elements interact or cooperate with each other to form a relatively independent organizational structure in the entire system or jointly accomplish relatively independent system functions. These are of great significance to the understanding of the characteristics of the entire network topology as well as the mining of the module functions of the various organizational networks.

This paper focuses on mathematical modeling of nodes, node sets and connection relationships in complex networks, and proposes recursive merged community detection algorithm (RMC), it includes three basic entities: one is a node, which is the most primitive node in a complex network; the other is a node cluster, similar node cluster formed by the merger of entities; third is the community. These three are progressive relationship, the node merged to form a node cluster, and then the community formed by the node cluster, the community is the final node cluster when the node clusters cannot be merged with each other (Fig. 2).

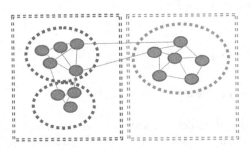

Fig. 2. Community structure based on node cluster

Since a node can be regarded as a node cluster formed by a single node, the relationship between nodes or nodes and node clusters is not considered separately, and only the relationship between the node clusters can be considered.

3.1 Node Cluster Attributes and Parameters

The main attributes of a node cluster include: the state of a node cluster, the set of neighbor nodes of a node cluster, and the set of nodes of a node cluster. The state of a node cluster mainly refers to how the node cluster interacts with other node clusters. It contains four states: a. the default state, b. the closed state, c. the following state, and d. the expanded state. When a single-node cluster is initialized, it is the default state, and the state of the node cluster is restored to its default state after the expansion of one node cluster occurs. The closed state indicates that the interior of a node cluster is relatively closely connected, does not follow other node clusters, or merges with other node clusters. In this case, the node cluster is already the final community. When the current node is the default state, if the most similar other node cluster C is found, and the most similar node cluster of the node cluster C is not the current node cluster, the current node cluster is in a following state and no other node clusters are combined, When a cluster merges, it is merged with C's following list, or after C follows other cluster, C is merged with C when it is merged. When the node cluster is the default state, and it is the most similar node cluster with the node cluster C, then the state is converted to an expansion state, it will merge C.

The index parameters of node cluster include: similarity, average connection density within clusters, and average connection density among clusters.

Similarity is used to describe the degree of similarity between the cluster of nodes, and its measurement is to examine the proportion of the same neighbor nodes of two cluster nodes in all neighbor nodes of two node clusters. Formula as shown in Eq. 4:

$$S = \left(N_i \cap N_j\right)/\left(N_i \cup N_j\right) \tag{4}$$

Among them, N_i is the neighbor node set of node cluster i.

The average connection density in a cluster is actually the average of the proportion of all nodes in a cluster in the cluster in the proportion of all connections. The formula is shown in Eq. 5:

$$\rho = \sum\nolimits_{i=1}^{N} \frac{C_{i1}}{C_{i2}}/N \tag{5}$$

Where "C_{i1}" is the number of connections in the node cluster of node i, "C_{i2}" is the total number of connections of node i, and N is the total number of nodes in the node set of the node cluster.

The average connection density between clusters refers to the average of the average connection density of clusters in each node cluster. As shown in Eq. 6:

$$q = \left(\sum\nolimits_{i=1}^{M} \rho_i\right)/M \tag{6}$$

Where ρ_i is the average intra-cluster connection density of the number i node cluster, and M is the total number of current node clusters.

3.2 Community Classification Quality Indicators

Modularity, first proposed by Mark Newman, is a commonly used method to measure the strength of the structure of the network community by comparing the difference in connectivity calculated between the existing network and the reference network under the same community classification to evaluate the advantages and disadvantages of online community.

The definition of module degree is as follows:

$$Q = \frac{1}{2m} * \sum_{ij} \left[A_{ij} - \frac{k_i * k_j}{2m} \right] \delta(C_i, C_j) \tag{7}$$

A_{ij} indicates whether the i node and the j node are adjacent, and if adjacent, $A_{ji} = 1$, otherwise $A_{ij} = 0$. In the corresponding network, the probability that one edge (i, j) exists is $\frac{k_i * k_j}{2m}$, m is the total number of connections, k_i represents the degree of node i, $\delta(C_i, C_j)$ is a two-valued function whose value is 1 when i node and j node belong to the same community, otherwise the value is 0.

The formula means the difference between the proportion of edges within the same community in the network and the expected proportion of the proportion of the inner edges of the reference network under the same community structure. If the module value is higher, then the effect of dividing the social network in a complex network is better.

3.3 Process Description

Proc1:

Step1: Initialize each node as a separate node cluster, and set its status to the default state.
Step2: Calculate the similarity between each node cluster, and record the most similar cluster of each node cluster.
Step3: Perform Proc2 on each node cluster.
Step4: If the status of each node cluster is closed, end; otherwise, execute step2.

Proc2:

Step1: If the cluster status of the node is closed, then the end.
Step2: End if the cluster status of the node is following.
Step3: If the node cluster and the node cluster C are most similar to each other, then the cluster of nodes is merged with C, and then the following list of the cluster of nodes and the cluster of nodes C is searched, and all the cluster of nodes on the linked list are merged together, Judge the closure conditions, if the conditions are true, the update is closed, otherwise update to the default state.
Step4: If the cluster with the most similar node is C, but the cluster with the most similar to C is not the cluster, then the cluster becomes a following state and adds it to C's following chain.

The algorithm is shown in Table 1 below.

Table 1. RMC algorithm

```
RMC():
1:  give initial unique label for each vertex;
2:  sort all vertices in the order of their degrees from large to
    small in vSet;
3:  foreach vertex i in sorted vSet:
4:      if k_i ≥ 3 and i.free = true:
5:          find vertex j which has the largest degree in N(i)
            and j.free = true;
6:          if j ≠ null:
7:              add vertices i and j to a new core;
8:              commNeiber ← N(i) ∩ N(j);
9:              sort commNeiber in the order of vertex degree
                from small to large;
10:             while commNeiber ≠ null:
11:                 foreach vertex h in sorted commNeiber:
12:                     add h to the core;
13:                     delete vertices not in N(h) from commNeiber;
14:                     delete vertex h from commNeiber;
15:         if sizeof(core) ≥ 3:
16:             add core to cores;
17: return cores;
```

4 Experimental Results and Analysis

4.1 Experimental Data

In order to test the performance of the community detection algorithm, many scholars in the field of sociology abstract a lot of network topographies with typical community structure. Such networks are the analysis of actual social networks, so their community structure often has definite practical significance. The development of social networks has provided large-scale network data for the research of the community detection algorithm. The scholars have collected and sorted the data information of many social networks as the test data of the community classification. In this paper, three data sets commonly used in the field of complex network analysis are selected: American Football Network [23], Karate Club [24] and Dolphins [25] Network as experimental data.

The node attributes of the dataset are shown in Table 2 below.

Table 2. Node properties of different network

Dataset	Nodes	Edges	Max_d	Min_d	Ave_d
Football	115	616	12	7	10
Karate club	34	78	17	1	4
Dolphins	62	159	12	1	5

4.2 Experimental Results

We chose label propagation algorithm, cellular machine learning algorithm and compared with the node cluster algorithm. The computation processes of the three algorithms in the American Football Club, Karate Club Network, and Dolphin Internet Society are shown in Fig. 3, Fig. 4, and Fig. 5. The horizontal axis is the number of iterations and the vertical axis is the number of nodes. Different colors represent different algorithms. The evaluation criteria of the community detection results adopted are modularity indicators. The higher the module degree, the better the effect of community classification.

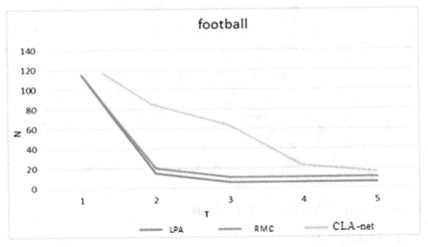

Fig. 3. Illustration of football based on different algorithm

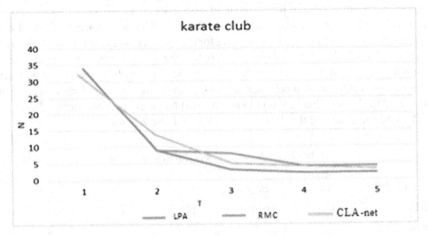

Fig. 4. Illustration of club based on different algorithm

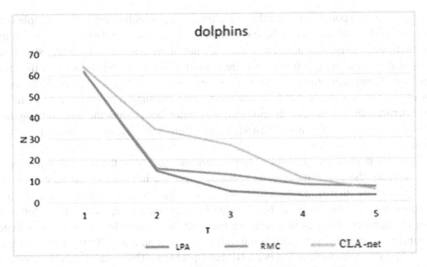

Fig. 5. Illustration of dolphins based on different algorithm

As can be seen from Fig. 3, Fig. 4 and Fig. 5, the node clustering algorithm iterative speed slightly better than the label propagation algorithm, convergence speed is also faster than the label algorithm. Cellular automata-based algorithm is essentially a multi-objective optimization algorithm, relatively complex, the convergence speed and iteration speed is slower. Three algorithms experimental results are shown in Table 3.

Table 3. Results of the Experiment based on Different Algorithm

Dataset	LPA		CLA-net		RMC	
	Q	CLUSTER	Q	CLUSTER	Q	CLUSTER
Football	0.547	5	0.605	13	0.988	11
Karate club	0.525	2	0.545	4	0.594	4
Dolphins	0.537	3	0.540	5	0.810	7

It can be seen from Table 3 that the modularity of node clustering algorithm is higher than that of tag propagation algorithm and cellular machine learning algorithm.

4.3 Experimental Analysis

The algorithm takes the set of node clusters formed by a single node as a starting set, calculates the similarity between two nodes, builds the similarity following linked list, recursively merges the existing node clusters step by step, and uses the preset closed condition as the node cluster whether to continue the conditions for the merger, ultimately rapidly makes division of society. During the initial experiment, due to the randomness in the iterative process of label algorithm, there are many kinds of community structures that meet the stop condition when the tags are updated asynchronously for multiple runs of the same data set. However, these community structures are relatively similar. In the process of node clustering algorithm, there are not many kinds of community structures that satisfy the ending condition, so there is no disadvantage of instability of the label algorithm. Comparing the experimental process in Fig. 3, Fig. 4 and Fig. 5, the node cluster algorithm is stable and fast. The modularity evaluation index is used to analyze the experimental results, and its modularity is much better than the label propagation algorithm. In general, the node clustering algorithm yields the best results on all three datasets.

Modulus degree can quantitatively measure the quality of the network community segmentation, the closer its value is 1, the stronger the intensity of the network community structure, that is, the better the quality of segmentation. Therefore, the optimal partitioning of the network can be obtained by maximizing the modularity. However, since the number of nodes in a network that may be divided by the network is huge, the number of possible partitions is a number with n as an index. Therefore, finding an optimal partition in partitioning is an NP-hard problem. This is the future direction of research.

5   Conclusion and Outlook

In this paper, a new recursive merged community detection algorithm based on node clusters is proposed and compared with the current typical label propagation algorithm and cellular machine learning algorithm. The label propagation algorithm predicts the label information of unlabeled nodes by marking the labeled information of the nodes, and depicts the structural features of the community from multiple sides based on the cellular machine algorithm. Both of them obtain the social grouping result by optimizing the module coefficients.

However, the recursive merged community detection algorithm based on node clusters gradually merges the existing node clusters by judging whether the cluster of nodes meets the preset merge conditions and achieving rapid division of communities.

In contrast to the experimental results on the common Zachary network, Dolphin network and American Football Club network, the node clustering algorithm obtains a more stable and rapid community classification than the label propagation algorithm and the cellular machine learning algorithm, and obtains on all three data sets the highest module value. It can be seen that the application of node clustering algorithm for community classification is more effective than the other two algorithms.

Acknowledgements. This work was supported in part by the China Scholarship Council and the National Natural Science Foundation of China under Grant No. 61472272.

References

1. Kernighan, B.W., Lin, S.: An efficient heuristic procedure for partitioning graphs. Bell Syst. Tech. J. **49**(2), 291–307 (1970)

2. Fiedler, M.: A property of eigenvectors of non-negative symmetric matrices and its application to graph theory. Czechoslov. Math. J. **25**(4), 619–633 (1975)
3. Newman, M.: Fast algorithm for detecting community structure in networks. Phys. Rev. E **69**(6), 066133 (2004)
4. Shi, C., Yan, Z.Y., Wang, Y.I., Cai, Y.A., Wu, B.: A genetic algorithm for detecting communities in large-scale complex networks. Adv. Complex Syst. **13**(1), 3–17 (2010)
5. Shi, C., Yu, P.S., Yan, Z., Huang, Y., Wang, B.: Comparison and selection of objective functions in multiobjective community detection. Comput. Intell. **30**(3), 562–582 (2014)
6. Filatovas, E., Kurasova, O., Sindhya, K.: Synchronous R-NSGA-II: an extended preference-based evolutionary algorithm for multi-objective optimization. Informatica **26**(1), 33–50 (2015)
7. Palla, G., Derényi, I., Farkas, I., Vicsek, T.: Uncovering the overlapping community structure of complex networks in nature and society. Nature **435**(7043), 814–818 (2005)
8. Kumpula, J.M., Kivelä, M., Kaski, K., Saramäki, J.: Sequential algorithm for fast clique percolation. Phys. Rev. E: Stat. Nonlinear Soft Matter Phys. **78**(2), 1815–1824 (2008)
9. Xie, J., Kelley, S., Szymanski, B.K.: Overlapping community detection in networks: the state-of-the-art and comparative study. ACM Comput. Surv. **45**(4), 115–123 (2011)
10. Shi, C., Cai, Y., Fu, D., Dong, Y., Wu, B.: A link clustering based overlapping community detection algorithm. Data Knowl. Eng. **87**(9), 394–404 (2013)
11. Su, Y.-J., Hsu, W.-L., Wun, J.-C.: Overlapping community detection with a maximal clique enumeration method in MapReduce. In: Pan, J.-S., Snasel, V., Corchado, E.S., Abraham, A., Wang, S.-L. (eds.) Intelligent Data analysis and its Applications, Volume I. AISC, vol. 297, pp. 367–376. Springer, Cham (2014). https://doi.org/10.1007/978-3-319-07776-5_38
12. Rosvall, M., Bergstrom, C.T.: Maps of random walks on complex networks reveal community structure. Proc. Natl. Acad. Sci. U.S.A. **105**(4), 1118–1123 (2008)
13. Kim, Y., Jeong, H.: Map equation for link communities. Phys. Rev. E: Stat. Nonlinear Soft Matter Phys. **84**(2), 1402–1409 (2011)
14. Jin, S., Li, A., Yang, S., Lin, W., Deng, B., Li, S.: A MapReduce and information compression based social community structure mining method. In: IEEE International Conference on Computational Science & Engineering, pp. 971–980. IEEE Computer Society, Washington (2013)
15. Mej, N., Leicht, E.A.: Mixture models and exploratory analysis in networks. Proc. Natl. Acad. Sci. U.S.A. **104**(23), 9564–9569 (2007)
16. Yu, L., Wu, B., Bai, W.: LBLP: link-clustering-based approach for overlapping community detection. Tsinghua Sci. Technol. **18**(4), 387–397 (2013)
17. Shen, H., Cheng, X., Cai, K., Maobin, H.: Detect overlapping and hierarchical community structure. Phys. A **388**(8), 1706–1712 (2008)
18. Rees, B.S., Gallagher, K.B.: Overlapping community detection using a community optimized graph swarm. Soc. Netw. Anal. Min. **2**(4), 405–417 (2014)
19. Qi, J., Jiang, F., Wang, X., Xu, B., Sun, Y.: Community clustering algorithm in complex networks based on microcommunity fusion. Math. Probl. Eng. 1–8 (2015)
20. Huang, J., Sun, H., Han, J., Feng, B.: Density-based shrinkage for revealing hierarchical and overlapping community structure in networks. Phys. A Stat. Mech. Appl. **390**(11), 2160–2171 (2011)
21. Raghavan, U.N., Albert, R., Kumara, S.: Near linear time algorithm to detect community structures in large-scale networks. Phys. Rev. E **76**(3), 036106 (2007)
22. Zhao, Y., Jiang, W., Li, S., et al.: A cellular learning automata based algorithm for detecting community structure in complex networks. Nerocomputing **151**, 1216–1226 (2015)
23. Girvan, M., Newman, M.E.: Community structure in social and biological networks. Proc. Natl. Acad. Sci. U.S.A. **99**(12), 7821–7826 (2002)

24. Zachary, W.W.: An information flow model for conflict and fission in small groups. J. Anthropol. Res. **33**(4), 452–473 (1977)
25. Lusseau, D., Schneider, K., Boisseau, O.J., et al.: The bottlenose dolphin community of Doubtful Sound features a large proportion of long-lasting associations (2003)

Purchase Preferences - Based Air Passenger Choice Behavior Analysis from Sales Transaction Data

Xinghua Li[1], Suixiang Gao[1], Wenguo Yang[1], Yu Si[1], and Zhen Liu[2(✉)]

[1] School of Mathematical Sciences, University of Chinese Academy of Sciences,
Beijing 100049, China
{lixinghua18,siyu191}@mails.ucas.ac.cn,
{sxgao,yangwg}@ucas.edu.cn
[2] UStrategy Technology Inc., Beijing, China
zliu@ustrategytech.com

Abstract. Travel providers such as airlines are becoming more and more interested in understanding how passengers choose among alternative products, especially the purchasing preferences of passengers. Getting information of air passenger choice behavior helps them better display and adapt their offer. Discrete choice models are appealing for airline revenue management (RM). In this paper, we apply latent class multinomial logit model (LC-MNL) to passenger choice behavior. The analysis based on actual sales transaction data reveals the purchase preferences of different passenger types. According to the distribution of the market, we divide passengers into three groups: low-price oriented, high-price oriented and no specific price preference. The low-price oriented passengers only choose products from the set consisting of the lowest price cabin classes of all flights while the high-price oriented passengers only choose products from the set consisting of the highest price cabin classes of all flights. Considered the passenger types in the transaction sales data are unknown, the latent class passenger choice model can better represent their heterogeneous purchasing preference. A developed EM algorithm is applied to solve the LC-MNL. In the developed EM algorithm, an indicator function containing the type of passengers and first choice information in period t is devised, the iterative process of the EM algorithm is more effective consequently. The proposed model and algorithm are evaluated on actual aviation sales transaction data in China. Experimental results show that the passenger choice behavior analysis based on the specific purchasing preferences performs well on actual aviation sales transaction data.

Keywords: Passenger choice behavior · Specific purchasing preference · LC-MNL · Developed EM algorithm

1 Introduction

Travel providers such as airlines and on-line travel agents are becoming more and more interested in understanding how passenger choose among alternative

W. Wu and H. Du (Eds.): AAIM 2021, LNCS 13153, pp. 303–314, 2021.
https://doi.org/10.1007/978-3-030-93176-6_26

products. Understanding passenger behavior and their purchasing preference is important in the travel industry. It can be used, for example, to meet the demand and market shares better in the context of dynamic markets.

We mainly focus on the choice behavior of air passengers based on specific purchasing preferences. Firstly, the choice set changes over time in the analysis of air passengers' choice behavior, that is, the flight opening cabin will change over time. When making a choice, passengers either choose a product from the choice set to buy or not. As the choice set is not fixed, the analysis of passenger choice behavior must take no-purchase data (unrecorded) into account. Secondly, for each sales transaction data, the corresponding passenger type information cannot be obtained. Without loss of generality, there will be different types of passengers in the market, and different types of passengers have different preferences for products, so it is necessary to analyze the types of passengers. Finally, in some aviation markets, there will be passengers with specific purchasing preferences. Some passengers will only choose high-priced tickets for certain reasons, while others are completely price-sensitive and will only choose low-priced tickets. Our study proposes the latent class multinomial logit model (LC-MNL). In summary, our contributions are as follows.

- First, we establish an LC-MNL model based on specific purchasing preferences to analyze the choice behavior of air passengers. Dividing the passengers into three groups: low-price oriented, high-price oriented and no specific price preference, and describe their respective expressions of the choice probability of choosing each product from choice set. Simultaneously, considering the data of no-purchase, it can be divided into two situations: arrival with no-purchase or no-arrival, and finally establish a passenger choice model: LC-MNL.
- A developed EM algorithm is tailored to solve the LC-MNL model. The developed EM algorithm devises an indicator function containing the type of passengers and first choice information in period t, and makes the iteration of the EM algorithm more efficient.
- Finally, we evaluate the efficiency of the proposed model and algorithm on the actual aviation transaction sales data set including flights departing from Chendu and arriving in Beijing. The experimental results show that our model and algorithm perform well on the actual aviation sales transaction data set.

The following paper is arranged as below. We view the related work in Sect. 2. Section 3 describes the process of setting a model through LC-MNL. The developed EM estimation algorithm is described in Sect. 4. In Sect. 5, we analyzed the experimental results on the actual data set. Finally, our work is concluded in Sect. 6.

2   Related Work

The passenger choice behavior attracts much attention due to its important role in revenue management (RM). Especially, discrete choice model has played an important role in RM in recent years due to its ability to account for passenger preferences.

The passenger discrete choice model (DCM) is widely used in industrial applications due to its simplicity and general good performance [1–4]. The DCM is an important area of research in diverse fields such as economics [9], marketing [10], and artificial intelligence [11]. The DCM framework was originally proposed by Nobel Prize winner Daniel McFadden [5], and has been the basis for all the subsequent research in the field. In the seminal work, McFadden introduces the Multinomial Logit model (MNL), which is widely used in industrial applications. In particular, the MNL is the most popular approach for air travel itinerary choice prediction [6–8, 20].

The latent class multinomial logit model (LC-MNL) was presented by Kamakura and Russell [12], also was described in Wedel and Kamakura [13]. They assumed that "latent classes" of multinomial logits represent segments of consumers who are relatively homogeneous in their preferences and response to marketing mix variables. Lee H et al. [14] proposes a latent class multinomial logit model to discover heterogeneous consumer groups. The expectation-maximization (EM) method is developed to estimate the parameters of the choice model. Pancras J et al. [15] contrasts the generalized multinomial logit model and the widely used latent class logit model which approaches for studying the heterogeneity in consumer purchases.

Some literature uses machine learning methods to study choice model. Mottini A et al. [16] presents a new choice model based on Pointer Networks. Given an input sequence, this type of deep neural architecture combines Recurrent Neural Networks with the Attention Mechanism learns the conditional probability of an output. Osogami T et al. [17] describes a model based on Restricted Boltzmann Machines. Hruschka H et al. [18] proposes to modify the MNL model by reformulating the utility equation with a feed-forward multilayer neural network.

3   Model Description

We assume that there are different types of passengers in the market. When passengers make ticket booking choices, different types of passengers will have different preferences. In this paper, we study passenger groups with heterogeneous preferences. Passengers can be divided into three groups: low-price oriented, high-price oriented and no specific price preference.

We study sales transaction data in T periods. Time periods are indexed by $t = 1, \cdots, T$. Specifically, define F as a set of products which consists of all cabin classes under all flights departing on the same day (in other words, a cabin class of a flight that departing on a certain day is a product), and the set of available products on period t is denoted by C_t, satisfying $C_t \subseteq F$. We defined F_1 as the set of products of passenger which is low-price oriented, while F_2 is high-price oriented. $F_1 \subseteq F$ consists of the lowest price cabin class of all flights in F, $F_2 \subseteq F$ consists of the highest price cabin class of all flights in F. We assume a discrete time and homogeneous Bernoulli arrival process for consumer arrival.

This means that in any period a passenger arrives with probability $0 \leqslant \lambda \leqslant 1$ (λ is called the arrival rate).

A passenger arriving in period t chooses either a product from available set C_t or no-purchase. We denote the no-purchase option by '0', which is normalized to have utility of zero. It is assumed that passengers are segmented into M distinct types based on their preference for the product. In this study, M can be arbitrarily given a value not less than 3. Passenger types are indexed by $\sigma = \{1, \cdots, M\}$, which $m = 1$ represents the passenger type is low-price oriented, $m = 2$ represents the passenger type is high-price oriented, $m = (3, \cdots, M)$ represents different passenger types of no specific price preference. The share of the population in type m is w_m. In this paper, We employ the latent class multinomial logit model (LC-MNL) [19].

The passenger choice model can be established on the basis that passengers are utility maximizers. Let U_{mf} be the utility of passengers with type m to alternative $f \in F$. Without loss of generality, we can decompose utility into two parts, u_{mf} represents the expected utility to alternative f and ϵ_{mf} represents random variable which are independent and follow a Gumbel distribution.

$$U_{mf} = u_{mf} + \epsilon_{mf} \tag{1}$$

The utility of no-purchase is:

$$U_{m0} = u_{m0} + \epsilon_{m0} \tag{2}$$

Without loss of generality, $u_{m0} = 0$. The preference value of passengers with type m for alternative f is denoted as $v_{mf} = e^{u_{mf}}$. Hence, $v_{m0} = 1$. The expression of v_{mf} is a standard result in the discrete choice theory [19]. The vector of preference weight is indexed by $\boldsymbol{v}_m = (v_{mf}, \forall f \in F)$.

3.1   No Specific Price Preference

In period t, for passengers with no specific price preference, he will purchase from the entire choice set. Hence, the probability of passengers with type $m(m = 3, \cdots, M)$ who choose product $f \in C_t \cup 0$ is:

$$P_f(C_t, \boldsymbol{v}_m) = \frac{v_{mf}}{\sum_{l \in C_t \cup \{0\}} v_{ml}} \tag{3}$$

3.2   Low-Price Oriented

To consider the purchasing preference of certain passengers in the market and the strong substitutability of domestic air travel comprehensively, we assume that passengers who are low-price oriented ($m = 1$) only choose either a product from ($C_t \cap F_1$) or no-purchase. In period t, low-price oriented passengers will only purchase products at the intersection of the current choice set C_t and low-price set F_1, and the rest will not purchase. Hence, the probability of passengers with type $m(m = 1)$ who purchase product $f \in C_t \cup \{0\}$ is:

$$P_f(C_t, \boldsymbol{v}_1) = \begin{cases} \frac{v_{1f}}{\sum_{l \in C_t \cup \{0\}} v_{1l}}, & f \in ((C_t \cap F_1) \cup \{0\}) \\ 0, & others \end{cases} \tag{4}$$

3.3 High-Price Oriented

Same as above, we assume that passengers who are high-price oriented $(m = 2)$ only choose either a product from $(C_t \cap F_2)$ or no-purchase. In period t, high-price oriented passengers will only purchase products at the intersection of the current choice set C_t and high-price set F_2, and the rest will not purchase. Hence, the probability of passengers with type $m(m = 2)$ who purchase product $f \in C_t \cup \{0\}$ is:

$$P_f(C_t, \boldsymbol{v}_2) = \begin{cases} \frac{v_{2f}}{\sum_{l \in C_t \cup \{0\}} v_{2l}}, & f \in ((C_t \cap F_2) \cup \{0\}) \\ 0, & others \end{cases} \tag{5}$$

We denote the vector of preference value as $\boldsymbol{v} = (\boldsymbol{v}_1, \cdots, \boldsymbol{v}_m)$ and the vector of population shares as $\boldsymbol{w} = (w_1, \cdots, w_m)$. In period t, the probability of purchasing product $f \in C_t \cup \{0\}$ as follows.

$$P_f(C_t; M, \boldsymbol{w}, \boldsymbol{v}) = \sum_{m=0}^{M} w_m \cdot P_f(C_t; \boldsymbol{v}_m) \tag{6}$$

Define f_t as the sales transaction data in period t. If passengers purchase f in period t, $f_t = f$; if there is no arrival or no purchase, $f_t = 0$. In this paper, we assume that there is at least one time period t, $f_t = f$, $\forall f \in F \cup \{0\}$.

4   Estimation-Based Algorithm Design

In this paper, we can't obtain the specific types of passengers. Simultaneously, it is difficult to distinguish the period of no arrival and the period of an arrival but no purchase.

4.1 The Complete Data Log-Likelihood Function

To supplement incomplete data, we define a_t as passengers' arrival and no arrival in period $\bar{P}$. If $a_t = 1$, $t \in \bar{P}_A$; if $a_t = 0$, $t \in \bar{P}_{\bar{A}}$.

We devise an indicator function $I(\cdot)$ [14] containing the type of passengers and the first choice information in period t. First choice is the most preferred choice in the set of all products (regardless of alternative options when it is out of stock). $\sigma_t = m$ represents that the passenger type in period t is m and

$\rho_t = f$ represents that first choice of the passenger in period t is product f. The complete data log-likelihood function is as follows.

$$\mathcal{L}_C(M, \boldsymbol{w}, \boldsymbol{v}.\lambda)$$

$$= \sum_{m=0}^{M} \sum_{t \in P} \sum_{f \in F \cup \{0\}} I(\sigma_t = m, \rho_t = f) \cdot (log\lambda + logw_m + logP_f(F; \boldsymbol{v}_m))$$

$$+ \sum_{m=0}^{M} \sum_{t \in \bar{P}} \sum_{f \in F \cup \{0\}} I(a_t = 1, \sigma_t = m, \rho_t = f) \cdot (log\lambda + logw_m + logP_f(F; \boldsymbol{v}_m))$$

$$+ \sum_{t \in \bar{P}} I(a_t = 0) \cdot log(1 - \lambda)$$

$$\tag{7}$$

4.2   Developed EM Algorithm

In this paper, expectation maximization (EM) algorithm operates on the complete data log-likelihood function (7), which is constructed assuming that all the arrivals, purchases and no purchases can be observed. The developed EM algorithm introduces an operator indicating the passenger type and the first choice in the framework of the traditional EM algorithm, making the iteration of the algorithm more effective. Note that the value of M is given in EM algorithm.

In the calculation process, the preference values of high-price oriented and low-price oriented passengers are processed as follows.

For low-price oriented passengers ($m = 1$), we assume that passengers with this type only choose the lowest-priced product for the flight (i.e. $f \in F_1$), for $f \notin F_1$, v_{1f} cannot be calculated directly. We use the passengers' price sensitivity to each flight to calculate $v_{1f}(f \notin F_1)$. The calculation equation is as follows. In this equation, f_1 represents the lowest price product belonging to the same flight as product f; a_{1f_1} is the sensitivity coefficient of low-price oriented passengers for the flight discount. b_1 is a hyperparameter.

$$v_{1f} = a_{1f_1}(-log(discount(j)) + b_1) \tag{8}$$

Similarly, for high-price oriented passengers ($m = 2$), we assume that passengers with this type only choose the highest-priced product for the flight (i.e. $f \in F_2$), for $f \notin F_2$, v_{2f} cannot be calculated directly. We use the passengers' price sensitivity to each flight to calculate $v_{2f}(f \notin F_2)$. The calculation equation is as follows. In this equation, f_2 represents the highest price product belonging to the same flight as product f; a_{2f_2} is the sensitivity coefficient of high-price oriented passengers for the flight discount. b_2 is a hyperparameter.

$$v_{2f} = a_{2f_2}(log(discount(j) + 1) + b_2) \tag{9}$$

According to the Eqs. (8, 9), the preference of low-price oriented passengers will decrease as the discount increases (see Fig. 1(a)), while the preference of high-price oriented passengers will increase as the discount increases (see Fig. 1(b)).

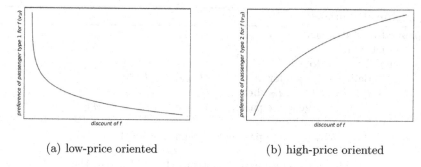

(a) low-price oriented   (b) high-price oriented

Fig. 1. The value of v with varying discount

As in the research of passenger choice preference, we pay more attention to the relative value of passengers' preference for each product rather than the absolute value, so this setting is reasonable. In this study, we design the following algorithm based on the EM algorithm framework.

Algorithm 1. Developed EM algorithm

1: **procedure** EM(M, w, v, λ)
2:     **Step 0.** Set the termination condition of EM algorithms ϵ and MAX_I and hyperparameter b_1, b_2.
3:     **Step 1.** (E-Step) Set iter=1. Update E_{mft} and $\hat{a}_t$ according to equation(15,17,11)(see appendix)
4:     Calculate $N_{mf} = \sum_{t \in P \cup \overline{P}} X_{mft}$ and $A = \sum_{t \in \overline{P}} a_t$.
5:     **Step 2.** (M-Step) $w^{new}, v^{new}, \lambda^{new} = M\_STEP(M, N, A)$.
6:     **Step 3.** (EM Stop Criterion)
7:     $Max\_diff = max\{max|w^{new} - w|, max|v^{new} - v|, \lambda^{new} - \lambda\}$.
8:     **if** $Max\_diff \leq \epsilon$ or $iter \geq MAX_I$ **then**
9:         Set $\mathcal{L}_C = \mathcal{L}_C(M, w, v, \lambda)$ according to equation(7).
10:        Terminate the program.
11:    **else**
12:        Set $w = w^{new}, v = v^{new}, \lambda = \lambda^{new}$. return Step1.(E-Step) and $iter = iter+1$
13: **return** $(w, v, \lambda.\mathcal{L}_C)$

1: **procedure** M_STEP(M, N, A)
2:     $w_m = \dfrac{\sum_{f \in F \cup \{0\}} N_{mf}}{\sum_{n=1}^{M} \sum_{f \in F \cup \{0\}} N_{nf}}$,
3:     $v_{mf} = \dfrac{N_{mf}}{N_{m0}}$
4:     $v_{1f} = -a_{0f_0}(log(discount(j)) + b_1)$, for $f \in \{F \backslash F_1\}$
5:     $v_{2f} = a_{1f_1}(log(discount(j) + 1) + b_2)$, for $f \in \{F \backslash F_2\}$
6:     $\lambda = \dfrac{|P|+A}{T}$
7: **return** (w, v, λ)

Algorithm 2. Initialize

1: **procedure** INITIALIZE(M, α)
2:     Initialize M.
3:     **for** $t = 1, ..., T$ **do**
4:         Initialize $X_{mft} = 0$ for $m = 1, ..., M$ and $f \in F \cup \{0\}$
5:         **if** $f_t \in F_1$ and $f_t \in F_2$ **then**
6:             Randomly generate passenger type n from $(1, \cdots, M)$;
7:         **else if** $f_t \in F_1$ **then**
8:             Randomly generate passenger type n from $(1, 3, \cdots, M)$;
9:         **else if** $f_t \in F_2$ **then**
10:             Randomly generate passenger type n from $(2, 3, \cdots, M)$;
11:         **else**
12:             Randomly generate passenger type n from $(3, \cdots, M)$.
13:         **if** $f_t = 0$ **then** $u \sim unif(0, 1)$. $X_{n0t} = 1$ if $u < \alpha$
14:         **else** $X_{njt} = 1$ if $j = j_t$
15:     Set $N_{mf} = \sum_{t \in P \cup \overline{P}} X_{mft}, A = |\overline{P}|, w, v, \lambda =M_STEP(M, N, A)$
16: **return** (w, v, λ)

Algorithm 3. Main Program

1: **procedure** MAIN
2:     **Step 0.** Set $\alpha \in [0, 1]$.
3:     **Step 1.**(Initial parameter setting) (w, v, λ)=INITIALIZE(M, α).
4:     **Step 2.**(Developed EM Algorithm) $w, v, \lambda, \mathscr{L}_C$ =EM(M, w, v, λ).

---

5   Real World Case Study

5.1   Dataset

In this section, some actual sales transaction data sets of air passenger are used to illustrate the applicability of the proposed model and algorithm. The dataset collected sales transaction data which the origin is Chengdu Shuangliu International Airport and the destination is Beijing Capital International Airport, and it contains information in terms of trip origin, trip destination, flight cabin class, data collection time, departure date, departure time and sales of each flight cabin class.

5.2   Data Processing

In this paper, we can obtain sales transaction data five days before the flight's departure for each flight. Without loss of generality, we assume that for a certain passenger, the departure date of the chosen product is fixed. Sales transaction data of flights departing on September 17,2020, September 18,2020 and October 15,2020 are used for analysis respectively.

Data Processing on sales transaction is as follows.

- Divide the time period according to data collection time, according to actual data, product sales transactions are mainly concentrated from 8 a.m. to 11 p.m. (see Fig. 2), so we use the data which is collected from 8 a.m. to 11 p.m. for analysis.
- Expand the no-purchase data of the remaining time periods based on the largest sales (we cannot obtain data about no-purchase, so expand no-purchase option by this processing).
- Determine the choice set corresponding to each piece of sales transaction data, namely, the opening cabin of each flight at the current time.
- For low-price oriented and high-price oriented passengers, their total choice set is the set of the lowest-priced product and the set of the highest-priced product of each flight respectively.

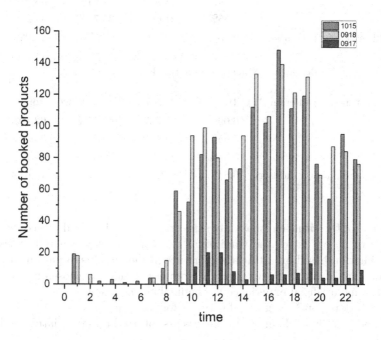

Fig. 2. Total number of booked products by collection time

5.3   Model Implementation and Experimental Results

We conduct experiments on three data sets after the above processing respectively. Using the above algorithm to solve the sales transaction data, we can obtain the preference value of each type of passengers for each product (i.e. v_{mf}). As we only have sales transaction data and choice set, in order to illustrate the performance of the model, we have to evaluate the model by using top-N accuracy.

Table 1. Top-N accuracy for the data on Sept.17.2020

Proposed model (different M)	Top-1 acc.	Top-5 acc.	Top-10 acc.	Top-15 acc.
M = 3	10.05%	30.65%	54.77%	67.84%
M = 4	8.04%	31.16%	60.30%	74.87%
M = 5	7.54%	27.64%	48.74%	66.83%
Traditional aviation industry	2.17%	10.87%	21.74%	32.61%

Table 2. Top-N accuracy for the data on Sept.18.2020

Proposed model (different M)	Top-1 acc.	Top-5 acc.	Top-10 acc.	Top-15 acc.
M = 3	6.27%	19.99%	35.56%	47.42%
M = 4	5.79%	18.95%	34.94%	48.31%
M = 5	5.65%	21.64%	35.77%	47.28%
Traditional aviation industry	0.93%	4.67%	9.35%	14.02%

Table 3. Top-N accuracy for the data on Oct.15.2020

Proposed model (different M)	Top-1 acc.	Top-5 acc.	Top-10 acc.	Top-15 acc.
M = 3	6.69%	26.36%	43.47%	62.10%
M = 4	5.65%	24.28%	43.23%	58.36%
M = 5	5.73%	22.45%	38.30%	56.21%
Traditional aviation industry	0.84%	4.20%	8.40%	12.61%

In view of the situation that only sales transaction data and choice set information can be obtained, the traditional aviation industry generally uses random estimation based on artificial experience in the analysis of passenger choice behavior, that is, for each passenger, a product is randomly chosen in the choice set with some artificial experience as the final choice result.

We calculate top-N accuracy for three, four and five passenger types respectively, and compare them with the results of traditional aviation industry method.

The data set of flights departing on September 17,2020 contains a total of 46 products, the top-N accuracy of the different amounts of passenger types in our study and the top-N accuracy of the traditional aviation industry method are shown in Table 1. In our research, the highest top-1 accuracy is 10.05% ($M = 3$), the lowest top-1 accuracy is 7.54% ($M = 5$), while the top-1 accuracy of the traditional aviation industry method is only 2.17%. The data set of flights departing on September 18, 2020 contains a total of 107 products, the top-N accuracy of the different number of passenger types in our study and the top-N accuracy of the traditional aviation industry method are shown in Table 2. In our research, the highest top-5 accuracy is 21.64% ($M = 5$), the lowest top-1 accuracy is 18.95% ($M = 4$). The data set of flights departing on October

15, 2020 contains a total of 119 products, the top-N accuracy of the different number of passenger types in our study and the top-N accuracy of the traditional aviation industry method are shown in Table 3. In our research, the highest top-15 accuracy is 62.10% ($M = 3$), the lowest top-1 accuracy is 56.21% ($M = 5$), while the top-15 accuracy of the traditional aviation industry method is only 12.61%.

According to the comparison of the above results, the top-N accuracy of our study far exceeds the random estimation in the traditional aviation industry, indicating that our study performs well in actual data sets and practical applications. Furthermore, we can also find that as the number of products in the data set increases, the advantages of our study are more significant.

6   Conclusions

In this paper, we focus on the booking choice behavior of air passengers with different purchasing preferences. Considering the different types of passengers in the market and the different purchasing preferences of each type, we consider passenger classifications with heterogeneous preferences and the market shares of different types, and establish a latent class multinomial logit model (LC-MNL). Then, in order to fit the air passengers' booking choice habits in the current market better, we define the choice behavior of air passengers with specific purchasing preferences, and divide the customers into three groups: low-price oriented, high-price oriented, and no specific price preference for analysis. Our method only utilizes the available sales transaction data and the corresponding choice set. Finally, we propose a developed EM algorithm to solve the LC-MNL. The algorithm devises an indicator function containing the type of passengers and first choice information in period t to make the EM iteration more effective. Simultaneously, formulas are introduced to determine the low-price oriented and high-price oriented passengers' preference for products in the calculation process. Experimental results show that our model and algorithm perform well on the actual aviation sales transaction data sets in China, and they can accurately describe the booking choice behavior of air passengers.

References

1. Papola, A.: Some developments on the cross-nested logit model. Transp. Res. Part B **38**(9), 833–851 (2004)
2. Wen, C.H., Wang, W.C., Fu, C.: Latent class nested logit model for analyzing high-speed rail access mode choice. Transp. Res. Part E Logist. Transp. Rev. **48**(2), 545–554 (2012)
3. Li, B.: The multinomial logit model revisited: a semi-parametric approach in discrete choice analysis. Transp. Res. Part B **45**(3), 461–473 (2011)
4. Vulcano, G., Van Ryzin, G., Chaar, W.: Om practice-choice-based revenue management: an empirical study of estimation and optimization. Manuf. Serv. Oper. Manag. **12**(3), 371–392 (2010)

5. Mcfadden, D.L.: Condition logit analysis of qualitative choice behavior. Front. Econometrics 105–142 (1974)
6. Busquets, J.G., Evans, A.D., Alonso, E.: Application of data mining to forecast air traffic: a 3-stage model using discrete choice modeling. In: AIAA Aviation Technology, Integration, and Operations Conference (2015)
7. Coldren, G.M., Koppelman, F.S., Kasturirangan, K., Mukherjee, A.: Modeling aggregate air-travel itinerary shares: logit model development at a major US airline. J. Air Transp. Manag. **9**(6), 361–369 (2003)
8. Warburg, V., Bhat, C., Adler, T.: Modeling demographic and unobserved heterogeneity in air passengers' sensitivity to service attributes in itinerary choice. Transp. Res. Rec. J. Transp. Res. Board **676**, 7–16 (2006)
9. Azadeh, S.S., Hosseinalifam, M., Savard, G.: The impact of customer behavior models on revenue management systems. CMS **12**(1), 99–109 (2014). https://doi.org/10.1007/s10287-014-0204-z
10. Chandukala, S.R., Kim, J., Otter, T., Rossi, P.E.: Choice models in marketing: economic assumptions, challenges and trends. Found. Trends Mark. **2**(2), 97–184 (2013)
11. Zhen, Y., Rai, P., Zha, H., Carin, L.: Cross-modal similarity learning via pairs, preferences, and active supervision. In: Twenty-Ninth AAAI Conference on Artificial Intelligence (2015)
12. Kamakura, W.A., Russell, G.J.: A probabilistic choice model for market segmentation and elasticity structure. J. Mark. Res. **26**(4), 379–390 (1989)
13. Wedel, M., Kamakura, W.: Market Segmentation: Conceptual and Methodological Foundations, 2nd edn. Kluwer Academic Publishers, Boston (2000)
14. Lee, H., Eun, Y.: Discovering heterogeneous consumer groups from sales transaction data. Eur. J. Oper. Res. **280**(1), 338–350 (2020)
15. Pancras, J., Dey, D.K.: A comparison of generalized multinomial logit and latent class approaches to studying consumer heterogeneity with some extensions of the generalized multinomial logit model. Appl. Stoch. Model. Bus. Ind. **27**(6), 567–578 (2011)
16. Mottini, A., Acuna-Agost, R.: Deep choice model using pointer networks for airline itinerary prediction. In: Proceedings of the 23rd ACM SIGKDD International Conference on Knowledge Discovery and Data Mining, pp. 1575–1583 (2017)
17. Osogami, T., Otsuka, M.: Restricted Boltzmann machines modeling human choice. In: Advances in Neural Information Processing Systems, vol. 27, pp. 73–81 (2014)
18. Hruschka, H., Fettes, W., Probst, M.: Analyzing purchase data by a neural net extension of the multinomial logit model. In: Dorffner, G., Bischof, H., Hornik, K. (eds.) ICANN 2001. LNCS, vol. 2130, pp. 790–795. Springer, Heidelberg (2001). https://doi.org/10.1007/3-540-44668-0_110
19. Train, K.E.: Discrete Choice Methods with Simulation. Cambridge University Press, Cambridge (2009)
20. Rusmevichientong, P., Shen, Z.J.M., Shmoys, D.B.: Dynamic assortment optimization with a multinomial logit choice model and capacity constraint. Oper. Res. **58**(6), 1666–1680 (2010)

Blockchain, Logic, Complexity and Reliability

A Multi-window Bitcoin Price Prediction Framework on Blockchain Transaction Graph

Xiao Li[1]([✉])[iD] and Linda Du[2]

[1] The University of Texas at Dallas, Richardson, TX 75080, USA
xiao.li@utdallas.edu
[2] The University of Texas at Austin, Austin, TX 78712, USA
yingfan.du@mccombs.utexas.edu

Abstract. Bitcoin, as one of the most popular cryptocurrency, has been attracting increasing attention from investors. Consequently, bitcoin price prediction is a rising academic topic. Existing bitcoin prediction works are mostly based on trivial feature engineering, that is, manually designed features or factors from multiple areas. Feature engineering not only requires tremendous human effort, but the effectiveness of the intuitively designed features can not be guaranteed. In this paper, we aim to mine the abundant patterns encoded in Bitcoin transactions, and propose k-order transaction graphs to reveal patterns under different scopes. We propose features based on a transaction graph to automatically encode the patterns. The Multi-Window Prediction Framework is proposed to train the model and make price predictions, which can take advantage of patterns from different historical periods. We further demonstrate that our proposed prediction method outperforms the state-of-art methods in the literature.

Keywords: Bitcoin · Blockchain · Transaction · Machine learning

1 Introduction

Bitcoin blockchain [24][1], the first application of blockchain, has been attracting increasing attention from various areas. *Bitcoin* is the cryptocurrency traded in the Bitcoin blockchain, which is a reward to the miners for successfully appending a block. Bitcoin can be traded with regular currency in financial markets like many other financial products, e.g. stocks, gold and crude oil [28]. Different from other products, bitcoin has highly volatile prices [1,5]. This provides investors with a great opportunity to earn a fortune from the striking difference in prices. Thus, bitcoin is becoming a popular financial asset, and attracts huge amounts of investment [30].

---

[1] In this paper, the terms "Bitcoin blockchain" or "Bitcoin" refer to the whole Bitcoin blockchain system and "bitcoin" refers to the cryptocurrency.

This work is partially supported by NSF 1907472.

Bitcoin price forecasting models are eagerly desired to provide the suggestions on whether the bitcoin price will rise or fall [10,29] to help investors decide whether and when they should buy or sell bitcoins. However, bitcoin price forecasting models usually require well-designed features to reveal the reason of bitcoin price change, which is a challenging task. The basic features of blockchain are the indexes reflecting the transaction information of Bitcoin blockchain, such as average degree of addresses, number of new addresses and total coin amount transferred in transactions [2]. Maesa et al. try to analyze the latent features of Bitcoin blockchain from the perspective of users transferring graph [20]. Mallqui et al. [21] include international economic indicators that were used to reflect the features of the global financial market, such as S&P500 future, NASDAQ future, and DAX index, which are features from a financial perspective. CerdaR et al. [8] and Yao et al. [29] introduce public opinion features into bitcoin price prediction through mining the sentiment from social media like Twitter and news articles.

Existing work has created features covering many aspects, including blockchain network, financial market information, and even public opinions. However it is still unclear what features or factors are useful, and how these features impact the price of bitcoin. Manually discovering or creating the features not only relies on heuristics but also consumes huge labour resource. In this paper, we try to develop a bitcoin prediction model that can directly learn features from the Bitcoin blockchain transactions without directly incorporating tedious information outside the blockchain, e.g. financial market information, and public sentiment. Instead, if the external factors beyond the Bitcoin blockchain, such as public sentiment or news, contribute to the bitcoin price change, they will eventually be reflected by the changes in the transactions and structure of the Bitcoin blockchain. In other words, if the external factors influence the action of users, the different actions taken by users will be reflected by the changes in the transactions in the Bitcoin blockchain. In this paper, we argue that the structure of Bitcoin blockchain encodes abundant transaction pattern information that can interpret the factors behind the bitcoin price change.

To capture these transaction patterns, we propose a blockchain transaction graph.

The blockchain transaction graph encodes the patterns of transactions which reflects market trend and status. As mentioned in [4], if the input addresses of a transaction is more than the output addresses, then the transaction is gathering bitcoins, indicating some users are buying bitcoins. On the other hand, if the input addresses of a transaction is less than the output addresses, then the transaction is splitting the bitcoins, indicating some users are selling bitcoins. Therefore by discovering these transaction patterns with a Bitcoin transaction graph and proposed prediction framework, we can leverage valuable information that can hardly be managed by manual feature engineering.

The main contributions of the paper can be summarized as follows:

- We propose a *k-order Transaction Subgraph* based on a transaction graph, to represent the transaction feature of blockchain.
- We proposed a transaction graph based feature to encode the implicit patterns behind the transactions, which is further fed to a novel machine learning

based **Multi-Window Prediction Framework** that can effectively learn the features of different historical periods.

- We evaluate the proposed method empirically using historical bitcoin prices and the results demonstrate superiority over recent state-of-the art methods.

The remainder of this paper is organized as follows: In Sect. 2, we review related recent literature. Section 3 proposes a transaction graph and describes how the subgraph feature is extracted. Next, in Sect. 4, we propose the Multi-Window Prediction Framework. In Sect. 5 we evaluate the proposed feature and the prediction framework. Finally, in Sect. 6, we conclude.

2   Related Work

The key issue of bitcoin price prediction is to discover and analyze determinants of bitcoin price. Various determinants have been studied including Google Trends [16,22], Wikipedia [16], Bitcoin tweets [6,22], social media or public opinions [7,8,29], and so on. Some papers consider both traditional features in the market as well as economical features of a digital currency [3,11]. Pieters and Vivanco [26] study the 11 bitcoin markets and present that standard financial regulations can have a non-negligible impact on the market for Bitcoin. Both Georgoula et al. [13] and Kristoufek [17] study the difference between long-term and short-term impact of the determinants on bitcoin price. Kristoufek [17] stresses that both time and frequency are crucial factors for bitcoin price dynamics since the price of bitcoin evolves overtime.

The structural information of the Bitcoin blockchain has also been used to mine determinants of the price of bitcoin. Akcora et al. [4] propose a Bitcoin graph model, upon which chainlets is proposed to represent graph structures in the Bitcoin.

In their further work [2], they propose occurrence matrix and amount matrix to encode the topological features of chainlets. In this paper, we also adopt the concept of occurrence matrix to encode the topological features. However, we design a different graph representation model to reveal the topological features of the Bitcoin blockchain.

There are also several theoretical [18,19,27] and empirical studies [5,15,23] that have looked at Bitcoin transactions focusing on the volume-return causality in the Bitcoin market. These studies focus on trading volumes or number of unique Bitcoin transactions and employ regression techniques. In this paper, we take our analyses further and extract patterns from the Bitcoin transactions using graph models.

Various machine learning methods can be adopted to learn the patterns from the features and forecast the price of bitcoin [10,31]. Felizardo et al. [9,12] compare several popular machine learning methods adopted in the bitcoin price prediction task. Methods include using a Hidden Markov Models to tackle the volatility of cryptocurrencies and predicting future movements with Long Short Term Memory networks (LSTM) [14] and using hybrid methods between AutoRegressive Integrated Moving Average (ARIMA) and machine learning [25].

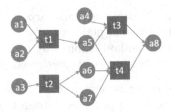

Fig. 1. A simple transaction graph

3   Transaction Graph and Subgraph Occurrence Pattern

In order to mine the blockchain transaction features, we define *transaction graph* to extract the blockchain transaction information.

Definition 1. (Transaction Graph): *A transaction graph is a directed graph $G = (A, T, E)$, where A is the set of addresses in the blockchain, T is the set of transactions in the blockchain, and E is the set of direct links from a_i to t_k, indicating a_i is one of the inputs of t_k, or from t_k to a_j, indicating a_j is one of the outputs of t_k, where $a_i, a_j \in A$ and $t_k \in T$.*

Figure 1 presents an example of a transaction graph, which contains 8 addresses and 4 transactions.

3.1   k-Order Transaction Subgraph

To specify characteristics of each transaction in the transaction graph, we define the k-order transaction subgraph of each transaction. The k-order transaction subgraph of a transaction t_i is a graph $G_{t_i}^k$ that contains only t_i and the transactions that spend the output of t_i in the next $k - 1$ steps, along with the corresponding addresses that connect to these transactions. The formal definition is given as Definition 2.

Definition 2. (K-order transaction subgraph): *The K-order transaction subgraph of a transaction t_i is a graph $G_{t_i}^k = (A^k, T^k, E^k)$, where $T^k = \{t_j | \exists a_n \in A^{k-1}, (a_n, t_j) \in E$ and $\exists (t_l, a_n) \in E^{k-1}$ for $t_l \in T^{k-1}\}$, $A^k = \{a_n | a_n \in A^{k-1}$ or $(t_j, a_n) \in E$ where $t_j \in T^k\}$. Specially, if $k = 1$, $G_{t_i}^1 = (A^1, T^1, E^1)$, where $A^1 = \{a_n | (a_n, t_i) \in E$ or $(t_i, a_n) \in E\}$, $T^1 = \{t_i\}$ and $E^1 = \{(a_n, t_i)$ or $(t_i, a_n) | a_n \in A^1\}$.*

If $k = 1$, then the k-order transaction subgraph of t_i contains only t_i along with its input addresses and output addresses. When k increases, the k order transaction subgraph will trace further along the bitcoin flow output by transaction t_i. Figure 2(a) and 2(b) shows the 1-order and 2-order transaction subgraph of transaction t_1 in Fig. 1, respectively.

The k-order transaction subgraphs have different patterns. Here we consider different patterns as different numbers of inputs and outputs addresses of the k-order transaction subgraphs.

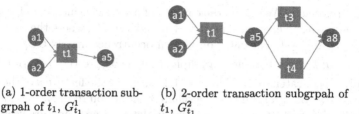

(a) 1-order transaction sub-grpah of t_1, $G^1_{t_1}$

(b) 2-order transaction subgrpah of t_1, $G^2_{t_1}$

Fig. 2. The 1 order nd 2-order transaction subgraph of t_1 in Fig. 1

The input addresses of a k-order transaction subgraph $G^k_{t_i}$ are the addresses that input to the first transaction in $G^k_{t_i}$. The output addresses of $G^k_{t_i}$ are the addresses that accepts the outputs of the last transactions in $G^k_{t_i}$. The input and output addresses are formally defined in Definition 3.

Definition 3. (Input and Output addresses of K-order transaction subgraph): *The input and output addresses of K-order transaction subgraph $G^k_{t_i}$ is $\mathcal{I}_{G^k_{t_i}}$ and $\mathcal{O}_{G^k_{t_i}}$, respectively. $\mathcal{I}_{G^k_{t_i}} = \{a_n | \exists (a_n, t_j) \in E^k, t_j \in T^k \text{ and } \forall t_k \in T^k, (t_k, a_n) \notin E^k\}$. $\mathcal{O}_{G^k_{t_i}} = \{a_n | \exists (t_k, a_n) \in E^k, t_k \in T^k \text{ and } \forall t_j \in T^k, (a_n, t_j) \notin E^k\}$.*

In Fig. 2(a), the addresses a_1 and a_2 are the input addresses of $G^1_{t_1}$, and the address a_5 is the output address of $G^1_{t_1}$. For higher order transaction subgraphs, the input and output addresses may be more complicated. For example, in Fig. 2(b), the input addresses of $G^2_{t_1}$ are $\{a_1, a_2\} = \mathcal{I}_{G^2_{t_1}}$, and the output addresses are $\{a_8\} = \mathcal{O}_{G^2_{t_1}}$.

Based on the concept of $\mathcal{I}_{G^k_{t_i}}$ and $\mathcal{O}_{G^k_{t_i}}$, we now further define the ***pattern*** of a transaction subgraph. The pattern of a k-order transaction graph of transaction t_i is denoted as $G^k_{(m,n)} = \{G^k_{t_i} || \mathcal{I}_{G^k_{t_i}} | = m, |\mathcal{O}_{G^k_{t_i}}| = n\}$, where m and n are the number of input addresses and output addresses of $G^k_{t_i}$ respectively.

For a given transaction graph generated from a blockchain transaction record during a specific period T, we can obtain a k order transaction subgraph $G^k_{t_i}$ of each transaction $t_i \in T$. The obtained transaction subgraphs may belong to different patterns. For the example in Fig. 2, $G^2_{t_1}$ belongs to the pattern $G^2_{(2,1)}$, while $G^2_{t_2}$ belongs to the pattern $G^2_{(1,1)}$.

We believe these patterns contain valuable information revealing the characteristics of each corresponding blockchain transaction in a period. In addition, the patterns obtained under different order k can reveal different levels of latent information. The benefit of denoting the pattern based on the number of input addresses and out addresses is that the patterns can be easily encoded into matrices, and therefore can be adopted as the features of the current transaction graph.

By summarizing the patterns of all k-order transaction graph $G^k_{t_i}$ of every transaction t_i in a transaction graph G, two key characteristics can be obtained 1) what kinds of patterns occur in the transaction graph, and 2) how many times these patterns occur. We extend the concept of occurrence matrix in literature [2] to a k order pattern occurrence matrix, denoted as OC^k, where the entry of OC^k is $OC^k_{(m,n)} = |G^k_{(m,n)}|$. The entry of pattern occurrence matrix $OC^k_{(m,n)}$ denotes the number of k-order transaction graphs that belong to the pattern $G^k_{(m,n)}$.

Finally we concatenate OC^k for $k = 1, 2, 3, .., s$ as the feature v of the transaction graph G we obtain from the blockchain transaction record. Now the *Bitcoin Price Prediction* problem can be specified in detail: use the feature vector v that is calculated from the transaction graph based on Bitcoin historical data in time period $[t-i, t]$, to predict bitcoin price at some future time $t+h$, P_{t+h}. Formally, we define the *price prediction task* as Definition 4.

Definition 4. (Bitcoin Price Prediction): *Given time $t' = t + \Delta t$, where $\Delta t \geq 0$, and Bitcoin historical data in time period $[t - s, t]$, where $s \in N^+$. Let P_t denote the price of bitcoin at time t. the bitcoin price prediction problem is to predict the price at time t', e.g. $P_{t'}$.*

4   Multi-window Prediction Framework

Transactions in the blockchain are time sequential, meaning the blockchain may shows different patterns at different periods of time. How much the future price is influenced by historical patterns and how far back we should look to discover these patterns are empirical questions. To answer these questions more systematically, we propose the Multi-Window Prediction Framework. This framework uses the features from different lengths of historical data to construct different submodels and incorporates the results from every submodel to form a final result. By taking advantage of all the submodels, this framework can boost the accuracy of our predictions.

Figure 3 illustrates the *Multi − Window Prediction Framework*. M_1 to M_s are s submodels that are trained separately on different windows of time with length s. For example, M_1 is the model trained by the features extracted from the past 1 day, and M_2 is the model trained by the features extracted from the past 2 days. When making price forecasts for a specific day $t' = t + \Delta t$ ($\Delta t \geq 0$),

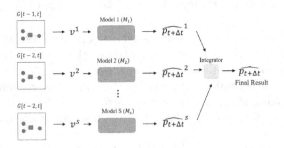

Fig. 3. Overview of the Multi-window Prediction Framework

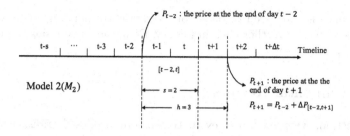

Fig. 4. Illustration of settings for submodel 2 (M_2) to predict P_{t+1}

each submodel will first output its individual prediction. The integrator will then combine the results into one final result.

The accuracy of the final result depends on the performance of each submodel. Next, we describe how each model is trained and makes future price predictions. In this paper, we predict the daily end price of bitcoin. The end price of day t' is denoted as $P_{t'}$. After extracting features from a historical period, say $[t-s,t]$, it is natural to directly predict $P_{t'}$. However, it is more reasonable to predict the price difference between $P_{t'}$ and P_{t-s}, denoted as $\Delta P_{[t-s,t']}$, and then derive the predicted P_t as $\hat{P}_t = P_{t-s} + \Delta P_{[t-s,t']}$. The reason is twofold: 1) we know the historical price P_{t-s}, and it should be considered to improve the prediction; 2) whatever features extracted from $[t-s,t]$ represents the characteristics only during $[t-s,t]$ in the bitcoin market, and these are the characteristics that bring changes to the price. Thus, it is more reasonable to use the features to interpret the price change rather than the exact price. Therefore, in this paper, we construct data sample pairs as $(\mathbf{x}, y)$, where $\mathbf{x}$ is the feature vector extracted from a historical period $[t-s,t]$, and $y = \Delta P_{[t-s,t']} = P_{t'} - P_{t-s}$. Each submodel will be retrained if it aims to predict a different future time. We denote the distance from the future time to be predicted as $h = t' - (t-s)$. Figure 4 illustrates an example of the parameters setting for submodels making predictions.

The integrator will combine the results from each submodel with different weights, which can be a simple linear function as follows:

$$\hat{P}_{t'} = r_1 * \hat{P}_{t'}^1 + r_2 * \hat{P}_{t'}^2 + ... + r_s * \hat{P}_{t'}^s \tag{1}$$

where $r_1 + r_2 + ... + r_s = 1$.

In this paper, we elaborately design the weights. Let $W_i = [r_1, r_2, r_3, ..., r_i]$. Specially, if the historical window size is 1, which indicates that we only employ one model to make the prediction, $W_1 = [r_1] = [1.0]$. As the historical window size increases, $i > 1$, W_i is defined as Eq. 2:

$$W_{i+1}[k] = W_i[k] \ (k = 1, ..., i-1)$$
$$W_{i+1}[i] = W_i[i] * \alpha \tag{2}$$
$$W_{i+1}[i+1] = W_i[i] * (1-\alpha)$$

where α controls the speed of decay of weights corresponding to results from submodels with data further back in history. Equation 2 maintains the property that $\sum_{r_j \in W_i} r_j = 1$ for $i > 0$.

5   Experimental Results

In this section, we present the evaluation of our proposed transaction graph based blockchain feature and Multi-Window Prediction Framework.

5.1   Data Preparation

To conduct the bitcoin price prediction task, we collect Bitcoin blockchain historical data and bitcoin price historical data. The Bitcoin blockchain data is downloaded from Google Bigquery public dataset crypto_Bitcoin[2] whose data is exported using Bitcoin etl tool[3]. The bitcoin price data is collected from Coindesk[4].

We select two historical periods for the experiments.

- **Interval 1:** From August 19th, 2013 to July 19th, 2016. The timestamps are divided daily. This period contains 1065 days, the first 80% days are used to train the model and the latter 20% is reserved for testing.
- **Interval 2:** From April 1st, 2013 to April 1st, 2017. The timestamps are divided daily. This period contains 1461 days, the first 70% days are used to train the model and the latter 30% is reserved for testing.

Interval 1 and *Interval 2* are identical to the datasets used in the literature [21], which will be used as a benchmark in the next sections. In this paper, we predict bitcoin daily closing price during the above periods.

For the evaluation metric, we adopt Mean Absolute Percentage Error (MAPE) to show the error between predicted prices and real prices. The MAPE is defined as $MAPE = \frac{1}{N} \sum_{i=1}^{N} \frac{|\hat{p_i} - p_i|}{p_i}$, where $\hat{p_i}$ is the predicted bitcoin price, while p_i is the real realized price.

5.2   Performance of Difference Submodels

Table 1 shows each submodel, M_1 to M_4, where each submodel adopts the same training strategy and machine learning prediction model. They only differ by the length of the historical window of time used when extracting the features. s is the length of the window of time, where $s = 1$ means the model extracts features from the past 1 day. h is the future time that the model aims to predict, where $h = 1$ means the model predicts the price the next day. Due space

---

[2] Dataset ID is bigquery-public-data: crypto_Bitcoin at https://cloud.google.com/bigquery.

[3] https://github.com/blockchain-etl/Bitcoin-etl.

[4] https://www.coindesk.com/.

Table 1. MAPE of Submodels (SVM Prediction) for Predicting Future Price

Submodels	Interval 1					Interval 2					Year 2017				
	$h=1$	$h=2$	$h=3$	$h=4$	$h=5$	$h=1$	$h=2$	$h=3$	$h=4$	$h=5$	$h=1$	$h=2$	$h=3$	$h=4$	$h=5$
M1 ($s=1$)	1.75%	2.59%	3.15%	3.77%	4.31%	1.74%	2.57%	3.21%	3.78%	4.29%	4.73%	7.09%	8.36%	10.30%	12.20%
M2 ($s=2$)	–	2.61%	3.16%	3.76%	4.29%	–	2.58%	3.20%	3.78%	4.29%	–	7.01%	8.36%	10.05%	11.90%
M3 ($s=3$)	–	–	3.17%	3.76%	4.29%	–	–	3.20%	3.76%	4.27%	–	–	8.24%	9.91%	11.70%
M4 ($s=4$)	–	–	–	3.75%	4.29%	–	–	–	3.78%	4.28%	–	–	–	9.85%	11.60%

constraints, we only show the results where each submodel adopts the Support Vector Machine (SVM) algorithm, which is the best in our record. We find that including more historical information in our models does not necessarily result in better performance in terms of MAPE. For example, M_2 at $h = 2$ obtains a worse prediction than M_1 at Interval 1 and Interval 2, despite the fact that M_2 considers one further day back than M_1. One can identify additional similar cases in Table 1. Therefore, we expect to achieve a higher MAPE by taking into consideration all the different submodels.

5.3   Performance of Combined Model

Figure 5 shows the effects of combining the submodels to produce the final prediction. M_1 means only submodel M_1 is adopted, $M_1 \sim M_2$ means the results from both submodels M_1 and M_2 were both used, $M_1 \sim M3$ means the results from submodels M_1, M_2 and M_3 were used, and so on. When $\alpha > 0.7$ in Interval 1, and $\alpha > 0.75$ in Interval 2, we can see the combined models outperform the single model (only M_1). When $\alpha = 0.85$ the Multi-Window Prediction Framework can produce the most accurate prediction with the lowest MAPE value.

Table 2 shows the specific MAPE values when $\alpha = 0.85$ and $h = 1$. We can observe that $M_1 \sim M_4$ produces the best results. Therefore we can conclude that 4-day historical information seems to be sufficient for predicting bitcoin price with our proposed Multi-Window Prediction Framework. The results reflect the high volatility of bitcoin price where the current price does not relate much to

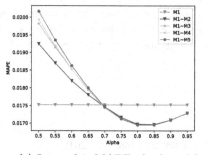
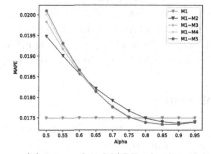

(a) Interval 1: MAPE of submodels   (b) Interval 2: MAPE of submodels

Fig. 5. MAPE of Multi-Window Prediction Framework when combining different submodels and alpha in Interval 1 and Interval 2 ($h = 1$), all using SVM

Table 2. MAPE of Multi-Window Prediction Framework when combining different submodels (SVM, $\alpha = 0.85$, $h = 1$)

Submodels incorporated	Interval 1	Interval 2
M1	1.75%	1.74%
M1+M2	1.70%	1.73%
M1+M2+M3	1.70%	**1.72%**
M1+M2+M3+M4	**1.69%**	1.72%
M1+M2+M3+M4+M5	1.70%	1.72%
Mallquietal. − SVM [21]	1.91%	1.81%

historical prices too far back and, instead, is more highly influenced by very recent characteristics in the Bitcoin blockchain.

5.4 Comparison with Benchmark

Mallqui et al. [21] study a similar bitcoin price prediction task. Mallqui et al. utilize several machine learning methods to forecast bitcoin price based on the proposed features including historical price, volume of trades and financial indicators. Since the SVM model performs the best in [21], we adopt the SVM prediction model for comparison, denoted as *Mallquietal. − SVM*. The result of *Mallquietal. − SVM* on both Interval 1 and Interval 2 are shown in Table 2. Our proposed combined model $M1 \sim M4$ outperforms *Mallquietal. − SVM*.

6 Conclusion

In this paper, we propose a transaction graph based machine learning method to forecast the price of bitcoin. The k-order transaction graphs of the transactions are proposed to reveal the transaction patterns in the Bitcoin blockchain. The occurrence matrix is then defined to encode the information patterns and we further represent them as features of the Bitcoin blockchain. We also propose the Multi-Window prediction framework to learn the transaction patterns from multiple blockchain historical periods. Results of comparative experiments show that the method we propose outperforms recent state-of-art methods, further demonstrating the effectiveness of our method.

References

1. Aalborg, H.A., Molnár, P., de Vries, J.E.: What can explain the price, volatility and trading volume of bitcoin? Financ. Res. Lett. **29**, 255–265 (2019)
2. Abay, N.C., et al.: ChainNet: learning on blockchain graphs with topological features. In: Wang, J., Shim, K., Wu, X. (eds.) 2019 IEEE International Conference on Data Mining, ICDM 2019, Beijing, China, 8–11 November 2019, pp. 946–951. IEEE (2019)

3. Aggarwal, A., Gupta, I., Garg, N., Goel, A.: Deep learning approach to determine the impact of socio economic factors on bitcoin price prediction. In: 2019 Twelfth International Conference on Contemporary Computing, IC3 2019, Noida, India, 8–10 August 2019, pp. 1–5. IEEE (2019)

4. Akcora, C.G., Dey, A.K., Gel, Y.R., Kantarcioglu, M.: Forecasting bitcoin price with graph chainlets. In: Phung, D., Tseng, V.S., Webb, G.I., Ho, B., Ganji, M., Rashidi, L. (eds.) PAKDD 2018. LNCS (LNAI), vol. 10939, pp. 765–776. Springer, Cham (2018). https://doi.org/10.1007/978-3-319-93040-4_60

5. Balcilar, M., Bouri, E., Gupta, R., Roubaud, D.: Can volume predict bitcoin returns and volatility? A quantiles-based approach. Econ. Model. **64**, 74–81 (2017)

6. Balfagih, A.M., Keselj, V.: Evaluating sentiment classifiers for bitcoin tweets in price prediction task. In: 2019 IEEE International Conference on Big Data (Big Data), Los Angeles, CA, USA, 9–12 December 2019, pp. 5499–5506. IEEE (2019)

7. Burnie, A., Yilmaz, E.: An analysis of the change in discussions on social media with bitcoin price. In: Piwowarski, B., Chevalier, M., Gaussier, É., Maarek, Y., Nie, J., Scholer, F. (eds.) Proceedings of the 42nd International ACM SIGIR Conference on Research and Development in Information Retrieval, SIGIR 2019, Paris, France, 21–25 July 2019, pp. 889–892. ACM (2019)

8. Cerda, G.C., Reutter, J.L.: Bitcoin price prediction through opinion mining. In: Amer-Yahia, S., et al. (eds.) Companion of The 2019 World Wide Web Conference, WWW 2019, San Francisco, CA, USA, 13–17 May 2019, pp. 755–762. ACM (2019)

9. Chen, C., Chang, J., Lin, F., Hung, J., Lin, C., Wang, Y.: Comparison of forcasting ability between backpropagation network and ARIMA in the prediction of bitcoin price. In: 2019 International Symposium on Intelligent Signal Processing and Communication Systems, ISPACS 2019, Taipei, Taiwan, 3–6 December 2019, pp. 1–2. IEEE (2019)

10. Chen, Z., Li, C., Sun, W.: Bitcoin price prediction using machine learning: an approach to sample dimension engineering. J. Comput. Appl. Math. **365** (2020)

11. Ciaian, P., Rajcaniova, M., d'Artis Kancs: The economics of bitcoin price formation. Appl. Econ. **48**(19), 1799–1815 (2016). https://doi.org/10.1080/00036846.2015.1109038

12. Felizardo, L., Oliveira, R., Del-Moral-Hernandez, E., Cozman, F.: Comparative study of bitcoin price prediction using WaveNets, recurrent neural networks and other machine learning methods. In: 6th International Conference on Behavioral, Economic and Socio-Cultural Computing, BESC 2019, Beijing, China, 28–30 October 2019, pp. 1–6. IEEE (2019)

13. Georgoula, I., Pournarakis, D., Bilanakos, C., Sotiropoulos, D.N., Giaglis, G.M.: Using time-series and sentiment analysis to detect the determinants of bitcoin prices. In: Proceedings of the 9th Mediterranean Conference on Information Systems, MCIS 2015, Samos, Greece, 2–5 October 2015, p. 20. AISeL (2015)

14. Hashish, I.A., Forni, F., Andreotti, G., Facchinetti, T., Darjani, S.: A hybrid model for bitcoin prices prediction using hidden Markov models and optimized LSTM networks. In: 24th IEEE International Conference on Emerging Technologies and Factory Automation, ETFA 2019, Zaragoza, Spain, 10–13 September 2019, pp. 721–728. IEEE (2019)

15. Koutmos, D.: Bitcoin returns and transaction activity. Econ. Lett. **167**, 81–85 (2018)

16. Kristoufek, L.: Bitcoin meets google trends and Wikipedia: quantifying the relationship between phenomena of the internet era. Sci. Rep. **3**(1), 1–7 (2013)

17. Kristoufek, L.: What are the main drivers of the bitcoin price? Evidence from wavelet coherence analysis. PloS One **10**(4), e0123923 (2015)

18. Kyle, A.S.: Continuous auctions and insider trading. Econometrica: J. Econometric Soc. 1315–1335 (1985)
19. Llorente, G., Michaely, R., Saar, G., Wang, J.: Dynamic volume-return relation of individual stocks. Rev. Financ. Stud. **15**(4), 1005–1047 (2002)
20. Maesa, D.D.F., Marino, A., Ricci, L.: Uncovering the bitcoin blockchain: an analysis of the full users graph. In: 2016 IEEE International Conference on Data Science and Advanced Analytics, DSAA 2016, Montreal, QC, Canada, 17–19 October 2016, pp. 537–546. IEEE (2016)
21. Mallqui, D.C.A., Fernandes, R.A.S.: Predicting the direction, maximum, minimum and closing prices of daily bitcoin exchange rate using machine learning techniques. Appl. Soft Comput. **75**, 596–606 (2019)
22. Mittal, A., Dhiman, V., Singh, A., Prakash, C.: Short-term bitcoin price fluctuation prediction using social media and web search data. In: 2019 Twelfth International Conference on Contemporary Computing, IC3 2019, Noida, India, 8–10 August 2019, pp. 1–6. IEEE (2019)
23. Naeem, M., Bouri, E., Boako, G., Roubaud, D.: Tail dependence in the return-volume of leading cryptocurrencies. Financ. Res. Lett. **36**, 101326 (2020)
24. Nakamoto, S.: Bitcoin: a peer-to-peer electronic cash system (2009)
25. Nguyen, D.-T., Le, H.-V.: Predicting the price of bitcoin using hybrid ARIMA and machine learning. In: Dang, T.K., Küng, J., Takizawa, M., Bui, S.H. (eds.) FDSE 2019. LNCS, vol. 11814, pp. 696–704. Springer, Cham (2019). https://doi.org/10.1007/978-3-030-35653-8_49
26. Pieters, G., Vivanco, S.: Financial regulations and price inconsistencies across bitcoin markets. Inf. Econ. Policy **39**, 1–14 (2017)
27. Schneider, J.: A rational expectations equilibrium with informative trading volume. J. Financ. **64**(6), 2783–2805 (2009)
28. Vassiliadis, S., Papadopoulos, P., Rangoussi, M., Konieczny, T., Gralewski, J.: Bitcoin value analysis based on cross-correlations. J. Internet Bank. Commer. **22**(S7), 1 (2017)
29. Yao, W., Xu, K., Li, Q.: Exploring the influence of news articles on bitcoin price with machine learning. In: 2019 IEEE Symposium on Computers and Communications, ISCC 2019, Barcelona, Spain, 29 June–3 July 2019, pp. 1–6. IEEE (2019)
30. Yermack, D.L.: Is bitcoin a real currency? An economic appraisal. Econ. Innov. eJournal (2013)
31. Yogeshwaran, S., Kaur, M.J., Maheshwari, P.: Project based learning: predicting bitcoin prices using deep learning. In: Ashmawy, A.K., Schreiter, S. (eds.) IEEE Global Engineering Education Conference, EDUCON 2019, Dubai, United Arab Emirates, 8–11 April 2019, pp. 1449–1454. IEEE (2019)

Sensitivity-Based Optimization
for Blockchain Selfish Mining

Jing-Yu Ma[1]($^{(\boxtimes)}$) (iD) and Quan-Lin Li[2]

[1] Bussiness School, Xuzhou University of Technology, Xuzhou 221018, China
[2] School of Economics and Management, Beijing University of Technology,
Beijing 100124, China
liquanlin@tsinghua.edu.cn

Abstract. In this paper, we provide a novel dynamic decision method
of blockchain selfish mining by applying the sensitivity-based optimization theory. Our aim is to find the optimal dynamic blockchain-pegged
policy of the dishonest mining pool. To study the selfish mining attacks,
two mining pools are designed by means of different competitive criterions, where the honest mining pool follows a two-block leading competitive criterion, while the dishonest mining pool follows a modification of
the two-block leading competitive criterion through using a blockchain-
pegged policy. To find the optimal blockchain-pegged policy, we set up
a policy-based continuous-time Markov process and analyze some key
factors. Based on this, we discuss monotonicity and optimality of the
long-run average profit with respect to the blockchain-pegged policy and
prove the structure of the optimal blockchain-pegged policy. We hope
the methodology and results derived in this paper can shed light on the
dynamic decision research on the selfish mining attacks of blockchain.

Keywords: Blockchain · Selfish mining · Blockchain-pegged policy ·
Sensitivity-based optimization · Markov decision process

1 Introduction

Blockchain is used to securely record a public shared ledger of Bitcoin payment
transactions among Internet users in an open P2P network. Though the security
of blockchain is always regarded as the top priority, it is still threatened by some
selfish mining attacks. In the PoW blockchain, the probability that an individual
miner can successfully mine a block becomes lower and lower, as the number of
joined miners increases. This greatly increases the mining risk of each individual
miner. In this situation, some miners willingly form a mining pool. Blockchain
selfish mining leads to colluding miners in dishonest mining pools, one of which

---

Supported by the National Natural Science Foundation of China under grant No.
71932002, and by the Beijing Social Science Foundation Research Base Project under
grant No. 19JDGLA004.

W. Wu and H. Du (Eds.): AAIM 2021, LNCS 13153, pp. 329–343, 2021.
https://doi.org/10.1007/978-3-030-93176-6_28

can obtain a revenue larger than their fair share. The existence of the selfish mining not only means unfair to solve PoW puzzles but also is a severe flaw in integrity of blockchain.

The existence of such selfish mining attacks was first proposed by Eyal and Sirer [4], they set up a Markov chain to express the dynamic of the selfish mining attacks efficiently. After then, some researchers extended and generalized such a similar method to discuss other attack strategies of blockchain. The newest work is Li et al. [13], which provided a new theoretical framework of pyramid Markov processes to solve some open and fundamental problems of blockchain selfish mining under a rigorously mathematical setting. Göbel et al. [7], Javier and Fralix [9] used two-dimensional Markov chain to study the selfish mining. Furthermore, some key research includes stubborn mining by Nayak et al. [16]; Ethereum by Niu and Feng [17]; multiple mining pools by Jain [8]; multi-stage blockchain by Chang et al. [3]; no block reward by Carlsten et al. [2]; power adjusting by Gao et al. [5].

In the study of blockchain selfish mining, it is a key to develop effective optimal methods and dynamic control techniques. However, little work has been done on applying Markov decision processes (MDPs) to set up optimal dynamic control policies for blockchain selfish mining. In general, such a study is more interesting, difficult and challenging. Based on Eyal and Sirer [4], Sapirshtein et al. [19] extended the underlying model for selfish mining attacks, and provided an algorithm to find ϵ-optimal policies for attackers within the model through MDPs. Furthermore, Wüst [20] provided a quantitative framework based on MDPs to analyse the security of different PoW blockchain instances with various parameters against selfish mining. Gervais et al. [6] extended the MDPs of Sapirshtein et al. [19] to determine optimal adversarial strategies for selfish mining. Recently, Zur et al. [24] presented a novel technique called ARR (Average Reward Ratio) MDPs to tighten the bound on the threshold for selfish mining in Ethereum.

The purpose of this paper is to apply the MDPs to set up an optimal parameterized policy (i.e., blockchain-pegged policy) for blockchain selfish mining. To do this, we first apply the sensitivity-based optimization theory in the study of blockchain selfish mining, which is an effective tool proposed for performance optimization of Markov systems by Cao [1]. Li [11] and Li and Cao [10] further extended and generalized such a method to a more general framework of perturbed Markov processes. A key idea in the sensitivity-based optimization theory is the performance difference equation that can quantify the performance difference of a Markov system under any two different policies. The performance difference equation gives a straightforward perspective to study the relation of the system performance between two different policies, which provides more sensitivity information. Thus, the sensitivity-based optimization theory has been applied to performance optimization in many practical areas. For example, the energy-efficient data centers by Xia et al. [21] and Ma et al. [14,15]; the inventory rationing by Li et al. [12]; the multi-hop wireless networks by Xia and Shihada [22]; the finance by Xia [23].

The main contributions of this paper are twofold. The first one is to apply the sensitivity-based optimization theory to study the blockchain selfish mining for the first time, in which we design a modification of the two-block leading competitive criterion for the dishonest mining pool. Different from previous works in the literature for applying an ordinary MDP to against the selfish mining attacks, we propose and develop an easier and more convenient dynamic decision method for the dishonest mining pool: the sensitivity-based optimization theory. Crucially, this sensitivity-based optimization theory may open a new avenue to the optimal blockchain-pegged policy of more general blockchain systems. The second contribution of this paper is to characterize the optimal blockchain-pegged policy of the dishonest mining pool. We analyze the monotonicity and optimality of the long-run average profit with respect to the blockchain-pegged policies under some restrained rewards. We obtain the structure of optimal blockchain-pegged policy is related to the blockchain reward. Therefore, the results of this paper give new insights on understanding not only competitive criterion design of blockchain selfish mining, but also applying the sensitivity-based optimization theory to dynamic decision for the dishonest mining pool. We hope that the methodology and results given in this paper can shed light on the study of more general blockchain systems.

The remainder of this paper is organized as follows. In Sect. 2, we describe a problem of blockchain selfish mining with two different mining pools. In Sect. 3, we establish a policy-based continuous-time Markov process and introduce some key factors. In Sect. 4, we discuss the monotonicity and optimality of the long-run average profit with respect to the blockchain-pegged policy by the sensitivity-based optimization theory. Finally, we give some concluding remarks in Sect. 5.

2   Problem Description

In this section, we give a problem description of blockchain selfish mining with two different mining pools. Also, we provide system structure, operational mode and mathematical notations.

Mining Pools: There are two different mining pools: honest and dishonest mining pools.

(a) The honest mining pool follows the Bitcoin protocol. If he mines a block, he will broadcast to whole community immediately. To avoid the 51% attacks, we assume the honest mining pool are majority in the blockchain system.

(b) The dishonest mining pool has the selfish mining attacks. When the dishonest mining pool mines a block, he can earn more unfair revenue. Such revenue will attract some rational honest miners to jump into the dishonest mining pool. We denote the efficiency-increased ratio of the dishonest mining pool and the net jumping's mining rate by τ and γ, respectively.

Selfish Mining Processes: We assume that the blocks mined by the honest and dishonest mining pools have formed two block branches forked at a tree root, and the growths of the two block branches are two Poisson processes with block-generating rates α_1 and α_2, respectively. In the honest mining pool, the block-generating rate α_1 is equal to the net mining rate, but the situation in the dishonest mining pool is a bit different. The block-generating rate for the dishonest mining pool is $\alpha_2 = \widetilde{\alpha}_2 (1 + \tau)$, where $\widetilde{\alpha}_2$ is regarded as the net mining rate when all the dishonest miners become honest. Following the protocol can not earn more rewards, the honest miners like to jump to the dishonest mining pool with the net jumping rate γ, the real mining rates of the honest and dishonest mining pools are given by $\lambda_1 = \alpha_1 - \gamma$ and $\lambda_2 = (\widetilde{\alpha}_2 + \gamma)(1 + \tau)$, respectively.

Note that mining costs of both mining pools contains two parts: (a) Power consumption cost. Let c_P be the power consumption price per unit of net mining rate and per unit of time. It is easy to see that the power consumption costs per unit of time with respect to the honest and dishonest mining pools are given by $c_P(\alpha_1 - \gamma)$ and $c_P(\widetilde{\alpha}_2 + \gamma)$, respectively. (b) Administrative cost. Let c_A be the administrative price per unit of real mining rate and per unit of time. Then the administrative costs per unit of time with respect to the honest and dishonest mining pools are given by $c_A(\alpha_1 - \gamma)$ and $c_A(\widetilde{\alpha}_2 + \gamma)(1 + \tau)$, respectively.

Competitive Criterions: In the blockchain selfish mining, the honest and dishonest mining pools compete fiercely in finding the nonces to generate the blocks, and they publish the blocks to make two block branches forked at a common tree root. For the two block branches, the longer block branch in the forked structure is called a *main chain*, which or the part of which will be pegged on the blockchain. Under the selfish mining attacks, such two mining pools follow the different competitive criterions.

(a) A two-block leading competitive criterion for the honest mining pool. The honest chain of blocks is taken as the main chain pegged on the blockchain, as soon as the honest chain of blocks is two blocks ahead of the dishonest chain of blocks.

(b) A modification of the two-block leading competitive criterion for the dishonest mining pool. Once the dishonest chain of blocks is the two blocks ahead of the honest chain of blocks, the dishonest chain of blocks can be taken as the main chain. To get more reward, the dishonest mining pool may prefer to keep its mined blocks secret, and continue to mine more blocks rather than broadcast all the mined information.

Since the dishonest miners are minority, their mining power is limited, the dishonest mining pool will not be extend infinitely. We assume that once the dishonest main chain contains m blocks, its part n blocks $(n \leq m)$ must be pegged on the blockchain immediately. In addition, the limitation of the dishonest main chain leads to that the honest main chain containing at most $n - 2$ blocks due to the two-block leading competitive criterion.

Blockchain-Pegged Processes: If the main chain is formed, then the mining processes are terminated immediately. The honest main chain or the part of the dishonest main chain is pegged on the blockchain, and the blockchain-pegged times are i.i.d. and exponential with mean $1/\mu$. The mining pool of the main chain can obtain an appropriate amount of reward (or compensation) from two different parts: A block reward r_B by the blockchain system and an average total transaction fee r_F in the block. At the same time, all the blocks of the other non-main chain become orphan and immediately return to the transaction pool without any new fee. Note that no new blocks are generated during the blockchain-pegged process of the main chain.

We assume that all the random variables defined above are independent of each other. Figure 1 provides an intuitive understanding for the two cases.

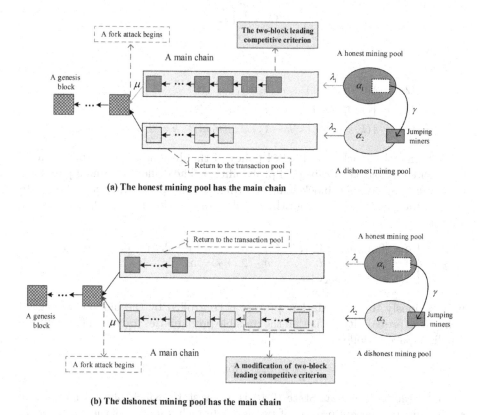

(a) The honest mining pool has the main chain

(b) The dishonest mining pool has the main chain

Fig. 1. A blockchain selfish mining with two different mining pools.

3   Optimization Model Formulation

In this section, we establish an optimization problem to find an optimal blockchain-pegged policy for the dishonest mining pool. To do this, we set up a policy-based continuous-time Markov process and introduce some key factors.

3.1   The States and Policies

To study the blockchain-pegged policy of the blockchain selfish mining with two different mining pools, we first define both 'states' and 'policies' to express such a stochastic dynamic.

Let $N_1(t)$ and $N_2(t)$ be the numbers of blocks mined by the honest and dishonest mining pools at time t, respectively. Then $(N_1(t), N_2(t))$ is regarded as the state of a Markov system at time t. Obviously, all the cases of State $(N_1(t), N_2(t))$ form a state space as follows:

$$\Omega = \bigcup_{k=0}^{m+2} \Omega_k,$$

where

$$\Omega_0 = \{(0,0), (0,1), \ldots, (0,m)\},$$
$$\Omega_1 = \{(1,0), (1,1), \ldots, (1,m)\},$$
$$\Omega_k = \{(k, k-2), (k, k-1), \ldots, (k, m)\}, k = 2, 3, \ldots, m+2.$$

Actually, the blockchain-pegged policy of the dishonest mining pool can be represented by blockchain-pegged probability p. The dishonest mining pool pegs the main chain on the blockchain according to the probability p at the state (n_1, n_2) for $(n_1, n_2) \in \Omega$. From the problem description in Sect. 2, it is easy to see that

$$p = \begin{cases} a \in [0,1], & n_1 = 0, 1, \ldots, m-3, \ n_2 = n_1 + 2, n_1 + 3, \ldots, m-1, \\ 1, & n_1 = 0, 1, \ldots, m-2, \ n_2 = m, \\ 0, & \text{otherwise.} \end{cases} \tag{1}$$

It is obviously that the Markov process is controlled by the blockchain-pegged policy (the probability p). Let all the possible probabilities p given in (1) compose a policy space as follows:

$$\mathcal{P} = \{p : p \in [0,1], \text{ for } (n_1, n_2) \in \Omega\}.$$

It is readily seen that State $(0,0)$ is a key state, which plays a key role in setting up the Markov process of two block branches forked at the tree root. In fact, State $(0,0)$ describes the tree root as the starting point of the fork attacks, e.g., see Fig. 2. If the Markov process enters State $(0,0)$, then the fork attack ends immediately, and the main chain is pegged on the blockchain.

Now, from Fig. 2, we provide an interpretation for the blockchain-pegged probability p as follows:

(1) In Part A-1, i.e., $n_1 = 0, 1, \ldots, m-3$ and $n_2 = n_1 + 2, n_1 + 3, \ldots, m-1$, the dishonest mining pool follows the modification of the two-block leading competitive criterion and forms the dishonest main chain, then the probability $p \in [0,1]$.

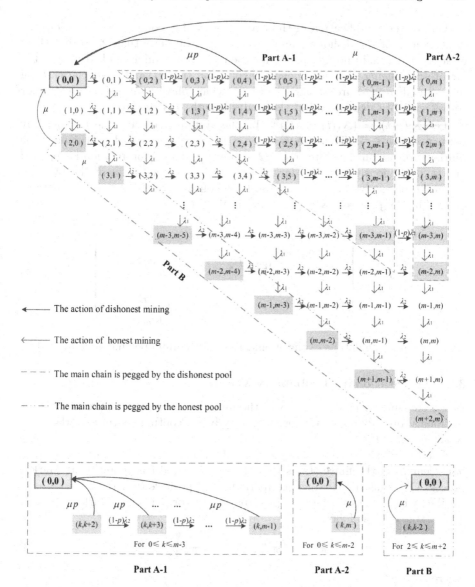

Fig. 2. The state transition relation of the Markov process.

(2) In Part A-2, i.e., $n_1 = 0, 1, \ldots, m - 2$ and $n_2 = m$, for the limitation of dishonest mining power, the dishonest main chain must be pegged on the blockchain, or there is a risk of getting no reward. It is easy to see that the probability p is taken as 1.

(3) In the rest of Fig. 2, it is the competitive process of honest and dishonest mining pools. Therefore, $p = 0$ for the dishonest main chain hasn't formed. In addition, the states in Part B mean that the honest main chain is formed.

Due to the modification of the two-block leading competitive criterion, the limitation m of the dishonest mining pool must be more than 2, so that there exist the blockchain-pegged policy for the dishonest mining pool. If $m \geq 5$, the infinitesimal generator has a general expression (note that the special cases of $m = 3$ and $m = 4$ are omitted here). In what follows, we assume $m \geq 5$ for convenience of calculation, but the analysis method is similar.

Let $\mathbf{X}^{(p)}(t) = (N_1(t), N_2(t))^{(p)}$ be the system state at time t under any given policy $p \in \mathcal{P}$. Then $\{\mathbf{X}^{(p)}(t) : t \geq 0\}$ is a policy-based continuous-time Markov process on the state space Ω whose state transition relation is depicted in Fig. 2. Obviously, such a Markov process is a special form of the pyramid Markov process given in Li et al. [13]. Based on this, the infinitesimal generator of the Markov process $\{\mathbf{X}^{(p)}(t) : t \geq 0\}$ is given by

$$
\mathbf{Q}^{(p)} = \begin{pmatrix}
Q_{0,0} & B_0 & & & & \\
Q_{1,0} & Q_{1,1} & B_1 & & & \\
Q_{2,0} & & Q_{2,2} & B_2 & & \\
\vdots & & & \ddots & \ddots & \\
Q_{m+1,0} & & & & Q_{m+1,m+1} & B_{m+1} \\
Q_{m+2,0} & & & & & Q_{m+2,m+2}
\end{pmatrix}. \tag{2}
$$

Here, we omit the details of the submatrices in the infinitesimal generator $\mathbf{Q}^{(p)}$.

3.2   The Stationary Probability Vector

Based on some special properties of the infinitesimal generator, we provide the stationary probability vector for the policy-based continuous-time Markov process $\{\mathbf{X}^{(p)}(t) : t \geq 0\}$.

For $n_1 = 0, 1, \ldots, m - 3$, $n_2 = n_1 + 2, n_1 + 3, \ldots, m - 1$ and $0 \leq p < 1$, it is clear from the finite states that the policy-based continuous-time Markov process $\mathbf{Q}^{(p)}$ must be irreducible, aperiodic and positive recurrent.

We write the stationary probability vector of the Markov process $\{\mathbf{X}^{(p)}(t) : t \geq 0\}$ as follows:

$$
\boldsymbol{\pi}^{(p)} = \left(\boldsymbol{\pi}_0^{(p)}, \boldsymbol{\pi}_1^{(p)}, \ldots, \boldsymbol{\pi}_{m+2}^{(p)} \right), \tag{3}
$$

where

$$
\boldsymbol{\pi}_0^{(p)} = \left(\pi^{(p)}(0,0), \pi^{(p)}(0,1), \ldots, \pi^{(p)}(0,m) \right),
$$

$$
\boldsymbol{\pi}_1^{(p)} = \left(\pi^{(p)}(1,0), \pi^{(p)}(1,1), \ldots, \pi^{(p)}(1,m) \right),
$$

$$
\boldsymbol{\pi}_k^{(p)} = \left(\pi^{(p)}(k,k-2), \pi^{(p)}(k,k-1), \ldots, \pi^{(p)}(k,m) \right), \quad 2 \leq k \leq m+2.
$$

Let

$$
\mathbf{D}_0 = 1,
$$

$$
\mathbf{D}_k = B_{k-1} \left(-Q_{k,k} \right)^{-1}, \quad k = 1, 2, \ldots, m+2. \tag{4}
$$

Then the following theorem provides an explicit expression for the stationary probability vector $\boldsymbol{\pi}^{(p)}$ by means of the system of linear equations: $\boldsymbol{\pi}^{(p)} Q^{(p)} = \mathbf{0}$ and $\boldsymbol{\pi}^{(p)} e = 1$.

Theorem 1. *The stationary probability vector $\boldsymbol{\pi}^{(p)}$ of the Markov process $Q^{(p)}$ is given by*

$$\boldsymbol{\pi}_k^{(p)} = \boldsymbol{\pi}_0^{(p)} \prod_{l=1}^{k} \mathbf{D}_l, \tag{5}$$

where $\boldsymbol{\pi}_0^{(p)}$ is determined by the system of linear equations

$$\boldsymbol{\pi}_0^{(p)} \left(\sum_{k=0}^{m+2} \prod_{l=0}^{k} \mathbf{D}_l Q_{k,0} \right) = \mathbf{0},$$

$$\boldsymbol{\pi}_0^{(p)} \left(\sum_{k=0}^{m+2} \prod_{l=0}^{k} \mathbf{D}_l e \right) = 1.$$

3.3   The Reward Function

A reward function of the dishonest mining pool with respect to both states and policies is defined as a profit rate (i.e., the total revenues minus the total costs per unit of time).

Let $R = r_B + r_F$ and $C = (\tilde{\alpha} + \gamma) [c_P + c_A (1 + \tau)]$. Then R and C denote the blockchain-pegged reward and the mining cost for the dishonest mining pool, respectively. According to Fig. 2, the reward function at State $(N_1(t), N_2(t))^{(p)}$ under the blockchain-pegged policy p is defined as follows:

$$f^{(p)}(n_1, n_2) = \begin{cases} n_2 R \mu p - C, & \text{if } 0 \le n_1 \le m - 3 \text{ and } n_1 + 2 \le n_2 \le m - 1, \\ m R \mu - C, & \text{if } 0 \le n_1 \le m - 2 \text{ and } n_2 = m, \\ -C, & \text{otherwise.} \end{cases}$$

We futher define a column vector $\boldsymbol{f}^{(p)}$ composed of the elements $f^{(p)}(n_1, n_2)$ as

$$\boldsymbol{f}^{(p)} = \left(\left(\boldsymbol{f}_0^{(p)} \right)^T, \left(\boldsymbol{f}_1^{(p)} \right)^T, \ldots, \left(\boldsymbol{f}_{m+2}^{(p)} \right)^T \right)^T, \tag{6}$$

where

$$\boldsymbol{f}_0^{(p)} = \left(f^{(p)}(0,0), f^{(p)}(0,1), \ldots, f^{(p)}(0,m) \right)^T,$$

$$\boldsymbol{f}_1^{(p)} = \left(f^{(p)}(1,0), f^{(p)}(1,1), \ldots, f^{(p)}(1,m) \right)^T,$$

$$\boldsymbol{f}_k^{(p)} = \left(f^{(p)}(k,k-2), f^{(p)}(k,k-1), \ldots, f^{(p)}(k,m) \right)^T, \quad k = 2, 3, \ldots, m+2.$$

In the remainder of this section, the long-run average profit of the dishonest mining pool under a blockchain-pegged policy p is defined as

$$\eta^p = \lim_{T \to +\infty} E\left\{ \frac{1}{T} \int_0^T f^{(p)}\left((N_1(t), N_2(t))^{(p)} \right) dt \right\} = \boldsymbol{\pi}^{(p)} \boldsymbol{f}^{(p)}, \qquad (7)$$

where $\boldsymbol{\pi}^{(p)}$ and $\boldsymbol{f}^{(p)}$ are given by (5) and (6), respectively.

3.4   The Performance Potential

The sensitivity-based optimization theory has a fundamental quantity called performance potential by Cao [1], which is defined as

$$g^{(p)}(n_1, n_2) = E\left\{ \int_0^{+\infty} \left[f^{(p)}\left(\mathbf{X}^{(p)}(t) \right) - \eta^p \right] dt \middle| \mathbf{X}^{(p)}(0) = (n_1, n_2) \right\}, \qquad (8)$$

where η^p is defined in (7). For any blockchain-pegged policy $p \in \mathcal{P}$, $g^{(p)}(n_1, n_2)$ quantifies the contribution of the initial State (n_1, n_2) to the long-run average profit of the dishonest mining pool. Here, $g^{(p)}(n_1, n_2)$ is also called the relative value function or the bias in the traditional MDP theory, see, e.g., Puterman [18]. We further define a column vector $\boldsymbol{g}^{(p)}$ as

$$\boldsymbol{g}^{(p)} = \left(\left(\boldsymbol{g}_0^{(p)} \right)^T, \left(\boldsymbol{g}_1^{(p)} \right)^T, \dots, \left(\boldsymbol{g}_{m+2}^{(p)} \right)^T \right)^T, \qquad (9)$$

where

$$\boldsymbol{g}_0^{(p)} = \left(g^{(p)}(0,0), g^{(p)}(0,1), \dots, g^{(p)}(0,m) \right)^T,$$

$$\boldsymbol{g}_1^{(p)} = \left(g^{(p)}(1,0), g^{(p)}(1,1), \dots, g^{(p)}(1,m) \right)^T,$$

$$\boldsymbol{g}_k^{(p)} = \left(g^{(p)}(k, k-2), g^{(p)}(k, k-1), \dots, g^{(p)}(k,m) \right)^T, \quad k = 2, 3, \dots, m+2.$$

A similar computation to that in Ma et al. [14,15] is omitted here, we can provide an expression for the vector $\boldsymbol{g}^{(p)}$, i.e.,

$$\boldsymbol{g}^{(p)} = R\boldsymbol{a} + \boldsymbol{b}, \qquad (10)$$

where $\boldsymbol{a}$ and $\boldsymbol{b}$ can be given by $\mathbf{Q}^{(p)}$, $\boldsymbol{\pi}^{(p)}$ and $\boldsymbol{f}^{(p)}$. It is seen that all the entries $g^{(p)}(n_1, n_2)$ in $\boldsymbol{g}^{(p)}$ are the linear functions of R. Therefore, our objective is to find the optimal blockchain-pegged policy p^* such that the long-run average profit of the dishonest mining pool η^p is maximize, that is,

$$p^* = \arg\max_{p \in \mathcal{P}} \{\eta^p\}. \qquad (11)$$

However, it is very challenging to analyze some interesting structure properties of the optimal blockchain-pegged policy p^*. In the remainder of this paper, we will apply the sensitivity-based optimization theory to study such an optimal problem.

4   Monotonicity and Optimality

In this section, we use the sensitivity-based optimization theory to discuss monotonicity and optimality of the long-run average profit of the dishonest mining pool with respect to the blockchain-pegged policy. Based on this, we obtain the optimal blockchain-pegged policy of the dishonest mining pool.

In an MDP, system policies will affect the element values of infinitesimal generator and reward function. That is, if the policy p changes, then the infinitesimal generator $\mathbf{Q}^{(p)}$ and the reward function $\boldsymbol{f}^{(p)}$ will have their corresponding changes. To express such a change mathematically, we take two different policies $p, p' \in \mathcal{P}$, both of which correspond to their infinitesimal generators $\mathbf{Q}^{(p)}$ and $\mathbf{Q}^{(p')}$, and to their reward functions $\boldsymbol{f}^{(p)}$ and $\boldsymbol{f}^{(p')}$.

The following lemma provides the performance difference equation for the difference $\eta^{p'} - \eta^p$ of the long-run average performances for any two blockchain-pegged policies $p, p' \in \mathcal{P}$. Here, we only restate it without proof, while readers may refer to Cao [1] and Ma et al. [14] for more details.

Lemma 1. *For any two blockchain-pegged policies $p, p' \in \mathcal{P}$, we have*

$$\eta^{p'} - \eta^p = \boldsymbol{\pi}^{(p')} \left[\left(\mathbf{Q}^{(p')} - \mathbf{Q}^{(p)} \right) \boldsymbol{g}^{(p)} + \left(\boldsymbol{f}^{(p')} - \boldsymbol{f}^{(p)} \right) \right]. \tag{12}$$

Therefore, to find the optimal blockchain-pegged policy p^*, we consider such two blockchain-pegged policies $p, p' \in \mathcal{P}$. Suppose the blockchain-pegged policy is changed from p to p', which corresponding the states (n_1, n_2) for $n_1 = 0, 1, \ldots, m-3$ and $n_2 = n_1 + 2, n_1 + 3, \ldots, m-1$, i.e., Part A-1 of Fig. 2.

Using Lemma 1, we examine the sensitivity of blockchain-pegged policy on the long-run average profit of the dishonest mining pool. Substituting (2) and (6) into (12), we have

$$\eta^{p'} - \eta^p$$
$$= \boldsymbol{\pi}^{(p')} \left[\left(\mathbf{Q}^{(p')} - \mathbf{Q}^{(p)} \right) \boldsymbol{g}^{(p)} + \left(\boldsymbol{f}^{(p')} - \boldsymbol{f}^{(p)} \right) \right]$$
$$= (p'-p) \sum_{n_1=0}^{m-3} \sum_{n_2=n_1+2}^{m-1} \pi^{(p')}(n_1, n_2) \left[\mu - (\mu - \lambda_2) g^{(p)}(n_1, n_2) - \lambda_2 g^{(p)}(n_1, n_2+1) + n_2 R \mu \right].$$
$$\tag{13}$$

With the difference (13), we can easily obtain the following equation

$$\frac{\triangle \eta^p}{\triangle p} = \sum_{n_1=0}^{m-3} \sum_{n_2=n_1+2}^{m-1} \pi^{(p')}(n_1, n_2) \left[\mu - (\mu - \lambda_2) g^{(p)}(n_1, n_2) - \lambda_2 g^{(p)}(n_1, n_2+1) + n_2 R \mu \right],$$
$$\tag{14}$$

where $\triangle \eta^p = \eta^{p'} - \eta^p$ and $\triangle p = p' - p$. As $p' \to p$,

$$\left. \frac{\mathrm{d}\eta^p}{\mathrm{d}p} \right|_{\triangle p \to 0} = \lim_{\triangle p \to 0} \frac{\eta^{p'} - \eta^p}{\triangle p},$$

we derive the following derivative equation

$$\frac{\mathrm{d}\eta^p}{\mathrm{d}p} = \sum_{n_1=0}^{m-3} \sum_{n_2=n_1+2}^{m-1} \pi^{(p)}(n_1, n_2) \left[\mu - (\mu - \lambda_2) g^{(p)}(n_1, n_2) - \lambda_2 g^{(p)}(n_1, n_2+1) + n_2 R\mu \right].$$

(15)

According to (10), $g^{(p)}(n_1, n_2)$ and $g^{(p)}(n_1, n_2 + 1)$ are both linear functions w.r.t. R. Thus, we denote $g^{(p)}(n_1, n_2)$ and $g^{(p)}(n_1, n_2 + 1)$ as $a_{n_1,n_2} R + b_{n_1,n_2}$ and $a_{n_1,n_2+1} R + b_{n_1,n_2+1}$, respectively. Substituting into (15), we have

$$\frac{\mathrm{d}\eta^p}{\mathrm{d}p} = \bar{a}R + \bar{b},$$

(16)

where

$$\bar{a} = \sum_{n_1=0}^{m-3} \sum_{n_2=n_1+2}^{m-1} \pi^{(p)}(n_1, n_2) \left[(\lambda_2 - \mu) a_{n_1,n_2} - \lambda_2 a_{n_1,n_2+1} + n_2\mu \right],$$

$$\bar{b} = \sum_{n_1=0}^{m-3} \sum_{n_2=n_1+2}^{m-1} \pi^{(p)}(n_1, n_2) \left[(\lambda_2 - \mu) b_{n_1,n_2} + \lambda_2 a_{n_1,n_2+1} b_{n_1,n_2+1} + \mu \right].$$

It is clear that $\frac{\mathrm{d}\eta^p}{\mathrm{d}p}$ is also a linear function w.r.t. R, and depends only on the current policy.

Remark 1. *It is seen from (16) that we only need to know the sign of $\frac{\mathrm{d}\eta^p}{\mathrm{d}p}$, instead of its precise value. The estimation accuracy of a sign is usually better than that of a value. Therefore, this feature can help us find the optimal blockchain-pegged policy effectively. Moreover, we see that we do not have to know some prior system information. Thus, the complete system information is not required in our approach and this is an advantage during the practical application.*

Remark 2. *The key idea of the sensitivity-based optimization theory is to utilize the performance sensitivity information, such as the performance difference, to conduct the optimization of stochastic systems. Therefore, even if the competition criteria become more complicated, it does not affect the applicability of our method.*

The following theorems discuss monotonicity and optimality of the long-run average profit η^p of the dishonest mining pool with respect to the blockchain-pegged policy p.

Theorem 2. *If $R > -\bar{b}/\bar{a}$, then the long run average profit η^p is strictly monotone increasing with respect to each decision element $p \in [0, 1]$, and the optimal blockchain-pegged policy $p^* = 1$.*

This theorem follows directly (16). It is seen that the optimal blockchain-pegged policy $p^* = 1$ just corresponding to any State $(n_1, n_1 + 2)$ in Part A-1 of Fig. 2, and the state transition has changed. In this case, the dishonest chain of blocks is the only two blocks ahead of the honest chain of blocks, the dishonest mining pool should peg on the blockchain, also follows the two-block leading competitive criterion.

Therefore, when the blockchain-pegged reward is higher with $R > -\bar{b}/\bar{a}$, it is seen that the dishonest miners become honest, all miners will follow the PoW protocol and broadcast to the whole community. In this case, the selfish mining attacks should be invalid.

Theorem 3. *If $0 \leq R < -\bar{b}/\bar{a}$, then the long run average profit η^p is strictly monotone decreasing with respect to each decision element $p \in [0,1]$, and the optimal blockchain-pegged policy $p^* = 0$.*

Simlar to Theorem 2, this theorem also follows directly (16). It is seen that the optimal blockchain-pegged policy $p^* = 0$ corresponding to any State (n_1, n_2) in Part A-1 of Fig. 2.

In the blockchain selfish mining, if the dishonest mining pool makes decision not to peg on the blockchain, i.e., $p^* = 0$, the main chain is detained to continue mining more blocks so that it is not broadcasted in the blockchain network, until the number of blocks reaches m for the limited mining bound. In this case, the dishonest mining pool prefer to obtain more mining profit through winning on mining more blocks, rather than peg on the blockchain prematurely.

Therefore, when the blockchain-pegged reward is lower with $0 \leq R < -\bar{b}/\bar{a}$, it is seen that the dishonest mining pool follows the n-block ($2 \leq n \leq m$) leading competitive criterion under the selfish mining attacks.

Theorem 4. *If $R = -\bar{b}/\bar{a}$, then the change of blockchain-pegged policy p no longer improve the long-run average profit η^p.*

With Theorem 4, the dishonest miners don't care about when the main chain is pegged on the blockchain, thus the blockchain-pegged policy can be chosen randomly in set $[0,1]$.

5 Concluding Remarks

In this paper, we propose a novel dynamic decision method by applying the sensitivity-based optimization theory to study the optimal blockchain-pegged policy of blockchain selfish mining with two different mining pools.

We describe a more general blockchain selfish mining with a modification of the two-block leading competitive criterion, which is related to the blockchain-pegged policies. To find the optimal blockchain-pegged policy of the dishonest mining pool, we analyze the monotonicity and optimality of the long-run average profit with respect to the blockchain-pegged policy under some restrained blockchain-pegged rewards. We prove the structure of optimal blockchain-pegged policy with respect to the blockchain-pegged rewards. Different from those previous works in the literature on applying the traditional MDP theory to the blockchain selfish mining, the sensitivity-based optimization theory used in this paper is easier and more convenient in the optimal policy study of blockchain selfish mining.

Along such a research line of applying the sensitivity-based optimization theory, there are a number of interesting directions for potential future research, for example:

- Extending to the blockchain selfish mining with multiple mining pools, for example, a different competitive criterion, no space limitation of the dishonest pool and so on;
- analyzing non-Poisson inputs such as Markovian arrival processes (MAPs) and/or non-exponential service times, e.g., the PH distributions;
- discussing the long-run average performance is influenced by some concave or convex reward (or cost) functions; and
- studying individual or social optimization for the blockchain selfish mining from a perspective of combining game theory with the sensitivity-based optimization.

References

1. Cao, X.R.: Stochastic Learning and Optimization–A Sensitivity-Based Approach. Springer, New York (2007)
2. Carlsten, M., Kalodner, H.A., Weinberg, S.M., et al.: On the instability of bitcoin without the block reward. In: ACM SIGSAC Conference on Computer and Communications Security, pp. 154–167. Association for Computing Machinery, New York (2016)
3. Chang, D., Hasan, M., Jain, P.: Spy based analysis of selfish mining attack on multi-stage blockchain. IACR Cryptol, ePrint 2019/1327, pp. 1–34 (2019)
4. Eyal, I., Sirer, E.G.: Majority is not enough: bitcoin mining is vulnerable. In: Christin, N., Safavi-Naini, R. (eds.) FC 2014. LNCS, vol. 8437, pp. 436–454. Springer, Heidelberg (2014). https://doi.org/10.1007/978-3-662-45472-5_28
5. Gao, S., Li, Z., Peng, Z., et al.: Power adjusting and bribery racing: novel mining attacks in the bitcoin system. In: ACM SIGSAC Conference on Computer and Communications Security, pp. 833–850. Association for Computing Machinery, New York (2019)
6. Gervais, A., Karame, G.O., Wüst, K., et al.: On the security and performance of Proof of Work blockchains. In: ACM SIGSAC Conference on Computer and Communications Security, pp. 3–16. Association for Computing Machinery, New York (2016)
7. Göbel, J., Keeler, H.P., Krzesinski, A.E., et al.: Bitcoin blockchain dynamics: the selfish-mine strategy in the presence of propagation delay. Perform. Eval. **104**, 23–41 (2016)
8. Jain, P.: Revenue generation strategy through selfish mining focusing multiple mining pools. Bachelor thesis, Computer Science & Applied Mathematics, Indraprastha Institute of Information Technology, India (2019)
9. Javier, K., Fralix, B.: A further study of some Markovian Bitcoin models from Göbel et al.. Stochast. Models **36**(2), 223–250 (2020)
10. Li, Q.L., Cao, J.: Two types of RG-factorizations of quasi-birth-and-death processes and their applications to stochastic integral functionals. Stoch. Model. **20**(3), 299–340 (2004)

11. Li, Q.L.: Constructive Computation in Stochastic Models with Applications: The RG Factorizations. Springer, Heidelberg (2010). https://doi.org/10.1007/978-3-642-11492-2
12. Li, Q.L., Li, Y.M., Ma, J.Y., et al.: A complete algebraic transformational solution for the optimal dynamic policy in inventory rationing across two demand classes. arXiv:1908.09295v1 (2019)
13. Li, Q.L., Chang, Y.X., Wu, X., et al.: A new theoretical framework of pyramid Markov processes for blockchain selfish mining. J. Syst. Sci. Syst. Eng. 1–45 (2021)
14. Ma, J.Y., Xia, L., Li, Q.L.: Optimal energy-efficient policies for data centers through Sensitivity-based optimization. Discret. Event Dyn. Syst. **29**(4), 567–606 (2019)
15. Ma, J.Y., Li, Q.L., Xia, L.: Optimal asynchronous dynamic policies in energy-efficient data centers. arXiv:1901.03371 (2019)
16. Nayak, K., Kumar, S., Miller, A., et al.: Stubborn mining: generalizing selfish mining and combining with an eclipse attack. In: IEEE European Symposium on Security and Privacy, Saarbruecken, pp. 305–320. IEEE (2016)
17. Niu, J., Feng, C.: Selfish mining in Ethereum. arXiv:1901.04620 (2019)
18. Puterman, M.L.: Markov Decision Processes: Discrete Stochastic Dynamic Programming. Wiley, Hoboken (1994)
19. Sapirshtein, A., Sompolinsky, Y., Zohar, A.: Optimal selfish mining strategies in bitcoin. In: Grossklags, J., Preneel, B. (eds.) FC 2016. LNCS, vol. 9603, pp. 515–532. Springer, Heidelberg (2017). https://doi.org/10.1007/978-3-662-54970-4_30
20. Wüst, K.: Security of blockchain technologies. Master thesis, Department of Computer Science, ETH Zürich, Switzerland (2016)
21. Xia, L., Zhang, Z.G., Li, Q.L.: A c/μ-rule for job assignment in heterogeneous group-server queues. Prod. Oper. Manag. 1–18 (2021)
22. Xia, L., Shihada, B.: A Jackson network model and threshold policy for joint optimization of energy and delay in multi-hop wireless networks. Eur. J. Oper. Res. **242**(3), 778–787 (2015)
23. Xia, L.: Risk-sensitive Markov decision processes with combined metrics of mean and variance. Prod. Oper. Manag. **29**(12), 2808–2827 (2020)
24. Zur, R.B., Eyal, I., Tamar, A.: Efficient MDP analysis for selfish-mining in blockchains. In: The 2nd ACM Conference on Advances in Financial Technologies, pp. 113–131. Association for Computing Machinery, New York (2020)

Design and Implementation of List and Dictionary in XD-M Language

Yajie Wang, Nan Zhang$^{(\boxtimes)}$, and Zhenhua Duan$^{(\boxtimes)}$

Institute of Computing Theory and Technology, and ISN Laboratory,
Xidian University, Xi'an 710071, China
wangyj@stu.xidian.edu.cn, nanzhang@xidian.edu.cn,
zhhduan@mail.xidian.edu.cn

Abstract. This paper presents the design and implementation of two data types, List and Dictionary, in the programming language XD-M. XD-M is an interpreted language with dynamic data types and its syntax is similar to Python. It is developed from the Modeling Simulation and Verification Language called MSVL. In the main part of the paper, we will discuss data structures, syntax and algorithms of List and Dictionary in XD-M, as well as their abstract syntax trees. Finally, an example is given to illustrate how to use List and Dictionary in XD-M programming.

Keywords: Programming language · List · Dictionary · Data structure · Abstract syntax tree

1 Introduction

Modeling Simulation and Verification Language (MSVL) [1–3] is a temporal logic programming language [1] derived from Projection Temporal Logic (PTL) [4, 5]. MSVL can perform modeling, simulation and verification of software and hardware systems so as to improve the correctness and reliability of systems. XD-M was developed by simplifying the MSVL syntax [6–8] so as to omit type declaration. The syntax of XD-M is similar to Python [9,10], and both support dynamic data types.

Although the Python [9,10] language is attractive because of many advantages, such as simplicity, portability, vast libraries, open source, and scalability, there are still some inconveniences. For instance, there are few Python tools supporting formal verification. Accordingly, we are motivated to develop a Python-like programming language and meanwhile make it have the potential of program verification. XD-M programs can be verified through formal specification [11]

This work was supported in part by the National Key Research and Development Program of China under Grant 2018AAA0103202, in part by the National Natural Science Foundation of China under Grants 62172322 and 61732013, and in part by the Key Science and Technology Innovation Team of Shaanxi Province under Grant 2019TD-001.

W. Wu and H. Du (Eds.): AAIM 2021, LNCS 13153, pp. 344–355, 2021.
https://doi.org/10.1007/978-3-030-93176-6_29

and verification methods based on MSVL [12–19]. This characteristic of the language is of great help to improve the correctness and reliability of software and hardware systems.

XD-M is an interpreted language that employs dynamic data types. The data types permitted in XD-M currently are integer, floating point, character, string, array and structure. The language interpreter can be used to simulate, model and verify XD-M programs.

In order to make the XD-M language more powerful and convenient use, it is necessary to enhance the interpreter to support two data types, namely, List and Dictionary in Python. Therefore, the following part of this paper will focus on the detailed implementation of List and Dictionary in XD-M.

This paper is organized as follows. In the next section, the design and implementation of List and Dictionary are presented, including the data structure, syntax and syntactic interpretation of List and Dictionary. In Sect. 3, an example is given to show how to use List and Dictionary in XD-M programs. In Sect. 4, conclusions are drawn.

2   Design and Implementation of List and Dictionary

First of all, we will introduce the execution flow of the XD-M interpreter. As shown in Fig. 1, lexical analysis, syntactic analysis, and semantic analysis are performed on the source program to get the output. The purpose of lexical analysis and syntactic analysis is to generate an abstract syntax tree, which is implemented with the help of Flex [20] and Bison [21].

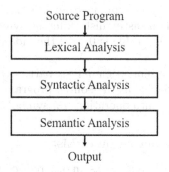

Fig. 1. Execution flow of the interpreter

Figure 2 shows the data structure of symbol, symbol table and abstract syntax tree. The entire data structure is not shown here, only some key member variables are selected.

The "CSyntaxNode" class represents a node in the abstract syntax tree, and "NodeType" is an enumerated type, which represents the type of the current node. An abstract syntax tree node contains a variable named "nodeType" of

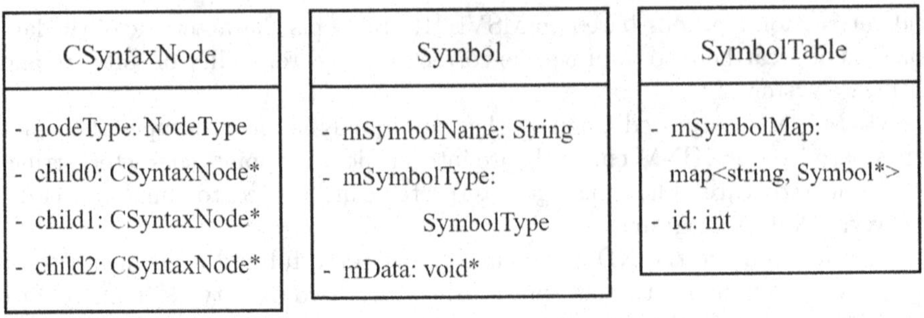

CSyntaxNode
- nodeType: NodeType
- child0: CSyntaxNode*
- child1: CSyntaxNode*
- child2: CSyntaxNode*

Symbol
- mSymbolName: String
- mSymbolType:
SymbolType
- mData: void*

SymbolTable
- mSymbolMap:
map<string, Symbol*>
- id: int

Fig. 2. Data structure of symbol, symbol table and abstract syntax tree

type "NodeType" and three pointers to child nodes. When the interpreter parses the abstract syntax tree, it will perform different actions according to the value of "nodeType".

The "Symbol" class contains a String variable "mSymbolName", which represents the name of the "Symbol", and a pointer variable "mData", which is used to point to data. In addition, there is a "SymbolType" variable "mSymbolType", which is an enumerated variable and similar to the "nodeType" in the "CSyntaxNode" class. It is used to indicate the type of the current "Symbol". We can understand that there is a one-to-one correspondence between a variable in XD-M program and a "Symbol". For example, if there is a floating-point variable named "f" in a XD-M program, this variable will correspond to a "Symbol" whose "mSymbolName" is "f", and "mSymbolType" is "SYMBOLFLOAT" (a constant).

The "SymbolTable" class contains an "mSymbolMap" variable and an "id" variable. "mSymbolMap" is a mapping from a String to a "Symbol" pointer. The "id" variable is used to distinguish different symbol tables. For example, the "id" of the global symbol table is 0, and the "id" of the local symbol table is 1 or other values.

Generally speaking, the semantic analysis part of the XD-M language interpreter is to parse the variables in the source program into symbols, and parse the operations on variables into operations on symbols.

2.1   Design and Implementation of List

Data Structure of List: List is defined as an ordered sequence of several elements, and the elements in List can be of any data type. For example, a List type variable can store a person's personal information, including name, age, height, weight, etc. List needs to support access, addition, deletion and modification of list elements. And the time complexity of element access is required to be O(1).

As shown in Fig. 3, a class named "MList" is defined to implement the function of List. In "MList" class, a variable named "list" is used to store data, which is an array of "Symbol" pointers. There are also two unsigned integers, one for capacity of List and the other for length of List. The initial value of "capacity" is set to 16, in which case "list" is an array of length 16 and each element is null pointer.

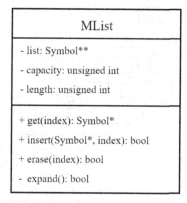

Fig. 3. Data structure of List

The public function "insert (Symbol*, index)" is to add a "Symbol" pointer to the "index" position in the list, and the elements after "index" will be moved back. As the number of list elements increase, "length" will also increase. When "length" increases to the same as "capacity", the expansion operation will be triggered, that is, the private function "expand ()" will be called.

Function "expand ()" will reapply for a memory space, and the space size is twice "capacity" multiplied by the number of pointer bytes. Copy the value in the original "list" to a new memory space, then the original memory will be released and "list" will point to the new memory space, and finally the value of "capacity" is doubled. In general, the average time complexity of the insert operation is $O(n)$.

Function "erase (index)" is used to delete the "Symbol" at the "index" position in the list and move the following elements forward. Its time complexity is also $O(n)$.

The "get (index)" function is simple. For a valid "index", it directly returns the "Symbol" pointer at the "index" position in the list. Therefore, the time complexity of element access is $O(1)$. If the "index" is invalid, an error warning will be output.

Syntax and Syntactic Interpretation of List: Because XD-M language uses dynamic data types, there is no statement for data type declarations. If we want to use List, we directly use the assignment statement to assign a List to a variable.

For example:

$$a =< 1, 1.762, \ 'c', \ "hello", [1, 3, 6], < 2, 7 >>$$

In this statement, a pair of angle brackets is used to represent a List initialization, and the elements in the brackets are separated by commas. The initialized list is assigned to a variable named "a", and then the data type of "a" is List.

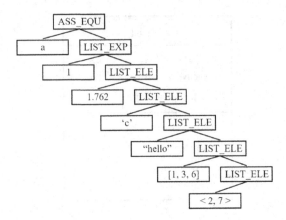

Fig. 4. Abstract syntax tree of List initialization statement

Figure 4 shows the abstract syntax tree constructed by the above statement. Each box represents a node of the abstract syntax tree. We use the value of the element to fill in the box instead of the abstract syntax tree of each element in the list. The purpose of this is to let us pay more attention to the parsing process of List initialization statement.

The type of the root node is "ASS_EQU", which means that the interpreter will parse an assignment statement. Root node has two child nodes. The first child node represents the variable in the front of the equal sign, that is "a". The second child node represents the List initialization expression behinds the equal sign. Its node type is "LIST_EXP", and the interpreter will first parse "LIST_EXP" node and its subtree.

When the interpreter parses the "LIST_EXP" node, it will instantiate an "MList" object, and then parse the child nodes of "LIST_EXP" node. The first child node represents an element in the list. The interpreter will interpret first child node to get a "Symbol" object, whose "mSymbolType" is "SYMBOLINT", and its "mData" is a pointer to an integer "1". This "Symbol" object will be added to the previously instantiated "MList" object. Then the interpreter will parse the second child node, whose node type is "LIST_ELE", indicating that there are still elements in the list that need to be parsed. Similarly, whenever the interpreter parsing an element, it adds a "Symbol" object to "MList" object. When the second child node of the last "LIST_ELE" node is null pointer, it means that all elements have been parsed. Next, the interpreter will back to the "LIST_EXP" node and instantiate a "Symbol" object, whose "mSymbolType" will set to "SYMBOLLIST", and its "mData" will point to "MList" object, and its "mSymbolName" will set to "a". Finally, this "Symbol" object will be added to the symbol table.

In addition to initializing a List variable, we also need some built-in functions related to List variables. The code snippet in Fig. 5 describes the syntax for operating list variables in the XD-M language, where "a" is a List variable,

a.at(index)	// get an element at the "index" position
a.append(anytype_data)	// append an element to the list
a.insert(anytype_data, index)	// insert an element to the "index" position
a.del(index)	// delete an element at the "index" position
a.size()	// return the size of list
a.clear()	// delete all element in list
a.head()	// return the first element in list
a.tail()	// return the last element in list
a.sub(index1, index2)	// return the sub-list and the index between "index1" and "index2"

Fig. 5. Built-in functions for List

"anytype_data" is a variable or constant of any data type, and "index" is an integer variable or constant.

2.2   Design and Implementation of Dictionary

Data Structure of Dictionary: Dictionary is defined as a collection of key-value pairs. The key is unique, which means that one key can only correspond to one value. For example, a Dictionary type variable can store an address book, where the key of each key-value pair is a person's name and the value is the corresponding address. Considering that the dictionary may require frequent access operations, hash table is chosen to store the data in Dictionary.

As shown in Fig. 6, a class named "MDict" is defined to implement the function of Dictionary. There are three private variables in "MDict" class. "mDictHashTable" represents the hash table, "capacity" represents the capacity of the hash table, and "size" represents the number of key-value pairs in the dictionary. The hash table is an array of "MDictSlot" type.

It can also be seen from the figure that "MDictSlot" contains two pointers to "Symbol", namely "key" and "value", and "hashRes" represents a hash value, which is stored in an array of "uint8_t" type. "hashRes" corresponds to the symbol "key" and is obtained from the private function "hash(Symbol*)" in "MDict" class. The input of this hash function is the address of a "Symbol", and its output is a 256-bit hash value (stored in an array of "uint8_t"). The SHA256 algorithm is used inside the hash function, mainly to take advantage of the low-collision of the algorithm. Of course, other hash function or combinations of hash functions are also possible. In this hash function, only symbols of integer, floating-point numbers, characters, and strings can be accepted. It is to ensure the immutability and uniqueness of the keys in the dictionary. But there is no restriction on the values, which can be a constant or variable of any data type.

Function "insert (Symbol*, Symbol*)" is used to complete the insertion operation of key-value pair. The input is two symbols, which represent key and value in turn. When the function is executed, it will first determine whether the "size" reaches three-quarters of the "capacity", and if the "size" reaches three-quarters of the "capacity", function "expand ()" will be called. By the way, the initial

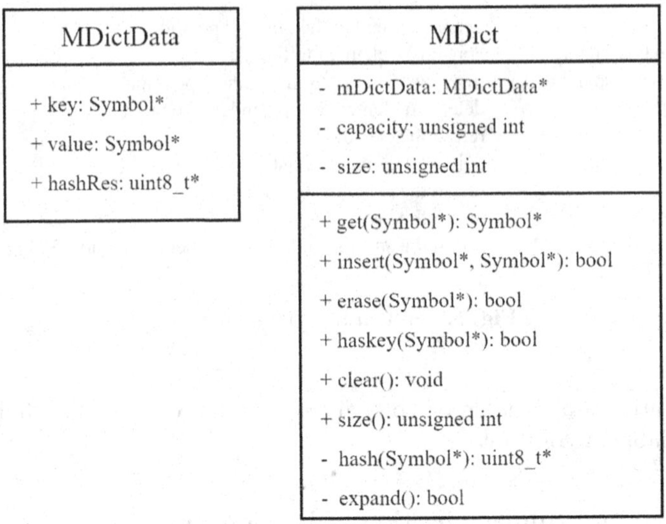

MDictData
+ key: Symbol*   + value: Symbol*   + hashRes: uint8_t*

MDict
- mDictData: MDictData*   - capacity: unsigned int   - size: unsigned int
+ get(Symbol*): Symbol*   + insert(Symbol*, Symbol*): bool   + erase(Symbol*): bool   + haskey(Symbol*): bool   + clear(): void   + size(): unsigned int   - hash(Symbol*): uint8_t*   - expand(): bool

Fig. 6. Data structure of Dictionary

value of "capacity" will be set to 16. Then the hash function is called on the key to get a 256-bit hash value, and the first $\log_2 capacity$ bits of the hash value are used as an index. The value of the index is between 0 and $capacity - 1$. And then the key, value and hash value will be stored at the "index" position of the hash table. If there is already a key-value pairs at the "index" position, it will compare the stored hash value ("hashRes") is equal to the hash value just calculated. If they are equal, the value will be updated directly, otherwise a hash conflict will occur. The method to solve the hash conflict is the open addressing method, which will find an empty slot and insert the key-value pair into the hash table.

Function "expand ()" is mentioned in the insert operation, which is used to expand the hash table. When "expand ()" is executed, a new hash table is applied for, and its size is twice the old one. Traverse the old hash table, reinsert each key-value pair into the new hash table, and finally update the capacity value.

Function "get (Symbol*)" is used to find a value corresponding to a given key. The process of querying the hash table is similar to that in the insert operation.

Function "haskey (Symbol*)" is to determine whether a "Symbol" is a key in the dictionary.

Function "erase (Symbol*)" is used to delete a key-value pair which is located by input parameter.

Syntax and Syntactic Interpretation of Dictionary: Similar to List, if we want to use Dictionary, we directly use the assignment statement to assign a Dictionary to a variable.

For example:

$$a = \{1 : \text{``hello''}, \text{``array''} : [1, 3, 5]\}$$

In this statement, a pair of curly braces are used to indicate the initialization of a dictionary, and a key-value pair is represented in the form of "key : value". The key and value are separated by a colon, and each key-value pairs are separated by a comma. Finally, this dictionary is assigned to the variable "a", and "a" becomes a dictionary type variable.

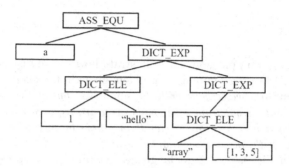

Fig. 7. Abstract syntax tree of Dictionary initialization statement

Figure 7 shows the abstract syntax tree of Dictionary initialization statement. The node type of the root node is "ASS_EQU", which means that an assignment statement will be parsed.

When the interpreter parses the first "DICT_EXP" node, it will instantiate an "MDict" object. The second and subsequent "DICT_EXP" nodes all mean that there are "DICT_ELE" nodes that need to be parsed. Each "DICT_ELE" node has two child nodes, and the interpreter parses them to get two "Symbol" objects, the first represents a key and the second represents a value. And then the interpreter will call the "insert" function in "MDict" object with two "Symbol" object as parameters. After the interpreter has parsed all "DICT_ELE" nodes, it will back to the first "DICT_EXP" node and instantiate a "Symbol" object, whose "mSymbolType" set to "SYMBOLDICT", and its "mData" will point to "mDict" object, and its "mSymbolName" will set to "a". Finally, this "Symbol" object will be added to the symbol table.

In addition to initializing a Dictionary variable, we also need some built-in functions related to Dictionary variables. The code snippet in Fig. 8 describes the syntax for operating Dictionary variables in the XD-M language, where "a" is a Dictionary variable, "key" is a variable or constant, whose type is limited to integer, floating point, character and string. "value" is a variable or constant of any data type. For an invalid statement, the interpreter will output an error message and skip this statement.

a.at(key)	// return a value, which corresponds to the key
a.append(key, value)	// append a key-value pair to the dictionary
a.del(key)	// delete a key-value pair
a.haskey(key);	// determine whether there is a key in the dictionary
a.size()	// return the size of the dictionary
a.clear()	// delete all key-value pairs in the dictionary

Fig. 8. Built-in functions for Dictionary

3   Example

In this section, an example is given to illustrate how to use List and Dictionary data types in XD-M programming. A problem given in the example is described as follows: *There are twelve balls with the same size, but one of them called bad ball has a different weight from other balls. You are required to find out the bad ball and confirm it is lighter or heavier than others by means of weighing them only three times with one scale without tick mark.*

As shown in Fig. 9, an XD-M language program is developed to solve the problem of finding the bad ball.

In the fourth line of the program, "balls" is a Dictionary type variable, which contains twelve key-value pairs, and each key-value pair represents a ball and its weight. In lines 2 to 3 in the program, the user is required to randomly choose two different integers and use them to instantiate the values of 12 key-value pairs in the dictionary. The first integer is randomly assigned to the value of a key-value pair and the second integer is assigned to values of the rested 11 key-value pairs one by one.

The next step is to find the bad ball by comparing just three times. In lines 11 to 13, "A", "B" and "C" are three List type variables. The balls are divided into three groups and stored in "A", "B", "C". The four balls in group A are numbered 1, 2, 3, 4, similarly, the four balls in group B are numbered 5, 6, 7, 8, and the four balls in group C are numbered 9, 10, 11, 12. The comparison processes of the program are explained below. To show the running process more intuitively, Fig. 10 lists all possible situations.

In the first comparison, the total weights of groups A and B are compared.

(1) If the weights of group A and group B are equal, the bad ball is among the four balls in group C and the bad ball can be easily found out by comparing twice.

(2) As shown in lines 46 to 63 in the program, if the weight of group A is greater than the weight of group B, there must be a ball in group A whose weight is greater than the normal weight or the weight of a ball in group B is less than the normal weight. Keep this case in mind, the program continues doing the following:

The second comparison can be processed as follows. Place the balls numbered 3, 4, 5, 6, 7 on the left hand side of the scale and the balls numbered 8, 9, 10, 11, 12 on the right hand side.

```
 1 frame(norWeight, badWeight, balls, A, B, C) and (
 2    badWeight = 0; mscan(badWeight);
 3    norWeight = 0; mscan(norWeight);
 4    balls = {"ball-1" :norWeight, "ball-2" :norWeight, "ball-3" :norWeight,
 5             "ball-4" :norWeight, "ball-5" :norWeight, "ball-6" :norWeight,
 6             "ball-7" :norWeight, "ball-8" :norWeight, "ball-9" :norWeight,
 7             "ball-10":norWeight, "ball-11":norWeight, "ball-12":norWeight};
 8    badBallNum = random(11) + 1;           // Take a random number between 1 and 12
 9    ballName = "ball-" + to_string(badBallNum);
10    balls.at(ballName) = badWeight;
11    A = <"ball-1","ball-2","ball-3","ball-4">;      //1, 2, 3, 4
12    B = <"ball-5","ball-6","ball-7","ball-8">;      //5, 6, 7, 8
13    C = <"ball-9","ball-10","ball-11","ball-12">;   //9,10,11,12
14    lWeight1 = 0; rWeight1 = 0; lWeight2 = 0; rWeight2 = 0; lWeight3 = 0; lWeight3 = 0;
15    lSide = <>; rSide = <>;
16    for(i = 0; i < 4; i:=i+1){
17       lWeight1 := lWeight1 + balls.at(A.at(i)); rWeight1 := rWeight1 + balls.at(B.at(i))};
18    if(lWeight1 == rWeight1) then{            // Equal weight means bad ball in C
19       lSide.append(A.at(0)); lSide.append(A.at(1)); lSide.append(A.at(2));
20       rSide.append(C.at(0)); rSide.append(C.at(1)); rSide.append(C.at(2));
21       for(i = 0; i < 3; i:=i+1) {
22          lWeight2:=lWeight2+balls.at(lSide.at(i));rWeight2:=rWeight2+balls.at(rSide.at(i)));
23          if(lWeight2 == rWeight2) then{       // Bad ball is 12(C3)
24             lWeight3 = balls.at(C.at(3)); rWeight3 = balls.at(A.at(0));
25             if(lWeight3>rWeight3) then{mout("Bad ball is ",C.at(3),", which is lighter than other balls\n")}
26             else{mout("Bad ball is ",C.at(3),", which is heavier than other balls\n")}}
27          else{
28             if(lWeight2 > rWeight2) then{      // Bad ball is in 9, 10, 11(lighter)
29                lWeight3 = balls.at(C.at(0)); rWeight3 = balls.at(C.at(1));
30                if(lWeight3==rWeight3) then{mout("Bad ball is ",C.at(2),", which is lighter than other balls\n")}
31                else{
32                   if(lWeight3>rWeight3) then{mout("Bad ball is ",C.at(1),", which is lighter than other balls\n")}
33                   else{mout("Bad ball is ",C.at(0),", which is lighter than other balls\n")}}}
34             else{                              // Bad ball is in 9, 10, 11(heavier)
35                lWeight3 = balls.at(C.at(0)); rWeight3 = balls.at(C.at(1));
36                if(lWeight3==rWeight3) then{mout("Bad ball is ",C.at(2),", which is heavier than other balls\n")}
37                else{
38                   if(lWeight3>rWeight3) then{mout("Bad ball is ",C.at(0),", which is heavier than other balls\n")}
39                   else{mout("Bad ball is ",C.at(1),", which is heavier than other balls\n")}}}}}
40    else {
41       lSide = <>; rSide = <>; lSide.append(A.at(2)); lSide.append(A.at(3));
42       lSide.append(B.at(0));lSide.append(B.at(1)); rSide.append(C.at(2)); rSide.append(B.at(3));
43       rSide.append(C.at(0));rSide.append(C.at(1));rSide.append(C.at(2));rSide.append(C.at(3));
44       for(i = 0; i < 5; i:=i+1){
45          lWeight2:=lWeight2+balls.at(lSide.at(i)); rWeight2:=rWeight2+balls.at(rSide.at(i))};
46       if(lWeight1 > rWeight1) then{            // The heavier ball in A or the lighter ball in B
47          if(lWeight2 == rWeight2) then{         // Bad ball is in 1, 2(heavier)
48             lWeight3 = balls.at(A.at(0)); rWeight3 = balls.at(A.at(1));
49             if(lWeight3>rWeight3) then{mout("Bad ball is ",A.at(0),", which is heavier than other balls\n")}
50             else{mout("Bad ball is ",A.at(1),", which is heavier than other balls\n")}}
51          else{
52             if(lWeight2 > rWeight2) then{       // Bad ball is in 3, 4(heavier), 8(lighter)
53                lWeight3 = balls.at(A.at(2)); rWeight3 = balls.at(A.at(3));
54                if(lWeight3==rWeight3) then{mout("Bad ball is ",B.at(3),", which is lighter than other balls\n")}
55                else{
56                   if(lWeight3>rWeight3) then{mout("Bad ball is ",A.at(2),", which is heavier than other balls\n")}
57                   else{mout("Bad ball is ",A.at(3),", which is heavier than other balls\n")}}}
58             else{                               // Bad ball is in 5, 6, 7(lighter)
59                lWeight3 = balls.at(B.at(0)); rWeight3 = balls.at(B.at(1));
60                if(lWeight3==rWeight3) then{mout("Bad ball is ",B.at(2),", which is lighter than other balls\n")}
61                else{
62                   if(lWeight3>rWeight3) then{mout("Bad ball is ",B.at(1),", which is lighter than other balls\n")}
63                   else{mout("Bad ball is ",B.at(0),", which is lighter than other balls\n")}}}}}
64       else{                                    // The heavier ball in B or the lighter ball in A
65          if(lWeight2 == rWeight2) then{         // Bad ball is in 1, 2(lighter)
66             lWeight3 = balls.at(A.at(0)); rWeight3 = balls.at(A.at(1));
67             if(lWeight3>rWeight3) then{mout("Bad ball is ",A.at(1),", which is lighter than other balls\n")}
68             else{ mout("Bad ball is ",A.at(0),", which is lighter than other balls\n")}}
69          else{
70             if(lWeight2 > rWeight2) then{       // Bad ball is in 5, 6, 7(heavier)
71                lWeight3 = balls.at(B.at(0)); rWeight3 = balls.at(B.at(1));
72                if(lWeight3==rWeight3) then{mout("Bad ball is ",B.at(2),", which is heavier than other balls\n")}
73                else{
74                   if(lWeight3>rWeight3) then{mout("Bad ball is ",B.at(0),", which is heavier than other balls\n")}
75                   else{ mout("Bad ball is ", B.at(1), ", which is heavier than other balls\n")}}}
76             else{                               // Bad ball is in 3, 4(lighter), 8(heavier)
77                lWeight3 = balls.at(A.at(2)); rWeight3 = balls.at(A.at(3));
78                if(lWeight3==rWeight3) then{mout("Bad ball is ",B.at(3),", which is heavier than other balls\n")}
79                else{
80                   if(lWeight3>rWeight3) then{mout("Bad ball is ",A.at(3),", which is lighter than other balls\n")}
81                   else{ mout("Bad ball is ",A.at(2),", which is lighter than other balls\n")}}}}}}}
```

Fig. 9. Example program

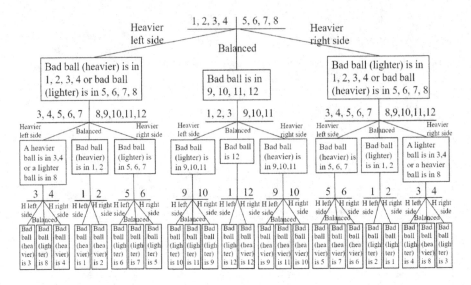

Fig. 10. Execution flow of the example program

1) If the left hand side of the scale is heavier than the right hand side, there must be a heavier ball in No. 3 and No. 4 or the lighter one is ball No. 8. Thirdly, No. 3 and No. 4 balls are compared. If they are equal, the No. 8 ball is a lighter one. Otherwise, the bad ball No. 3 or No. 4 is the heavier one.

2) If the left hand side of the scale is equal to the right hand side, there must be a heavier ball in No. 1 and No. 2.
   Then thirdly No. 1 and No. 2 balls are compared, and the bad ball No. 1 or No. 2 is the heavier one.

3) If the left hand side of the scale is lighter than the right hand side, there must be a lighter ball in No. 5, No. 6, and No. 7.
   Then thirdly, No. 5 and No. 6 balls are compared. If they are equal, the No. 7 ball is the lighter ball. Otherwise, the bad ball No. 5 or No. 6 is the lighter one.

(3) On the other hand, in lines 64 to 81 of the program, if the weight of group A is lighter than the weight of group B, the explanation of the program running process to find out the bad ball is similar to the previous case (2).

4   Conclusion

This paper introduces the implementation of two data types, List and Dictionary in XD-M language, including data structures, algorithms, and the interpretation process of abstract syntax trees. In the future, more operations and functions on List and Dictionary need to be added and implemented so as to make XD-M more powerful and friendly.

References

1. Duan, Z.: Temporal Logic and Temporal Logic Programming. Science Press, Beijing (2005)
2. Duan, Z., Tian, C.: A practical decision procedure for propositional projection temporal logic with infinite models. Theor. Comput. Sci. **554**, 169–190 (2014)
3. Duan, Z., Tian, C., Zhang, N.: A canonical form based decision procedure and model checking approach for propositional projection temporal logic. Theor. Comput. Sci. **609**, 544–560 (2016)
4. Rosner, R., Pnueli, A.: A choppy logic. In: Proceedings of the Symposium on Logic in Computer Science, Cambridge, pp. 306–313 (1986)
5. Bowman, H., Thompson, S.: A decision procedure and complete axiomatization of finite interval temporal logic with projection. J. Logic Comput. **13**, 195–239 (2003)
6. Wang, X., Duan, Z., Zhao, L.: Formalizing and implementing types in MSVL. In: Proceedings of the International Workshop of Structured Object-Oriented Formal Language and Method, Queenstown, pp. 62–75 (2013)
7. Duan, Z., Yang, X., Koutny, M.: Framed temporal logic programming. Sci. Comput. Program. **70**, 31–61 (2008)
8. Zhang, N., Duan, Z., Tian, C.: A mechanism of function calls in MSVL. Theor. Comput. Sci. **654**, 11–25 (2016)
9. Mark, H., Andy, R.: Python Programming on Win32. O'Reilly & Associates Inc., Sebastopol (2000)
10. Python Homepage. https://www.python.org/
11. Duan, Z., Tian, C., Zhang, N.: Normal form expressions of propositional projection temporal logic. In: Cai, Z., Zelikovsky, A., Bourgeois, A. (eds.) COCOON 2014. LNCS, vol. 8591, pp. 84–93. Springer, Cham (2014). https://doi.org/10.1007/978-3-319-08783-2_8
12. Valmari, A.: A stubborn attack on state explosion. Form. Method Syst. Des. **1**, 297–322 (1992). https://doi.org/10.1007/BF00709154
13. Burch, J., Clarke, E., McMillan, K., et al.: Symbolic model checking: 10^{20} states and beyond. Inform. Comput. **98**, 142–170 (1992)
14. Duan, Z., Tian, C.: A unified model checking approach with projection temporal logic. In: Liu, S., Maibaum, T., Araki, K. (eds.) ICFEM 2008. LNCS, vol. 5256, pp. 167–186. Springer, Heidelberg (2008). https://doi.org/10.1007/978-3-540-88194-0_12
15. Biere, A., Cimati, A., Clarke, E., et al.: Bounded model checking. Adv. Comput. **58**, 117–148 (2003)
16. Zhang, N., Duan, Z., Tian, C.: Model checking concurrent systems with MSVL. Sci. China Inf. Sci. **59**(11), 1–3 (2016). https://doi.org/10.1007/s11432-015-0882-6
17. Clarke, E., Grumberg, O., Long, D.: Model checking and abstraction. ACM Trans. Progr. Lang. Syst. **16**, 1512–1542 (1992)
18. Tian, C., Duan, Z., Duan, Z.: Making CEGAR more efficient in software model checking. IEEE Trans. Softw. Eng. **40**, 1206–1223 (2014)
19. Godefroid, P., Wolper, P.: A partial approach to model checking. Inform. Comput. **110**, 305–326 (1994)
20. Flex Homepage. https://github.com/westes/flex
21. GNU Bison Homepage. http://www.gnu.org/software/bison/

Reliable Edge Intelligence Using JPEG Progressive

Haobin Luo$^{(\boxtimes)}$, Xiangang Du, Luobing Dong⬤, Guowei Su, and Ruijie Chen

Xidian University, No. 2 South Taibai Road, Xi'an 710071, Shaanxi, China

Abstract. Edge intelligence (EI), the interdisciplinary field of edge computing and artificial intelligence (AI), aims at achieving time-critical AI services. Existing work mainly focuses on task offloading and model optimization to realize low latency as well as low power consumption. However, to guarantee low latency in lossy mobile network, user datagram protocol (UDP) is preferred to the transmission control protocol (TCP). Although image files can tolerate a small portion of corruption, latent packet loss and out-of-order can damage the overall service reliability. To improve it, we found that progressive encoding in joint photographic expert group (JPEG) standard tends to keep low-frequency coefficients in the first place compared to the commonly used baseline encoding. This paper studies whether JPEG Progressive achieves more reliable EI than JPEG Baseline in unreliable network environment. In our experiments, including scenarios of single packet loss, multiple packet loss and random packet loss, JPEG Progressive achieves significantly higher inference performance, with huge benefit in latency due to UDP.

Keywords: Edge intelligence · Edge computing · Artificial intelligence · JPEG · Reliability

1 Introduction

The confluence of artificial intelligence (AI) and edge computing gives birth to edge intelligence (EI) [1]. So far, it is not simply the combination of edge computing and artificial intelligence but a collaboration of edge and cloud [2]. The main objectives of edge computing are to reduce both the latency and the bandwidth [3]. Introducing edge computing to AI enables AI applications in time-critical scenarios.

In mobile edge intelligence, where end devices are wirelessly connected, problem caused by unstable channels arises, especially packet loss and out-of-order. To guarantee reliable transmission, the typical solution is to use transmission control protocol (TCP) at transport layer [6]. However, due to its default additive increase and multiplicative decrease (AIMD) strategy [5], the throughput under lossy network will be greatly limited [4]. Also, the retransmission is not essential for loss-tolerable data like images and video streams. The above two characteristics make TCP not preferable, in contradiction to latency decrease, which is the major goal of edge computing. In this scenario, user datagram protocol (UDP) becomes a more suitable choice. Work has been done to

© Springer Nature Switzerland AG 2021
W. Wu and H. Du (Eds.): AAIM 2021, LNCS 13153, pp. 356–368, 2021.
https://doi.org/10.1007/978-3-030-93176-6_30

improve the service reliability of edge intelligence based on UDP in the scenario of image classification [7].

Although pictures can abide corruptions, image quality will be inevitably impaired [11]. Figure 1 exhibits a picture of spatial distortion, channel misplacing and early termination, three types of possible perturbations. Spatial distortion is caused by other pixels taking the places of prior pixels. Potential channel misplacing arises with spatial distortion, so that the remainder of the picture shows a biased color. Besides, in early termination, missed pixels at the bottom are filled with gray color.

Fig. 1. A corrupted picture with: 1) spatial distortion 2) channel misplacing and 3) early termination.

In AI applications, image distortion results in inference degradation, especially in object detection. Some objects are split into multiple, bringing confusion to the model. Also, the coordination shift caused by packet loss may result in taking the wrong response. Therefore, it is necessary to minimize the impact of transmission loss.

Improving EI reliability in unreliable network can be a tradeoff between transmission latency and inference performance if only network protocol is replaced. Since most pictures are organized in jpg or jpeg format [22], we wonder whether a sophisticated encoding can realize a more reliable EI. JPG and JPEG pictures are encoded according to Joint Photographic Expert Group (JPEG) standard. Among 4 types of encodings defined in JPEG [8], JPEG Sequential is most prevalently used, especially one realization called JPEG Baseline. In this encoding method, the whole image is rendered in one scan. JPEG

Progressive, instead, renders the picture by multiple scans so that users have a rough preview before the picture is fully downloaded. Its principle is to put low-frequency coefficients closer to the beginning of the file. We assume this characteristic makes pictures more resistant to corruptions.

In this paper, we research whether the 'rough preview' of JPEG Progressive will help improve the service reliability of Image-based Edge Intelligence. Section 2 introduces the related work regarding JPEG Progressive and reliable edge intelligence. Work in both EI and JPEG encoding areas are impressive, but we are the first to attempt to improve the reliability of EI with the help of JPEG Progressive. Section 3 designs the experiment and Sect. 4 discusses the results. Our experiment takes object detection as an example and shows the JPEG Progressive improves the EI reliability in lossy environment.

2 Related Work

2.1 JPEG Progressive

The JPEG still picture compression standard [8] defined four encoding modes of still pictures. This paper focuses on the JPEG Sequential, JPEG Progressive for their prevalent application. Specifically, we discuss baseline encoding and spectrum selection, two typical realization of each mode. In the rest of this paper, baseline and spectrum selection can be considered equivalent to the JPEG Sequential and JPEG Progressive respectively for simplicity. The difference between them is that in entropy encoding phase, progressive encoding tends to keep the global low-frequency coefficients in the first place. In contrast, the baseline encoding only puts the low-frequency coefficients first inside each 8 * 8 block.

M. Mody et al. [9] proposed a decoder design for JPEG Baseline decoders to decode JPEG Progressive pictures and it largely reduces the memory requirements. Andrew Louie et al. [12] put forward an JPEG Progressive encoder design for real-time systems to minimize latency. Yan et al. [10] provided an image compression method for storage based on JPEG Progressive. Wiseman [11] showed that JPEG Progressive helps alleviate JPEG inaccuracy appearance. Their work paved the way of our work.

2.2 Reliable Edge Intelligence

Deng et al. [1] provided an insight into the interdisciplinary field edge intelligence by dividing EI into two categories 'AI for edge' and 'AI on edge', the former emphasizes on edge computing while the latter focuses more on AI. To formally define edge intelligence, Zhou et al. [2] proposed six levels of edge intelligence, from cloud-edge co-inference (level 1) to all on-device (level 6).

To satisfy the time-critical requirements of EI, 'AI for edge' approaches focus on partitioning computationally intensive components from the resource-constrained devices. Kang et al. [13], Hu et al. [14] and Dong et al. [20] optimized the latency, power consumption as well as network throughput by partitioning AI models. Furthermore, Dong et al. [15] formulated the above solution into a n-fold integer programing problem by combinatorial optimization. Alternatively, [16] and [17] considered reducing latency by image compression, therefore reduces both transmission cost and inference cost. On the other hand, 'AI on edge' approaches [18, 19] adopt early-exit inference to reduce the inference latency.

Above work considered the network transmission to be lossless and mainly focused on optimizing latency, throughput and energy consumption. However, in the practice of mobile edge computing, noisy network causes a great increase in the transmission latency. Liu and Zhang [7] found the UDP-based offloading can improve the normalized service reliability by up to 70% for time-critical services under lossy network. Lee et al. [21] discussed the approaches to design robust AI platforms against input perturbations. Work to improve EI reliability under such scenario remains insufficient.

3   Experiment Setup

To support our assumption that the progressive encoding helps improve the EI reliability in lossy network environments, the experiment is designed so that pictures in JPEG Baseline and JPEG Progressive suffer same extents of corruption in each scenario. The EI reliability can be quantized by the model considering both inference performance and latency.

3.1   Experiment Environment

Hardware Platform and Network Environment. As Fig. 2 suggests, we consider a simple edge intelligence scenario where one user equipment (UE) and one edge server (ES) are deployed. In our experiments, the UE of the system is Raspberry Pi 4B with 2 GB RAM, a general embedded system for image collection. The ES is a web server simulated by a laptop with i7 11800H CPU and RTX3060 GPU. For a better comparison, UE and ES are connected under the same Wi-Fi LAN with RTT 3 ms. The latency and packet loss are realized by TC commands at the UE or dropping packets at the ES.

Fig. 2. Experiment model

Data Organization. Like most web applications, our data are JSON serialized. Pictures are base64 encoded and each packet takes 1 KB of the base64 content. To truncate each file efficiently in lossy network, late arrived packets will be discarded by the ES. In order to do so, 3 other fields, filename, sequence number and total sequence, are provided in each packet.

AI Models and Datasets. To investigate whether and how much JPEG Progressive encoding can improve the service reliability in a lossy network, the pretrained YOLOv5s is used as an example of objection detection applications. It is a lightweight model with 7266973 parameters in state-of-the-art YOLOv5 series. And the model is evaluated on COCO128 of different corruption by the metric mAP (0.5:0.95), also refered as mAP_{val}.

3.2 System Model

The EI reliability can be generalized with respect to average inference reliability and average time reliability, where $C_1, C_2 and p_1, p_2$ are weights to adjust in different scenarios:

$$\overline{R_{EI}} = C_1\overline{R_{inference}}^{p1} \times C_2\overline{R_{time}}^{p2}$$

The inference reliability can be measured by a specific metric. Specifically, we use mAP (0.5:0.95) for $R_{inference}$ in this paper. The unreliability in time, on the other hand, is mainly caused by timeout so that a picture is missed or fails to be computed in time by ES. Similar to [7], our time reliability is defined as:

$$\overline{R_{time}} = 1 - \overline{F_{timeout}}$$

$\overline{F_{timeout}}$ in the above equation refers to the average timeout probability. For image-based EI applications, the system latency can be written as the sum of the transmission latency and the computation latency:

$$T_{total} = T_{transmission} + T_{computation}$$

For each sample picture i, the timeout state is either 1 or 0, in the below equation where ε is the Heaviside function and δ refers to the threshold:

$$F^i_{timeout} = \varepsilon(T^i_{total} - \delta)$$

With enough number of samples, the average timeout probability can be derived as:

$$\overline{F_{timeout}} = \lim_{I \to \infty} [\frac{1}{I}\sum_{i=1}^{I} \varepsilon\left(T^i_{total} - \delta\right)]$$

Overall, the EI reliability can be expressed as:

$$\overline{R_{EI}} = C_1C2 \cdot mAP_{val}^{p1}(1 - \lim_{I \to \infty} [\frac{1}{I}\sum_{i=1}^{I} \varepsilon\left(T^i_{transmission} + T^i_{computation} - \delta\right)])^{p2}$$

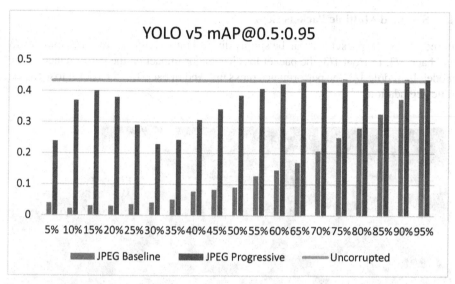

Fig. 3. The mAP (0.5:0.95) of YOLO v5 with corrupted input pictures under single packet loss at difference positions of a file.

4 Results and Discussion

4.1 Specified Single Packet Loss

This experiment is to find out the relationship between the position of single packet loss and the inference reliability. Since the first packet contains a not disposable file header and each file size of YOLO128 varies from 20 KB to over 200 KB, specifying lost packet number has different influence on images. Therefore, one packet is dropped at different positions of every picture, ranging from the position at 5% to 95%.

As Fig. 3 shows, image distortion in baseline encoded pictures is destructive. For a single packet loss of 1 KB taking place in first 35% of the file, the mAP_{val} drops dramatically down to below 0.05. The mAP_{val} of a single packet loss at 85% reaches 0.327, very close to 0.348, the mAP_{val} of YOLOv3-tiny. For baseline encoded pictures, single packet loss later than 85% has minor effect on inference reliability.

In contrary, JPEG Progressive behaves more robust to single packet loss, with nearly all mAP_{val} values over 0.3. The performance is much more ideal than the baseline encoding. However, the mAP_{val} experiences a drop for a packet loss around 30%. Figure 4 demonstrates an example of the above scenario. Most contours in the picture, possibly encoded around that position of the picture, suffer a shift, that explains the reason of underperformance at mid-frequency.

4.2 Specified Multiple Packets Loss

Based on 4.1, the packet loss can be simply divided into 3 regions, front 1/3, middle 1/3 and back 1/3. In front 1/3, the packet loss has major impact on the performance of the model. In middle 1/3, the performance picks up. And in back 1/3, the packet loss causes little degradation.

Fig. 4. A corrupted JPEG Progressive picture with single packet loss

The aim of this part is to investigate whether image corruption in JPEG Progressive still outperforms JPEG Baseline in inference when more packets are lost. We divide the loss position into 3 regions: front 1/3, middle 1/3 and back 1/3, with both consecutive and random packet loss specified by random module in numpy package. For comparison, the number of lost packets is set to be 2, 4, 6 and 8, corresponding to 2 KB, 4 KB, 6 KB and 8 KB loss respectively.

Table 1 indicates that in both scenarios of consecutive and random packet loss, JPEG Progressive outperforms the JPEG Baseline. Especially in the front 1/3, With consecutive loss of 6 KB, the mAP_{val} remains above 0.3, while this matric of JPEG Baseline already drops below 0.05.

The result follows a similar pattern when the packet loss happens closer to the end of the file, with a less severe impact. In middle 1/3, the mAP_{val} of baseline encoding reaches slightly above 0.1 for consecutive loss. If the lost packets are discrete, the performance is still poor. In most cases in middle 1/3, JPEG Progressive remains mAP_{val} above 0.3 except for 4 KB random losses. Packet loss in the back 1/3 has little influence on the inference performance.

Table 1. The mAP (0.5:0.95) values of YOLO v5 under different cases of multiple packet loss

Loss type	Encoding	Loss size	Front 1/3	Middle 1/3	Back 1/3
Lossless	Baseline	0 KB	0.434	0.434	0.434
Consecutive		2 KB	0.0413	0.107	0.296
		4 KB	0.0377	0.103	0.315
		6 KB	0.0443	0.126	0.315
		8 KB	0.0261	0.12	0.311
	Progressive	2 KB	0.314	0.376	0.436
		4 KB	0.328	0.369	0.437
		6 KB	0.305	0.382	0.443
		8 KB	0.263	0.374	0.443
Random	Baseline	2 KB	0.0235	0.0717	0.274
		4 KB	0.0156	0.0652	0.244
		6 KB	0.0166	0.0593	0.233
		8 KB	0.019	0.049	0.22
	Progressive	2 KB	0.208	0.332	0.441
		4 KB	0.119	0.317	0.441
		6 KB	0.0562	0.267	0.432
		8 KB	0.0421	0.266	0.438

Another key observation is that for same size of packet loss, random packet loss causes more damage to the inference performance. Both encodings satisfy this observation, which coincides with [7]. One explanation is that consecutive packet loss is just one packet loss of larger size while random packet loss cannot be merged.

The experiment explores the scenarios where multiple packets are lost, with respect to the loss position and whether the losses are consecutive. JPEG Progressive still outperforms the Baseline encoding, especially early packet loss, resulting in severe distortion in Baseline, the progressive encoding keeps the most low-frequency components in the right position.

4.3   Random Loss Simulation

Section 4.2 compares JPEG Baseline and JPEG Progressive in extreme scenarios when more than 1 packet of 1 KB are lost. However, the random numbers generated in Sect. 4.3 are pseudo-random numbers. To simulate more practical scenarios, tc (for traffic control) commands in Linux are applied. The lossy network is simulated by setting packet loss rate to 0.5%, 1% and 2% respectively and repeatedly sending the pictures from UE to ES, and the results are shown in Fig. 5.

Compared to the original mAP_{val} 0.434 under reliable data transmission, the mAP_{val} of JPEG Baseline drop dramatically to 0.12 when the packet loss is 0.5%, equivalent to one packet loss for images of size 200 KB. For higher loss rates, most JPEG Baseline pictures become not recognizable and it is meaningless to take them for object detection.

And in total random scenarios, the JPEG Progressive still behaves more robust to packet loss. Under loss rate 0.5%, the mean mAP_{val} remains above 0.3, which is highly usable. And even in the extreme scenario with 2% loss rate, the resulting mAP_{val} outperforms the JPEG Baseline under loss rate 0.5%.

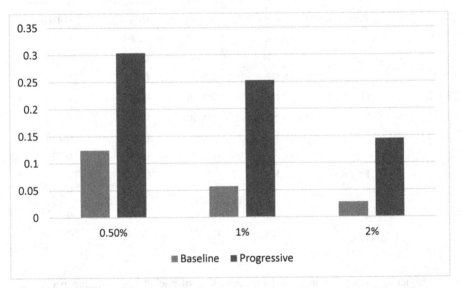

Fig. 5. The mAP (0.5:0.95) of YOLOv5 with corrupted input pictures under different loss rate

4.4   Time Reliability Analysis

In practice, the $\overline{R_{time}}$ in Sect. 3.2 is calculated according to the number of result pictures after inference within the time threshold. The time threshold δ in our experiment is 100 ms.

Table 2 compares the computation latency of baseline encoding with the progressive encoding. The difference in computation latency is trivial except the additional latency in re-encoding at UE. The overall average computation latency is less than 20 ms. However, if the inference is implemented on the UE, the inference latency can easily exceed the threshold.

Table 2. The difference in computation latency (ms)

Encoding method	Re-encoding latency	Preprocessing time	Inference time	NMS time
Baseline	/	2.0	6.3	8.3
Progressive	3.7	2.1	6.1	7.5

Then we consider the transmission latency, which is linear to file size and inversely linear to network throughput. The original baseline encoded COCO128 dataset takes 21.3 MB while the progressive version (100% quality) takes 3.5 MB more. The difference in size is not significant, making the comparison in throughput between TCP and UDP decisive.

TCP guarantees reliable data transmission so that $\overline{R_{inference}}$ is only determined by the inference model, at the cost of low $\overline{R_{time}}$. UDP has higher $\overline{R_{time}}$ but lower $\overline{R_{inference}}$. We compare the network throughput in scenarios described in Sect. 4.3, with round trip time (RTT) 3 ms and bandwidth 100 Mbps.

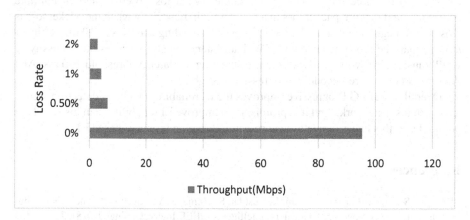

Fig. 6. TCP throughput in lossy network

Figure 6 suggests that TCP throughput drops dramatically in a lossy network. For example, the file size of progressive encoded first picture in COCO128 is 235 KB. With 1% loss rate, it takes 19 ms under UDP while TCP requires more than 400 ms, which is not acceptable for time-critical tasks. As a result, few pictures finish the task in time under TCP. Under UDP, the system outputs 119 result pictures on average, and the primary reason for picture loss is encoding corruption.

5 Future Work

This paper implements the preliminary research to improve the EI reliability by JPEG Progressive encoding in lossy network.

To generalize our results, further experiments on EfficientDet and even in image classification should be implemented. Besides, finding the inner connection between DCT and characteristics of the neural network is the key to answer the fundamental reason of the improvement in inference reliability. Also, in Sect. 4.1, the high frequency coefficients in progressive encoding have little contribution to the inference reliability, which can be a clue to design an optimized DCT-based image encoding for EI.

6 Conclusion

This paper considers the scenario of edge intelligence in noisy network environment. Since pictures can tolerate small distortions, to further reduce the transmission latency, UDP is preferred to TCP. However, the unreliable network transmission is likely to cause packet loss. The resulting image distortion causes damage to EI reliability, especially in object detection applications. To improve EI performance in lossy environment, we consider JPEG Progressive encoding is a possible solution.

In our experiments, progressive encoding achieves higher inference reliability (mAP_{val}) in all three scenarios of fixed single packet loss, pseudo-random multiple packet loss and random packet loss in a real network. Especially when the packet loss is close to the beginning of the file, the Progressive encoding achieves 5–10 times higher mAP_{val} than the Baseline encoding. The later latency analysis proves the necessity of UDP transmission, with nearly all pictures meeting the latency threshold 100 ms. And the time expense to re-encode in progressive is slight.

Overall, the JPEG Progressive improves the EI reliability in object detection significantly in lossy network. And it is promising to improve EI reliability in industries. Last but not least, this paper indicates new research direction in image-based EI.

References

1. Deng, S., Zhao, H., Fang, W., Yin, J., Dustdar, S., Zomaya, A.: Edge intelligence: the confluence of edge computing and artificial intelligence. IEEE Internet Things J. **7**(8), 7457–7469 (2020). https://doi.org/10.1109/JIOT.2020.2984887
2. Zhou, Z., Chen, X., Li, E., Zeng, L., Luo, K., Zhang, J.: Edge intelligence: paving the last mile of artificial intelligence with edge computing. Proc. IEEE **107**(8), 1738–1762 (2019). https://doi.org/10.1109/JPROC.2019.2918951
3. What is Edge Computing: The Network Edge Explained. https://www.cloudwards.net/what-is-edge-computing/. Accessed 27 Sept 2021
4. Measuring Network Performance: Links Between Latency, Throughput and Packet Loss. https://accedian.com/enterprises/blog/measuring-network-performance-latency-throughput-packet-loss/. Accessed 27 Sept 2021
5. Additive increase/multiplicative decrease. https://en.wikipedia.org/wiki/Additive_increase/multiplicative_decrease. Accessed 09 Sept 2021

6. Zimmermann, H.: OSI reference model - the ISO model of architecture for open systems inter-connection. IEEE Trans. Commun. **28**(4), 425–432 (1980). https://doi.org/10.1109/TCOM.1980.1094702

7. Liu, J., Zhang, Q.: To improve service reliability for ai-powered time-critical services using imperfect transmission in MEC: an experimental study. IEEE Internet Things J. **7**(10), 9357–9371 (2020). https://doi.org/10.1109/JIOT.2020.2984333

8. Wallace, G.K.: The JPEG still picture compression standard. IEEE Trans. Consum. Electron. **38**(1), xviii–xxxiv (1992). https://doi.org/10.1109/30.125072

9. Mody, M., Paladiya, V., Ahuja, K.: Efficient progressive JPEG decoder using JPEG baseline hardware. In: 2013 IEEE Second International Conference on Image Information Processing (ICIIP-2013), pp. 369–372 (2013). https://doi.org/10.1109/ICIIP.2013.6707617

10. Yan, E., Zhang, K., Wang, X., Strauss, K., Ceze, L.: Customizing progressive {JPEG} for efficient image storage. In: 9th {USENIX} Workshop on Hot Topics in Storage and File Systems (HotStorage 17) (2017)

11. Wiseman, Y.: Alleviation of JPEG inaccuracy appearance. Int. J. Multimed. Ubiquitous Eng. **11**(3), 133–142 (2016). https://doi.org/10.14257/ijmue.2016.11.3.13

12. Louie, A., Cheng, A.M.K.: Work-in-progress: designing a server-side progressive JPEG encoder for real-time applications. In: 2020 IEEE Real-Time Systems Symposium (RTSS), pp. 379–382 (2020). https://doi.org/10.1109/RTSS49844.2020.00043

13. Kang, Y., et al.: Neurosurgeon: collaborative intelligence between the cloud and mobile edge. ACM SIGARCH Comput. Archit. News **45**(1), 615–629 (2017). https://doi.org/10.1145/3093337.3037698

14. Hu, C., Bao, W., Wang, D., Liu, F.: Dynamic adaptive DNN surgery for inference acceleration on the edge. In: IEEE INFOCOM 2019 - IEEE Conference on Computer Communications, pp. 1423–1431 (2019). https://doi.org/10.1109/INFOCOM.2019.8737614

15. Dong, L., Wu, W., Guo, Q., Satpute, M.N., Znati, T., Du, D.Z.: Reliability-aware offloading and allocation in multilevel edge computing system. IEEE Trans. Reliab. **70**(1), 200–211 (2021). https://doi.org/10.1109/TR.2019.2909279

16. Li, H., Hu, C., Jiang, J., Wang, Z., Wen, Y., Zhu, W.: JALAD: joint accuracy-and latency-aware deep structure decoupling for edge-cloud execution. In: 2018 IEEE 24th International Conference on Parallel and Distributed Systems (ICPADS), pp. 671–678 (2018). https://doi.org/10.1109/PADSW.2018.8645013

17. Eshratifar, A.E., Esmaili, A., Pedram, M.: BottleNet: a deep learning architecture for intel-ligent mobile cloud computing services. In: 2019 IEEE/ACM International Symposium on Low Power Electronics and Design (ISLPED), pp. 1–6 (2019). https://doi.org/10.1109/ISLPED.2019.8824955

18. Teerapittayanon, S., McDanel, B., Kung, H.T.: BranchyNet: fast inference via early exiting from deep neural networks. In: 2016 23rd International Conference on Pattern Recognition (ICPR), pp. 2464–2469 (2016). https://doi.org/10.1109/ICPR.2016.7900006

19. Li, E., Zhou, Z., Chen, X.: Edge intelligence: on-demand deep learning model co-inference with device-edge synergy. In: Proceedings of the 2018 Workshop on Mobile Edge Com-munications (MECOMM'18), pp. 31–36. Association for Computing Machinery, New York (2018). https://doi.org/10.1145/3229556.3229562.

20. Dong, L., Satpute, M.N., Shan, J., Liu, B., Yu, Y., Yan, T.: Computation offloading for mobile-edge computing with multi-user. In: 2019 IEEE 39th International Conference on Dis-tributed Computing Systems (ICDCS), pp. 841–850 (2019). https://doi.org/10.1109/ICDCS.2019.00088

21. Lee, M., She, X., Chakraborty, B., Dash, S., Mudassar, B., Mukhopadhyay, S.: Reliable edge intelligence in unreliable environment, In: 2021 Design, Automation & Test in Europe Conference & Exhibition (DATE), pp. 896–901 (2021). https://doi.org/10.23919/DATE51 398.2021.9474097
22. Liu, X., Lu, W., Huang, T., Liu, H., Xue, Y., Yeung, Y.: Scaling factor estimation on JPEG compressed images by cyclostationarity analysis. Multimed. Tools Appl. **78**(7), 7947–7964 (2018). https://doi.org/10.1007/s11042-018-6411-9

A Game-Theoretic Analysis of Deep Neural Networks

Chunying Ren[1], Zijun Wu[2(✉)], Dachuan Xu[1], and Wenqing Xu[1]

[1] Department of Operations Research and Information Engineering,
Beijing University of Technology, Pingleyuan 100, Beijing 100124, China
renchunying@emails.bjut.edu.cn, {xudc,xwq}@bjut.edu.cn
[2] Institute for Applied Optimization, Department of Artificial Intelligence
and Bigdata, Hefei University, Jinxiu 99, Hefei 230601, Anhui, China
wuzj@hfuu.edu.cn

Abstract. As an important part of machine learning, deep learning has been intensively used in various fields relevant to data science. Despite of its popularity in practice, it is still of challenging to compute the optimal parameters of a deep neural network, which has been shown to be NP-hard. We devote the present paper to an analysis of deep neural networks with nonatomic congestion games, and expect that this can inspire the computation of optimal parameters of deep neural networks. We consider a deep neural network with linear activation functions of the form $x + b$ for some biases b that need not be zero. We show under mild conditions that learning the weights and the biases is equivalent to computing the social optimum flow of a nonatomic congestion game. When the deep neural network is for classification, then the learning is even equivalent to computing the equilibrium flow. These results generalize a recent seminar work by [18], who have shown similar results for deep neural networks of linear activation functions with *zero* biases.

Keywords: Deep neural networks · Game theory · Wardrop equilibria · Local minima

1 Introduction

In 1980, the first International Symposium on Machine Learning was held in Carnegie Mellon (CMU) in the United States, which has made machine learning be known all over the world. Since then, machine learning has been widely used, see, e.g., [1,5,13]. With the development of machine learning, its application scenarios have become more complex, and people have consciously associated game theory with machine learning. In fact, the idea of game theory has always existed in many machine learning exploration processes such as the classic *support vector machine* (SVM) or *reinforcement learning*.

One can regard the classic SVM as finding a (pure) *Nash equilibrium* (NE) for a two-person zero-sum game, since both of the two classes (players) want to minimize their cost (classification errors), and an optimal classifier would then

© Springer Nature Switzerland AG 2021
W. Wu and H. Du (Eds.): AAIM 2021, LNCS 13153, pp. 369–379, 2021.
https://doi.org/10.1007/978-3-030-93176-6_31

be a hyperplane that perfectly balances the cost of the two classes. This has been realized by [12] for Network Intrusion Detection (NID), and obtained an SVM classifier with better learning ability and generalized performance. Moreover, reinforcement learning can be seen as a game between the player and the environment. [10] has designed a zero-sum random game reinforcement learning algorithm. Then [11] have shown the convergence of that algorithm. [8] have proposed a reinforcement learning method under the framework of general-sum random games, in which some highly restrictive assumptions on the form of stage games are made to ensure the convergence.

While hybridizing machine learning with games has attracted much attention, only few researchers have considered the combination of deep learning and games. Deep learning is a machine learning method with a deep architecture. It was formally proposed by [7] in 2006. Recently, it has attracted much mainstreamed attention since it overcomes the shortcomings of manual design features rooted in traditional algorithms. So far, deep learning has successfully applied to computer vision, pattern recognition, speech recognition, natural language processing and recommendation systems, and others.

Despite of its popularity in application, deep learning itself is NP-hard. [14] have proved in 1987 that finding the global minimum of a general nonconvex function is NP-complete. With deep learning, one might expect that the function induced by the deep model has a certain structure to make the resulting nonconvex optimization easier to handle. Unfortunately, [3] have shown that training a simple 3-layer neural network is already NP-hard. Hence, commonly used solution methods for deep learning are still these convex optimization techniques such as various gradient descent approaches. [4] have used the property of strict saddle point to prove that stochastic gradient descent at an arbitrary starting point converges to a local minimum in polynomial iterations. This is the first work to a global convergence guarantee for stochastic gradient descent of nonconvex functions with exponentially many local minima and saddle points. As a continuation, [9] have proved that the gradient descent method with disturbance deformation converges to the second-order stationary point, and the required number of iterations depends on the logarithm of the dimension. For the case where all saddle points are nondegenerate, then all second-order stationary points would be local minima, and thus the results of [9] show that the perturbed gradient descent method can escape saddle points almost for free.

We consider in this paper an alternative way for deep learning, which aims at a possible paradigm for learning deep neural networks with *nonatomic congestion games*, see [16]. As indicated by [6], training a deep neural network requires optimizing a non-convex and non-concave objective function. Even in the case of a linear activation function, it may lead to any local minimum far away from the global minimum. Therefore, it's very useful to theoretically analyze deep neural networks with linear activation functions and we restrict our study only to such networks.

A seminar work towards this direction has been done recently by [18]. They have studied a linear deep neural network with activation functions of the form

$f(x) = x$. They showed that the corresponding deep neural network can be reformulated to be a nonatomic congestion game, and establish the equivalence relationship between the optimal weights of the deep neural networks and the *Wardrop equilibria* [19] of that nonatomic congestion game. Beyond this work, studies towards deep learning with other games have also attracted certain attention, see, e.g., [2,17] and others. [17] have shown an approach to formulate a deep neural network for supervised learning with differentiable convex gates as a simultaneous move two-person zero-sum game, which allowed a bijection between Karush-Kuhn-Tucker (KKT) points and Nash equilibria. [2] has proposed to optimize the deep network using game methods, and studied deep neural networks with nondifferentiable activation functions. In particular, [2] has explained from the perspective of game theory why the convex optimization design method in modern convolutional networks with nonconvex loss functions guarantees convergence.

We continue the work of [18] to consider *general* linear activation functions, that is, $f(x) = x + b$ for certain biases b that might not be zero. We show that the results of [18] carry over to this general case. While the results are similar, the proofs are highly nontrivial due to the influence of the extra biases. In particular, the nonatomic congestion game constructed by [18] does not apply in this general case. Compared with the work of [18], we consider nonatomic congestion games with more players, with each layer adding an O/D pair. Hence, we have proposed a new constructive method for the nonatomic congestion game, and then prove the results under this new environment.

The paper is organized as follows: Sect. 2 introduces the involved notions; Sect. 3 shows the construction and reports our main results; Sect. 4 shows a short summary of the whole paper. To improve the readability, we provide the proofs in an Appendix.

2   Model and Preliminaries

As we consider both deep neural networks and nonatomic congestion games, we devote this section to reviewing the relevant concepts.

2.1   Deep Neural Networks

We represent a *deep (feed-forwarding) neural network* (DNN) symbolically by a tuple $\mathcal{N} = (V, E, L, I, O, F, W)$ with components defined as follows.

N1) V is the *neuron set* of $\mathcal{N}$. We assume, w.l.o.g., that V is finite and nonempty, i.e., $\mathcal{N}$ is not an empty DNN.

N2) E is the *link set* of $\mathcal{N}$. We assume, w.l.o.g., that neurons on the same layer of $\mathcal{N}$ are *disconnected*, and that two consecutive layers of $\mathcal{N}$ are *fully connected*. In other words, $\mathcal{N}$ is a *fully connected* DNN.

N3) $L \geq 2$ is the *layer number* of $\mathcal{N}$, i.e., $\mathcal{N}$ has $L - 2$ *hidden layers*. For each $l = 1, \ldots, L$, we denote by n_l the *number* of neurons on the l-th layer.

N4) I is the neuron set of the first layer of $\mathcal{N}$, i.e., the *input layer* of $\mathcal{N}$. According to N3), the input layer has n_1 neurons.

N5) O is *the neuron set* of the last layer of $\mathcal{N}$, i.e., the *output layer* of $\mathcal{N}$. According to N3), the output layer has n_L neurons.

N6) $F = (f_v)_{v \in V \setminus I}$ is the *activation function vector* of $\mathcal{N}$, i.e., $f_v : \mathbb{R} \to \mathbb{R}$ is the *activation function* of neuron $v \in V \setminus I$. Note that the neurons of input layer are responsible for receiving information from outsides, and so need not have activation functions.

N7) Every link $e \in E$ has a *weight* w_e. We denote by $W = (w_e)_{e \in E}$ the *weight vector* of $\mathcal{N}$. As $\mathcal{N}$ is feed-forwarding, every link $e \in E$ connects two neurons belonging respectively to consecutive layers. We thus write the weight w_e explicitly as $w_{ij}^{(l)}$ when the link e connects the i-th neuron $v_i^{(l)}$ of the l-th layer and the j-th neuron $v_j^{(l-1)}$ of the $(l-1)$-th layer. Then W is written equivalently as $(w^{(l)})_{l=2,\dots,L}$, where $w^{(l)} := (w_{ij}^{(l)})_{1 \le i \le n_l,\ 1 \le j \le n_{l-1}}$ is the *matrix of weights* connecting the layer $l-1$ and the layer l for each $l = 2, \dots, L$.

As two consecutive layers are fully connected, a neuron $v_i^{(l)}$ of layer l receives information from all neurons $v_j^{(l-1)}$ of layer $l - 1$ with weights $w_{ij}^{(l)}$ for each $l = 2, \dots, L$. Suppose that $g_{v_j^{(l-1)}}$ is the *output* of the neuron $v_j^{(l-1)}$ of layer $l-1$ for each $j = 1, \dots, n_{l-1}$ and each $l = 2, \dots, L$. Then we have the *feed-forwarding recursion* (2.1),

$$g_{v_i^{(l)}} = f_{v_i^{(l)}} \left(\sum_{j=1}^{n_{l-1}} g_{v_j^{(l-1)}} \cdot w_{ij}^{(l)} \right), \quad i = 1, \dots, n_l, \ l = 2, \dots, L, \qquad (2.1)$$

where $f_{v_i^{(l)}}(\cdot)$ is the activation function of neuron $v_i^{(l)}$.

We focus on *linear* activation functions with biases, i.e., $f_v(x) = x + b_v$ for some bias $b_v \ge 0$ and each neuron $v \in V \setminus I$. We denote by $b = (b_v)_{v \in V \setminus I}$ the bias vector. With respect to these linear activation functions, the recursion (2.1) can be written explicitly as

$$g_{v_i^{(l)}} = \sum_{j=1}^{n_{l-1}} g_{v_j^{(l-1)}} \cdot w_{ij}^{(l)} + b_{v_i^{(l)}}, \quad i = 1, \dots, n_l, \ l = 2, \dots, L. \qquad (2.2)$$

Then the weights $w_{ij}^{(l)}$ together with the biases $b_{v_i^{(l)}}$ control the outputs of $\mathcal{N}$ when the inputs of $\mathcal{N}$ are given. A typical supervised learning task is then to compute weights $w_{ij}^{(l)}$ and biases $b_{v_i^{(l)}}$ for $\mathcal{N}$ such that the resulting outputs of $\mathcal{N}$ match the *labels* of given training samples as good as possible.

Consider now an arbitrary training set $\mathcal{H} = \{(X^1, Y^1), (X^2, Y^2), \dots, (X^N, Y^N)\}$, where $X^n = (x_{v_j^{(1)}}^n)_{j=1,\dots,n_1} \in \mathbb{R}^{n_1}$ is an n_1-dimensional *feature vector* and $Y^n = (y_{v_i^{(L)}}^n)_{i=1,\dots,n_L} \in \mathbb{R}^{n_L}$ is an n_L-dimensional *label* for each $n = 1, 2, \dots, N$. When the n-th feature vector X^n is fed to $\mathcal{N}$, the input layer of $\mathcal{N}$ sends directly X^n to the second layer, i.e., the j-th neuron $v_j^{(1)}$ of the first layer outputs $g_{v_j^{(1)}}(X^n) := x_{v_j^{(1)}}^n$ for each $j = 1, \ldots, n_1$. This follows since the neurons of input layer are only responsible for receiving information from outsides. With the recursion (2.2), this received information X^n will be iteratively perceived and eventually transformed to an n_L-dimensional output $O(X^n) = \big(g_{v_i^{(L)}}(X^n)\big)_{i=1,\ldots,n_L}$. Here, we use symbols $g_{v_i^{(L)}}(X^n)$ to indicate the dependence of the outputs of $\mathcal{N}$ on the input feature vector X^n.

We hope that these output labels $O(X^n)$ are *close* enough to the true labels Y^n of feature vectors X^n. Their "distance" is usually measured by a specified *loss* function $\ell : \mathbb{R}^{n_L} \times \mathbb{R}^{n_L} \to \mathbb{R}_+ := [0, \infty)$. Then the "quality" of weights $w_{ij}^{(l)}$ and biases $b_{v_i^{(l)}}$ could be quantified by the resulting *average loss* $loss(W, b) = \frac{1}{N} \sum_{n=1}^{N} \ell(O(X^n), Y^n)$, where the notation $loss(W, b)$ indicates the dependence of the average loss on the weights $w_{ij}^{(l)}$ and the biases $b_{v_i^{(l)}}$, w.r.t. the given training set $\mathcal{H}$. In our study, we will assume that the weights $w_{ij}^{(l)}$ satisfy the conditions

$$w_{ij}^{(l)} \geq 0, \quad \forall i = 1, \ldots, n_l, \ \forall j = 1, \ldots, n_{l-1}, \ \forall l = 2, \ldots, L,$$

$$\sum_{i=1}^{n_l} w_{ij}^{(l)} = 1, \quad \forall j = 1, \ldots, n_{l-1}, \ \forall l = 2, \ldots, L, \tag{2.3}$$

and that the biases $b_{v_i^{(l)}}$ fulfill the conditions

$$b_{v_i^{(l)}} \geq 0 \text{ and } \sum_{j=1}^{n_l} b_{v_j^{(l)}} = 1, \quad \forall i = 1, \ldots, n_l, \ \forall l = 2, \ldots, L. \tag{2.4}$$

Then an *optimal* pair (W^*, b^*) of weights and biases for $\mathcal{N}$, w.r.t. a given training set $\mathcal{H}$ will fulfill conditions (2.3)–(2.4) and *minimize* the above average loss.

2.2 Nonatomic Congestion Games

A *nonatomic congestion game* (NCG) is written as a tuple $\Gamma = (\mathcal{G}, \mathcal{K}, c, \mathcal{D}, \mathcal{P}, r)$, see also [15], whose components are defined as follows.

G1) $\mathcal{G} = (\mathcal{V}, \mathcal{A})$ is a directed graph (*routing network*) with a finite non-empty arc set $\mathcal{A}$ and a finite non-empty vertex set $\mathcal{V}$.

G2) $\mathcal{K} = \{(o_1, t_1), \ldots, (o_K, t_K)\} \subseteq \mathcal{V} \times \mathcal{V}$ is a finite set of travel *origin-destination* (O/D) pairs defined on the routing network $\mathcal{G}$, where $K = |\mathcal{K}| > 0$ is the constant denoting the number of O/D pairs of Γ. In the sequel, we shall identify an O/D pair (o_k, t_k) with its index k for simplifying notation.

G3) $c = (c_a)_{a \in \mathcal{A}}$ is a vector of functions with each $c_a(\cdot)$ denoting a flow-dependent *non-negative, continuous* and *non-decreasing* cost (i.e., traversal latency) of an arc $a \in \mathcal{A}$.

G4) $\mathcal{D} = (d_k)_{k \in \mathcal{K}}$ is a real-valued vector with each $d_k > 0$ denoting the amount of traffic of O/D pair (o_k, t_k) for each $k \in \mathcal{K}$.

G5) *Feasible strategies* of an O/D pair $k \in \mathcal{K}$ are simple paths leading from o_k to t_k. We denote the set of all feasible strategies of O/D pair k by $\mathcal{P}_k$ for each $k \in \mathcal{K}$. Then $\mathcal{P} := \bigcup_{k \in \mathcal{K}} \mathcal{P}_k$ is the set of all strategies of Γ.

G6) Associated with a strategy $p \in \mathcal{P}_k$ and an arc $a \in p$ is a *nonnegative* rate $r_{p,a}$ of consumption. We put $r_{p,a} = 0$ when $a \notin p$. Then $r := (r_{p,a})_{p \in \mathcal{P}, a \in \mathcal{A}}$ is the *consumption* matrix of Γ.

Players in Γ are assumed to be infinitesimal, and each of them controls only a negligible fraction of demands. They need to route their traffic *simultaneously* and *independently*, which results in a *traffic flow* $z = (z_p)_{p \in \mathcal{P}}$ such that $z_p \geq 0$ is the *flow value* of path $p \in \mathcal{P}$ and $\sum_{p \in \mathcal{P}_k} z_p = d_k$ for all $k \in \mathcal{K}$. This flow $z = (z_p)_{p \in \mathcal{P}}$ in turn induces an *arc flow* $(z_a)_{a \in \mathcal{A}}$, where $z_a := \sum_{k \in \mathcal{K}} \sum_{p \in \mathcal{P}_k : a \in p} r_{p,a} \cdot z_p$ is the resulting flow value of arc $a \in \mathcal{A}$. Then an arc $a \in \mathcal{A}$ has a cost of $c_a(z_a)$, w.r.t. $z = (z_p)_{p \in \mathcal{P}}$. We assume that the cost is *linearly aggregated* along a path, and so a path $p \in \mathcal{P}$ has a cost of $c_p(z) := \sum_{a \in \mathcal{A} : a \in p} r_{p,a} \cdot c_a(z_a)$, w.r.t. flow z. The *social cost* $SC(z)$, w.r.t. flow z is then computed by $SC(z) = \sum_{k \in \mathcal{K}} \sum_{p \in \mathcal{P}_k} c_p(z) \cdot z_p = \sum_{a \in \mathcal{A}} c_a(z_a) \cdot z_a$. Furthermore, a *social optimum* (SO) of Γ is a flow which minimizes this social cost of Γ.

While Γ has SO flows, they are usually hard to be attained in practice, since players are selfish and only want to follow their own cheapest paths. This selfish behavior leads to an equilibrium state of Γ. Formally, a traffic flow z of Γ is called a *Wardrop equilibrium* (WE) of Γ if $c_p(z) \leq c_{p'}(z)$ for each O/D pair $k \in \mathcal{K}$ and any two paths $p, p' \in \mathcal{P}_k$ with $z_p > 0$, see [19]. Since players in Γ are infinitesimal and the cost functions c_a are nondecreasing, nonnegative and continuous, a WE of Γ corresponds to a pure *Nash equilibrium* and all WE have the same social cost, see, e.g., [16] and [19].

Note that WE flows may deviate significantly from SO flows, see [16]. The *worst-case* ratio of the social cost of a WE flow over that of an SO flow is known as the *Price of Anarchy* (PoA), which quantifies the *inefficiency* of WE flows, see [16]. Formally, we define the PoA of Γ as

$$\rho(\Gamma) := \max_{\tilde{z}, \, z^*} \frac{SC(\tilde{z})}{SC(z^*)} = \frac{SC(\tilde{z})}{SC(z^*)}, \tag{2.5}$$

where $\tilde{z}$ is an arbitrary WE flow of Γ and z^* is an arbitrary SO flow of Γ. We used in (2.5) the fact that all WE flows of Γ have the same social cost under our assumption.

3   Main Results

Before we show our main results, we first make some assumptions on the DNN $\mathcal{N}$ and its training set $\mathcal{H}$ as follows.

A1) We assume that every feature vector X^n in $\mathcal{H}$ has only *nonnegative* features, i.e., $x^n_{v_j^{(1)}} \geq 0$ for each $n = 1, \ldots, N$ and each $j = 1, \ldots, n_1$. Note that this is not restrictive, as we can make all features of $\mathcal{H}$ nonnegative by subtracting the *minimum* feature value $\min \{x^n_{v_j^{(1)}} : n = 1, \ldots, N, j = 1, \ldots, n_1\}$.

A2) The activation functions $f_v(\cdot)$ of $\mathcal{N}$ are linear with biases satisfying conditions (2.4).

A3) The weights $w_{ij}^{(l)}$ of $\mathcal{N}$ satisfy conditions (2.3).

A4) All hidden layers are wider than the output layer, i.e., $n_l \geq n_L$ for all $l \geq 2$.

A5) The loss function $\ell : \mathbb{R}^{n_L} \times \mathbb{R}^{n_L} \to \mathbb{R}_+$ of $\mathcal{N}$ is *linearly separable*, or *linearly decomposable*, that means, $\ell(O(X^n), Y^n) = \sum_{i=1}^{n_L} \ell_1(g_{v_i^{(L)}}(X^n), y^n_{v_i^{(L)}})$ for each $n = 1, \ldots, N$, where $\ell_1 : \mathbb{R} \times \mathbb{R} \to \mathbb{R}_+$ is a "similarity" measure on $\mathbb{R}$. Then we have

$$loss(W, b) = \frac{1}{N} \sum_{n=1}^{N} \ell(O(X^n), Y^n) = \frac{1}{N} \sum_{n=1}^{N} \sum_{i=1}^{n_L} \ell_1(g_{v_i^{(L)}}(X^n), y^n_{v_i^{(L)}}). \quad (3.1)$$

3.1   From DNNs to NCGs

Consider now an arbitrary training set $\mathcal{H}$ satisfying A1) and a DNN $\mathcal{N} = (V, E, L, I, O, F, W)$ satisfying A2)–A5). We now construct an NCG $\Gamma(\mathcal{N}, \mathcal{H}) = (\mathcal{G}, \mathcal{K}, c, \mathcal{D}, \mathcal{P}, r)$ in accordance with the training set $\mathcal{H}$ and the DNN $\mathcal{N}$. We first build the routing network $\mathcal{G}$ of $\Gamma(\mathcal{N}, \mathcal{H})$ as follows.

1) Each neuron of $\mathcal{N}$ becomes a vertex of $\mathcal{G}$, and each link $e \in E$ of $\mathcal{N}$ becomes an arc of $\mathcal{G}$.

2) For each $l \geq 2$, we add an *auxiliary* vertex $v^{(l)}$ and n_l auxiliary arcs $(v^{(l)}, v_j^{(l)})$ leading from $v^{(l)}$ to the neurons $v_j^{(l)}$ on layer l. We denote the set of these $L - 1$ auxiliary vertices $v^{(l)}$ by $\mathcal{V}^{(L)}$, and the set of the $\sum_{l=2}^{L} n_l$ auxiliary arcs $(v^{(l)}, v_j^{(l)})$ by $\mathcal{A}^{(L)}$.

3) We then add to the resulting $\mathcal{G}$ a *common destination vertex* (i.e., sink point) J.

4) For each $i = 1, \ldots, n_L$, we insert $N - 1$ *auxiliary* vertices $v_i^{(L+1)}, \ldots, v_i^{(L+N-1)}$ between the i-th neuron $v_i^{(L)}$ of the output layer and the sink point J, and connect them consecutively by N arcs $a_i^{(1)} = (v_i^{(L)}, v_i^{(L+1)})$, $a_i^{(2)} = (v_i^{(L+1)}, v_i^{(L+2)})$, $\ldots$, $a_i^{(N)} = (v_i^{(L+N-1)}, J)$. We put $\mathcal{V}^{(N)} := \{v_i^{(L+n)} : i = 1, \ldots, n_L, n = 1, \ldots, N-1\}$ and $\mathcal{A}^{(N)} := \{a_i^{(n)} : i = 1, \ldots, n_L, n = 1, \ldots, N\}$.

5) Finally, the vertex set of $\mathcal{G}$ is $\mathcal{V} := V \cup \mathcal{V}^{(L)} \cup \{J\} \cup \mathcal{V}^{(N)}$, and the arc set of $\mathcal{G}$ is $\mathcal{A} := E \cup \mathcal{A}^{(L)} \cup \mathcal{A}^{(N)}$.

With this routing network $\mathcal{G} = (\mathcal{V}, \mathcal{A})$, we define $n_1 + L - 1$ O/D pairs and their traffic demands as follows.

6) To avoid ambiguity, we denote by $\mathcal{V}^{(1)}$ the set of vertices in $\mathcal{V}$ that are neurons on the input layer of $\mathcal{N}$, i.e., $\mathcal{V}^{(1)} = I$. We then put $\tilde{\mathcal{V}} = \mathcal{V}^{(1)} \cup \mathcal{V}^{(L)}$. For each

vertex $v \in \tilde{\mathcal{V}}$, we build an O/D pair (v, J) with *unit* traffic demand. Then we obtain a set $\mathcal{K} = \{(v, J) : v \in \tilde{\mathcal{V}}\}$ of $n_1 + L - 1$ O/D pairs and a vector $\mathcal{D} = (d_v)_{v \in \tilde{\mathcal{V}}}$ of demands, where $d_v \equiv 1$ is the traffic demand of O/D pair (v, J) for each $v \in \tilde{\mathcal{V}}$.

As the routing network $\mathcal{G}$ and the O/D pair set $\mathcal{K}$ have already been determined, the strategy set $\mathcal{P} = \cup_{v \in \tilde{\mathcal{V}}} \mathcal{P}_v$ then follows accordingly. Here, we note that the strategy set $\mathcal{P}_v$ of each O/D pair (v, J) is not empty, since consecutive layers of $\mathcal{N}$ are fully connected.

7) We define the rate of consumption as follows. For each origin $v \in \tilde{\mathcal{V}}$, each path $p \in \mathcal{P}_v$, and each arc $a \in p$, we put $r_{p,a}$ as 1 if $a \in E \cup A^{(L)}$, as x_v^n if $v = v_j^{(1)} \in \mathcal{V}^{(1)} = I$ for $j = 1, \ldots, n_1$, and $a = a_i^{(n)} \in \mathcal{A}^{(N)}$ for $i = 1, \ldots, n_L$ and $n = 1, \ldots, N$, and as 1 if $v \in \mathcal{V}^{(L)}$ and $a = a_i^{(n)} \in \mathcal{A}^{(N)}$. Here, we recall that $x_{v_j^{(1)}}^n \geq 0$ is the j-th feature of the n-th observation X^n of $\mathcal{H}$, which is a constant, w.r.t. the given training set $\mathcal{H}$.

8) Finally, we define the arcs cost functions as below,

$$c_a(\eta) := \begin{cases} 0, & \text{if } a \in E \cup A^{(L)}, \text{ or } \eta = 0, \\ \ell_1(\eta, y_{v_i^{(L)}}^n)/\eta, & \text{if } \eta > 0 \text{ and } a = a_i^{(n)} \in \mathcal{A}^{(N)}, \end{cases} \quad \forall \eta \geq 0 \; \forall a \in \mathcal{A}.$$

Here, we recall that $y_{v_i^{(L)}}^n \geq 0$ is the i-th entry of n-th label Y^n of $\mathcal{H}$, which is a constant, w.r.t. the given training set $\mathcal{H}$.

With all above, we have finished the construction of $\Gamma(\mathcal{N}, \mathcal{H})$. We illustrate it with Example 1 below.

Example 1. Consider now a deep neural network as shown in Fig. 1(a). Suppose that we are given a training set $\mathcal{H} = \{(X^1, Y^1), (X^2, Y^2), (X^3, Y^3)\}$. We first extend the network by introducing two new vertices for the biases, which then results in Fig. 1(b). We further extend the network by adding the common destination J, and the auxiliary vertices and arcs connecting the output layer and J, see Fig. 1(c), which then forms the desired network of the nonatomic congestion game.

3.2   Learning $\mathcal{N}$ With $\Gamma(\mathcal{N}, H)$

We aim now to show that learning $\mathcal{N}$ for $\mathcal{H}$ is equivalent to compute SO flows of $\Gamma(\mathcal{N}, \mathcal{H})$. This generalizes the recent work of [18], who showed similar results for DNNs with linear activation functions and *zero* biases.

Theorem 1 below shows that there is certain correspondence between flows of $\Gamma(\mathcal{N}, \mathcal{H})$ and pairs (W, b) of weights and biases of $\mathcal{N}$. In particular, it shows that the average loss $loss(W, b)$ of $\mathcal{N}$, w.r.t. a pair (W, b) of weights and biases coincides exactly with $\frac{1}{N} \cdot SC(z)$ for a flow z of $\Gamma(\mathcal{N}, \mathcal{H})$, and vice versa. This means that an optimal pair (W^*, b^*) of weights and biases of $\mathcal{N}$ for $\mathcal{H}$ corresponds to an SO flow z^* of $\Gamma(\mathcal{N}, \mathcal{H})$.

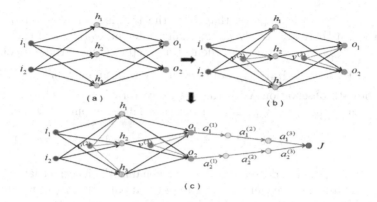

Fig. 1. An example of how a deep neural network (DNN) can be transformed into a nonatomic congestion game (NCG) network.

Theorem 1. *Consider an arbitrary training set $\mathcal{H}$ satisfying A1), and a DNN $\mathcal{N}$ satisying A2)–A5). Let $\Gamma(\mathcal{N}, \mathcal{H})$ be the NCG defined as above. Then, for each pair (W, b) of weights and biases of $\mathcal{N}$, there is a flow z of $\Gamma(\mathcal{N}, \mathcal{H})$ such that $loss(W, b) = \frac{1}{N} SC(z)$. Moreover, for each flow z of $\Gamma(\mathcal{N}, \mathcal{H})$, there is a pair (W, b) of weights and biases of $\mathcal{N}$ such that $loss(W, b) = \frac{1}{N} SC(z)$.*

In the proof of Theorem 1, we have shown effective methods for the transformations between flows of $\Gamma(\mathcal{N}, \mathcal{H})$ and the pairs (W, b) of weights and biases of $\mathcal{N}$. Hence, we may obtain an optimal pair (W^*, b^*) of $\mathcal{N}$ for $\mathcal{H}$ by computing an SO flow z^* of $\Gamma(\mathcal{N}, \mathcal{H})$.

Theorem 2 below shows that each *locally* optimal pair (W, b) of weights and biases of $\mathcal{N}$ corresponds to a WE flow $\tilde{z}$ of $\Gamma(\mathcal{N}, \mathcal{H})$, and vice versa, when $\ell_1(\eta, y^n_{v_i^{(L)}})$ is additionally of the form $A^n_{v_i^{(L)}} \cdot \eta^\beta$ for some constants $A^n_{v_i^{(L)}} \geq 0$ depending on $y^n_{v_i^{(L)}}$ and $\beta \geq 1$ for each $i \in \{1, \ldots, n_L\}$ and $n \in \{1, \ldots, N\}$.

Theorem 2. *Consider an arbitrary training set $\mathcal{H}$ satisfying A1), and a DNN $\mathcal{N}$ satisying A2)–A5). Suppose that $\ell_1(\eta, y^n_{v_i^{(L)}}) = A^n_{v_i^{(L)}} \cdot \eta^\beta$ for some constants $A^n_{v_i^{(L)}} \geq 0$ dependent of $y^n_{v_i^{(L)}}$ and $\beta \geq 1$ for each $i = 1, \ldots, n_L$ and each $n = 1, \ldots, N$. Then, each locally optimal pair (W, b) of weights and biases of $\mathcal{N}$ for $\mathcal{H}$ corresponds to a WE flow $\tilde{z}$ of $\Gamma(\mathcal{N}, \mathcal{H})$, and vice versa.*

Theorem 2 follows from Theorem 1 and the convexity of the resulting function $loss(W, b)$. [20] have shown that nonatomic congestion games with monomial cost functions of the same degree are well designed and thus have the PoA of 1. This applies to $\Gamma(\mathcal{N}, \mathcal{H})$ when the deep neural network $\mathcal{N}$ has loss functions of the particular form as in Theorem 2. In other words, WE flows and SO flows of $\Gamma(\mathcal{N}, \mathcal{H})$ coincide when $\mathcal{N}$ has loss functions of the specific form.

While $\ell_1(\eta, y^n_{v_i^{(L)}})$ of Theorem 2 is seemingly restrictive, it applies when the labels Y^n of $\mathcal{H}$ are *binary*, i.e., $y^n_{v_i^{(L)}} \in \{0, 1\}$ for all $n = 1, \ldots, N$ and all $i =$

$1, \ldots, n_L$. A typical example is that $\mathcal{H}$ is the training set of a *classification* task, in which there are totally n_L mutually different *objects*, and each $X^n = (x^n_{v_j^{(1)}})_{j=1,\ldots,n_1}$ is an observation of the feature vector of one of these n_L objects. The labels $Y^n = (y^n_{v_i^{(L)}})_{i=1,\ldots,n_L}$ are then binary with an entry $y^n_{v_i^{(L)}} = 1$ if and only if the n-th observation X^n is the feature vector of the i-th object. For this case, we can define $\ell_1(\cdot)$ by penalizing *incorrect* labels output by $\mathcal{N}$, i.e.,

$$\ell_1(\eta, y^n_{v_i^{(L)}}) = A^n_{v_i^{(L)}} \cdot \eta^\beta \text{ with } A^n_{v_i^{(L)}} := 1 - y^n_{v_i^{(L)}}, \quad \forall i = 1, \ldots, n_L, \, \forall n = 1, \ldots, N,$$

where $\beta \geq 1$ reflects the degree, to which we penalize the incorrect labels. Then Theorem 2 states that we can also find an optimal pair (W^*, b^*) of weights and biases of $\mathcal{N}$ by computing the WE flows of $\Gamma(\mathcal{N}, \mathcal{H})$ when $\mathcal{N}$ is a DNN for classification.

4   Conclusion

In this article, we have considered more general deep linear neural networks $\mathcal{N}$ with activation function of form $f_v(x) = x + b_v$. Similar with that of [18], we are able to show that there is a nonatomic congestion game $\Gamma(\mathcal{N}, \mathcal{H})$ such that learning the weights and biases of $\mathcal{N}$ for $\mathcal{H}$ is equivalent to computing the SO flows of $\Gamma(\mathcal{N}, \mathcal{H})$. In particular, we show that the learning task is even equivalent to computing the WE flows of $\Gamma(\mathcal{N}, \mathcal{H})$ when $\mathcal{N}$ is a deep neural network for classification. While the theory has been there, we may still need a computation to verify the results, which we leave as a future work. Moreover, this paper considers loss functions that are linearly decomposable, which then resulted in nonatomic congestion games with separable arc cost. One may then wonder if the results still hold when the loss functions are not decomposable. For this more general case, we can also construct a nonatomic congestion game for the deep neural network. However, the game will then have nonseparable cost, which would be very difficult to analyze. We thus leave this also as a future work.

Acknowledgement. The first and third authors are supported by Beijing Natural Science Foundation Project No. Z200002 and National Natural Science Foundation of China (No. 12131003); The second author is supported by National Natural Science Foundation of China (No. 61906062), Natural Science Foundation Project of Anhui Science and Technology Department (No. 1908085QF262), and Natural Science Foundation Project of Anhui Education Department (No. KJ2019A0834).

References

1. Alpaydin, E.: Introduction to Machine Learning (Adaptive Computation and Machine Learning Series). The MIT Press, Cambridge (2008)
2. Balduzzi, D.: Deep online convex optimization with gated games. arXiv preprint, arXiv:1604.01952 (2016)

3. Blum, A.L., Rivest, R.L.: Training a 3-node neural network is np-complete. Neural Netw. **5**(1), 117–127 (1992)
4. Ge, R., Huang, F., Jin, C., Yuan, Y.: Escaping from saddle points - online stochastic gradient for tensor decomposition. In: Proceedings of the Conference on Learning Theory, vol. 40, pp. 797–842 (2015)
5. Goldberg, D.E., Holland, J.H.: Genetic algorithms and machine learning. Mach. Learn. **3**, 95–99 (1988)
6. Goodfellow, I., Bengio, Y., Courville, A.: Deep Learning. The MIT Press, Cambridge (2016)
7. Hinton, G.E., Salakhutdinov, R.R.: Reducing the dimensionality of data with neural networks. Science **313**(5786), 504–507 (2006)
8. Hu, J., Wellman, M.P.: Nash q-learning for general-sum stochastic games. J. Mach. Learn. Res. **4**, 1039–1069 (2003)
9. Jin, C., Ge, R., Netrapalli, P., Kakade, S.M., Jordan, M.I.: How to escape saddle points efficiently. In: Proceedings of the 34th International Conference on Machine Learning, vol. 70, pp. 1724–1732. PMLR (2017)
10. Littman, M.L.: Markov games as a framework for multi-agent reinforcement learning. In: Proceedings of the Eleventh International Conference on Machine Learning, pp. 157–163. Morgan Kaufmann (1994)
11. Littman, M.L., Szepesvári, C.: A generalized reinforcement-learning model: convergence and applications. In: Proceedings of the Thirteenth International Conference on Machine Learning, pp. 310–318. Morgan Kaufmann (1996)
12. Liu, Y., Pi, D.: A novel kernel SVM algorithm with game theory for network intrusion detection. Trans. Internet Inf. Syst. **11**(8), 4043–4060 (2017)
13. Kollovieh, M., Bani-Harouni, D.: Machine learning. Der Hautarzt **72**(8), 719–719 (2021). https://doi.org/10.1007/s00105-021-04834-0
14. Murty, K.G., Kabadi, S.N.: Some NP-complete problems in quadratic and nonlinear programming. Math. Program. **39**(2), 117–129 (1987)
15. Roughgarden, T.: Routing games. Algorithmic Game Theory **18**, 459–484 (2007)
16. Roughgarden, T., Tardos, É.: How bad is selfish routing? J. ACM **49**(2), 236–259 (2002)
17. Schuurmans, D., Zinkevich, M.: Deep learning games. In: Proceedings of the Annual Conference on Neural Information Processing Systems, vol. 29, pp. 1678–1686 (2016)
18. Vesseron, N., Redko, I., Laclau, C.: Deep neural networks are congestion games: From loss landscape to Wardrop equilibrium and beyond. In: Proceedings of the 24th International Conference on Artificial Intelligence and Statistics, vol. 130, pp. 1765–1773. PMLR (2021)
19. Wardrop, J.G.: Road paper. Some theoretical aspects of road traffic research. In: Proceedings of the Institution of Civil Engineers, vol. 1, pp. 325–362 (1952)
20. Wu, Z., Möhring, R.H., Chen, Y., Xu, D.: Selfishness need not be bad. Oper. Res. **69**(2), 410–435 (2021)

Energy Complexity of Satisfying Assignments in Monotone Circuits: On the Complexity of Computing the Best Case

Janio Carlos Nascimento Silva[1,2(✉)], Uéverton S. Souza[1], and Luiz Satoru Ochi[1]

[1] Instituto de Computação, Universidade Federal Fluminense, Niterói, RJ, Brazil
{ueverton,satoru}@ic.uff.br
[2] Instituto Federal do Tocantins, Campus Porto Nacional, Porto Nacional, TO, Brazil
janio.carlos@ifto.edu.br

Abstract. Measures of circuit complexity are usually analyzed to ensure the computation of Boolean functions with economy and efficiency. One of these measures is the *energy complexity*, which is related to the number of gates that output true in a circuit for an assignment. The idea behind energy complexity comes from the counting of 'firing' neurons in a natural neural network. The initial model is based on threshold circuits, but recent works also have analyzed the energy complexity of traditional Boolean circuits. In this work, we discuss the time complexity needed to compute the best case energy complexity among satisfying assignments of a monotone Boolean circuit, and we call such a problem as $\textsc{MinEC}_M^+$. In the $\textsc{MinEC}_M^+$ problem, we are given a monotone Boolean circuit C, a positive integer k and asked to determine whether there is a satisfying assignment X for C such that $EC(C, X) \leq k$, where $EC(C, X)$ is the number of gates that output true in C according to the assignment X. We prove that $\textsc{MinEC}_M^+$ is NP-complete even when the input monotone circuit is planar. Besides, we show that the problem is W[1]-hard but in XP when parameterized by the size of the solution. In contrast, we show that $\textsc{MinEC}_M^+$ on bounded genus circuits is FPT.

Keywords: Energy complexity · Monotone circuit · Genus · FPT

1 Introduction

Circuit Complexity is a research field that aims to study bounds for measures (such as size and depth) of circuits that compute Boolean functions. The *size* of a circuit is its number of logic gates, and *depth* is the largest path from any input to the output gate. A circuit complexity analysis can provide precise lower/upper

Supported by CAPES, CNPq and FAPERJ.

W. Wu and H. Du (Eds.): AAIM 2021, LNCS 13153, pp. 380–391, 2021.
https://doi.org/10.1007/978-3-030-93176-6_32

bounds on circuits classes that represent classic decision problems besides the possibility to design efficient Boolean circuits according to specific properties (see [9,13,21]). In addition, some important bounds are described to deal with different definitions of size [1]. From a combinatorial point of view, several optimization problems address the minimization of measures in some circuits classes, such as the notorious WEIGHTED CIRCUIT SATISFIABILITY problem, where the weight measures the amount of true values assigned to the input variables. Despite this 'zoo of measures', optimizing properties like size and depth does not always guarantee an 'efficient' design of a specific circuit class. Depending on the purpose, a circuit with small size (either considering gates or wires) or depth can be inappropriate; such a situation was identified in [19].

When faced with threshold circuits used as an artificial neural network, it is possible to observe a contrast with neurons of the human brain. The authors in [19] (based in neuroscience literature) argue that the activation of neurons in a human brain happens sparsely. It was shown in [12] that the metabolic cost of a single spike in cortical computation is very high in a way that approximately 1% of the neurons can be activated simultaneously. This phenomenon happens due to the asymmetric energy cost between neurons activated and non-activated in natural cases. From the other side, digital circuits, when satisfied (outputting true), on average activate 50% of the gates. Under different perspectives, 'energy' (or 'power') of a circuit is a measure that has a lot of attention in the literature. Due to multiple models (from biology, electronics, or purely theoretical), several works address different ways of analyzing the energy of a circuit. In [11], the energy consumption of a circuit considers the switching energy consumed by wires (edges) and gates of VLSI circuits. In [3] and [4], it is analyzed the *voltage-to-energy* consumed by the gates, taking into consideration the *failure-to-energy*. Other different models are explored in [16] and [5], such works try to explore concepts of energy too intrinsic to the design of practical circuits on electronics.

In this paper, we deal with a circuit complexity measure called *energy complexity* (EC). The idea behind this measure is to evaluate the number of gates in a circuit that returns true for an assignment. A similar concept called 'power of circuit' was studied by [20]. The term *energy complexity* was introduced in [19] as an alternative to the dilemma *artificial vs natural* described above. In [19], the authors prove that the minimization of circuit energy complexity obtains a different structure from the minimization of previously considered circuit complexity measures and potentially closer to the structure of neural networks in the human brain. The authors proved initial lower bounds for energy complexity and other circuit complexity measure called *entropy*.

With a different perspective, this work dedicates attention to optimization and decision problems related to energy complexity. More precisely, we consider the problem of determining the satisfying assignment with minimum energy consumption in monotone circuits, i.e., the best case energy complexity of a satisfying assignment in the class of monotone circuits – MINEC^+_M. The minimization of energy complexity potentially can help the design of circuits with sparse activation, hence, more similar to biological models. Our focus is on time (parameterized) complexity of the henceforth defined MINEC^+_M.

In [2], a measure called *certification-width* was described, which is the size of a minimum subset of edges that are enough to certificate a satisfying assignment. Such edges form a structure called *succinct certificate* that can be seen as a minimal map of edges to be followed in order to activate the output gate. Similar structures can also be found in [17,18]. Note that there are similarities between certification-width and energy complexity. Both measures indicate saturation levels of circuits, but while certification-width focuses on edges, energy complexity is about the activation of gates. However, energy complexity presents two additional challenges: (i) *EC* ignores the 'firing' of input gates; (ii) *EC* counts activated gates even if its signal does not reach the output gate (due to unsatisfied gates). These two issues forbid rushed conclusions about *EC* based on what we know about certification-width. Nevertheless, the study in [2] also motivates the study of the complexity of computing the best case energy complexity of satisfying assignments in monotone circuits.

Note that in energy complexity problems in addition to working with the gates needed to satisfy the circuit, it is still necessary to handle gates that assignments may collaterally activate. Such behavior makes working with energy complexity problems more challenging than typical satisfying problems where the focus is only on the minimal set of inputs, gates, or wires/edges sufficient to satisfy the circuit.

Preliminaries

Given a Boolean circuit C and an assignment X, the Energy Complexity of X into C, $EC(C, X)$, is defined as the number of gates that output true in C according to the assignment X. The (Worst-Case) Energy Complexity of C (denoted by $EC(C)$) is the maximum $EC(C, X)$ among all possible assignments X (See [7]). Analogously, the Best-Case Energy Complexity of C (denoted by $MinEC(C)$) is the minimum $EC(C, X)$ among all possible assignments X.

While computing the worst-case energy complexity of satisfying assignments in monotone circuits is trivial (activate all inputs), the problem of computing the best-case energy complexity among all satisfying assignments in monotone circuits seems a challenge. Therefore, in this work, we address this particular case where the circuit is monotone, focusing on the following decision problem:

BEST-CASE ENERGY COMPLEXITY OF SATISFYING ASSIGNMENTS IN MONOTONE CIRCUITS – MinEC_M^+
Instance: A monotone Boolean circuit C and a positive integer k.
Question: Is there a satisfying assignment X for C such that $EC(C, X) \le k$?

Besides, we denote by $k\text{-MinEC}_M^+$ the parameterized version of MinEC_M^+ where k is taking as the parameter.

2   Computational Complexity Analysis

In this section, we present our (parameterized) complexity results regarding MinEC_M^+. Using a reduction from PLANAR VERTEX COVER, similar to that employed in [2], we are able to show that $\textsc{MinEC}_M^+$ is NP-complete even when restrict to planar circuits.

Theorem 1. $\textsc{MinEC}_M^+$ *is NP-complete even when restricted to planar circuits.*

Next, we investigate the parameterized complexity of $\textsc{MinEC}_M^+$.

Theorem 2. k-$\textsc{MinEC}_M^+$ *is in XP.*

Proof. Let $C = (V, E)$ be a circuit with $V = I \cup G \cup \{v_{out}\}$, where I is the set of inputs of C, G is the set of gates and v_{out} is the output gate. If C has a satisfying assignment X such that $EC(C, X) \leq k$ then we can find X as follows:

1. Suppose that X is the satisfying assignment with $EC(C, X) \leq k$ having minimum weight (i.e., minimum number of inputs assigned as `true`);

2. First, we "guess" the set T of gates that should be activated by X, that is, in $n^{O(k)}$ time, we enumerate each subset T of gates such that $|T| \leq k$ and check each one in a new branch;

3. For each T we can check in polynomial time whether it is consistent, that is:
   - $v_{out} \in T$;
   - For each `OR`-gate v in T either it has an in-neighbor in T or it has an in-neighbor in I, and for each `AND`-gate v in T its in-neighborhood is contained in $T \cup I$;
   - Conversely, each `OR`-gate $w \notin T$ has no in-neighbor in T, and for each `AND`-gate $w \notin T$ has at least one in-neighbor that is not in T;
   - Also, no input is mutually in-neighbor of an `AND`-gate in T and an `OR`-gate not in T;
   - Note that, if T is the set of gates activated by X then it holds that: any input i that is in-neighbor of an `AND`-gate in T should be set as `true` in X; any input i that is in-neighbor of an `OR`-gate not in T should be set as `false` in X. Let X' be such a partial assignment;
   - Therefore, each `OR`-gate v in T having no in-neighbor in T has at least one in-neighbor in I that is not set as `false` by X', and each `AND`-gate $w \notin T$ having no in-neighbor in $G \setminus T$ has at least one in-neighbor in I that is not set as `true` in X'.

4. Since X has minimum weight, from a given consistent set T, in order to extend X' into a satisfying assignment X with $EC(C, X) \leq k$ (if any), it is enough to "guess" the minimal set of inputs that should be set as `true` to activate the `OR`-gates in T having no in-neighbor in T. As $|T| \leq k$, such subset of inputs is also bounded by k, thus, in $n^{O(k)}$ time, we can enumerate (if any) each assignment X'' extending X' by setting at most k additional inputs as `true` in such a way that each `OR`-gates in T has at least one in-neighbor activated. At this point, from the guessed set T we obtain the assignment X if there is some X'' for which each `AND`-gate $w \notin T$ having no in-neighbor in $G \setminus T$ has at least one in-neighbor in I that is set as `false`.

Note that for any satisfying assignment X of C the set of activated gates must satisfy the properties described in step 3. Since steps 2 and 4 check in $n^{O(k)}$ time all possibilities, it holds that $\textsc{MinEC}_M^+$ is XP-time solvable.   □

Now, we show the W[1]-hardness of k-MinEC_M^+ using a reduction from MUL-TICOLORED CLIQUE.

MULTICOLORED CLIQUE
Instance: A graph Q with a vertex-coloring $\ell : V(G) \rightarrow \{1, 2, \ldots, c\}$.
Parameter: A positive integer c.
Question: Does Q have a c-clique containing all c colors?

Theorem 3. k-MinEC_M^+ *is W[1]-hard.*

Proof. Let (Q, c) be an instance of MULTICOLORED CLIQUE and let $V_1, V_2, \ldots, V_c$ be the color classes of Q. Without loss of generality, we consider that each vertex in V_i has at least one neighbor in $V_j (i \neq j)$. We construct an instance (C, k) of $\text{MinEC}_M^+(k)$ as follows:

1. create an output gate v_{out} in C and set $f(v_{out}) = \text{AND}$;
2. for each color c_i of Q, create a gate w_i with $f(w_i) = \text{OR}$ and add an edge from w_i to v_{out};
3. for each color class V_i of Q, create copies V_i^1, V_i^2, V_i^3 and V_i^4 in C;
4. add edges from each vertex in V_i^4 to w_i;
5. let v^1, v^2, v^3 and v^4 be the copies of a vertex $v \in V(Q)$; add edges $(v^1, v^2), (v^2, v^3)$ and (v^3, v^4) to G; set V_i^1 as the input set; and assign $f(v^2) = f(v^3) = \text{OR}$ and $f(v^4) = \text{AND}$;
6. for each vertex $v^4 \in V_i^4 (1 \leq i \leq c)$, create $c - 1$ new OR-in-neighbors $a_{v^4}^j (1 \leq j \leq c$ and $i \neq j)$, and for each $u^2 \in V_j^2$ such that $vu \in E(Q)$ create an AND-vertex b_{vu}^j and the following edges: $(b_{vu}^j, a_{v^4}^j), (u^2, b_{vu}^j)$ and (v^2, b_{vu}^j);
7. finally, set $k = 2c^2 + 2c + 1$.

If Q contains a multicolored clique K such that $|K| = c$, then it is possible to find a satisfying assignment of C that consumes k energy by mapping the set S of gates/vertices that must be activated (outputs true) as follows: (a) v_{out} and all of its in-neighbors must belong to S; (b) for each OR-gate $w_i \in S$, we want include in S exactly the in-neighbor $v^4 \in V_i^4$ such that $v \in K$, therefore, we set $f(v^1) = \text{true}$ if and only if $v \in K$ (At this point, by construction, for each $v \notin K$ holds that every vertex between v^1 and v^4 will be inactivated); (c) for each $v^4 \in S$, all of its in-neighbors must be in S, and for each $a_{v^4}^j$ in S, its unique in-neighbor in S must be the AND-gate b_{vu}^j such that $f(v^1) = f(u^1) = \text{true}$ (recall that K has exactly one vertex per color). (d) finally, a vertex $v^2 \in V_2$ belongs to S if and only if its in-neighbor v^1 outputs true. Through a simple count one can conclude that $|S| = 2c^2 + 2c + 1$. Thus, the defined assignment satisfies C by consuming k energy as required.

Conversely, if C has a satisfying assignment X with energy complexity at most $k = 2c^2 + 2c + 1$ then it is possible to obtain a multicolored clique K of Q as follows: a vertex v of Q belongs to K if and only if v^2 outputs true. Since, by construction, any satisfying assignment of C activates at least $2c^2 + 2c + 1$ gates in $V(C) \setminus (V_2 \cup V_1)$, the assignment X activates at most c gates in V_2. Besides, the construction also implies that at least one input per color must activated in order to satisfy C. So, X activates exactly c gates in V_2 (one per color). Therefore, K has exactly one vertex per color. Now, to show that K induces a clique is enough to observe the if v_2 and u_2 are activated in X into C and $vu \notin E(G)$, then for the color j of u holds that b_{vr}^j is inactivated by X into C for any neighbor r of v with color j. Thus, v_4 and w_i are also inactivated, where i is the color of v, which contradicts the fact that X satisfies C. Therefore, K induces a clique. □

2.1   On Monotone Circuits with Bounded Genus

A graph G has *genus* at most g if it can be drawn on a surface of genus g (a sphere with g handles) without edge intersections. We consider the genus of a circuit as the genus of its underlying undirected graph.

In this section, we show that k-MINEC_M^+ on bounded genus circuits can be reduced to k-MINEC_M^+ on bounded treewidth circuits.

Definition 1. *A* tree decomposition *of a undirected graph G is a pair $\mathcal{T} = (T, \{X_t\}_{t\in V(T)})$, such that T is a tree where each node t is assigned to a set of vertices $X_t \subseteq V(G)$, called* bags, *according to the following conditions:*

- $\bigcup_{t\in V(T)} X_t = V(G)$;
- *For each $uv \in E(G)$ there is a node t such that $\{u, v\} \subseteq X_t$;*
- *For each $v \in V(G)$, the set $T_v = \{t \in V(T) : v \in X_t\}$ spans a subtree of T.*

The *width* of a tree decomposition $\mathcal{T}$ is the size of its largest bag minus one. The treewidth of G is the minimum width among all tree decompositions of G.

Definition 2. *A graph H is a* minor *of a graph G if H can be constructed from G by deleting vertices or edges, and contracting edges.*

Theorem 4 (Excluded Grid Theorem [14]). *Let t be a non-negative integer. Then every planar graph G of treewidth at least $9t/2$ contains a grid $t \times t$ as a minor.*

From the Excluded Grid Theorem, it is easy to see that there is a connection between the diameter of a planar graph and its treewidth. In [15], Robertson and Seymour presented a bound for the treewidth of a planar graph with respect to its radius, which also implies a bound regarding the diameter.

Definition 3. *For every face F of a planar embedding M, we define $d(F)$ to be the minimum value of r such that there is a sequence $F_0, F_1, \ldots, F_r$ of faces of M, where F_0 is the external face, $F = F_r$, and for $1 \leq j \leq r$ there is a vertex v incident with both F_{j-1} and F_j. The radius $\rho(M)$ of M is the minimum value r such that $d(F) \leq r$ for all faces F of M. The radius of a planar graph is the minimum of the radius of its planar embeddings.*

Theorem 5 (Radius Theorem [15]). *If G is planar and has radius at most r then its treewidth is at most $3r + 1$.*

Using Theorem 5 we are able to either solve MinEC_M^+ on planar circuits or outputs an equivalent instance C' with treewidth bounded by a function of k.

Lemma 1. *Let (C, k) be an instance of MinEC_M^+. There is an algorithm that in polynomial time either solves (C, k) or outputs an equivalent instance (C', k) of MinEC_M^+ where each vertex is at distance at most $2k + 1$ from v_{out} in the underlying undirected graph of C'.*

Proof. From an instance (C, k) of MinEC_M^+, we apply the following reduction rules to obtain C':

1. Delete every input vertex which is at a distance greater than k to v_{out};
2. Delete every vertex which is at a distance greater than $k + 1$ from its nearest input vertex;
3. Delete any AND-vertex which lost one of its in-neighbors;
4. Delete any OR-vertex in which its in-degree became equal to 0;
5. Repeat steps 1 to 4 as long as possible;
6. If $C' = \emptyset$ then we conclude that (C, k) is a *no*-instance of MinEC_M^+.

We now discuss the safety of the previous reduction rules: if an input vertex v is at a distance greater than k from v_{out}, since C is monotone, then v is not useful to satisfy v_{out} in any assignment X with $EC(C, X) \leq k$, thus we can assume that v outputs `false` and given the monotonicity of C we can safely remove v (Rule 1). Similarly, gates that are at a distance greater than k from its nearest input vertex must output `false` in an assignment X; otherwise, X consumes energy greater than k. Note that vertices at a distance exactly $k + 1$ from its nearest input vertex can be useful to show that a given assignment consumes energy greater than k. However, gates at a distance of at least $k + 2$ from its nearest input vertex can be removed once its neighbors are sufficient to certify the negative answer (Rule 2). Besides, if for any assignment X with $EC(C, X) \leq k$ holds that some (resp. every) in-neighbor of an AND(resp. OR)-vertex v must output `false`, then v must output `false` as well. Thus, Rule 3 and Rule 4 are safe. From the safety of rules 1–4, it follows that Rule 5 and Rule 6 are safe. Finally, if $C' \neq \emptyset$ then C' has only vertices at a distance at most $2k + 1$ from v_{out} in the underlying undirected graph of C'.   □

Note that the underlying undirected graph of the circuits obtained from Lemma 1 have diameter bounded by $4k + 2$. Therefore, contrasting with the W[1]-hardness for the general case, Corollary 1 holds.

Corollary 1. *MinEC_M^+ is fixed-parameter tractable when restricted to monotone circuits having bounded maximum in-degree.*

Also, notice that a gate with large in-degree can always be replaced by a binary tree using only binary gates, but for or-gates it makes a relevant difference in the energy complexity. Therefore, replacing large in-degree gates is not a useful strategy for dealing with k-MinEC_M^+. On the other hand, Lemma 1 also implies that if C' is planar then it also has bounded radius , thus, by Theorem 5, it follows that the underlying undirected graph of C' has treewidth bounded by a function of k. We extend the previous reasoning for bounded genus circuits.

Given a vertex-set $S \subseteq V(G)$ of a simple graph G such that the subgraph of G induced by S, denoted $G[S]$, is connected, contracting S means contracting the edges between the vertices in S to obtain a single vertex at the end. We say that a graph H is an *s-contraction* of a graph G if H can be obtained after applying to G a (possibly empty) sequence of edge contractions.

The following is a construction presented in [8] and [10]. Consider an $(r \times r)$-grid. A corner vertex of the grid is a vertex of the grid of degree 2. By Γ_r we denote the graph obtained from the $(r \times r)$-grid as follows : construct first the Γ_r' by triangulating all internal faces of the $(r \times r)$-grid such that all internal vertices of the grid are of degree 6, and all non-corner external vertices of the grid are of degree 4 (Γ_r' is unique up to isomorphism). Two of the corners of the initial grid have degree 2 in Γ_r'; let x be one of them. Γ_r obtained from Γ_r' by adding all the edges having x as an endpoint and a vertex of the external face of the grid that is not already a neighbor of x as the other endpoint. Observe again that Γ_r' is unique up to isomorphism. The following is a lemma from [10] implied from Lemma 6 in [8].

Lemma 2 (Lemma 4.5 in [10]). *Let G be a graph of genus g, and let r be any positive integer. If G excludes Γ_r as an s-contraction, then the treewidth of G is at most $(2r + 4) \cdot (g + 1)^{3/2}$.*

Lemma 3. *Let C' be the circuit obtained from Lemma 1. It holds that C' has treewidth at most $(8k + 14) \cdot (g + 1)^{3/2}$, where g is the genus of C'.*

Proof. First, notice that for each vertex u of a Γ_{4k+5} there is another vertex v such that the distance between u and v is at least $2k + 2$. Now, suppose that C' has Γ_{4k+5} as an s-contraction, and let u be a vertex of a Γ_{4k+5} such that u is either v_{out} or a vertex obtained by contracting S containing v_{out}. Since there is a vertex v such that the distance between u and v is at least $2k + 2$, it holds that C' has a vertex at distance greater than $2k + 1$ from v_{out}, which is a contradiction (see Lemma 1). Thus, by Lemma 2 we have that the treewidth of C' is at most $(8k + 14) \cdot (g + 1)^{3/2}$. $\qquad\qquad\square$

3   Dynamic Programming on Bounded Treewidth Circuits

From Lemma 3, in order to solve k-MinEC_M^+ in FPT-time on bounded genus instances, it is enough to present an FPT algorithm parameterized by the treewidth of the input. To design a dynamic programming on tree decompositions, without loss of generality, we may consider that we are given a tree decomposition that is a rooted *extended nice tree decomposition* (see [6] for details).

Theorem 6. MinEC_M^+ *can be solved in time $2^{O(tw)} \cdot n$, where tw is the treewidth of the underlying undirected graph of the input.*

Proof. Let $C = (I, G, v_{out})$ be a monotone circuit where I is the set of inputs of C, G is the set of gates with out-degree greater than 0 and v_{out} is a single output vertex. Let $\mathcal{T} = (T, \{X_t\}_{t \in V(T)})$ be a rooted extended nice tree decomposition of C. Consider also T_t as the subtree of T rooted by node t (bag X_t) and C_t be the graph/circuit having $\mathcal{T}_t = (T_t, \{X_i\}_{i \in V(T_t)})$ as tree decomposition. For convenience, we add the vertex v_{out} to every bag of T; thus, the width of $\mathcal{T}$ is increased by at most one.

Now, note that an assignment X satisfies a monotone circuit C if and only if it induces an activation set $\mathcal{S}_X$ such that:

1. $v_{out} \in \mathcal{S}_X$;
2. for each $v \in \mathcal{S}_X$ holds that:
   - if $f(v) = $ AND then every in-neighbor of v is in $\mathcal{S}_X$;
   - if $f(v) = $ OR then at least one among the in-neighbors of v is in $\mathcal{S}_X$;
3. for each $v \notin \mathcal{S}_X$ holds that:
   - if $f(v) = $ AND then at least one among the in-neighbor of v is not in $\mathcal{S}_X$;
   - if $f(v) = $ OR then every in-neighbor of v is not in $\mathcal{S}_X$;

Properties 1 and 2 describe the necessary and sufficient conditions for a set $\mathcal{S}_X$ of activated gates to certify a satisfying assignment. Property 3 ensures that $\mathcal{S}_X$ is maximal regarding the property of having been activated by X.

Therefore, the problem of finding a satisfying assignment X which minimizes $EC(C, X)$ can be seen as the problem of finding a satisfying assignment X which minimizes $|\mathcal{S}_X \setminus I|$. Thus, we define $c[t, S, \mathcal{B}^{OR}, \mathcal{B}^{AND}]$ as the cardinality of a minimum set of gates $\mathcal{S}_t$ (if any) of C_t such that:

- $v_{out} \in S$ and $S = X_t \cap \mathcal{S}_t$; (we say that $X_t \setminus S = \overline{S}$)
- for each $v \in V(C_t) \setminus X_t$ properties 2 and 3 holds with respect to $\mathcal{S}_t$;
- for each $v \in S$ such that $f(v) = $ AND, all in-neighbors of v in C_t are in $\mathcal{S}_t$;
- The set $\mathcal{B}^{OR}$ is the subset of OR-gates in S already having in-neighbors in $\mathcal{S}_t$;
- for each $v \in \overline{S}$ such that $f(v) = $ OR, all in-neighbors of v in C_t are not in $\mathcal{S}_t$;
- The set $\mathcal{B}^{AND}$ is the subset of AND-gates in $\overline{S}$ already having in-neighbors that are not in $\mathcal{S}_t$;

Furthermore, the optimal solution of the main problem can be found either at $c[r, \{v_{out}\}, \{v_{out}\}, \{\}]$, if $f(v_{out}) = $ OR, or at $c[r, \{v_{out}\}, \{\}, \{\}]$, if $f(v_{out}) = $ AND, where r is the root of the tree decomposition $\mathcal{T}$. Recall that, for any node t we assume that $v_{out} \in S$.

In order to solve MINEC_M^+, the counting of gates that output **true** (in the solution) are made in introduce vertex nodes. Note that in the introduce node of an OR-vertex v, it can not be simultaneously in S and $\mathcal{B}^{OR}$ because when a vertex is introduced then it is isolated in C_t (we are considering an extended nice tree decomposition, i.e. the edges are introduced in introduce edge nodes).

Leaf Node. Let t be a leaf node, then $X_t = \{v_{out}\}$. Since v_{out} must be in S, then $\mathcal{B}^{AND} = \emptyset$. Thus, we have three subproblems in Eq. (1).

$$c[t, \{v_{out}\}, \mathcal{B}^{OR}, \mathcal{B}^{AND}] = \begin{cases} 1, & \text{if } f(v_{out}) = \text{AND} \\ 1, & \text{if } f(v_{out}) = \text{OR and } v_{out} \notin \mathcal{B}^{OR} \\ \infty, & \text{if } f(v_{out}) = \text{OR and } v_{out} \in \mathcal{B}^{OR} \end{cases} \quad (1)$$

Introduce Vertex Node. Let t be an introduce vertex node with exactly one child t' such that $X_t = X_{t'} \cup \{v\}$. In the graph C_t, v is an isolated vertex; consequently, as in the leaf nodes, there is infeasibility whenever v belongs to $\mathcal{B}^{\mathrm{OR}}$ or $\mathcal{B}^{\mathrm{AND}}$. Besides, we have the possibility of v be an input vertex ($f(v) \notin \{\mathrm{AND}, \mathrm{OR}\}$) or $v \notin S$, such situations only rescue previous subproblems without increment the current subsolution. On the other hand, we increment the subsolution by 1 whenever $v \in S$. All possibilities are covered in Eqs. (2), (3) and (4).

- If $f(v) \notin \{\mathrm{AND}, \mathrm{OR}\}$ then

$$c[t, S, \mathcal{B}^{\mathrm{OR}}, \mathcal{B}^{\mathrm{AND}}] = c[t', S \setminus \{v\}, \mathcal{B}^{\mathrm{OR}}, \mathcal{B}^{\mathrm{AND}}] \tag{2}$$

- If $f(v) = \mathrm{OR}$ then

$$c[t, S, \mathcal{B}^{\mathrm{OR}}, \mathcal{B}^{\mathrm{AND}}] = \begin{cases} c[t', S, \mathcal{B}^{\mathrm{OR}}, \mathcal{B}^{\mathrm{AND}}], & \text{if } v \notin S \\ c[t', S \setminus \{v\}, \mathcal{B}^{\mathrm{OR}}, \mathcal{B}^{\mathrm{AND}}] + 1, & \text{if } v \in S \text{ and } v \notin \mathcal{B}^{\mathrm{OR}} \\ \infty, & \text{if } v \in S \text{ and } v \in \mathcal{B}^{\mathrm{OR}} \end{cases} \tag{3}$$

- If $f(v) = \mathrm{AND}$ then

$$c[t, S, \mathcal{B}^{\mathrm{OR}}, \mathcal{B}^{\mathrm{AND}}] = \begin{cases} c[t', S \setminus \{v\}, \mathcal{B}^{\mathrm{OR}}, \mathcal{B}^{\mathrm{AND}}] + 1, & \text{if } v \in S \\ c[t', S, \mathcal{B}^{\mathrm{OR}}, \mathcal{B}^{\mathrm{AND}}], & \text{if } v \notin S \text{ and } v \notin \mathcal{B}^{\mathrm{AND}} \\ \infty, & \text{if } v \notin S \text{ and } v \in \mathcal{B}^{\mathrm{AND}} \end{cases} \tag{4}$$

Introduce Edge Node. Let t be an introduce edge node and t' its child such that $X_t = X_{t'}$, which introduces the directed edge uv such that $\{u, v\} \subseteq X_t$. Now, by including an edge, we can evaluate each subproblem concerning the sets $\mathcal{B}^{\mathrm{OR}}$ and $\mathcal{B}^{\mathrm{AND}}$; so, for each OR-gate $v \in S$, at least one in-neighbor also must be in S; so, either uv attend this demand or another already introduced edge satisfied that. We apply the same reasoning for AND-gates: considering an AND-gate $v \in \overline{S}$, then at least one in-edge of v need comes to another vertex in $\overline{S}$; if uv do not attend this requirement, the current subproblem is assigned to a previous subproblem where $v \in \mathcal{B}^{\mathrm{AND}}$. All these conditions are handled in Eqs. (5) and (6). Recall that we are introducing the directed edge uv.

- If $f(v) = \mathrm{OR}$ then $c[t, S, \mathcal{B}^{\mathrm{OR}}, \mathcal{B}^{\mathrm{AND}}]$ is equal to

$$\begin{cases} c[t', S, \mathcal{B}^{\mathrm{OR}}, \mathcal{B}^{\mathrm{AND}}], & \text{if } u \notin S \\ \infty, & \text{if } u \in S \text{ and } v \notin S \cap \mathcal{B}^{\mathrm{OR}} \\ \min\{c[t', S, \mathcal{B}^{\mathrm{OR}}, \mathcal{B}^{\mathrm{AND}}], c[t', S, \mathcal{B}^{\mathrm{OR}} \setminus \{v\}, \mathcal{B}^{\mathrm{AND}}]\}, & \text{if } u \in S \text{ and } v \in S \cap \mathcal{B}^{\mathrm{OR}} \end{cases} \tag{5}$$

- If $f(v) = \text{AND}$ then $c[t, S, \mathcal{B}^{\text{OR}}, \mathcal{B}^{\text{AND}}]$ is equal to

$$\begin{cases} c[t', S, \mathcal{B}^{\text{OR}}, \mathcal{B}^{\text{AND}}], & \text{if } u \in S \\ \infty, & \text{if } u \notin S \text{ and } v \in S \\ \infty, & \text{if } \{u, v\} \subseteq \overline{S} \text{ and } v \notin \mathcal{B}^{\text{AND}} \\ \min\left\{c[t', S, \mathcal{B}^{\text{OR}}, \mathcal{B}^{\text{AND}}], c[t', S, \mathcal{B}^{\text{OR}}, \mathcal{B}^{\text{AND}} \setminus \{v\}]\right\}, & \text{if } \{u, v\} \subseteq \overline{S} \text{ and } v \in \mathcal{B}^{\text{AND}} \end{cases}$$

$$(6)$$

Forget Node. Let t be a forget node and t' be its child such that $X_t = X_{t'} \setminus v$. In this case, we verify the best among either selecting or not v in current subproblem. If v is an input vertex, then this verification is trivial (it is enough to rescue the minimum subsolution varying only the membership of v in S). For OR-gates and AND-gates, the same verification are made but considering the feasibility of v through its membership in $\mathcal{B}^{\text{OR}}$ and $\mathcal{B}^{\text{AND}}$. Equations (7), (8) and (9) summarize these three scenarios.

- If $f(v) \neq \{\text{AND}, \text{OR}\}$ then

$$c[t, S, \mathcal{B}^{\text{OR}}, \mathcal{B}^{\text{AND}}] = \min\left\{c[t', S, \mathcal{B}^{\text{OR}}, \mathcal{B}^{\text{AND}}], c[t', S \cup \{v\}, \mathcal{B}^{\text{OR}}, \mathcal{B}^{\text{AND}}]\right\} \quad (7)$$

- If $f(v) = \text{OR}$ then

$$c[t, S, \mathcal{B}^{\text{OR}}, \mathcal{B}^{\text{AND}}] = \min\left\{c[t', S, \mathcal{B}^{\text{OR}}, \mathcal{B}^{\text{AND}}], c[t', S \cup \{v\}, \mathcal{B}^{\text{OR}} \cup \{v\}, \mathcal{B}^{\text{AND}}]\right\} \quad (8)$$

- If $f(v) = \text{AND}$ then

$$c[t, S, \mathcal{B}^{\text{OR}}, \mathcal{B}^{\text{AND}}] = \min\left\{c[t', S, \mathcal{B}^{\text{OR}}, \mathcal{B}^{\text{AND}} \cup \{v\}], c[t', S \cup \{v\}, \mathcal{B}^{\text{OR}}, \mathcal{B}^{\text{AND}}]\right\} \quad (9)$$

Join Node. Let t be a join node with two children t_1 and t_2. For tabulation of the join nodes, we need to combine two partial solutions – one originating from C_{t_1} and another from C_{t_2} – in such a way that the merging is a feasible solution. Recall that G is acyclic so we don't need to care about cycles. Also, if a gate is activated in C_t it must be activated in both children, so we must subtract duplicity. However, since each edge of C_t is in either C_{t_1} or C_{t_2}, the feasibility of merging children's solutions is guaranteed assuming that whether $v \in \mathcal{B}^{\text{OR}}/\mathcal{B}^{\text{AND}}$ then it is also in the respective set of one of the children, as described in Eq. 10.

$$c[t, S, \mathcal{B}^{\text{OR}}, \mathcal{B}^{\text{AND}}] = \min_{\mathcal{B}_1^{\text{OR}}, \mathcal{B}_1^{\text{AND}}, \mathcal{B}_2^{\text{OR}}, \mathcal{B}_2^{\text{AND}}} \left\{c[t_1, S, \mathcal{B}_1^{\text{OR}}, \mathcal{B}_1^{\text{AND}}] + c[t_2, S, \mathcal{B}_2^{\text{OR}}, \mathcal{B}_2^{\text{AND}}]\right\} - |S \setminus I|$$

$$(10)$$

where $\mathcal{B}^{\text{OR}} = \mathcal{B}_1^{\text{OR}} \cup \mathcal{B}_2^{\text{OR}}$ and $\mathcal{B}^{\text{AND}} = \mathcal{B}_1^{\text{AND}} \cup \mathcal{B}_2^{\text{AND}}$.

Every bag of $\mathcal{T}$ has at most $tw + 2$ vertices (including v_{out}) and v_{out} is fixed in the solution, thus each bag has at most 2^{tw+1} possible subsets S, there are at most 2^{tw+2} possible sets $\mathcal{B}^{\text{OR}}$, and there are at most 2^{tw+1} sets $\mathcal{B}^{\text{AND}}$. Therefore, the entire matrix has size $2^{O(tw)} \cdot n$. As each entry of the table can be computed in $2^{O(tw)}$ time, it holds that the algorithm performs in time $2^{O(tw)} \cdot n$. □

Corollary 2. MinEC_M^+ *can be solved in time* $2^{O(k \cdot (g+1)^{3/2})} \cdot n^{O(1)}$, *where g is the genus of the input.*

References

1. Allender, E., Koucký, M.: Amplifying lower bounds by means of self-reducibility. J. ACM (JACM) **57**(3), 1–36 (2010)
2. Alves, M.R., de Oliveira Oliveira, M., Nascimento Silva, J.C., Souza, U.S.: Succinct certification of monotone circuits. Theoret. Comput. Sci. **889**, 1–13 (2021). https://doi.org/10.1016/j.tcs.2021.07.032
3. Antoniadis, A., Barcelo, N., Nugent, M., Pruhs, K., Scquizzato, M.: Energy-efficient circuit design. In: Proceedings of the 5th Conference on Innovations in Theoretical Computer Science, pp. 303–312 (2014)
4. Barcelo, N., Nugent, M., Pruhs, K., Scquizzato, M.: Almost all functions require exponential energy. In: Italiano, G.F., Pighizzini, G., Sannella, D.T. (eds.) MFCS 2015. LNCS, vol. 9235, pp. 90–101. Springer, Heidelberg (2015). https://doi.org/10.1007/978-3-662-48054-0_8
5. Blake, C.G., Kschischang, F.R.: On the VLSI energy complexity of LDPC decoder circuits. IEEE Trans. Inf. Theory **63**(5), 2781–2795 (2017)
6. Cygan, M., et al.: Parameterized Algorithms. Springer, Cham (2015). https://doi.org/10.1007/978-3-319-21275-3
7. Dinesh, K., Otiv, S., Sarma, J.: New bounds for energy complexity of Boolean functions. Theoret. Comput. Sci. **845**, 59–75 (2020)
8. Fomin, F.V., Golovach, P., Thilikos, D.M.: Contraction obstructions for treewidth. J. Comb. Theory, Ser. B **101**(5), 302–314 (2011)
9. Hastad, J.: Almost optimal lower bounds for small depth circuits. In: Proceedings of the Eighteenth Annual ACM Symposium on Theory of Computing, pp. 6–20 (1986)
10. Kanj, I., Thilikos, D.M., Xia, G.: On the parameterized complexity of monotone and antimonotone weighted circuit satisfiability. Inf. Comput. **257**, 139–156 (2017)
11. Kissin, G.: Measuring energy consumption in VLSI circuits: a foundation. In: Proceedings of the Fourteenth Annual ACM Symposium on Theory of Computing, pp. 99–104 (1982)
12. Lennie, P.: The cost of cortical computation. Curr. Biol. **13**(6), 493–497 (2003)
13. Razborov, A.A.: Lower bounds on the size of bounded depth circuits over a complete basis with logical addition. Math. Notes Acad. Sci. USSR **41**(4), 333–338 (1987). https://doi.org/10.1007/BF01137685
14. Robertson, N., Seymour, P., Thomas, R.: Quickly excluding a planar graph. J. Comb. Theory Ser. B **62**(2), 323–348 (1994)
15. Robertson, N., Seymour, P.D.: Graph minors. III. Planar tree-width. J. Comb. Theory Ser. B **36**(1), 49–64 (1984)
16. Šíma, J.: Energy complexity of recurrent neural networks. Neural Comput. **26**(5), 953–973 (2014)
17. Souza, U.S., Protti, F.: Tractability, hardness, and kernelization lower bound for and/or graph solution. Discret. Appl. Math. **232**, 125–133 (2017). https://doi.org/10.1016/j.dam.2017.07.029
18. Souza, U.S., Protti, F., Dantas da Silva, M.: Revisiting the complexity of and/or graph solution. J. Comput. Syst. Sci. **79**(7), 1156–1163 (2013). https://doi.org/10.1016/j.jcss.2013.04.001
19. Uchizawa, K., Douglas, R., Maass, W.: On the computational power of threshold circuits with sparse activity. Neural Comput. **18**(12), 2994–3008 (2006)
20. Vaintsvaig, M.N.: On the power of networks of functional elements. In: Doklady Akademii Nauk. vol. 139–2, pp. 320–323. Russian Academy of Sciences (1961)
21. Williams, R.: Nonuniform ACC circuit lower bounds. J. ACM (JACM) **61**(1), 1–32 (2014)

Miscellaneous

The Independence Numbers of Weighted Graphs with Forbidden Cycles

Ye Wang[1] and Yan Li[2(✉)]

[1] Harbin Engineering University, Harbin 150001, China
[2] University of Shanghai for Science and Technology, Shanghai 200093, China
li_yan0919@usst.edu.cn

Abstract. Let $G = (V, E, w)$ be a weighted graph. The independence number of a weighted graph G is the maximum $w(S)$ taken over all independent sets S of G. By a well-known method, we can show that the independence number of G is at least $\sum_{v \in V} \frac{w(v)}{1+d(v)}$, where $d(v)$ is the degree of v. In this paper, we consider the independence numbers of weighted graphs with forbidden cycles. For a graph G, the odd girth of G is the smallest length of an odd cycle in G. We show that if G is a weighted graph with odd girth $2m + 1$ for $m \geq 3$, then the independence number of G is at least $c \left(\sum_v w(v) d(v)^{1/(m-2)} \right)^{(m-2)/(m-1)}$, where c is a constant.

Keywords: Independence number · Random algorithm · Weighted graph · Odd girth

1 Introduction

Let $G = (V, E)$ be a graph with order N and degree sequence $\{d(v) : v \in V\}$, and average degree d. Let $\alpha(G)$ be the independence number of G. A well-known lower bound for $\alpha(G)$ is Turán theorem that $\alpha(G) \geq N/(1 + d)$. This result was strengthened by Caro [4] and Wei [13] as

$$\alpha(G) \geq \sum_{v \in V} \frac{1}{1 + d(v)}. \tag{1}$$

Alon and Spencer [3] showed an elegant probabilistic proof for this result.

A graph $G = (V, E)$ is called weighted if each vertex $v \in V$ is assigned a weight $w(v) \in [0, \infty)$. We shall write the weighted graph G as (V, E, w). For a subset S of V, set $w(S) = \sum_{v \in S} w(v)$. The weighted independence number of G, denoted by $\alpha_w(G)$, is the maximum $w(S)$ taken over all independent sets S of G. Note that $\alpha_w(G)$ can always achieve its value at a maximal independent

Supported in part by NSFC (12101156).

W. Wu and H. Du (Eds.): AAIM 2021, LNCS 13153, pp. 395–399, 2021.
https://doi.org/10.1007/978-3-030-93176-6_33

set, but the set is not necessarily maximum. It is easy to extend the proof of (1) of Alon and Spencer to show that

$$\alpha_w(G) \geq \sum_{v \in V} \frac{w(v)}{1 + d(v)}. \tag{2}$$

When G is locally sparse, the Turán's bound can be improved greatly. The most studied such graphs are triangle-free graphs. (G is $H-$free if G contains no H as a subgraph.) A result of Ajtai, Komlós and Szemerédi [1] states that

$$\alpha(G) \geq cN\frac{\log d}{d} \tag{3}$$

for any triangle-free graph G, where $c > 0$ is a constant. The improvements have been obtained by Griggs [6] and Shearer [10,11]. Li, Rousseau and Zang [7–9] generalized the result for graphs with sparse neighborhoods. Alon [2] gave that for graphs whose chromatic number of neighborhood of each vertex is bounded.

We shall have a more general criteria for local sparseness of a graph G. For a graph G, the odd girth of G is the smallest length of an odd cycle in G. Shearer [12] showed that if the odd girth of G is $2m + 1$, then

$$\alpha(G) \geq \frac{1}{2^{(m-2)/(m-1)}} \left(\sum_{v \in V} d(v)^{1/(m-2)} \right)^{(m-2)/(m-1)}. \tag{4}$$

In particular, if G is $d-$regular, than $\alpha(G) \geq cN^{1-1/(m-1)}d^{1/(m-1)}$, where c is a constant. When $d \gg N^{1/m}(\log N)^{(m-1)/m}$, this lower bound is much better than that in (3).

The propose of this paper is to generalize (4) for weighted graphs in a manner similar to that from (1) to (2). Let $\mathcal{M}$ be the family of all maximum independent sets of G, and let $\overline{w} = \max\{w(I)/|I| : I \in \mathcal{M}\}$. Note that $|I| = \alpha(G)$ for any $I \in \mathcal{M}$. It is trivial to see

$$\alpha_w(G) = \max_{I \in \mathcal{M}} w(I) = \overline{w}\alpha(G).$$

Theorem 1. Let $G = (V, E, w)$ be a weighted graph. If the odd girth of G is $2m + 1$, where $m \geq 3$, then

$$\alpha_w(G) \geq c \left(\sum_{v \in V} w(v) d(v)^{1/(m-2)} \right)^{(m-2)/(m-1)},$$

where $c = \frac{\overline{w}^{1/(m-1)}}{2^{(m-2)/(m-1)}}$ is a constant.

Remark: The above result is sharp for $K_{n,n}$ when both sums of weights of vertices in two classes are equal. We are not expecting that the coefficient c is independent of w. In fact, for any $t > 0$, we have $\alpha_w(G) = \frac{1}{t}\alpha_{tw}(G)$. If c is independent of w, then the quantity

$$\frac{c}{t}\left(\sum_{v \in V} tw(v)\,d(v)^{1/(m-2)}\right)^{(m-2)/(m-1)} = \frac{c}{t^{1/(m-1)}}\left(\sum_{v \in V} w(v)\,d(v)^{1/(m-2)}\right)^{(m-2)/(m-1)},$$

which can be arbitrarily large as t tends to zero. However, it is reasonable to try to replace $\overline{w}$ in the coefficient by some quantity that is independent of $\mathcal{M}$. The minimum value of $w(v)$ is a trivial one.

2   Proof of Main Result

Given a graph $G = (V, E, w)$, let $N_i(v)$ be the set of all the vertices of distance i from vertex v in G, and $d_i(v) = |N_i(v)|$. Thus $N_0(v) = \{v\}$. To show that $G = (V, E, w)$ contains an independent set I with $w(I)$ as desired, we may apply the following algorithm: Initially set $I = \emptyset$. Let v be a vertex to be chosen, set $I = I \cup \{v\}$ and $G = G \setminus \{v \cup N_1(v)\}$. Repeat the process until G contains no vertex. The criteria for choosing v is random.

Lemma 1. *Let $G = (V, E, w)$ be a triangle-free weighted graph. Then*

$$\alpha_w(G) \geq \sum_{v \in V} \frac{d_1(v)}{1 + d_1(v) + d_2(v)} w(v).$$

Proof. Let us randomly label the vertices of G with a permutation of integers $1, 2, \ldots, N$, where N is the order of G. Let X be the set of all the vertices v such that the smallest label on $\{v\} \cup N_1(v) \cup N_2(v)$ is on a vertex in $N_1(v)$. Then the probability that v belongs to X is $\frac{d_1(v)}{1 + d_1(v) + d_2(v)}$, and thus the expectation of $w(X)$ is $\sum_{v \in V} \frac{d_1(v)}{1 + d_1(v) + d_2(v)} w(v)$. It follows that there is some fixed permutation of integers $1, 2, \ldots, N$ such that

$$w(X) \geq \sum_{v \in V} \frac{d_1(v)}{1 + d_1(v) + d_2(v)} w(v).$$

To this end, we shall show that X is an independent set. Suppose to the contrary that there is an edge uv in G with $u, v \in X$. Let u_1 be the vertex with the smallest label on $\{u\} \cup N_1(u) \cup N_2(u)$, and v_1 the vertex with the smallest label on $\{v\} \cup N_1(v) \cup N_2(v)$. Clearly $u_1 \neq v_1$ as G is triangle-free. Since $v_1 \in N_2(u)$, by the definition of X we see that the label of u_1 is less than that of v_1. Similarly, as $u_1 \in N_2(v)$, we know that the label of v_1 is less than that of u_1, which is a contradiction. So we have

$$\alpha_w(G) \geq w(X) \geq \sum_{v \in V} \frac{d_1(v)}{1 + d_1(v) + d_2(v)} w(v),$$

completing the proof.   □

Lemma 2. *Let $G = (V, E, w)$ be a weighted graph with odd girth $2m + 1$, where $m \geq 2$. Then for any $2 \leq \ell \leq m$,*

$$\alpha_w(G) \geq \frac{1}{2} \sum_{v \in V} \frac{1 + d_1(v) + \cdots + d_{\ell-1}(v)}{1 + d_1(v) + \cdots + d_\ell(v)} w(v).$$

Proof. The proof is similar to that of Lemma 1, so we only give a sketch. Randomly label the vertices of G. Define X as the set of all vertices v such that the smallest label on $\cup_{j=0}^{\ell} N_j(v)$ is on a vertex in $\cup_{j=0}^{\ell-1} N_j(v)$. By consider the expectation of $w(X)$, we assert that there is an X such that

$$w(X) \geq \sum_{v \in V} \frac{1 + d_1(v) + \cdots + d_{\ell-1}(v)}{1 + d_1(v) + \cdots + d_\ell(v)} w(v).$$

The next step is to show that X is bipartite. If uv is an edge in G with $u, v \in X$, let u_1 be the vertex with the smallest label on $\cup_{j=0}^{\ell} N_j(u)$, and v_1 the vertex with the smallest label on $\cup_{j=0}^{\ell} N_j(v)$. Similarly, we can conclude that $u_1 = v_1$. Thus X can be partitioned into $X_1, X_2, \cdots, X_p$, such that there is no edge between X_i and X_j for $i \neq j$, and all the vertices in X_i share the same vertex v_i with the smallest label. Furthermore, the distance between each vertex in X_i and v_i is at most $\ell - 1$. Partition X_i into X_i' and X_i'' by distance of any vertex $v \in X_i$ to v_i odd and even, respectively. As G contains no $C_{2\ell-1}$, both $\cup_i X_i'$ and $\cup_i X_i''$ are independent sets, and hence X is bipartite. □

Lemma 3. *Let $G = (V, E, w)$ be a weighted graph with odd girth $2m + 1$, where $m \geq 3$. Then*

$$\alpha_w(G) \geq \frac{1}{2^{(m-3)/(m-2)}} \sum_{v \in V} w(v) \left(\frac{d_1(v)}{1 + d_1(v) + \cdots + d_{m-1}(v)} \right)^{1/(m-2)}.$$

Proof. Applying Lemma 1 and Lemma 2 repeatedly, we have

$$\alpha_w(G) \geq \sum_{v \in V} \frac{w(v)}{m-2} \left(\frac{d_1(v)}{1 + d_1(v) + d_2(v)} + \frac{1 + d_1(v) + d_2(v)}{2(1 + d_1(v) + d_2(v) + d_3(v))} \right.$$

$$\left. + \cdots + \frac{1 + d_1(v) + \cdots + d_{m-2}(v)}{2(1 + d_1(v) + \cdots + d_{m-1}(v))} \right)$$

$$\geq \frac{1}{2^{(m-3)/(m-2)}} \sum_{v \in V} w(v) \left(\frac{d_1(v)}{1 + d_1(v) + \cdots + d_{m-1}(v)} \right)^{1/(m-2)},$$

where the last inequality is based on the fact that the arithmetic mean is at least the geometric mean. □

Proof of Theorem 1. For any vertex v in G and any $i \leq m-1$, as G contains no C_{2i+1}, we have $N_i(v)$ is an independent set. So both $\cup_{even\ i} N_i(v)$ and $\cup_{odd\ i} N_i(v)$ are independent sets hence $\cup_{i=0}^{m-1} N_i(v)$ is bipartite. So we obtain

$$2\alpha(G) \geq 1 + d_1(v) + \cdots + d_{m-1}(v).$$

The fact that $\alpha_w(G) = \overline{w}\alpha(G)$ implies

$$\frac{2}{\overline{w}}\alpha_w(G) \geq 1 + d_1(v) + \cdots + d_{m-1}(v).$$

Plugging this into the inequality in Lemma 3, we have

$$\alpha_w(G)^{(m-1)/(m-2)} \geq \frac{\overline{w}^{1/(m-2)}}{2} \sum_{v \in V} w(v)d(v)^{1/(m-2)},$$

and the desired bound follows immediately. □

References

1. Ajtai, M., Komlós, J., Szemerédi, E.: A note on Ramsey numbers. J. Combin. Theory Ser. A **29**(3), 354–360 (1980)
2. Alon, N.: Independence numbers of locally sparse graphs and a Ramsey type problem. Random Struct. Algorithms **9**(3), 271–278 (1996)
3. Alon, N., Spencer, J.: The Probabilistic Method. Wiley, New York (1992)
4. Caro, Y.: New results on the independence number. Technical report. Tel-Aviv University (1979)
5. Erdős, P., Faudree, R., Rousseau, C., Schelp, R.: On cycle-complete Ramsey numbers. J. Graph Theory **2**(1), 53–64 (1978)
6. Griggs, J.R.: An upper bound on the Ramsey number $R(3, n)$. J. Combin. Theory Ser. B **35**(2), 145–153 (1983)
7. Li, Y., Rousseau, C.: On book-complete graph Ramsey numbers. J. Combin. Theory Ser. B **68**(1), 36–44 (1996)
8. Li, Y., Rousseau, C., Zang, W.: Asymptotic upper bounds for Ramsey functions. Graphs Combin. **17**(1), 123–128 (2001)
9. Li, Y., Zang, W.: The independence number of graphs with a forbidden cycle and Ramsey numbers. J. Combin. Optim. **7**(4), 353–359 (2003)
10. Shearer, J.: A note on the independence number of triangle-free graphs. Discrete Math. **46**(1), 83–87 (1983)
11. Shearer, J.: Lower bounds for small diagonal Ramsey numbers. J. Combin. Theory Ser. A **42**(2), 302–304 (1986)
12. Shearer, J.: The independence number of dense graphs with large odd girth. Electron. J. Combin. **2**, #N2 (1995)
13. Wei, V.: A lower bound on the stability number of a simple graph. Bell Laboratories Technical Memorandum. No. 81-11217-9 (1981)

Wegner's Conjecture on 2-Distance Coloring

Junlei Zhu[1]($\boxtimes$), Yuehua Bu[2,3], and Hongguo Zhu[2]

[1] College of Data Science, Jiaxing University, Jiaxing 314001, China
[2] Department of Mathematics, Zhejiang Normal University,
Jinhua 321004, Zhejiang, China
[3] Xingzhi College, Zhejiang Normal University, Jinhua 321004, Zhejiang, China

Abstract. A 2 distance k-coloring of a graph G is a function $f : V(G) \rightarrow \{1, 2, \ldots, k\}$ such that $|f(u) - f(v)| \geq 1$ if $1 \leq d(u, v) \leq 2$, where $d(u, v)$ is the distance between the two vertices u and v. The 2-distance chromatic number of G, written $\chi_2(G)$, is the minimum k such that G has such a coloring. In this paper, we show that $\chi_2(G) \leq 5\Delta - 7$ holds for planar graphs G with maximum degree $\Delta \geq 5$, which improves a result due to Zhu and Bu (J. Comb. Optim. 36:55–64, 2018).

Keywords: Planar graph · 2-distance coloring · Maximum degree · Girth · Wegner's Conjecture

1 Introduction

A 2 distance k-coloring of a graph G is a function $f : V(G) \rightarrow \{1, 2, \ldots, k\}$ such that $|f(u) - f(v)| \geq 1$ if $1 \leq d(u, v) \leq 2$, where $d(u, v)$ is the distance between the two vertices u and v. The 2-distance chromatic number of G, written $\chi_2(G)$, is the minimum k such that G has such a coloring.

The research of this problem can be traced back to 1977. In [10], Wegner first proposed the 2-distance coloring of graphs and showed that $\chi_2(G) \leq 8$ holds for planar graphs G with maximum degree 3. He further conjectured that the upper bound 8 can be reduced to 7, which has been confirmed by Thomasse [9]. In [6], Montassier and Raspaud showed that the upper bound 7 is tight. For planar graphs with maximum degree $\Delta \geq 4$, Wenger posed the following conjecture.

Conjecture 1.1. *Let G be a planar graph with maximum degree Δ. If $4 \leq \Delta \leq 7$, then $\chi_2(G) \leq \Delta + 5$. If $\Delta \geq 8$, then $\chi_2(G) \leq \lfloor \frac{3\Delta}{2} \rfloor + 1$.*

Conjecture 1.1 is still open. However, several upper bounds in terms of maximum degree Δ have been proven as follows.

1. $\chi_2(G) \leq 2\Delta + 25$ [4].
2. $\chi_2(G) \leq 2\Delta + 16$ if $\Delta \geq 8$ [8].
3. $\chi_2(G) \leq \lceil \frac{9}{5}\Delta \rceil + 1$ if $\Delta \geq 47$ [2].
4. $\chi_2(G) \leq \lfloor \frac{9}{5}\Delta \rfloor + 2$ if $\Delta \geq 749$ [1].

© Springer Nature Switzerland AG 2021
W. Wu and H. Du (Eds.): AAIM 2021, LNCS 13153, pp. 400–405, 2021.
https://doi.org/10.1007/978-3-030-93176-6_34

5. $\chi_2(G) \leq \lceil \frac{5\Delta}{3} \rceil + 78$ and $\chi_2(G) \leq \lceil \frac{5}{3}\Delta \rceil + 25$ if $\Delta \geq 241$ [5].
6. $\chi_2(G) \leq 20$ if $\Delta \leq 5$ and $\chi_2(G) \leq 5\Delta - 7$ if $\Delta \geq 6$ [11].

In this paper, we improved the upper bound on $\chi_2(G)$ for planar graph G with maximum degree at most 5 to 18 by induction on the number of vertices and edges.

Theorem 1.1. *If G is a planar graphs with maximum degree $\Delta \leq 5$, then $\chi_2(G) \leq 18$.*

2 Notations

All graphs considered here are simple and finite. For a planar graph G, we denote its vertex set, edge set, face set, maximum degree and minimum degree by $V(G)$, $E(G)$, $F(G)$, $\Delta(G)$ and $\delta(G)$ respectively. For $x \in V(G) \cup F(G)$, let $d_G(x)$ denote the degree of x in G. We drop G in $d_G(v)$ when G is clear from the context. A vertex of degree k (resp. at least k, at most k) is called a k-vertex (resp. k^+-vertex, k^--vertex). A face of degree k (resp. at least k, at most k) is called a k-face (resp. k^+-face, k^--face). In a 2-connected graph, each 5^--face is a cycle. Let $F(v)$ be the set of colors cannot be used for vertex v. Let $t(v)$ be the number of 3-faces incident with vertex v. Let $n_3(f)$ be the number of 3-vertices incident with face f. A $[v_1 v_2 \cdots v_k]$-face is a k-face with vertices $v_1, v_2, \cdots v_k$ on its boundary. A $(x_1, x_2, \cdots, x_k)$-face is a k-face with vertices of degree $x_1, x_2, \cdots, x_k$.

For an edge uv, let G/uv be the graph obtained from G by contracting the edge uv. After the operation G/uv, the following properties of the vertex degree was formulated.

Proposition 2.1. *Let $H = G/uv$ and v' be the new vertex in H obtained by contacting the edge uv. Then we have*

(1) $d_H(w) \leq d_G(w)$ for each vertex $w \in V(H) \setminus \{v'\}$ and $d_H(v') = d_G(u) + d_G(v) - 2 - t_G(uv)$, where $t_G(uv)$ is the number of 3-faces incident with the edge uv.

(2) For any vertices $w, w' \in V(H) \setminus \{v'\}$, $d_H(w, w') \leq d_G(w, w')$ and $d_H(w, v') \leq d_G(w, u)$.

3 Planar Graphs with Maximum Degree at Most 5

We prove Theorem 1.1 by contradiction. Let G be a minimal counterexample with the smallest number of vertices and edges. Then G is connected and $\chi_2(G) > 18$. By the minimality of G, for any planar graph H with $\Delta(H) \leq 5$ and $|V(H)| + |E(H)| < |V(G)| + |E(G)|$, we have that $\chi_2(H) \leq 18$. We first establish structural properties of G. Let $C = \{1, 2, \cdots, 18\}$.

Lemma 3.1. *G is 2-connected.*

Proof. If G is disconnected, then each component has a 2-distance 18-coloring, contradiction. Assume that G is connected and v is a cut vertex of G. Let $H_1, H_2, \cdots H_t(t \geq 2)$ be the components of $G - v$ and let $G_i = G[V(H_i \cup \{v\})]$. By the minimality, $\chi_2(G_i) \leq 18$. Permutate the color of v in each G_i such that v is colored by the same color, denote the current coloring by φ. If no two vertices of $N(v)$ are colored by the same color, φ is a 2-distance 18-coloring of G, a contradiction. Without loss of generality, let $v_1 \in G_1, v_2 \in G_2$ and $\varphi(v_1) = \varphi(v_2)$. There is at least one color $\alpha \notin C - N_\varphi(v) \cup \{\varphi(v)\}$(here $N_\varphi(v)$ is a color set of vertices in $N(v)$ under the coloring φ). Permutate two colors, $\varphi(v_2)$ and α, in G_2. If there are other cases we can do the same procedure to get a 2-distance 18-coloring of G, a contradiction. $\square$

Lemma 3.2. $\delta(G) \geq 3$.

Proof. Assume that v is a 2-vertex of G and $N(v) = \{u, w\}$, then we contact the edge uv to a new vertex v'. Obviously, the obtained graph G/uv is also a planar graph with $\Delta(G) \leq 5$, by the minimality, $\chi_2(G/uv) \leq 18$. Let φ be a 2-distance 18-coloring of G'. Color the vertex u by $\varphi(v')$. The remaining vertices keep their colors. Since $|F(v)| \leq 2 \times 5 = 10$, we can color v by a color $\alpha \in C - \{\varphi(x) | x \in V(G), d_G(v, x) \leq 2\}$. Thus, φ can be extended to a 2-distance 18-coloring of G, a contradiction. $\square$

Lemma 3.3. *Every 3-vertex is adjacent to three 5-vertices.*

Proof. Assume that 3-vertex v has a 4^--neighbor u. Contact the edge uv to a new vertex v'. By the minimality, $\chi_2(G/uv) \leq 18$. Let φ be a 2-distance 18-coloring of G/uv. Color the vertex u by $\varphi(v')$. The remaining vertices keep their colors. Since $|F(v)| \leq 2 \times 5 + 4 = 14$, φ can be extended to a 2-distance 18-coloring of G, a contradiction. $\square$

Lemma 3.4. *Every 3-face in G is either a $(4, 5, 5)$-face or a $(5, 5, 5)$-face.*

Proof. Assume that 3-face $[uvw]$ is incident with a 3-vertex v and v_1 is another neighbor of v. Let $G' = G - v + uv_1$. By the minimality, $\chi_2(G') \leq 18$. Let φ be a 2-distance 18-coloring of G'. Note that the colors on vertices v_1, u and w are distinct and $|F(v)| \leq 5 + 4 + 4 = 13$, φ can be extended to a 2-distance 18-coloring of G. Assume that 3-face $[uvw]$ is incident with two 4-vertices u and v. By the minimality, $\chi_2(G - uv) \leq 18$. Erase the colors on vertices u and v. Since $|F(u) \leq 16, |F(v)| \leq 16$, we can recolor vertices u and v, a contradiction. $\square$

Lemma 3.5. *Every 3-vertex is incident with at least two 5^+-faces.*

Proof. By Lemma 3.3, 3-vertex v is incident with 4^+-faces. Assume that 3-vertex v is incident with two 4-faces $[vv_1uv_2]$ and $[vv_2wv_3]$. Let $G' = G - v + v_1v_3$. By the minimality, $\chi_2(G') \leq 18$. Let φ be a 2-distance 18-coloring of G'. Note that the colors on vertices v_i are distinct for $i = 1, 2, 3$ and $|F(v)| \leq 4 + 5 + 4 = 13$, φ can be extended to a 2-distance 18-coloring of G, a contradiction. $\square$

Lemma 3.6. *Every 4-vertex is incident to at most one 3-face.*

Proof. Let v_1, v_2, v_3 and v_4 be four neighbors of 4-vertex v in clockwise. Let $v_1v_2 \in E(G)$. If $v_2v_3 \in E(G)$, then let $G' = G - v + v_2v_4$. By the minimality, $\chi_2(G') \leq 18$. Let φ be a 2-distance 18-coloring of G'. Note that the colors on vertices v_i are distinct for $i = 1, 2, 3, 4$ and $|F(v)| \leq 16$, φ can be extended to a 2-distance 18-coloring of G, a contradiction. If $v_3v_4 \in E(G)$, then let $G' = G - v + v_2v_3 + v_1v_4$. By the minimality, $\chi_2(G') \leq 18$. Let φ be a 2-distance 18-coloring of G'. Note that the colors on vertices v_i are distinct for $i = 1, 2, 3, 4$ and $|F(v)| \leq 16$, φ can be extended to a 2-distance 18-coloring of G, a contradiction. $\square$

Lemma 3.7. *Let v be a 5-vertex. Then*

(1) $t(v) \leq 3$.
(2) If $t(v) = 3$, then v is incident to at most one $(4, 5, 5)$-face.
(3) If v is incident to one $(4, 5, 5)$-face and two $(5, 5, 5)$-faces, then v is incident to two 5-faces.

Proof. (1) Assume that v is incident with four 3-faces $[v_1vv_2]$, $[v_2vv_3]$, $[v_3vv_4]$ and $[v_4vv_5]$. Let $G' = G - v + v_1v_5$. By the minimality, $\chi_2(G') \leq 18$. Note that the colors on vertices v_i for $i = 1, 2, 3, 4, 5$ are distinct and $|F(v)| \leq 4 + 3 \times 3 + 4 = 17$, we can color v, a contradiction.

(2) Let v_1, v_2, v_3, v_4 and v_5 be five neighbors of v in clockwise. Let $v_1v_2 \in E(G)$ and $d(v_1) = 4$, then by Lemma 3.6, $v_1v_5 \notin E(G)$. By the minimality, $\chi_2(G - vv_1) \leq 18$. Erase the colors on vertices v and v_1. If $v_2v_3, v_3v_4 \in E(G)$ or $v_2v_3, v_4v_5 \in E(G)$, then $|F(v)| \leq 2 + 3 + 3 + 4 + 5 = 17$, $|F(v_1)| \leq 5 + 5 + 3 + 3 = 16$. Thus, we can recolor v and v_1, contradiction. Now we can assume that $v_3v_4, v_4v_5 \in E(G)$. If $d(v_i) < 5$ for some $i = 3, 4, 5$, then $|F(v_1)| \leq 17$, $|F(v)| \leq 16$, we can recolor v_1 and v, contradiction.

(3) Let v_1, v_2, v_4, v_4 and v_5 be five neighbors of v in clockwise. By the analysis above, we can assume that $v_1v_2, v_3v_4, v_4v_5 \in E(G)$. By the minimality, $\chi_2(G - vv_1) \leq 18$. Erase the colors on vertices v and v_1. If v is incident to a $[vv_2uv_3]$-face or a $[vv_5wv_1]$-face, then $|F(v_1)| \leq 17$, $|F(v)| \leq 16$, we can recolor v_1 and v, contradiction. $\square$

Lemma 3.8. *Every 5-face has at most one 3-vertex.*

Proof. Let $f = [v_1v_2v_3v_4v_5]$ be a 5-face. By Lemma 3.2, $n_3(f) \leq 2$. Assume that $d(v_1) = d(v_3) = 3$. Contact vertices v_1 and v_3 to a new vertex v. Denote the the obtained graph by G'. Note that $d_{G'}(v) = 5$, by the minimality, $\chi_2(G') \leq 18$. Let φ be a 2-distance 18-coloring of G'. Note that the colors on vertices v_2, v_4 and v_5 are distinct, we color vertex v_3 by $\varphi(v)$. Since $|F(v_1)| \leq 3 \times 5 = 15$, φ can be extended to a 2-distance 18-coloring of G, a contradiction. $\square$

Proof of Theorem 1.1.
Since G is connected, we define a weight function w by $w(x) = d(x) - 4$ for $x \in V(G) \cup F(G)$. By Euler's formula $|V(G)| - |E(G)| + |F(G)| = 2$ and formula $\sum_{v \in V(G)} d(v) = 2|E| = \sum_{f \in F(G)} d(f)$, we can derive

$$\sum_{x \in V(G) \cup F(G)} w(x) = -8.$$

We then design appropriate discharging rules and redistribute weights accordingly. Once the discharging is finished, a new weight function w' is produced. During the process, the total sum of weights is kept fixed. It follows that

$$\sum_{x \in V(G) \cup F(G)} w'(x) = \sum_{x \in V(G) \cup F(G)} w(x) = -8.$$

However, we will show that after the discharging is complete, the new weight function $w'(x) \geq 0$ for all $x \in V(G) \cup F(G)$. This leads to the following obvious contradiction

$$0 \leq \sum_{x \in V(G) \cup F(G)} w'(x) = \sum_{x \in V(G) \cup F(G)} w(x) = -8 < 0.$$

Discharging Rules:

R1. Every 5-vertex gives $\frac{1}{2}$ to each incident $(4, 5, 5)$-face and $\frac{1}{3}$ to each incident $(5, 5, 5)$-face.

R2. Every 5^+-face gives $\frac{1}{2}$ to each incident 3-vertex and $\frac{1}{12}$ to each incident 5-vertex.

Checking $w'(v) \geq 0, v \in V(G)$.

By Lemma 3.2, $\delta(G) \geq 3$.

Case $d(v) = 3$.

By Lemma 3.5 and R2, v receives at least $\frac{1}{2} \times 2 = 1$ from its incident 5^+-faces. Thus $w'(v) \geq 3 - 4 + 1 = 0$.

Case $d(v) = 4$.

Since v does not give out or receive any charge of v and thus $w'(v) = w(v) = 0$.

Case $d(v) = 5$.

By Lemma 3.7(1), $t(v) \leq 3$. Assume that $t(v) = 3$, then by Lemma 3.7(2), v is incident to at most one $(4, 5, 5)$-face. If v is incident to one $(4, 5, 5)$-face, then by Lemma 3.7(3) and R1, R2, $w'(v) = 5 - 4 - \frac{1}{2} - \frac{1}{3} \times 2 + \frac{1}{12} \times 2 = 0$. Otherwise, $w'(v) \geq 5 - 4 - \frac{1}{3} \times 3 = 0$ by R1. If $t(v) \leq 2$, then by R1, $w'(v) \geq 5 - 4 - \frac{1}{2} \times 2 = 0$.

Checking $w'(f) \geq 0, f \in F(G)$.

Case $d(f) = 3$.

By Lemma 3.4, f is a $(4, 5, 5)$-face or a $(5, 5, 5)$-face. If f is a $(4, 5, 5)$-face, then $w'(f) = 3 - 4 + \frac{1}{2} \times 2 = 0$. If f is a $(5, 5, 5)$-face, then $w'(f) = 3 - 4 + \frac{1}{3} \times 3 = 0$.

Case $d(f) = 4$.

Since no 4-face gives out or receives any charge, $w'(f) = w(f) = 0$.

Case $d(f) = 5$.

By Lemma 3.8, $n_3(f) \leq 1$. By R2, $w'(f) \geq 5 - 4 - \frac{1}{2} - \frac{1}{12} \times 4 > 0$.

Case $d(f) \geq 6$.

By Lemma 3.3, $n_3(f) \le \lfloor \frac{d(f)}{2} \rfloor$. By R2, $w'(f) \ge d(f) - \frac{1}{2} \lfloor \frac{d(f)}{2} \rfloor - \frac{1}{12}(d(f) - \lfloor \frac{d(f)}{2} \rfloor) > 0$.

By the analysis above, we proved that $w'(x) \ge 0$ for all $x \in V \cup F$ and thus we complete the proof of Theorem 1.1.

Acknowledgement. This research was supported by National Science Foundation of China under Grant Nos. 11901243, 11771403 and Zhejiang Provincial Natural Science Foundation of China under Grant No. LQ19A010005.

References

1. Agnarsson, G., Halldorsson, M.M.: Coloring powers of planar graphs. SIAM J. Discrete Math. **16**, 651–662 (2003)
2. Borodin, O.V., Broersma, H.J., Glebov, A., Heuvel, J.V.D.: Stars and bunches in planar graphs. Part II: General planar graphs and colourings, CDAM researches report 2002-05 (2002)
3. Griggs, J.R., Yeh, R.K.: Labelling graphs with a condition at distance 2. SIAM J. Discrete Math. **5**, 586–595 (1992)
4. van den Heuvel, J., McGuinness, S.M., Molloy, Salavatipour, M.: Coloring of the square of planar graph. J. Graph Theory **42**, 110–124 (2003)
5. Molloy, M., Salavatipour, M.: A bound on the chromatic number of the square of a planar graph. J. Comb. Theory Ser. B. **94**, 189–213 (2005)
6. Montassier, M., Raspaud, A.: A note on 2-facial coloring of plane graphs. Inf. Process. Lett. **98**, 235–241 (2006)
7. Roberts, F.S.: T-colorings of graphs: recent results and open problems. Discrete Math. **93**, 229–245 (1991)
8. Song, H.M., Lai, H.J.: Upper bounds of r-hued colorings of planar graphs. Discrete Appl. Math. **243**, 262–269 (2018)
9. Thomasse, C.: Applications of Tutte cycles, Technical report, Technical University of Denmark (2001)
10. Wegner, G.: Graphs with given diameter and a coloring problem. Technical report, University of Dortmund (1977)
11. Zhu, J., Bu, Y.: Minimum 2-distance coloring of planar graphs and channel assignment. J. Comb. Optim. **36**(1), 55–64 (2018). https://doi.org/10.1007/s10878-018-0285-7

An Efficient Oracle for Counting Shortest Paths in Planar Graphs

Ye Gong and Qian-Ping Gu$^{(\boxtimes)}$

School of Computing Science, Simon Fraser University, Burnaby, Canada
{yeg,qgu}@sfu.ca

Abstract. We propose an $O(\sqrt{n})$ query time and $O(n^{1.5})$ size oracle which, given a pair of vertices u and v in a planar graph G of n vertices, answers the number of shortest paths from u to v. Our oracle can answer a query whether there is a unique shortest path from u to v in $O(\log n)$ time. Bezáková and Searns [ISAAC 2018] give an $O(\sqrt{n})$ query time and $O(n^{1.5})$ size oracle for counting shortest paths in planar graphs. Applying this oracle directly, it takes $O(\sqrt{n})$ time to answer whether there is a unique shortest path from u to v. A key component in our oracle is to apply Voronoi diagrams on planar graphs to speed up the query time. Computational studies show that our oracle is faster to answer queries than the oracle of Bezáková and Searns for large graphs. Applying Voronoi diagrams on planar graphs, significant theoretical improvements have been made for distance oracles. Our studies confirm that Voronoi diagrams are efficient data structures for distance oracles in practice.

Keywords: Distance oracles · Voronoi diagrams on planar graphs · Computational study

1 Introduction

Computing shortest distances/paths is a most fundamental problem in graph algorithms and has numerous applications. To answer a query for the shortest distance $d(u,v)$ from a vertex u to a vertex v in a graph G, one approach is to use a single source shortest path (SSSP) algorithm to compute $d(u,v)$. Another approach is to precompute and store $d(u,v)$ from u to v for all pairs of u and v in a 2-dimensional array, and get $d(u,v)$ from the array. The former approach takes $t(n)$ time and $O(m)$ memory space to answer a query for a graph of n vertices and m edges, where $t(n)$ is the running time of the SSSP algorithm, while the latter one takes $O(1)$ time and $O(n^2)$ space. The $t(n)$ query time of an SSSP algorithm is considered inefficient and $O(n^2)$ space is too large for applications expecting a real time answer in large graphs. To reduce the $t(n)$ query time and $O(n^2)$ memory space, one more approach is to precompute and store some distance information in a data structure called oracle, and get $d(u,v)$ from the oracle. There is a tradeoff between the query time and memory space (oracle size). Distance oracles with better query time and size have been extensively studied, a survey on this topic can be found in [13].

W. Wu and H. Du (Eds.): AAIM 2021, LNCS 13153, pp. 406–417, 2021.
https://doi.org/10.1007/978-3-030-93176-6_35

Planar graphs are simple and elegant models for many real networks and distance oracles for planar graphs have received much attention. Distance oracles can be classified as approximate oracles and exact oracles. For a query on $d(u, v)$, an approximate oracle gives an answer at least $d(u, v)$ and at most $\alpha d(u, v)$, where $\alpha > 1$ is called a stretch factor. Near-constant query time and near-linear size approximate oracles on planar graphs have been obtained at a cost of $(1 + \epsilon)$ stretch for constant $\epsilon > 0$ (see [13, 16]). An exact oracle returns $d(u, v)$ as the answer. Polylogarithmic query time and near-linear size exact oracles on planar graphs are known (see [4, 11]). There are two major techniques used in exact oracles on a planar graph G: one is an r-division of G that recursively decomposes G into subgraphs by vertex-separators of small size [7, 10] and the other is Voronoi diagrams on planar graphs introduced by Cabello [3].

Given vertices u and v in G, counting the number $ns(u, v)$ of shortest paths from u to v in G is an important problem with many applications such as for counting minimum (s, t)-cut in planar graphs and route guidance systems [1, 5, 14]. Bezáková and Searns give an $O(\sqrt{n})$ query time and $O(n^{1.5})$ size oracle (BS oracle) for counting $ns(u, v)$ in a planar graph G of n vertices and positive edge lengths [2]. The BS oracle is based on a recursive decomposition of G by balanced separators: Divide the vertices of G into three disjoint subsets (A, B, C) such that $|C| = O(\sqrt{|V(G)|})$, A and B have similar size and there do not exist vertices $u \in A$ and $v \in B$ connected by an edge (u, v) or (v, u). C is called a separator for G. Each of A and B is viewed as a subgraph of G and divided recursively. For u and v in a subgraph R, let $d_R(u, v)$ be the shortest distance and $ns_R(u, v)$ be the number of shortest paths from u to v in R (initially $R = G$, $d_R(u, v) = d(u, v)$ and $ns_R(u, v) = ns(u, v)$). Let C be the separator for R in a recursive decomposition of G. A vertex s in C is called a *feasible site* for vertices u and v if $d(u, v) = d_R(u, s) + d_R(s, v)$. An outline of the BS oracle is as follows:

- Compute a recursive decomposition of G. For each subgraph R in the decomposition, compute and store $d_R(u, s)$, $d_R(s, v)$, $ns_R(u, s)$ and $ns_R(s, v)$ for every pair of u and v in R and every s in the separator C for R. For a feasible site s, $ns_R(u, s) \times ns_R(s, v)$ gives the number of shortest paths from u to v containing s in R. These $ns_R(u, s)$ and $ns_R(s, v)$ are computed in such a way that each shortest path from u to v is counted once in $ns_R(u, s) \times ns_R(s, v)$ for exactly one feasible site s. Then $ns(u, v) = \sum_s$ feasible site $ns_R(u, s) \times ns_R(s, v)$.
- To answer a query on $ns(u, v)$: Start from $R = G$, if u and v are in R then check vertices in C for R by enumeration to find all feasible sites in C, and recurse on the child of R containing u and v until u and v are in different subgraphs. Return $ns(u, v) = \sum_s$ feasible site $ns_R(u, s) \times ns_R(s, v)$.

The total number of vertices in C for every R containing u and v is $O(\sqrt{n})$. So, the BS oracle has $O(\sqrt{n})$ query time[1] and $O(n^{1.5})$ size. The preprocessing time of the oracle is $O(n^{1.5})$.

---

[1] It is assumed that every arithmetic operation takes $O(1)$ time. This paper also follows this assumption.

The BS oracle can be modified by replacing a recursive decomposition with an r-division of G below: Divide the edges of G into two subsets (subgraphs, called *pieces*) of similar size by a vertex-separator of size $O(\sqrt{n})$; and then divide each piece recursively. For a piece R of G, let $|R|$ be the size of R and $\partial(R)$ be the set of *boundary* vertices of R (each vertex is incident to an edge of R and an edge not in R). Then $|\partial(R)| = O(\sqrt{|R|})$. For vertices u and v in a piece R divided into pieces P and Q with u in Q and v in P, any path from u to v in G must contain a boundary vertex s in $\partial(P)$. For u in Q and v in P, a vertex s in $\partial(P)$ is a *feasible site* if $d(u,v) = d(u,s) + d(s,v)$. One can precompute and store $d(u,s)$, $d(s,v)$, ns(u,s) and ns(s,v) for each $s \in \partial(P)$. Then ns(u,v) can be computed from $\sum_s$ feasible site ns$(u,s) \times$ ns(s,v).

If G has a unique shortest path from u to v for every pair of u and v, then there is a unique feasible site s in $\partial(P)$ for u in Q and v in P. The unique feasible site can be found in $O(\log n)$ time by Voronoi diagrams on planar graphs [6,8]. Applying Voronoi diagrams to find the unique feasible site in $O(\log n)$ time, Gawrychowski et al. give an $O(\log n)$ query time and $O(n^{1.5})$ size oracle to answer queries on $d(u,v)$ [8]. It takes $O(n^2)$ time to compute the data structure for Voronoi diagrams and the preprocessing time for the oracle of [8] is $O(n^2)$.

In this paper, we propose a new oracle for counting shortest paths based on an idea of applying Voronoi diagrams to find multiple feasible sites, an extension of the application of Voronoi diagrams in [8]. Similar to the works in [2,8], our oracle uses an r-division of G, precomputes and stores the shortest distances and numbers of shortest paths in the r-division. Let P and Q be the two pieces from a piece R in an r-division of G. A new ingredient in our oracle is to apply Voronoi diagrams to find all feasible sites for $u \in Q$ and $v \in P$ instead of enumerating vertices of $\partial(P)$ as that in [2]. We develop an algorithm (Algorithm 1) which applies Voronoi diagrams to find all feasible sites. The running time of Algorithm 1 is $O(\sqrt{n})$ (the number of feasible sites can be $O(\sqrt{n})$ in a worst case) which dominates the query time of our oracle. It is expected that the worst cases rarely happen in practice. Computational studies show that Algorithm 1 runs faster than checking $\partial(P)$ by enumeration as that in [2] to find all feasible sites for large $|\partial(P)|$ in grid graphs. The results also confirm that the data structure for Voronoi diagrams in [8] is indeed efficient for exact distance oracles for large planar graphs in practice.

Our oracle can answer a query whether there is a unique shortest path from a vertex u to a vertex v in G in $O(\log n)$ time. The unique shortest path problem has many practical applications, for example, in route guidance system for path choices [5,14]. Finding all feasible sites in $\partial(P)$ by enumeration as that in [2], it takes $O(\sqrt{n})$ time to answer a query on the unique shortest path problem while our oracle answers the query in $O(\log n)$ time.

Our main contributions are summarized as follows:

- An oracle which, given any planar graph G with n vertices and positive edge lengths, and any two vertices u and v in G, gives the number ns(u,v) of shortest paths from u to v in $O(\sqrt{n})$ time. The oracle answers whether there is a unique shortest path from u to v or not in $O(\log n)$ time. The oracle has $O(n^{1.5})$ size and $O(n^2)$ preprocessing time.

- Perform computational studies which show that our oracle is faster than the previous one in [2] and confirm that Voronoi diagrams are efficient data structures for distance oracles in practice.

The rest of the paper is organized as follows. In the next section, we give the preliminaries of the paper. We describe our oracle in Sect. 3 and report the computational results in Sect. 4. The final section concludes the paper.

2   Preliminaries

We describe our oracle for directed planar graphs. The oracle can be applied to undirected planar graphs as well. Let $G(V, E)$ be a directed planar graph with vertex set V and edge set E. Each edge (u, v) from vertex u to vertex v is assigned a length $l(u, v) > 0$. A path in G is a sequence of vertices $v_1, .., v_k$ with (v_i, v_{i+1}) an edge of E for $1 \le i \le k - 1$. Path $v_1, .., v_k$ is called a path from v_1 to v_k, denoted as $v_1 \to v_k$ for convenience. A path is simple if each vertex of G appears in it at most once. The length of path $v_1 \to v_k$ is $\sum_{1 \le i \le k-1} l(v_i, v_{i+1})$. A path $u \to v$ in G is shortest if its length is the minimum among all paths $u \to v$. Since $l(u, v) > 0$ for every (u, v) in E, a shortest path $u \to v$ is a simple path. The shortest distance $d(u, v)$ from u to v is the length of a shortest path $u \to v$, $d(u, u)$ is defined 0.

Bezáková and Searns [2] give an algorithm for counting the shortest paths from a source vertex u to all other vertices in G: first compute a shortest path tree rooted at u by a single source shortest path (SSSP) algorithm; an edge (v_1, v_2) in G is tight if $d(u, v_2) - d(u, v_1) = l(v_1, v_2)$; then add all tight edges to the shortest path tree to get a directed acyclic graph (DAG) and count the shortest paths from u to other vertices by a topological sort on the DAG. The running time of the algorithm is dominated by the time of the SSSP algorithm.

We assume that G has a planar embedding on a plane: each vertex u of G is mapped to a point and each edge (u, v) is mapped to a simple curve (does not cross itself) connecting u and v in the plane and the curves are pairwise non-crossing. Two edges (u, v) and (v, u) are viewed as a bidirectional edge and embedded as one curve. The embedding of G is called a plane graph (also denoted by G). Faces of a plane graph G are the maximal regions of the plane after removing the curves of G. A face f is enclosed by a set E_f of edges. We say f and edges of E_f are incident to each other, and f and the end vertices of edges in E_f are incident to each other. We assume G is strongly connected and triangulated (each face of G is incident to three edges). If G has a face f incident to more than three edges, we can triangulate f by adding new bidirectional edges with infinite length in the interior of f connecting some vertices incident to f. An embedding and a triangulation of G can be computed in $O(n)$ time.

Given a plane graph G, the planar dual of G is a plane graph $G^*(V^*, E^*)$: for each face f in G, there is a vertex f^* in V^* that is a point in the interior of f; for any two faces f and g separated by an edge e of G, there is an edge

$e^* = (f^*, g^*)$ in E^* that is a simple curve which connects f^* and g^* in the plane, intersects e exactly once and does not intersect any edge of G other than e. A Jordan curve separator for a plane graph G is a closed curve in the plane that does not intersect itself and intersects G only at vertices. The size of a Jordan curve is the number of vertices of G it intersects.

An r-division $\mathcal{T}$ of a plane graph G of n vertices can be computed as follows [8, 10, 15]: divide G into two pieces by a Jordan curve separator, then divide each piece recursively until each piece has at most r vertices and $O(\sqrt{r})$ boundary vertices. A face f in a piece R of an r-division $\mathcal{T}$ is called a *hole* if f is not an original face of G. For each piece R of $\mathcal{T}$, recall that $\partial(R)$ is the set of boundary vertices of R. Each vertex of $\partial(R)$ is incident to some hole. $\mathcal{T}$ can be viewed as a rooted binary tree with G the root. For each node R of $\mathcal{T}$, the depth of R is the number of edges in the path from G to R in $\mathcal{T}$. Applying the result of Miller [12], an r-division $\mathcal{T}$ of G can be computed in $O(n)$ time with the properties [8, 10, 15]: each internal node (piece) R of $\mathcal{T}$ is divided into two pieces P and Q by a Jordan curve of size $O(\sqrt{|R|})$, each piece of $\mathcal{T}$ has $O(1)$ holes, the number of vertices of each piece of depth d is $O(n/c^{d/3})$, and the number of vertices of $\partial(R)$ for R of depth d is $O(\sqrt{n}/c^{d/3})$ for some constant $c > 1$. An r-division of a plane G naturally gives an r-division of the planar graph G.

We adopt the notions related to Voronoi diagrams from [8]. Assume that for every two vertices u and v in G, there is a unique shortest path $u \to v$. Given a piece P of G in $\mathcal{T}$ and a hole h of P, let S be the set of vertices incident to h. Assume that every face of P except h is triangulated. Each vertex s in S is called a *site* and assigned a weight $w(s) > 0$. For a site s in S and a vertex v in P, the weighted distance from s to v is defined as $\mathrm{wd}(s, v) = w(s) + d(s, v)$. The additive weighted Voronoi diagram $\mathrm{VD}(S, w)$ is a partition of $V(P)$ into pairwise disjoint sets, one set $\mathrm{Vor}(s)$ for each $s \in S$, called *Voronoi cell of s*, $s \in \mathrm{Vor}(s)$ and a vertex $v \in V(P) \setminus S$ is in $\mathrm{Vor}(s)$ if $\mathrm{wd}(s, v) < \mathrm{wd}(s', v)$ for any $s' \in S$ with $s' \neq s$. We simply call $\mathrm{VD}(S, w)$ a Voronoi diagram and denote it by VD when the context is clear. Let P^* be the planar dual of P and VD a Voronoi diagram on P. A dual representation VD^* for VD is a ternary tree with the following properties: VD^* has $|S|$ leaf vertices; each internal vertex f^* of VD^* is a vertex of P^* corresponding to a face f of P such that the three vertices of P incident to f are in three distinct Voronoi cells (we say these three Voronoi cells are incident to f^*); each leaf vertex is a point in the interior of h; and each Voronoi cell is incident to at least one internal vertex of VD^*. Let T^* be a subtree of VD^*. Removing an internal vertex f^* and replacing it with copies, one for each edge incident to f^*, decomposes T^* into three subtrees. A vertex f^* of T^* is a *centriod* of T^* if each of the three subtrees created by f^* has at most $(|V(T^*)| + 1)/2$ edges. There is a centroid in T^* if it has more than one edge. A centroid decomposition of VD^* is as follows: Start from $T^* = \mathrm{VD}^*$; find a centroid f^* of T^* to get three subtrees and decompose each subtree recursively until each subtree has a single edge.

For a centroid f^* of T^*, let T_0^*, T_1^*, T_2^* be the three subtrees created by f^*, y_0, y_1, y_2 be the three vertices of P incident to f^*, and $s(0), s(1), s(2)$ be the three sites of S such that y_i is in Vor($s(i)$) for $i = 0, 1, 2$, all are in the clockwise order as shown in Fig. 1 (a). Let p_i be the shortest path $s(i) \rightarrow y_i$ for $i = 0, 1, 2$. Let J be the Jordan curve separating P from hole h, and J_i be the segment of J from $s(i + 2)$ to $s(i)$ in clockwise direction for $i = 0, 1, 2$ (additions in this paragraph are modulo 3). Let Z_i be the region enclosed by path p_i, edge (y_i, y_{i+2}), path p_{i+2} and segment J_i (see Fig. 1 (a)). It is given in [8] a data structure which, given v in P and f^* of T^*, decides in $O(1)$ time one of the following: v is on path p_i for some $i \in \{0, 1, 2\}$, or v is in the interior of Z_i for some $i \in \{0, 1, 2\}$. In the former case v is in Vor($s(i)$), and in the latter case, the Voronoi cell containing v can be found recursively on T_i^*. At the bottom of the recursion, T^* has one edge e^* which is incident to two Voronoi cells Vor(s) and Vor(s'). By comparing wd(s, v) and wd(s', v), the Voronoi cell containing v can be found in $O(1)$ time. Since the depth of the recursion by centroids of T^* is $O(\log |S|)$ and $|S| \leq n$, the Voronoi cell containing v can be found in $O(\log n)$ time.

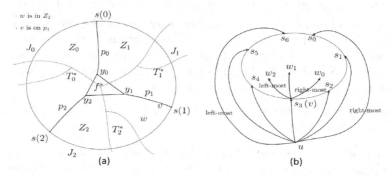

Fig. 1. (a) Decompose T^* into three subtrees by centroid f^*. (b) Clockwise order of sites, right-most and left-most edges.

A vertex u is in the hole h if the embedding of u is a point in the interior of h. For each piece R divided into P and Q, each hole h of P and each u in the hole h, the oracle of [8] assigns $w(s) = d(u, s)$ for every $s \in S$ and constructs a Voronoi diagram VD. Given u and v in G, the oracle finds the Voronoi cell Vor(s) containing v in $O(\log n)$ time, and thus $d(u, v) = \text{wd}(s, v) = d(u, s) + d(s, v)$ can be obtained in $O(\log n)$ time. Readers may refer to [8] for more details.

3 Our Oracle

Preprocessing. Let G be a planar graph of n vertices with positive edge lengths. We first compute an r-division T of G described in Sect. 2. For each internal node R of T, let P and Q be the two children of R separated by a Jordan curve. Let h be the region in which Q is embedded (h is a hole of P). Let S be the set of vertices incident to h. We compute and store the shortest distances and the number of shortest paths from u to s and from s to v (resp. from v to s and from s to u) for every u in Q, every s in S and every v in P. We only describe how to compute the distances and number of paths from u to s and from s to v. The computation for those from v to s and from s to u is symmetric with the roles of P and Q exchanged. The shortest distances $d(u, s)$ and $d(s, v)$ can be computed by shortest path algorithms in [9] with s in S as sources ($d(u, s)$ are computed with the orientation of edges of G reversed).

More technical details are involved in computing the number $ns(u, s)$ of shortest paths from u to s and $ns(s, v)$ from s to v. For $u \in Q$ and $v \in P$, let $F = \{s \in S | d(u, v) = d(u, s) + d(s, v)\}$ be the set of feasible sites. For an $s \in F$, $ns(u, s) \times ns(s, v)$ gives the number of shortest paths from u to v containing s. A shortest path $u \to v$ may contain multiple feasible sites and $\sum_{s \in F} ns(u, s) \times ns(s, v)$ may over count the number of shortest paths from u to v. We use an approach similar to that in [2] to compute $ns(u, s)$ and $ns(s, v)$ such that each shortest path $u \to v$ is counted once for exactly one $s \in F$. Let $\overleftarrow{G}$ be the graph obtained by reversing the direction of every edge of G. For each site $s \in S$, we compute a shortest path tree in G rooted at s and add the tight edges to form a DAG $D_G(s)$, and a shortest path tree in $\overleftarrow{G}$ rooted at s and add the tight edges to form a DAG $D_{\overleftarrow{G}}(s)$. We compute $ns(s, v)$ for every v in P on $D_G(s)$ and $ns(u, s)$ for every u in Q on $D_{\overleftarrow{G}}(s)$. Next we remove every edge (x, y) with $y \in S$ from $D_G(s)$ to get graph $D'_G(s)$, then compute the number of shortest paths $ns'(s, v)$ from s to v for every v in $D'_G(s)$.

Lemma 1. For $u \in Q$ and $v \in P$, if neither u nor v is in S then $\sum_{s \in F} ns(u, s) \times ns'(s, v)$ gives $ns(u, v)$.

Proof. We prove the statement that a shortest path p from u to v is counted once in $\sum_{s \in F} ns(u, s) \times ns'(s, v)$ for exactly one feasible site to get the lemma. If a shortest path $p : u \to v$ contains a single feasible site s, then the subpath path $s \to v$ of p is in $D'_G(s)$ and is counted in $ns(u, s) \times ns'(s, v)$, and the statement is true. Assume that p has multiple feasible sites $s(1), .., s(j)$ $(1 < j)$ of S with $s(j)$ closest to v. Then the subpath $s(j) \to v$ of p is in $D'_G(s(j))$ but for $i < j$, the subpath $s(i) \to v$ of p is not in $D'_G(s(i))$ because $s(i) \to v$ contains an edge $(x, s(j))$ and every edge $(x, s(j))$ is removed. Therefore, p is counted once in $ns(u, s(j)) \times ns'(s(j), v)$ only, and the statement holds. □

Following Lemma 1, in what follows, we say $s \in S$ is a *feasible site* for $u \in Q$ and $v \in P$ if $d(u, v) = d(u, s) + d(s, v)$ and $ns'(s, v) > 0$.

We view the hole h of P as the outer face of P. We assign each site of S an index and denote the sites of S as $s_0, s_1, .., s_k$ in their clockwise order on h with an arbitrary site as s_0 (see Fig. 1 (b)). For any vertex u in h and any vertex v in P, a path $u \to v$ must have a site s of S. For u in h, we remove all edges of Q, add edge (u, s_i) with length $d(u, s_i)$ for every $s_i \in S$, and remove every edge (x, s_i) in P for every x and every $s_i \in S$ to get a graph $\tilde{P}$. We find a shortest path tree rooted as u in $\tilde{P}$ and add the tight edges to form a DAG $D_{\tilde{P}}(u)$. We assume a planar embedding of $D_{\tilde{P}}(u)$. We identify edge (u, s_0) as the right-most edge from u. For a vertex v and an edge (x, v), the right-most edge from v w.r.t. edge (x, v) is the first edge (v, w) in the counter-clockwise order from (x, v) as they incident to v in their embedding (see Fig. 1 (b), (s_3, w_0) is the right-most edge from s_3 w.r.t. (u, s_3)). The right-most search is a depth-first search with the restriction that, for each vertex v visited, edges (v, w) from v are explored in the counter-clockwise order from the right-most edge (w.r.t. edge (x, v) by which v is first visited for $v \neq u$). For each vertex v in P, we find the unique right-most shortest path from u to v by the right-most search on $D_{\tilde{P}}(u)$. For $s \in S$ and $v \in P$, we assign $w(s) = d(u, s)$ and $\mathrm{wd}(s, v) = w(s) + \tilde{d}(s, v)$, where $\tilde{d}(s, v)$ is the shortest distance from s to v in $\tilde{P}$. Based on the unique right-most shortest paths, we construct an oracle of [8], called ORACLE_R, which returns the right-most feasible site (the feasible site of smallest index) and $d(u, v)$ for u in Q and v in P in $O(\log n)$ time. Notice that the oracle of [8] requires that $\mathrm{wd}(s_i, v)$ has a distinct value for each s_i. To get this property, we break the tie by defining $\mathrm{wd}(s_i, v)$ is smaller than $\mathrm{wd}(s_j, v)$ if $i < j$ when $\mathrm{wd}(s_i, v)$ and $\mathrm{wd}(s_j, v)$ have the same value for different s_i and s_j.

We also construct an oracle ORACLE_L of [8] based on the left-most shortest paths. We identify edge (u, s_k) as the left-most edge from u. For a vertex v and an edge (x, v), the left-most edge from v w.r.t. edge (x, v) is the first edge (v, w) in the clockwise order from (x, v) as they incident to v in their embedding (see Fig. 1 (b), (s_3, w_2) is the left-most edge from s_3 w.r.t. (u, s_3)). The left-most search is a depth-first search with the restriction that, for each vertex v visited, edges (v, w) from v are explored in the clockwise order from the left-most edge (w.r.t. edge (x, v) by which v is first visited for $v \neq u$). For each vertex v in P, we find the unique left-most shortest path from u to v by the left-most search on the DAG. Based on the unique left-most shortest paths, ORACLE_L returns the left-most feasible site (the feasible site of largest index) and $d(u, v)$ for u in Q and v in P in $O(\log n)$ time.

Answering Query. Given u and v, to get $\mathrm{ns}(u, v)$, we find the piece R of smallest depth in $\mathcal{T}$ such that R is divided into P and Q with u in Q and v in P. Then u is either in the hole h of P or in the set S of vertices incident to h. When u or v is in S, $\mathrm{ns}(u, v)$ has been stored and get $\mathrm{ns}(u, v)$ is trivial. The difficult part is to find $\mathrm{ns}(u, v)$ without enumeration on S when neither u nor v is in S. In this case, u is in h, and we say v is in P for simplicity. We give an algorithm to find all feasible sites for this case. Given u in h, let VD be the Voronoi diagram in ORACLE_R and VD^* be the dual representation of VD. For a subtree T^* of VD^* and a centroid f^* of T^*, let T_0^*, T_1^*, T_2^* be the three subtrees created by f^*, y_0, y_1, y_2 be the three vertices of P incident to f^*, $s(0), s(1), s(2)$ be the three sites of S such that y_i is in $\mathrm{Vor}(s(i))$ as shown in Fig. 1 (a). Let p_i be the right-most shortest path from $s(i)$ to y_i for $i = 0, 1, 2$. Notice that p_i is in the Voronoi cell $\mathrm{Vor}(s(i))$ of VD. For a site s of S, we denote the index of s by $\mathrm{ind}(s)$. Given u in h and v in P, let s^* be the right-most feasible site (v is in $\mathrm{Vor}(s^*)$) and s^{**} be the left-most feasible site. Then for any feasible site s, $\mathrm{ind}(s^*) \leq \mathrm{ind}(s) \leq \mathrm{ind}(s^{**})$. We use ORACLE_R to find s^* and ORACLE_L to find s^{**}. Let $\mathrm{rng}(s^*, s^{**}) = \{s \,|\, \mathrm{ind}(s^*) < \mathrm{ind}(s) < \mathrm{ind}(s^{**})\}$ be the range for searching feasible sites. Let $\mathrm{dep}(T^*)$ be the depth of the recursion on T^* by centroids f^*. If $|\mathrm{rng}(s^*, s^{**})| \leq a\,\mathrm{dep}(T^*)$ for some constant $a > 0$ then we check

Algorithm 1 Find-Feasible-sites
    Call $ORACLE_R$ to find s^* and $ORACLE_L$ to find s^{**};
    if $|rng(s^*, s^{**})| \leq adep(T^*)$ **then** /* $a > 0$ is a constant. */
        check each vertex s with $ind(s^*) < ind(s) < ind(s^{**})$ to find feasible sites and return;
    $T^* = VD^*$; Recurse-on(T^*);
Subroutine Recurse-on(T^*) /* addition operations in the subroutine are modulo 3. */
    if $rng(s^*, s^{**}) \cap rng(T^*) = \emptyset$ **then** return;
    if T^* has one edge e^* **then** find feasible sites from the two sites incident to e^* and return;
    Decompose T^* into T_0^*, T_1^*, T_2^* by centroid f^*;
    Find the location of v; /* v is either in the interior of Z_i or on p_i for some $i \in \{0, 1, 2\}$. */
    Case 1: v is in the interior of Z_i and $s^* \neq s(i + 2)$.
        Case 1.1: $wd(s^*, v) < wd(s(i), v)$. Recurse-on(T_i^*);
        Case 1.2: $wd(s^*, v) = wd(s(i), v) < wd(s(i + 1), v)$.
            Recurse-on(T_i^*); Recurse-on(T_{i+1}^*);
        Case 1.3: $wd(s^*, v) = wd(s(i), v) = wd(s(i + 1), v)$.
            Recurse-on(T_i^*); Recurse-on(T_{i+1}^*); Recurse-on(T_{i+2}^*);
    Case 2: v is in the interior of Z_i and $s^* = s(i + 2)$.
        Case 2.1 $wd(s^*, v) < wd(s(i), v)$. Recurse-on(T_i^*); Recurse-on(T_{i+2}^*);
        Case 2.2: $wd(s^*, v) = wd(s(i), v)$.
            Recurse-on(T_{i+1}^*); Recurse-on(T_{i+2}^*);
    Case 3: v is on p_i.
        Case 3.1: $wd(s^*, v) < wd(s(i + 1), v)$. Recurse-on$(T_i^*)$; Recurse-on$(T_{i+1}^*)$;
        Case 3.2: $wd(s^*, v) = wd(s(i + 1), v)$.
            Recurse-on(T_i^*); Recurse-on(T_{i+1}^*); Recurse-on(T_{i+2}^*);

Fig. 2. Algorithm for computing all feasible sites for u in the hole h and v in P.

every site s in $rng(s^*, s^{**})$, otherwise, we recurse on T^* to find all feasible sites. We use $rng(s^*, s^{**})$ to narrow down the recursion range. More specifically, let $rng(T^*)$ be the set of sites incident to T^*. Initially, $rng(T^*) = S$. For a subtree T_i^*, the sites incident to T_i^* are located from $s(i + 2)$ to $s(i)$ (additions in this paragraph are modulo 3) in the clockwise direction. If $ind(s(i + 2)) < ind(s(i))$ then $rng(T_i^*) = \{s | ind(s(i+2)) \leq ind(s) \leq ind(s(i))\}$. If $ind(s(i+2)) > ind(s(i))$ then $rng(T_i^*) = \{s | 0 \leq ind(s) \leq ind(s(i))$ or $ind(s(i + 2)) \leq ind(s) \leq k\}$. If $rng(s^*, s^{**}) \cap rng(T_i^*) = \emptyset$ then T_i^* will not be recursed. The pseudo code of the algorithm is given in Fig. 2.

Analysis for the Oracle

Lemma 2. *Algorithm 1 finds all feasible sites for u and v in $O(|S|)$ time.*

A proof for the lemma is omitted due to space limit. Our oracle for counting shortest paths consisting of $ORACLE_R$, $ORACLE_L$ and the precomputed numbers of shortest paths. Given u and v in G, the oracle finds the piece R on which we find $ns(u, v)$. If u or v is in S then $ns(u, v)$ can be obtained from the stored information. Otherwise, the oracle uses Algorithm 1 to find all feasible sites and return $\sum_{s \in F} ns(u, s) \times ns'(s, v)$ for $ns(u, v)$. Now we have our first result.

Theorem 1. *There is an oracle which, given any planar graph G with n vertices and positive edge lengths, and any two vertices u and v in G, gives the number* ns(u, v) of shortest paths from u to v in $O(\sqrt{n})$ time. The oracle has $O(n^{1.5})$ size and $O(n^2)$ preprocessing time.

Proof. It takes $O(\log n)$ time to find the piece R of $\mathcal{T}$ on which we find ns(u, v). If u or v is in S, then the oracle returns the correct value for ns(u, v) in $O(1)$ time. Otherwise, by Lemmas 1 and 2, the oracle also returns the correct value for ns(u, v). By Lemma 2, it takes $O(|S|)$ time to get ns(u, v). Since $|S| = O(\sqrt{n})$, our oracle takes $O(\sqrt{n})$ time to find ns(u, v). As shown in [8], each of ORACLE$_R$ and ORACLE$_L$ has $O(n^{1.5})$ size and can be computed in $O(n^2)$ time using the shortest path algorithms in [9] for computing the shortest path trees. When the shortest path trees are given, the number of shortest paths can be computed in $O(n^2)$ time. So, our oracle has $O(n^{1.5})$ size and $O(n^2)$ preprocessing time.   □

For vertices u and v in G let s^* be the right-most feasible site and s^{**} be left-most feasible site. If $s^* = s^{**}$ and ns$(u, s^*) \times$ ns$(s^*, v) = 1$ then there is a unique shortest path $u \to v$, otherwise there are multiple shortest paths $u \to v$. By ORACLE$_R$ and ORACLE$_L$, s^* and s^{**} can be found in $O(\log n)$ time and we obtain our next result.

Theorem 2. *There is an oracle which, given any planar graph G with n vertices and positive edge lengths, and any two vertices u and v in G, answers whether there is a unique shortest path from u to v or not in $O(\log n)$ time. The oracle has $O(n^{1.5})$ size and $O(n^2)$ preprocessing time.*

4   Computational Results

We compare the practical performance of the enumeration method which finds ns(u, v) by enumerating vertices in S and Algorithm 1. The comparison is focused on the time to find ns(u, v) for $u \in Q$ and $v \in P$ when P, Q and S are given. For this purpose, we use a grid graph as P. An $x \times y$ grid is a graph $H(V, E)$ with vertex set $V = \{v(i, j) | 1 \leq i \leq x, 1 \leq j \leq y\}$ and there is an edge $(v(i, j), v(i', j'))$ from $v(i, j)$ to $v(i', j')$ if $i = i'$ and $|j - j'| = 1$ or $|i - i'| = 1$ and $j = j'$. We have two settings for P, S and Q. In Setting (I), P is the subgraph of H induced by the vertices in the rows from 1 to $\lceil x/2 \rceil$; S is the set of vertices in the $(\lceil x/2 \rceil)$th row $(|S| = y)$ and Q is the subgraph of H induced by $V(H) \backslash V(P)$; and we randomly select 10 vertices of Q as u. In Setting (II), P is H; S is the set of vertices incident to the outer face (hole h) of P $(|S| = 2(x + y) - 4 \approx 2(x + y))$ and Q is in the interior of h; we use a point in h as u in Q and connect u to every site $s \in S$ by an edge (u, s) with a length specified later; we assign the edges (u, s) 10 sets of independently selected lengths to simulate 10 different u's in Q. In bothing settings, each edge of H is assigned a random length from $[1, 10]$. For every u, we run the algorithms to find ns(u, v) for every v in $V(P) \backslash S$. The average query time for a single pair (u, v) is the total query time for all (u, v) pairs divided by the number of pairs. We implemented the algorithms in C++ and run the implementations on a laptop with 2.0 GHz quad-core Intel Core i5 CPU, 16 GB memory.   We test the algorithms on $x \times x$ grids in Setting (I)

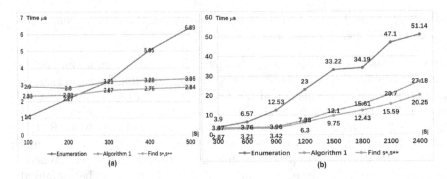

Fig. 3. (a) Average time to find $ns(u, v)$ in $x \times x$ grids with $|S| = 100, .., 500$. (b) Average time to find $ns(u, v)$ in $x \times 2x$ grids with $|S| = 300, 600, .., 2400$. Time unit is microsecond μs.

with $x = 100, .., 500$. The average query times of enumeration, Algorithm 1 and finding s^* and s^{**} are given in Fig. 3 (a). We test the algorithms on $x \times 2x$ grids P in Setting (II) with $x = 50, 100, .., 400$ ($|S| \approx 300, 600, .., 2400$). For each $x \times 2x$ grid and every u, we assign each edge (u, s) a length randomly selected from $[0.4x, 8x]$. The average query times for enumeration, Algorithm 1 and finding s^* and s^{**} for the grids in Setting (II) are given in Fig. 3 (b).

The results show that Algorithm 1 is faster than enumeration when $|S| > 300$. The time for finding s^* and s^{**} gives the query time to answer a query for the unique shortest path problem. The results show that applying Voronoi diagrams can reduce the query time significantly for counting shortest paths and answering whether there is a unique shortest path from u to v from enumeration. To find the shortest distance from u to v, one only needs to find s^* (or s^{**}) which takes about half of the time to find both s^* and s^{**}. This confirms that the data structure in [8] for the Voronoi diagrams on planar graphs is efficient in query time for distance oracles.

5   Conclusion

We proposed a new oracle for counting shortest paths in planar graphs, which uses Voronoi diagrams to speed up the query time. In worst cases, our oracle has the same $O(\sqrt{n})$ query time as that of the best known oracle [2] for G of n vertices. Computational results show that the worst cases rarely happen and the query time of our oracle is faster than the oracle in [2] for large grids. Our future works include proving an upper bound on the average query time of the oracle that is close to its practical performance. It is interesting to apply the oracles for counting shortest paths to real systems.

Acknowledgement. The authors thank anonymous reviewers for their constructive comments. The research was partially supported by NSERC discovery grant.

References

1. Bezáková, I., Friedlander, A.J.: Computing and sampling minimum (s, t)-cuts in weighted planar graphs in polynomial time. Theoret. Comput. Sci. **417**, 2–11 (2012)
2. Bezáková, I., Searns, A.: On counting oracles for path problems. In: Proceedings of the 29th International Symposium on Algorithms and Computation (ISAAC 2018), pp. 56:1–56:12 (2018)
3. Cabello, S.: Subquadratic algorithms for diameter and sum of pairwise distances in planar graphs. In: Proceedings of the 28th ACM/SIAM Symposium on Discrete Algorithms (SODA 2017), pp. 2143–2152 (2017)
4. Charalampopoulos, P., Gawrychowski, P., Mozes, S., Weimann, O.: Almost optimal distance oracles for planar graphs. In: Proceedings of the 51st Annual ACM Symposium on Theory of Computing (STOC 2019), pp. 138–151 (2019)
5. Chen, B.Y., Chen, X.-W., Chen, H.-P., Lam, W.H.K.: Efficient algorithm for finding k shortest paths based on re-optimization technique. Transp. Res. Part E: Logistics Transp. Rev. **133**, 101819 (2020)
6. Cohen-Addad, V., Dahlgaard, S., Wulff-Nilsen, C.: Fast and compact distance oracle for planar graphs. In: Proceedings of the 58th IEEE Annual Symposium on Foundations of Computer Science (FOCS2018), pp. 962–973 (2018)
7. Frederickson, G.N.: Fast algorithms for shortest paths in planar graphs. SIAM J. Comput. **16**(6), 1004–1022 (1987)
8. Gawrychowski, P., Mozes, S., Weimann, O., Wulff-Nilsen, C.: Better tradeoffs for exact distance oracles in planar graphs. In: Proceedings of the 29th ACM/SIAM Symposium on Discrete Algorithms (SODA 2018), pp. 515–529 (2018)
9. Henzinger, M.R., Klein, P.N., Rao, S., Subramanian, S.: Faster shortest-path algorithms for planar graphs. J. Comput. Syst. Sci. **55**(1), 3–23 (1997)
10. Klein, P.N., Mozes, S., Sommer, C.: Structured recursive separator decompositions for planar graphs. In: Proceedings of the 45th Annual ACM Symposium on Theory of Computing (STOC 2013), pp. 505–514 (2013)
11. Long, Y., Pettie, S.: Planar distance oracles with better time-space tradeoffs. In: Proceedings of the 32nd ACM/SIAM Symposium on Discrete Algorithms (SODA 2021), pp. 2517–2536 (2021)
12. Miller, G.L.: Finding small simple cycle separators for 2-connected planar graphs. J. Comput. Syst. Sci. **32**(3), 265–279 (1986)
13. Sommer, C.: Shortest-path queries in static networks. ACM Comput. Surv. **46**(4), 1–31 (2014)
14. Vanhove, S., Fack, V.: An efficient heuristic for computing many shortest path alternatives in road networks. Int. J. Geogr. Inf. Sci. **26**(6), 1031–1050 (2012)
15. van Walderveen, F., Zeh, N., Arge, L.: Multiway simple cycle separators and I/O-efficient algorithms for planar graphs. In: Proceedings of the 24th ACM/SIAM Symposium on Discrete Algorithms (SODA 2013), pp. 901–918 (2013)
16. Wulff-Nilsen, C.: Approximate distance oracles for planar graphs with improved time-space tradeoffs. In: Proceedings of the 27th ACM/SIAM Symposium on Discrete Algorithms (SODA 2016), pp. 351–362 (2016)

Restrained and Total Restrained Domination in Cographs

Xue-gang Chen[1] and Moo Young Sohn[2(✉)]

[1] Mathematics, North China Electric Power University, Beijing 102206, China
[2] Mathematics, Changwon National University, Changwon 641-773, Korea
mysohn@changwon.ac.kr

Abstract. Let $\gamma_r(G)$ and $\gamma_{tr}(G)$ denote the restrained domination number and total restrained domination number of G, respectively. The minimum total restrained domination problem is to find a total restrained dominating set of minimum cardinality. In this paper, we correct a minor error in Pandey et al. [5] and design a linear-time algorithm for finding the restrained domination number of cographs. Furthermore, we propose a linear-time algorithm to solve the minimum total restrained domination problem in cographs.

Keywords: Restrained domination number · Total restrained domination number · Cographs

1 Introduction

The concept of domination in graphs, with its many variations, is now well studied in graph theory. The literature on the subject has been surveyed and detailed in [8] and [9]. Let G be a simple and undirected graph. The vertex set and the edge set of G are denoted by $V(G)$ and $E(G)$, respectively. Let $n(G) = |V(G)|$. The degree, neighborhood and closed neighborhood of a vertex v in the graph G are denoted by $d_G(v)$, $N_G(v)$ and $N_G[v] = N_G(v) \cup \{v\}$, respectively. If the graph G is clear from context, we simply write $d(v)$, $N(v)$ and $N[v]$, respectively. The minimum degree and maximum degree of the graph G are denoted by $\delta(G)$ and $\Delta(G)$, respectively. Let $S \subseteq V(G)$, $N_G(S) = \bigcup_{v \in S} N_G(v)$ and $N_G[S] = N_G(S) \cup S$. The graph induced by $S \subseteq V$ is denoted by $G[S]$. The *diameter* of G, denoted by $diam(G)$, is the maximum distance among pairs of vertices in G.

A set $S \subseteq V$ in a graph G is called a *dominating set* if every vertex in $V(G) - S$ is adjacent to at least one vertex in S. The *domination number* $\gamma(G)$ equals the minimum cardinality of a dominating set in G. Moreover, a dominating set of G of cardinality $\gamma(G)$ is called a γ-set of G.

This research was supported by the Basic Science Research Program through the National Research Foundation of Korea (NRF) funded by the Ministry of Education (2020R1I1A3A04036669).

W. Wu and H. Du (Eds.): AAIM 2021, LNCS 13153, pp. 418–425, 2021.
https://doi.org/10.1007/978-3-030-93176-6_36

A dominating set S of a graph G is called a *restrained dominating set* if every vertex in $V - S$ is adjacent to a vertex in $V - S$. The *restrained domination number* of G, denoted by $\gamma_r(G)$, is the minimum cardinality of a restrained dominating set of G. A restrained dominating set of G of cardinality $\gamma_r(G)$ is called a γ_r-set of G. Telle and Proskurowski in [7] introduced restrained domination as a vertex partitioning problem. One possible application of the concept of restrained domination is that of prisoners and guards. Here, each vertex not in the restrained domination set corresponds to a position of a prisoner, and every vertex in the restrained dominating set correspond to a position of guard. Note that each prisoner's position is observed by a guard's (to effect security) while each prisoner's position is seen by at least one other prisoner's position (to protect the rights of prisoners). To be cost effective, it is desirable to place as few guards as possible (in the sense above).

A set $S \subseteq V$ is a *total restrained dominating set* of G if every vertex is adjacent to a vertex in S and every vertex in $V - S$ is adjacent to a vertex in $V - S$. The *total restrained domination number* of G, denoted by $\gamma_{tr}(G)$, is the minimum cardinality of a total restrained dominating set of G. A total restrained dominating set of G of cardinality $\gamma_{tr}(G)$ is called a γ_{tr}-set of G. The concept of total restrained domination in graphs was introduced in [6] and has been studied in [3]. For example, let $P_7 = v_1 v_2 \cdots v_7$ be the path with vertices set $\{v_i | i = 1, 2, \cdots, 7\}$. Then $\{v_1, v_4, v_7\}$ is a restrained dominating set of P_7 and $\gamma_r(P_7) = 3$. $\{v_1, v_2, v_5, v_6, v_7\}$ is a total restrained dominating set of P_7 and $\gamma_{tr}(P_7) = 5$.

For a graph G, any vertex of degree one is called a *leaf* and the neighbour of a leaf is called a *support vertex* of G.

Let G_1 and G_2 be two graphs such that $V(G_1) \cap V(G_2) = \emptyset$. Then the *union* $G = G_1 \cup G_2$ has $V(G) = V(G_1) \cup V(G_2)$ and $E(G) = E(G_1) \cup E(G_2)$. The *join* $G = G_1 + G_2$ has $V(G) = V(G_1) \cup V(G_2)$ and $E(G) = E(G_1) \cup E(G_2) \cup \{uv | u \in V(G_1), v \in V(G_2)\}$.

A cograph, or complement-reducible graph, or P_4-free graph, is a graph that can be generated from the single-vertex graph K_1 by complementation and disjoint union. That is, the family of cographs is the smallest class of graphs that includes K_1 and is closed under complementation and disjoint union.

Any cograph may be constructed using the following recursive construction :

1. any single vertex graph is a cograph;
2. if G is a cograph, so is its complement graph $\overline{G}$;
3. if G and H are cographs, so is their disjoint union $G \cup H$.

The cographs may be defined as the graphs that can be constructed using these operations, starting from the single-vertex graphs. Alternatively, instead of using the complement operation, one can use the join operation, which consists of forming the disjoint union $G \cup H$ and then adding an edge between every pair of a vertex from G and a vertex from H.

We often associate with a cograph G a rooted binary tree T_G called a cotree. In T_G, its leaves are in one to one correspondence with the vertices of G. For an internal vertex t_i of T_G, let V_i be the set of vertices in G that correspond to leaves in the subtree of T_G rooted at t_i, and we denote by G_i the subgraph of G induced by V_i.

Every internal vertex of T_G is labeled either 0 or 1, corresponding to the disjoint union and join operations, respectively in the following way. Let t_i be an internal vertex of T_G with children t_j and t_k. If t_i is labeled by 0, then $G_i = G_j \cup G_k$. If t_i is labeled by 1, then $G_i = G_j + G_k$. An internal vertex of T_G is called 0-*vertex* if it is labeled by 0, and 1-*vertex* if it is labeled by 1.

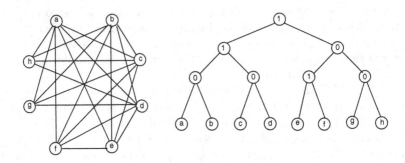

Fig. 1. A cograph G and its cotree T_G.

The restrained domination problem [2] is known to be NP-complete even for chordal graphs. A linear time algorithm to compute a minimum restrained dominating set of a tree has been proposed in [2]. In [4], Panda et al. posed a linear time algorithm to compute a minimum restrained dominating set of a proper interval graph. Pandey et al. [5] presented a polynomial time reduction that proves the NP-completeness of the restrained domination problem for undirected path graphs, chordal bipartite graphs, circle graphs.

The total restrained domination problem [6] is known to be NP-complete even for bipartite graphs and chordal graphs. Araki et al. in [1] proposed a linear-time algorithm for finding the secure domination number of cographs.

In this paper, we correct a minor error in [5] and design a linear-time algorithm for finding the restrained domination number of cographs. Furthermore, we propose a linear-time algorithm to solve the minimum total restrained domination problem in cographs.

2   Restrained and Total Restrained Domination Number in Cographs

For any cograph H, let $w(H)$ denote the connected component number of H. Suppose that $H_1, H_2, \cdots, H_{w(H)}$ be the connected components of H. Let $l(H) = |\{H_i | H_i \cong K_1, i = 1, 2, \cdots, w(H)\}|$ and $t(H) = \max\{|V(H_i)| : i = 1, 2, \cdots, w(H)\}$.

We first consider the values of γ, γ_r and γ_{tr} for the union of cographs G and H. The following result is obvious.

Lemma 1. *Let G and H be cographs.*

(1) $\gamma(G \cup H) = \gamma(G) + \gamma(H)$.
(2) $\gamma_r(G \cup H) = \gamma_r(G) + \gamma_r(H)$.
(3) $\gamma_{tr}(G \cup H) = \gamma_{tr}(G) + \gamma_{tr}(H)$.

Next we consider γ, γ_r and γ_{tr} for the join of cographs G and H. Pandey et al. [5] gave the following result.

Lemma 2 *([5]). Let G and H be cographs.*

$$\gamma_r(G + H) = \begin{cases} 2, & \text{if } n(G) = 1 \text{ and } n(H) = 1 \\ \min\{\gamma(G), \gamma(H), 2\}, & \text{if } n(G) \geq 2 \text{ and } n(H) \geq 2 \\ \min\{1 + l(G), \gamma(G)\}, & \text{if } n(G) \geq 2 \text{ and } n(H) = 1 \\ \min\{1 + l(H), \gamma(H)\}, & \text{if } n(G) = 1 \text{ and } n(H) \geq 2. \end{cases}$$

There is a minor error in the result. For example, let $G = K_1$ and H be the complement of K_t, then $\gamma_r(G + H) = t + 1$. However, $\min\{1 + l(H), \gamma(H)\} = t$. So $\gamma_r(G + H) \neq \min\{1 + l(H), \gamma(H)\}$. We give the following results.

Theorem 1. *Let H be a cograph. Then*

(1) $\gamma(K_1 + H) = 1$.
(2) $\gamma_r(K_1 + H) = l(H) + 1$.

Proof. (1) is easily obtained. So consider (2). Assume that $V(K_1) = \{u\}$. If $l(H) = 0$, then $\{u\}$ is a restrained dominating set of $K_1 + H$. Hence, $\gamma_r(K_1 + H) = 1$.

Suppose that $l(H) \geq 1$. Without loss of generality, we can assume that H_i is isomorphic to K_1 for $i = 1, 2, \cdots, l(H)$. It is obvious that $(\bigcup_{i=1}^{l(H)} V(H_i)) \cup \{u\}$ is a restrained dominating set of $K_1 + H$. Hence, $\gamma_r(K_1 + H) \leq l(H) + 1$. Let D be a γ_r-set of $K_1 + H$. It is obvious that $(\bigcup_{i=1}^{l(H)} V(H_i)) \subseteq D$ and $D \cap ((\bigcup_{i=l(H)+1}^{w(H)} V(H_i)) \cup \{u\}) \neq \emptyset$. So, $l(H) + 1 \leq |D| = \gamma_r(K_1 + H)$. Therefore, $\gamma_r(K_1 + H) = l(H) + 1$.

Theorem 2. *Let G and H be cographs with $n(G) \geq 2$ and $n(H) \geq 2$. Then*

$$\gamma(G + H) = \gamma_r(G + H) = \begin{cases} 1, & \text{if } \gamma(G) = 1 \text{ or } \gamma(H) = 1 \\ 2, & \text{if } gamma(G) \geq 2 \text{ and } gamma(H) \geq 2 \end{cases}$$

Proof. Suppose that $\gamma(G) = 1$. Assume that $\{u\}$ is a dominating set of G. It is obvious that $\{u\}$ is both a dominating set and a restrained dominating set of $G + H$. Hence, $\gamma(G + H) \leq 1$ and $\gamma_r(G + H) \leq 1$. Since $\gamma(G + H) \geq 1$ and $\gamma_r(G + H) \geq 1$, it follows that $\gamma(G + H) = 1$ and $\gamma_r(G + H) = 1$.

Suppose that $\gamma(G) \geq 2$ and $\gamma(H) \geq 2$. Let $u \in V(G)$ and $v \in V(H)$. It is obvious that $\{u, v\}$ is both a dominating set and a restrained dominating set of $G + H$. Hence, $\gamma(G + H) \leq 2$ and $\gamma_r(G + H) \leq 2$. Suppose that $\gamma(G + H) = 1$. Assume that $\{u\} \subseteq V(G)$ is a dominating set of $G + H$. Then $\{u\}$ is a dominating set of G. So $\gamma(G) \leq 1$, which is a contradiction with $\gamma(G) \geq 2$. Hence $\gamma(G + H) \geq 2$. By a similar proof, it follows that $\gamma_r(G + H) \geq 2$.

Theorem 3. *Let H be a cograph. Then*

$$\gamma_{tr}(K_1 + H) = \begin{cases} 3, & \text{if } l(H) = 0 \text{ and } t(H) = 2 \\ 2, & \text{if } l(H) = 0 \text{ and } t(H) \geq 3 \\ l(H) + 1, & \text{if } l(H) \geq 1. \end{cases}$$

Proof. Say $V(K_1) = \{u\}$. We discuss it from the following cases.

Case 1 $l(H) = 0$. Suppose that $t(H) \geq 3$. Without loss of generality, we can assume that $|V(H_1)| = t(H)$. Let H_1' be a spanning tree of H_1 and v be a leaf of H_1'. Then $\{u, v\}$ is a total restrained dominating set of $K_1 + H$. Hence, $\gamma_{tr}(K_1 + H) \leq 2$. Since $\gamma_{tr}(K_1 + H) \geq 2$, it follows that $\gamma_{tr}(K_1 + H) = 2$. Hence, we can assume that $t(H) = 2$.

If $w(H) = 1$, then $K_1 + H$ is isomorphic to K_3. It is obvious that $\gamma_{tr}(K_1 + H) = 3$. We can assume that $w(H) \geq 2$. Then $V(H_1) \cup \{u\}$ is a total restrained dominating set of $K_1 + H$. Hence, $\gamma_{tr}(K_1 + H) \leq |V(H_1) \cup \{u\}| = 3$.

Let D be a γ_{tr}-set of $K_1 + H$, where H is a cograph with $l(H) = 0$, $w(H) \geq 2$ and $t(H) = 2$. Then $u \in D$. Otherwise, suppose that $u \notin D$. Then $D = V(H)$ and $|D| \geq 4$. Then $V(H_1) \cup \{u\}$ is a total restrained dominating set of $K_1 + H$ with cardinality less than $|D|$, which is a contradiction. Hence, $u \in D$.

Since D is a total restrained dominating set of $K_1 + H$, there exists H_i such that $D \cap V(H_i) \neq \emptyset$. Without loss of generality, we can assume that $D \cap V(H_1) \neq \emptyset$. Since $|V(H_1)| = 2$, $V(H_1) \subseteq D$. Hence, $3 \leq |D| = \gamma_{tr}(K_1 + H)$. Hence, $\gamma_{tr}(K_1 + H) = 3$.

Case 2 $l(H) \geq 1$. Without loss of generality, we can assume that H_i is isomorphic to K_1 for $i = 1, 2, \cdots, l(H)$. It is obvious that $(\bigcup_{i=1}^{l(H)} V(H_i)) \cup \{u\}$ is a total restrained dominating set of $K_1 + H$. Hence, $\gamma_{tr}(K_1 + H) \leq l(H) + 1$. Let D be a γ_{tr}-set of $K_1 + H$. It is obvious that $(\bigcup_{i=1}^{l(H)} V(H_i) \cup \{u\}) \subseteq D$. So, $l(H) + 1 \leq |D| = \gamma_{tr}(K_1 + H)$. Therefore, $\gamma_{tr}(K_1 + H) = l(H) + 1$.

Theorem 4. *Let G and H be cographs with $n(G) \geq 2$ and $n(H) \geq 2$. Then $\gamma_{tr}(G + H) = 2$.*

Proof. Let $u \in V(G)$ and $v \in V(H)$. It is obvious that $\{u, v\}$ is a total restrained dominating set of $G + H$. Hence, $\gamma_{tr}(G + H) \leq 2$. Since $\gamma_{tr}(G + H) \geq 2$, it follows that $\gamma_{tr}(G + H) = 2$.

3   Algorithms

Our algorithms for finding the restrained domination number and total restrained domination number of a cograph G consists of two phases. In the first phase, construct a cotree T_G corresponding to G. In the second phase, compute the restrained and total restrained domination number of G_i for internal vertices of T_G by simple bottom-up calculations on the cotrees.

Let G be a cograph. Suppose that t_i is a vertex of T_G and G_i is the induced subgraph of G that is corresponding to t_i. Algorithm 1 and Algorithm 2 compute the restrained domination number and total restrained domination number of a cograph, respectively. If G has isolated vertex, define $\gamma_{tr}(G) = \infty$.

Algorithm 1: $RS(t_i)$: Computes the restrained domination number of a cograph G.

Input: a vertex t_i of the cotree T_G.
Output: the 4-tuple $(\gamma_r(G_i), \gamma(G_i), l(G_i), n(G_i))$, where G_i is the subgraph corresponding to t_i.

if t_i is a leaf **then return** $(1,1,1,1)$;
$t_l \leftarrow$ the left child of t_i; $t_r \leftarrow$ the right child of t_i;
$(s_l, g_l, l_l, n_l) \leftarrow RS(t_l)$;
$(s_r, g_r, l_r, n_r) \leftarrow RS(t_r)$;
if t_i is 0-vertex **then**
    return $(s_l + s_r, g_l + g_r, l_l + l_r, n_l + n_r)$;
else
    if $n_l = 1$ **then** $\gamma_r \leftarrow l_r + 1$;
    else if $n_r = 1$ **then** $\gamma_r \leftarrow l_l + 1$;
    else
        if $g_l = 1$ or $g_r = 1$ **then** $\gamma_r \leftarrow 1$;
        else $\gamma_r \leftarrow 2$;
    $\gamma \leftarrow \min\{g_l, g_r, 2\}$;
    return$(\gamma_r, \gamma, 0, n_l + n_r)$.

Algorithm 2: $TR(t_i)$: Computes the total restrained domination number of a cograph G

Input: a vertex t_i of the cotree T_G.
Output: the 4-tuple $(\gamma_{tr}(G_i), t(G_i), l(G_i), n(G_i))$, where G_i is the subgraph corresponding to t_i.

if t_i is a leaf **then return** $(\infty, 1, 1, 1)$;
$t_l \leftarrow$ the left child of t_i; $t_r \leftarrow$ the right child of t_i;
$(s_l, t_l, l_l, n_l) \leftarrow TR(t_l)$;
$(s_r, t_r, l_r, n_r) \leftarrow TR(t_r)$;
if t_i is 0-vertex **then**
    return $(s_l + s_r, \max\{t_l, t_r\}, l_l + l_r, n_l + n_r)$;
else
    if $n_l = 1$ **then**
        if $l_r = 0$ and $t_r = 2$ **then** $\gamma_{tr} \leftarrow 3$;
        else if $l_r = 0$ and $t_r \geq 3$ **then** $\gamma_{tr} \leftarrow 2$;
        else $\gamma_{tr} \leftarrow l_r + 1$;
    else if $n_r = 1$ **then**
        if $l_l = 0$ and $t_l = 2$ **then** $\gamma_{tr} \leftarrow 3$;
        else if $l_l = 0$ and $t_l \geq 3$ **then** $\gamma_{tr} \leftarrow 2$;
        else $\gamma_{tr} \leftarrow l_l + 1$;
    else $\gamma_{tr} \leftarrow 2$;
    return$(\gamma_{tr}, n_l + n_r, 0, n_l + n_r)$.

We can see easily that Algorithm 1 computes the restrained domination number of G_i from Lemma 1, Theorem 1 and Theorem 2. Algorithm 2 computes the total restrained domination number of G_i from Lemma 1, Theorem 3 and Theorem 4.

Theorem 5. *Given a cograph G of n vertices and m edges, the restrained and total restrained domination number of G can be found in $O(n+m)$ time, respectively.*

Proof. First a cotree T_G from G in time $O(n+m)$. Then we can find the restrained and total restrained domination number by calling $RS(t_{root})$ and $TR(t_{root})$, respectively, where t_{root} is the root vertex of T_G.

Since the cotree T_G has n leaves, it has $2n-1$ vertices. Hence the computation time for $RS(t_{root})$ and $TR(t_{root})$ is $O(n)$, respectively. Therefore, the total computation time is $O(n+m)$.

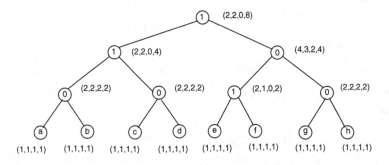

Fig. 2. An example of the execution of Algorithm 1.

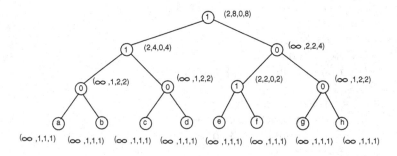

Fig. 3. An example of the execution of Algorithm 2.

Figure 2 is an example of the execution of Algorithm 1 for the cograph in Fig. 1, the 4-tuples near each node of the cotree means the values for $(\gamma_r(G_i), \gamma(G_i), l(G_i), n(G_i))$. Figure 3 is an example of the execution of Algorithm 2 for the cograph in Fig. 1, the 4-tuples near each node of the cotree means the values for $(\gamma_{tr}(G_i), t(G_i), l(G_i), n(G_i))$. Hence, $\gamma_r(G) = 2$ and $\gamma_{tr}(G) = 2$.

References

1. Araki, T., Yamanaka, R.: Secure domination in cographs. Discrete Appl. Math. (2019). in press. https://doi.org/10.1016/j.dam.2019.02.043
2. Domke, G.S., Hattingh, J.H., Hedetniemi, S.T., Laskar, R.C., Markus, L.R.: Restrained domination in graphs. Discrete Math. **203**(1–3), 61–69 (1999)
3. Joubert, E.: Total restrained domination in claw-free graphs with minimum degree at least two. Discrete Appl. Math. **159**, 2078–2097 (2011)
4. Panda, B.S., Pradhan, D.: A linear time algorithm to compute a minimum restrained dominating set in proper interval graphs. Discrete Math. Algorithms Appl. (2015). https://doi.org/10.1142/S1793830915500202
5. Pandey, A., Panda, B.S.: Some algorithmic results on restrained domination in graphs (2016). arXiv:1606.02340v1 [cs.DM]
6. Ma, D.X., Chen, X.G., Sun, L.: On total restrained domination in graphs. Czechoslov. Math. J. **55**(1), 165–173 (1999)
7. Telle, J.A., Proskurowski, A.: Algorithms for vertex partitioning problems on partial k-trees. SIAM J. Discrete Math. **10**, 529–550 (1997)
8. Haynes, T.W., Hedetniemi, S.T., Slater, P.J.: Fundamentals of Domination in Graphs. Marcel Dekker Inc., New York (1998)
9. Haynes, T.W., Hedetniemi, S.T., Slater, P.J.: Domination in Graphs: Advanced Topics. Marcel Dekker Inc., New York (1998)

An Order Approach for the Core Maintenance Problem on Edge-Weighted Graphs

Bin Liu$^{(\boxtimes)}$ ⓘ, Zhenming Liu, and Feiteng Zhang ⓘ

School of Mathematical Sciences, Ocean University of China, Qingdao 266100, China
binliu@ouc.edu.cn

Abstract. As a critical structure, a k-core is a maximal connected subgraph with the minimum degree $\delta \geq k$ of a simple unweighted graph, where integer $k \geq 0$. Define the core number of a vertex w as the maximum k such that w is contained in a k-core. There are two main problems: *the core decomposition problem* which is calculating the core numbers of all vertices in static graphs, and *the core maintenance problem* which is updating the core numbers in dynamic graphs. Although, core numbers can be updated by the core decomposition algorithms, only a small part of vertices' core numbers have changed after the change of a graph. Thus, it is necessary to update core numbers locally to reduce the cost. In this paper, we study the core maintenance problem on *edge-weighted graphs* by using the k-order that is a vertex sequence ordered by the order that the core decomposition algorithm removes vertices. We design the core maintenance algorithms for inserting one edge at a time and the method of updating the k-order, which reduce the searching range and the time cost evidently. For the removing case, we use the existing subcore algorithms to do the core maintenance and modify it with the method of updating k-order we design.

Keywords: Core maintenance · Edge-weighted graphs · k-core

1 Introduction

The graph is a simple and practical model. The application fields of graphs become increasingly widely, and include bioinformatics, social networks, communications technology and so on. With the development of large scale networks, graphs structure analysis has attracted much attention. People concentrate on the cohesive subgraph of a graph, which can explore the properties of graphs and solve some problems in its corresponding applications. The cohesive subgraphs include cliques, k-cores, k-trusses, to just name a few, and the most concerned subgraph structure is the k-core.

This work was supported in part by the National Natural Science Foundation of China (11971447, 11871442), and the Fundamental Research Funds for the Central Universities.

ⓒ Springer Nature Switzerland AG 2021
W. Wu and H. Du (Eds.): AAIM 2021, LNCS 13153, pp. 426–437, 2021.
https://doi.org/10.1007/978-3-030-93176-6_37

A k-core C_k is a maximal connected subgraph with the minimum degree $\delta_{C_k} \geq k$ of a simple undirected unweighted graph $G = (V, E)$, where k is a nonnegative integer, and is proposed in 1983 by Seidman [15]. As an critical subgraph structure, k-core has been used in community detection [2,12], bioinformatics [1,5], the analysis for structural properties of graphs [17] and so on.

The theoretical research of the k-core mainly includes *the core decomposition problem* and *the core maintenance problem*. Define the core number of a vertex $w \in V$ as the maximum k such that $w \in C_k$ on G. The core decomposition problem is that calculate the core numbers of all vertices in G. In 2003, Batagelj et al. proposed an $O(|E|)$ time core decomposition algorithm for an unweighted graph [3], which is the pioneering work of the k-core. Later, researchers considered different situations and did extensive works: the distribute k-core decomposition algorithms [7,11], core decomposition on uncertain graphs [4,13], the external-memory algorithm [6], to just name a few.

The core maintenance problem is that update the core numbers of vertices on dynamic graphs which are graphs changing over time. Although, using the core decomposition algorithms can update the core numbers for dynamic graphs, it costs too much for the reason that only a part of vertices' core numbers change after the change of a graph. Therefore, finding algorithms to update the core numbers locally becomes the focus of researchers. In [14], Sarıyüce et al. presented three core maintenance algorithms with $O(|E|)$ time. In [9], Li et al. designed similar core maintenance algorithms with [14], independently. Based on the above results, Jin et al. [8] proposed parallel core maintenance algorithms for inserting or removing multiple edges at a time; Zhang et al. [16] proposed a fast order-based core maintenance algorithm for inserting an edge; Liu et al. [10] solved the core maintenance problem for edge-weighted graphs. There are still many achievements, and we only list a few briefly.

In this paper, we address the core maintenance problem on *edge-weighted graphs* by the k-order which is firstly introduced in [16]. Let $G = (V, E, c)$ be an edge-weighed graph, and $c : E \rightarrow \{1, 2, \cdots, b\}$ is the edge weight function, where b is a positive integer. Next, we define the degree of vertex u, denoted by $d_G(u)$, is the sum of weights of all incident edges of u in G. A *(weighted) k-core* C_k is a maximal connected subgraph with the minimum degree $\delta_{C_k} \geq k$ of a simple undirected weighted graph G, where k is a nonnegative integer. And the definition of *core number*, denoted by $core(w)$ for $w \in V$, is the maximum k such that $w \in C_k$ on G. Then we define the *(weighted) k-order* which is the vertex sequence formed in the order in which the core decomposition algorithm accesses the vertices in [10] for weighted graphs. Consider inserting an edge $e = (u, v)$ with $core(u) \leq core(v)$ to G, and $O = O_0 O_1 O_2 \cdots O_{k-1} O_k O_{k+1} \cdots O_{k_{max}}$ is the k-order of G. The core maintenance algorithms search from u to the end of $O_{core(u)+c_e-1}$ layer by layer, and use the *indicative degree* we define to judge whether a vertex has a higher core number. Suppose that we are processing on the layer O_K. If a vertex's indicative degree is greater than K, then we put it into V' which consists of vertices whose core numbers may increase. If a vertex's indicative degree is less than or equal to K, then the new core number of this vertex is K and we put it to the end of O'_K which is the segment with new core number K of the k-order O' of $G + e$. In this way, updating the core numbers and the k-order will complete for inserting case until all layers that need to be processed is processed. It should be noted that the size of V' is reduced significantly, which improves the time complexity compared to [10]. As for the removing case, we use the subcore algorithms in [10] to solve this case, and modify it by adding the capacity of updating the k-order for the next core number maintenance. We summary our contributions as follows:

- We first introduce the weighted k-order of edge-weighted graphs, and find the properties of vertices whose core numbers may increase on k-order after inserting an edge.
- Using the k-order, we first propose the core maintenance algorithms for the dynamic edge-weighted graphs after inserting an edge, and presented the method of updating the k-order which is nested into the core maintenance algorithms. Our algorithms reduce the size of the search space V' compared with traversal algorithms in [10], so that the time cost is reduced when update the core numbers of vertices locally.
- For the removing case, we use the subcore algorithms in [10] to update core numbers and modify it with the method of updating the k-order we design.

The rest of this paper is organized as follows. Section 2 gives the fundamental definitions and results about weighted k-core and weighted k-order. Section 3 presents the core maintenance results that are already existing or first found by us on weighted graphs. Then, the core maintenance algorithms using the k-order as a tool are presented in Sect. 4, with the analyses of them. Lastly, Sect. 5 concludes the content and results about this paper.

2   Preliminaries

In this section, we present the fundamental definitions about (weighted) k-core and the properties of them. Next, the core decomposition algorithm for edge-weighted graphs designed in [10], which is the fundamental result for the core maintenance problem on edge-weighted graphs, is shown.

Let $G = (V, E, c)$ be a simple undirected edge-weighed graph, which is defined on the underlying graph $G = (V, E)$, meanwhile, $c : E \rightarrow \{1, 2, \cdots, b\}$ is the edge weight function, where b is a positive integer. For any vertex u in G, the neighborhood of u is defined as $\{v | (u, v) \in E\}$, denoted by $N_G(u)$. Next, we define the degree of vertex u, denoted by $d_G(u)$, is the sum of weights of all incident edges of u in G, i.e., $d_G(u) = \sum_{(u,v) \in E} c_{(u,v)}$, where c_e is the weight of edge e. Let $n = |V|$, $m = |E|$ and $\delta_G = \min\{d_G(u) | u \in G\}$.

Definition 1 (*(weighted) k-core*). *Let H be a subgraph of weighted graph $G = (V, E, c)$. If H satisfies the following three conditions, then H is a (weighted) k-core of G, denoted by C_k, where k is a nonnegative integer: (a) $\delta_H \geq k$; (b) H is a connected subgraph; (c) H is a maximal subgraph which satisfies the above two conditions.*

Property 1. For a vertex $u \in C_k$, then C_k is a unique k-core that contains u, denoted by C_k^u. For any k-core C_k, where k is an integer and $k \geq 1$, there is an unique C_{k-1} that contains C_k.

Definition 2 *((weighted) core number/coreness).* *In a weighted graph $G = (V, E, c)$, the (weighted) max-k-core of vertex u, denoted by C^u, is a k-core that contains u and has the maximum value of k. Then the value of k is defined as the (weighted) core number/coreness of u, denoted by $core(u)$.*

In 2003, Batagelj et al. [3] proposed a core decomposition algorithm with time complexity $O(m)$ for unweighted graphs. Based on it, Liu et al. [10] designed a core decomposition algorithm with time complexity $O(m)$ for edge-weighted graphs. We modify it to make it output the core numbers and the vertex sequence O ordered by the processing order of the algorithm.

Definition 3 *((weighted) k-order).* *A (weighted) k-order of a weighted graph $G = (V, E, c)$, denoted by O, is a vertex sequence which contains all vertices in G and produced from the core decomposition algorithm in [10]. For any pair of vertices u and v in G, if u is removed before v, then we use $u \preceq v$ to denote that u is in front of v in O.*

It is obviously that the k-order is not unique, since the non-decreasing order is not unique. But all k-orders of a certain graph have the following property.

Property 2. In a weighted graph $G = (V, E, c)$, u and v is any pair of vertices, O is a k-order of G. (a) If $core(u) < core(v)$, then $u \preceq v$. (b) If $core(u) = core(v)$ and u is removed before v, then $u \preceq v$. (c) All the vertices with the same core number k are on a continuous segment of O, denoted by O_k. Meanwhile, all those segments are ordered in a non-decreasing order by the core numbers of vertices in them.

Definition 4 *((weighted) remaining degree).* *Let O be a k-order of a weighted graph $G = (V, E, c)$. For any vertex $u \in V$, the (weighted) remaining degree of u about O, denoted by $d_G^+(u)$, is defined as $d_G^+(u) = \sum_{(u,v) \in E, \ u \preceq v} c_{(u,v)}$.*

According to Property 2, vertex sequence O can be written as $O_0 O_1 O_2 \cdots O_{k-1} O_k O_{k+1} \cdots O_{k_{max}}$ where k_{max} is the maximum core number of the weighted graph. Then, we get Lemma 1.

Lemma 1. *In a weighted graph $G = (V, E, c)$, let o_l be a vertex sequence consisting of all vertices with core numbers equal to l. Then the vertex sequence $o_0 o_1 o_2 \cdots o_{l-1} o_l o_{l+1} \cdots o_{k_{max}}$ is a k-order if and only if, for any l, $d_G^+(u) \leq l$ for any $u \in o_l$.*

Example 1. As shown in Fig. 1, the whole graph is a 2-core. The max-k-core of vertex u is a 6-core, thus $core(u) = 6$. Input this graph to the decomposition algorithm. Then we can get that a vertex sequence is $O = (w, y, x, s, v, z, q, t, u)$ and the core numbers are 2, 2, 2, 5, 6, 6, 6, 6, 6 respectively. The sequence O is a k-order of the graph, and $k_{max} = 6$, $O_2 = (w, y, x)$, $O_5 = (s)$, $O_6 = (v, z, q, t, u)$.

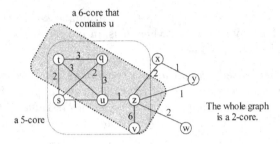

Fig. 1. An example of k-cores in a weighted graph.

Problem Statements: For unweighted graphs, the state-of-the-art core maintenance algorithm is designed in [16] in the case of inserting one edge at a time, which is operated on the k-order. In this paper, we use the k-order as a tool to solve the core maintenance problem on dynamic edge-weighted graphs in the case of inserting or removing one edge at a time. Meanwhile, we need to update the k-order after the change of the graph for the next core number updating.

3   Theoretical Findings

In this section, we present some theoretical results to support the design of the core maintenance problem on dynamic weighted problem. We only study the cases of inserting or removing one edge at a time. Then, there are three main challenges we are facing to update core numbers locally. The first is determining the changing value of the core number of a vertex after the change of a weighted graph. The second is finding the set of vertices whose core numbers change on the k-order, denoted by V^*, after the change. Lastly, we need to update the k-order of the new graph for the next round of core maintenance. Based on those three problems, we give the approaches to them and design core maintenance algorithms.

3.1   Previous Works

For any vertex u in G, we use $core'(u)$ and C'^u to denote the new core number and the new max-k-core of u on G' which is changed from G, respectively.

Liu et al. has proved theorems. One of them tells us that the absolute changing value of the core number of a vertex range from 0 to c_e after the insertion or deletion. Another two are used to search the set of vertices whose core numbers may change, named V' ($V' \supseteq V^*$), after inserting or removing an edge respectively. In this paper, we use those theorems to find the vertices whose core numbers will change on the k-order, which will reduce the size of V' significantly contrast to [10].

3.2   Inserting Case

Consider inserting an edge $e = (u, v)$ with $core(u) \le core(v)$ to a weighted graph $G = (V, E, c)$. We find the features of vertices whose core numbers will increase on a k-order after inserting e, by which the inserting core maintenance algorithm operates. They are described in the following theorem.

Theorem 1. *In a weighted graph* $G = (V, E, c)$, $u, v \in V$, $e = (u, v)$ *with* $core(u) \le core(v)$ *is not in* E. $O = O_0 O_1 O_2 \cdots O_{k-1} O_k O_{k+1} \cdots O_{k_{max}}$ *is the* k-order *of* G. *Insert an edge* $e = (u, v)$ *to* G, *then for any vertices* $w \in O_L$, *(a) if* $L < core(u)$, *then* $core'(w) = core(w)$; *(b) if* $L > core(u) + c_e - 1$, *then* $core'(w) = core(w)$; *(c) if* $L = core(u)$ *and* $w \preceq u$, *then* $core'(w) = core(w)$; *(d) if there is a path* $P = (u, u_1, u_2 \cdots w)$ *on* $G + e$, $u \preceq u_1 \preceq u_2 \preceq \cdots \preceq w$ *on* O *and any vertex* $w' \in P$ *satisfies that* $core(u) \le core(w') \le core(u) + c_e - 1$, *then* $core(w)$ *may increase.*

Consider a scenario that $core(w)$ may increase after insertion and $w \preceq w'$. If $core(w')$ locate in $[core(u), core(u) + c_e - 1]$, then w may be one vertex that help w' to have a high core number. Therefore, the definition *(weighted) candidate degree* is introduced to indicate this situation.

Definition 5 (*(weighted) candidate degree*). *Let* O *be a* k-order *of a weighted graph* $G = (V, E, c)$. *Insert an edge* $e = (u, v)$ *to* G. *For any vertex* $w \in V$, *the (weighted) candidate degree of* w, *denoted by* $d_G^*(w)$, *is defined as* $d_G^*(w) = \sum_{(w, w') \in E, \ w' \preceq w, \ core'(w') \ may \ be \ greater \ than \ core(w)} c_{(w, w')}$.

3.3   Removing Case

Consider removing an edge $e = (u, v)$ with $core(u) \le core(v)$ from a weighted graph $G = (V, E, c)$.

Definition 6 (*(weighted) maximum current degree*) [10]. *In a weighted graph* $G = (V, E, c)$, *for any vertex* $w \in V$, *define the (weighted) maximum current degree of* w, *denoted by* $WMD(w)$, *as* $WMD(w) = \sum_{\substack{(w, w') \in E \\ core(w') \ge core(w)}} c_{(w, w')}$.

Property 3. In a weighted graph $G = (V, E, c)$, $WMD(w) \ge core(w)$ *for any vertex* $w \in V$.

The maximum current degree of a vertex w is the part of degree of w that help w to have the core number $core(w)$. Conversely, if $\sum_{(w, w') \in E, \ core(w') \ge K} c_{(w, w')} < K$, then $core(w) < K$. It is modified and used as the indicator to judge the core number of a vertex after removing an edge.

Liu et al. [10] designed the (weighted) subcore core maintenance algorithms by the above results and principles in removing case. The subcore of a vertex u is a vertex set consisting of the vertices whose core numbers may change after removing e. In fact, the subcore of u is V' after removing e. In this paper, we quote these algorithms to solve the removing case and design new k-order updating method which is combined into the subcore algorithms.

4   Core Maintenance Algorithms

In this section, we propose the core maintenance algorithms for edge-weighted graphs, with the methods for updating the k-order after the change.

In the previous sections, the theoretical findings show what kind of vertices' core numbers may change and what the changing values may be. Then, there are new judge indicators we propose to judge whether a vertex $w \in V'$ is in V^* and determine its new core number. The algorithms we design are as follows.

4.1   Inserting Case

Algorithms Analysis and Process: Insert an edge $e = (u, v)$ with $core(u) \leq core(v)$ to weighted graph $G = (V, E, c)$, $O = O_0 O_1 O_2 \cdots O_{k-1} O_k O_{k+1} \cdots O_{k_{max}}$ is the k-order of G.

Different from the core maintenance problem on unweighted graphs [16], not only the core number of the vertex $w \in O_{core(u)}$ and $u \preceq w$ may change, but the vertex with the core number locating in $[core(u) + 1, core(u) + c_e - 1]$ also may have the change of the core number. Meanwhile, the changing values of the core numbers are range from 0 to c_e. Thus, the approach to judging whether the core number will change after inserting e to G directly in [16] is not work.

To solve the above question, a layer by layer judgment method operated on a k-order is proposed. In this method, we need to judge and calculate the new core number directly, rather than judge whose core number increases. We use the *indicative degree* $id(w) = d_G^+(w) + d_G^*(w)$, which is dynamic, as the indicator to judge the new core number of w, for any $w \in V'$.

Then, Algorithm 1 and 2 are designed to solve the core maintenance problem on weighted graphs for the edge insertion. Due to operated on the k-order, the size of V' (the vertices that have been added to V') is reduced and the time complexity is also improved compared to the traversal algorithm in [10].

To begin with, for a vertex w, the remaining degree $d_G^+(w)$ is prepared. Next, we start from the left of the search range on the k-order, i.e., vertex u, and process vertex by vertex to the right until the end of O_{c_e-1} is processed. In fact, we operate on $O_{core(u)}, O_{core(u)+1} \cdots O_{core(u)+c_e-1}$ in turn. Update core numbers and O'_K on each layer O_K in them, which is a round.

In each round, we use indicator $id(w)$ to judge whether a vertex w has a higher core number, when we get to w. Then, there are three cases divided by $id(w)$ and $d_G^*(w)$. Without loss of generality, assume the round we are processing is on the layer O_K. The process corresponds to the Algorithm 2.

The first case: $id(w) > K$. In this case, w is a vertex whose core number may be greater than K. Thus, w is added to V'. Meanwhile, for any w's neighbor w' satisfying $w' \in O_c = O_{core(u)} O_{core(u)+1} \cdots O_{core(u)+c_e-1}$ and $w \preceq w'$, $d_G^*(w') = d_G^*(w') + c_e$. Because w may have the ability to help w' to have a higher core number. It should be noted that if w is already stored in V' on the processed layers, it does not need to do the above operation again.

The second case: $id(w) \leq K$ and $d_G^*(w) = 0$. If w satisfies $id(w) \leq K$ and $d_G^*(w) = 0$, then w does not have high enough degree to increase its core number,

$core(w) = K$. Especially, if $V' \neq \emptyset$ and the initial vertex u of O_K has $id(u) \leq K$, then the core numbers of vertices in V' are K. Algorithm 1 is terminated.

The third case: $id(w) \leq K$ and $d_G^*(w) > 0$. Similar to the second case, w does not have high enough degree to increase its core number. Furthermore, for any w's neighbor w' satisfying $w' \in V'$ and $w' \preceq w$, $d_G^+(w') = d_G^+(w') - c_e$. If w is already stored in V' on the processed layers, for any w's neighbor w' satisfying $w' \in O_c$ and $w \preceq w'$, $d_G^*(w') = d_G^*(w') - c_e$. However, the decrease of $d_G^+(w')$ and $d_G^*(w')$ may lead to $id(w') \leq K$ for a vertex $w \in V'$. Define a vertex set R which contains the vertices x satisfying $x \in V'$, $id(x) \leq K$ and x is processed in this layer. For any $x \in R$, the new core number $core(x) = K$. Assume that w' satisfying the conditions of vertices in R, then put w' to R. Now, we should remove all the vertices in R from V' and decrease the $d_G^+()$ or $d_G^*()$ for some neighbors of $x \in R$. Specifically, there are three situations (take removing $x \in R$ from V' as an example).

(1) For each $x' \in V' \cap N_{G'}(x)$ and $x' \preceq x$, $d_G^+(x') = d_G^+(w) - c_{(x',x)}$. If $id(x') \leq K$, add x' to R.
(2) If $x \in V'$, then for each $x' \in N_{G'}(x) \cap V'$ and $x \preceq x' \preceq w$, $d_G^*(x') = d_G^*(x') - c_{(x,x')}$. If $id(x') \leq K$, add x' to R.
(3) If $x \in V'$, for each $x' \in N_{G'}(x) \cap O_c$ and $w \preceq x'$, $d_G^*(x') = d_G^*(x') - c_{(x,x')}$.

When processing $O_{core(u)+c_e-1}$ is finished and $V' \neq \emptyset$, the new core numbers of vertices in V' are $core(u) + c_e$.

K-Order and the Remaining Degree Update: The k-order and the remaining degrees are updated with the process of judging the core numbers of vertices. Without loss of generality, assume the round we are processing is on the layer O_K. The update for the segment O_K' of the k-order O' of G' completes in this round. Details are as follows.

Initially, we set O_K' as an empty sequence except $O'_{core(u)}$ which is the segment of $O_{core(u)}$ before u. When the core number of a vertex $w \in O_K$ is determined as K, the vertex w should be appended to the end of O_K'. Meanwhile, the remaining degree $d_{G'}^+(w) = d_G^+(w) + d_G^*(w)$ which is less than or equal to K. If $V' \neq \emptyset$ when this layer is finished, then we should append the vertices in V' to the beginning of O_{K+1} in the order of O_K to make the next round of judgment. When processing $O_{core(u)+c_e-1}$ is finished and $V' \neq \emptyset$, the combinant of V' and $O_{core(u)+c_e}$ is $O'_{core(u)+c_e}$. The remaining degrees of vertices that are not processed or assigned remaining degrees keep unchanged or are those output from the algorithms. Until now, the construction of k-order O' and remaining degree update are completed.

Theorem 2. *The sequence O' produced from Algorithm 1 is a k-order of $G' = G + e$.*

4.2 Removing Case

Algorithms Introduction: Remove an edge $e = (u, v)$ with $core(u) \leq core(v)$ from weighted graph $G = (V, E, c)$, $O = O_0 O_1 O_2 \cdots O_{k-1} O_k O_{k+1} \cdots O_{k_{max}}$ is

Algorithm 1. Inserting Case: Core Maintenance on a Weighted Graph by a k-order

---

Require: A weighted graph $G = (V, E, c)$; core numbers and $d_G^+(v)$ for each $v \in V$; a k-order O of G; the inserted edge $e = (u, v)$ with weight c_e

Ensure: New core number for each $v \in V'$ and new k-order O' of $G' = (V, E + e, c)$

1: $r = \operatorname{argmin}_{\{u,v\}} = \{core(u), core(v)\}$; $d_G^+(r) = d_G^+(r) + c_e$
2: **if** $d_G^+(r) \le core(r)$ **then**
3:     core numbers and O do not change
4: **end if**
5: $O_{core(r)} \leftarrow r$ and the segment of $O_{core(r)}$ after r
6: $O_c = O_{core(r)} O_{core(r)+1} \cdots O_{core(r)+c_e-1}$
7: $d_G^*(w) = 0$, for any $w \in O_c$
8: $V' = \emptyset$
9: **for** $K = core(r)$ to $core(r) + c_e - 1$ **do**
10:     $n' = |O_K|$, denote O_K as $(v_1 v_2 \cdots v_{n'})$
11:     **if** $K = core(r)$ **then**
12:         $O'_K \leftarrow$ the segment of $O_{core(r)}$ before r
13:     **else**
14:         $O'_K \leftarrow$ empty sequence
15:     **end if**
16:     $i = 1$
17:     Search and Judge on O_K
18:     **if** $V' = \emptyset$ **then**
19:         break   % *No subsequent vertex on O_c will increase the core number*
20:     **else**
21:         put the vertices in V' to the beginning of O_{K+1} in the order of O
22:     **end if**
23: **end for**
24: **for** each $w \in V'$ **do**
25:     $core(w) = core(r) + c_e$; $d_{G'}^+(w) = d_G^+(w)$
26: **end for**
27: $O' = O_1 O_2 \cdots O_{core(r)-1} O'_{core(r)} O'_{core(r)+1} \cdots O'_{core(r)+c_e-1} O'_{core(r)+c_e} \cdots O_{k'_{max}}$
    % *the O_K that is not operated will remain unchanged*
28: **return** $core(w), O'$

---

the k-order of G. In this part, we quote the (weighted) subcore algorithms [10] and modify them to update the k-order at the same time after removing e from G.

A dynamic indicator named *current degree* (*cd*) is proposed to calculate the new core numbers of vertices in V' (i.e., the subcore). And only the vertices in V' need to be assigned the *cd* values. For each vertex $w \in V'$, the current degree of w, denoted by $cd(w)$, is calculated initially as follows,

$$cd(w) = \sum_{(w,w') \in E \setminus \{e\}, \; core(w') \ge core(u) - c_e + 1} c_{(w,w')}.$$

To begin with, the subcore S_u and *cd* values should be prepared by BFS traversal. All vertices in S_u are ordered in a non-decreasing order by their *cd* values. H is the vertex-induce subgraph for S_u on $G - e$ which we need to search to update core numbers. Next, the new core numbers of vertices in S_u is calculated and judged layer by layer from $core(u) - c_e$ to $core(u)$. However, it

---

Algorithm 2. Search and Judge on O_K

1: $R \leftarrow$ empty queue
2: **while** $i \leq n'$ **do**
3:   **if** $d_G^+(v_i) + d_G^*(v_i) > K$ **then**
4:     **if** v_i is not in V' **then**
5:       $V' = V' \cup \{v_i\}$;
6:       **for each** $w \in N_{G'}(w) \cap O_c$ and $v_i \preceq w$ **do**
7:         $d_G^*(w) = d_G^*(w) + c_{(v_i,w)}$
8:       **end for**
9:     **end if**
10:     $i = i + 1$
11:   **else if** $d_G^*(v_i) = 0$ **then**
12:     **if** $v_i = r$ **then**
13:       **for each** $w \in V'$ **do**
14:         $core(w) = K$
15:       **end for**
16:       $V' = \emptyset$; $O'_k = O_K$; break
17:     **else**
18:       append v_i to O'_K; $i = i + 1$
19:     **end if**
20:   **else**
21:     R.enqueue(v_i)
22:     **while** $R \neq \emptyset$ **do**
23:       $x \leftarrow R$.dequque; $core(x) = K$
24:       append x to O'_K; $V' = V' - \{x\}$; $d_{G'}^+(x) = d_G^+(x) + d_G^*(x)$; $d_{G'}^*(x) = 0$
25:       **for each** $w \in V' \cap N_{G'}(x)$ and $w \preceq x$ **do**
26:         $d_G^+(w) = d_G^+(w) - c_{(w,x)}$
27:         **if** $d_G^+(w) + d_G^*(w) \leq K$ **then**
28:           R.enqueue(w)
29:         **end if**
30:       **end for**
31:       **if** $x \in V'$ **then**
32:         **for each** $w \in N_{G'}(x) \cap V'$ and $x \preceq w \preceq v_i$ **do**
33:           $d_G^*(w) = d_G^*(w) - c_{(x,w)}$
34:           **if** $d_G^+(w) + d_G^*(w) \leq K$ **then**
35:             R.enqueue(w)
36:           **end if**
37:         **end for**
38:         **for each** $w \in N_{G'}(x) \cap O_c$ and $v_i \preceq w$ **do**
39:           $d_G^*(w) = d_G^*(w) - c_{(x,w)}$
40:         **end for**
41:       **end if**
42:     **end while**
43:   **end if**
44: **end while**
45: **return** V', O'_K, vertices' $core()$, $d_{G'}^+()$ in O'_K, $d_G^+()$ and $d_G^*()$

---

may happen that no vertex has new core number K that locates in $[core(u) - c_e, core(u)]$. Then, the algorithms skips these layers.

In each round, i.e., a layer, processing each vertex w of S_u in order, the algorithm judge the core number of w. If $core(w)$ is determined, then remove w from H, and $cd(w') = cd(w') - c_{(w,w')}$ for any vertex $w' \in N_{G'}(w) \cap V(H)$. Reorder the rest vertices in H. Do the above operation until the cd value of the first vertex in order in H is greater than the core numbers of vertices in this layer. When H is a null graph, the algorithm is terminated.

K-Order and the Remaining Degree Update: At the beginning, all vertices in S_u are removed from the k-order O. In the round with core number K, append the vertex w that have just been assigned core number with K to the end of O'_K. In this way, a k-order of G' can be obtained after processing all vertices in H. Meanwhile, their cd values are assigned to the their remaining degrees when they are removed from H. The remaining degrees of other vertices are remain unchanged.

Theorem 3. *The sequence O' produced from the algorithm for removing case is a k-order of $G' = G - e$.*

5   Conclusion

In this paper, we propose the (weighted) k-order and design the core maintenance algorithms for edge-weighted graphs in inserting case by the k-order. Our algorithms update the core numbers of vertices on dynamic graphs locally, and reduce the size of V' which contains all vertices whose core numbers may change compared with traversal algorithms in [10]. These algorithms reduce the search scope and reduce the update time. For the removing case, we quote the (weighted) subcore algorithms [10]. Meanwhile, we also present the k-order updating methods for both insertion and deletion cases.

References

1. Bader, G.D., Hogue, C.W.V.: An automated method for finding molecular complexes in large protein interaction networks. BMC Bioinform. **4**(1), 1–27 (2003)
2. Barbieri, N., Bonchi, F., Galimberti, E., Gullo, F.: Efficient and effective community search. Data Min. Knowl. Disc. **29**(5), 1406–1433 (2015). https://doi.org/10.1007/s10618-015-0422-1
3. Batagelj, V., Zaversnik, M.: An O(m) algorithm for cores decomposition of networks. In: The Computing Research Repository (CoRR). arXiv: cs.DS/0310049 (2003)
4. Bonchi, F., Gullo, F., Kaltenbrunner, A., Volkovich, Y.: Core decomposition of uncertain graphs. In: Proceedings of the 20th ACM SIGKDD International Conference on Knowledge Discovery and Data Mining, pp. 1316–1325. ACM, New York (2014)

5. Cheng, Y., Lu, C., Wang, N.: Local k-core clustering for gene networks. In: 2013 IEEE International Conference on Bioinformatics and Biomedicine, pp. 9–15. IEEE, Shanghai (2013)
6. Cheng, J., Ke, Y., Chu, S., Özsu, M.T.: Efficient core decomposition in massive networks. In: 27th International Conference on Data Engineering (ICDE), pp. 51–62. IEEE, Hannover (2011)
7. Jakma, P., Orczyk, M., Perkins, C.S., Fayed, M.: Distributed k-core decomposition of dynamic graphs. In: Proceedings of the 2012 ACM Conference on CoNEXT Student Workshop, pp. 39–40. ACM, Nice (2012)
8. Jin, H., Wang, N., Yu, D., Hua, Q.S., Shi, X., Xie, X.: Core maintenance in dynamic graphs: a parallel approach based on matching. IEEE Trans. Parallel Distrib. Syst. **29**(11), 2416–2428 (2018)
9. Li, R.H., Yu, J.X., Mao, R.: Efficient core maintenance in large dynamic graphs. IEEE Trans. Knowl. Data Eng. **26**(10), 2453–2465 (2013)
10. Liu, B., Zhang, F.: Incremental algorithms of the core maintenance problem on edge-weighted graphs. IEEE Access **8**, 63872–63884 (2020)
11. Mandal, A., Al Hasan, M.: A distributed k-core decomposition algorithm on spark. In: 2017 IEEE International Conference on Big Data (Big Data), pp. 976–981. IEEE, Boston (2017)
12. Papadopoulos, S., Kompatsiaris, Y., Vakali, A., Spyridonos, P.: Community detection in social media. Data Min. Knowl. Disc. **24**(3), 515–554 (2012)
13. Peng, Y., Zhang, Y., Zhang, W., Lin, X., Qin, L.: Efficient probabilistic k-core computation on uncertain graphs. In: 2018 IEEE 34th International Conference on Data Engineering (ICDE), pp. 1192–1203. IEEE, Paris (2018)
14. Sariyüce, A.E., Gedik, B., Jacques-Silva, G., Wu, K.L., Çatalyürek, Ü.V.: Incremental k-core decomposition: algorithms and evaluation. VLDB J. **25**(3), 425–447 (2016)
15. Seidman, S.B.: Network structure and minimum degree. Soc. Netw. **5**(3), 269–287 (1983)
16. Zhang, Y., Yu, J.X., Zhang, Y., Qin, L.: A fast order-based approach for core maintenance. In: 2017 IEEE 33rd International Conference on Data Engineering (ICDE), pp. 337–348. IEEE, San Diego (2017)
17. Zhang, F., Zhang, Y., Qin, L., Zhang, W., Lin, X.: When engagement meets similarity: efficient (k, r)-core computation on social networks. Proc. VLDB Endow. **10**(10), 998–1009 (2017)

Fixed-Parameter Tractability for Book Drawing with Bounded Number of Crossings per Edge

Yunlong Liu[1,2]([🖂])(ⓘ), Yixuan Li[1](ⓘ), and Jingui Huang[1,2]([🖂])(ⓘ)

[1] College of Information Science and Engineering, Hunan Provincial Key Laboratory of Intelligent Computing and Language Information Processing, Hunan Normal University, Changsha 410081, People's Republic of China
{ylliu,yxlee,hjg}@hunnu.edu.cn
[2] Hunan Xiangjiang Artificial Intelligence Academy, Changsha, China

Abstract. The problem BOOK DRAWING with a bounded number of crossings per edge asks, given a graph $G = (V, E)$ and two integers k, b, whether there is a k-page book drawing of G such that the maximum number of crossings per edge is at most b. In this paper, we study this problem from a parameterized complexity viewpoint. Specifically, we investigate the problem parameterized by both the maximum number b of crossings per edge and the vertex cover number τ of G, and show that this parameterized problem admits a kernel of size $(3b+1) \cdot 2^{\mathcal{O}(\tau)}$ and admits a fixed-parameter tractable algorithm running in time $\mathcal{O}((3b+1) \cdot 2^{(3b+2) \cdot 2^{\mathcal{O}(\tau)}} + \tau \cdot |V|)$. Together with our previous result for FIXED-ORDER BOOK DRAWING with a bounded number of crossings per edge (COCOA 2020), our result provides a more complete answer to a question posed by Bhore et al. (J. Graph Alg. Appl. 2020).

1 Introduction

Book drawing is a main and popular paradigm for drawing graphs. Combinatorially, a *k-page book drawing* $\langle \prec, \sigma \rangle$ of a graph $G = (V, E)$ consists of a linear ordering $\prec$ of its vertices along a spine and an assignment σ of each edge to one of k pages, which are half-planes bounded by the spine [18]. The spine and the k pages construct a book. In particular, a k-page book drawing that is crossing free is called a *k-page book embedding*; if the ordering of vertices in $V(G)$ along the spine is specified, then a k-page book drawing (or embedding) is called fixed-order. These definitions are illustrated in Fig. 1. Book drawings have been extensively studied due to their wide range of applications including network visualization, combinatorial geometry, VLSI design, RNA folding, knot theory and others (see, e.g., [5, 11, 15]).

This research was supported in part by the National Natural Science Foundation of China under Grant No. 61572190 and Hunan Provincial Science and Technology Program under Grant No. 2018TP1018.

W. Wu and H. Du (Eds.): AAIM 2021, LNCS 13153, pp. 438–449, 2021.
https://doi.org/10.1007/978-3-030-93176-6_38

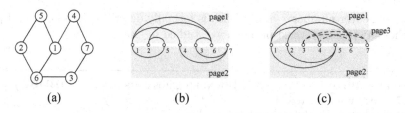

Fig. 1. (a) This graph G, with vertices $\{1, 2, 3, 4, 5, 6, 7\}$, has (b) a 2-page book drawing with at most 1 crossing per edge and (c) a 3-page fixed-order book drawing with at most 1 crossing per edge when vertices are specified in increasing order.

Algorithms on drawing graphs have attracted considerable attention and much interest as they provide geometric representations of abstract graphs. For problems whose algorithms show exponential growth, parameterized complexity theory seeks algorithms and analysis that confine the exponential dependence to some parameter of the input graph that can be hoped to be small in practice. This promising approach to deal with difficult graph drawing problems has led to several fixed-parameter tractable algorithms (see, e.g., [2–4,7,12,13,17,21]). A popular parameter has been the vertex cover number [2,7,17]. Recent studies on BOOK DRAWING parameterized by vertex cover number have arisen from the BOOK EMBEDDING problem and related works [6,19].

The BOOK EMBEDDING problem decides whether a given graph G admits a k-page book embedding. From the viewpoint of computational complexity, the 2-PAGE BOOK EMBEDDING problem, determining whether a graph can be embedded to two pages, is known to be **NP**-complete [10]. In view of the hardness of this problem, even for a small fixed number k, Bhore et al. [6] introduced the vertex cover number of the given graph as parameter and presented a fixed-parameter tractable algorithm. Furthermore, Bhore et al. [6] posed some open problems, e.g., investigating the parameterized complexity for (FIXED-ORDER) BOOK DRAWING in the setting where a bounded number of crossings per edge is given as part of the input. These open problems were thought to be interesting because some page in a k-page book drawing with a bounded number of crossings per edge may contain an unbounded number of crossings. Recently, we have shown that FIXED-ORDER BOOK DRAWING with a bounded number of crossings per edge is fixed-parameter tractable with respect to both the maximum number of crossings per edge and the vertex cover number of the input graph [20].

In this paper, we focus on the problem BOOK DRAWING with a bounded number of crossings per edge. This problem asks, given a graph $G = (V, E)$ and two integers k, b, whether there is a k-page book drawing $\langle \prec, \sigma \rangle$ of G such that the maximum number of crossings per edge is at most b. We denote by $\mathrm{bd}(G, b)$ the book drawing thickness of G (i.e., the minimum k such that (G, k, b) is a yes-instance of this problem). Since 2-PAGE BOOK EMBEDDING is **NP**-complete [10], there can be no fixed-parameter tractable algorithm for this problem parameterized only by the crossing number unless $\mathbf{P} = \mathbf{NP}$. Therefore, we study this problem parameterized by both the maximum number of crossings per edge and the vertex cover number τ of G, which is formally described as follows and just matches the parameterized problem we studied in [20].

---

k-PAGE BOOK DRAWING WITH BOUNDED CROSSINGS PER EDGE (**BDBC**)
Input: an undirected graph $G = (V, E)$, two integers k, b ;
Parameters: b, the vertex cover number τ ;
Question: does there exist a k-page book drawing $\langle \prec, \sigma \rangle$ of G such that the number of crossings *per edge* is at most b ?

Let (G, k, b, τ) be an instance of the BDBC problem. Observing that (G, k, b, τ) can be translated into $|V(G)|!$ instances of the problem FIXED-ORDER BOOK DRAWING with a bounded number of crossings per edge, we first explore a kernel for the BDBC problem by the following approach. A minimum vertex cover C of G is firstly computed and the vertices in $V(G) \backslash C$ are partitioned into $2^\tau - 1$ types. In each set of vertices of the same type, we focus on the subset of vertices that have at least two incident edges assigned to the same page and the subset of vertices that have at most one edge on each page and have at least one edge which produces crossing on some page. To estimate the size of these subsets in any yes-instance, we introduce two kinds of 1-page graphs, by which two upper bound functions of both τ and b can be derived respectively. Based on this approach, we obtain a kernel of size $(3b+1) \cdot 2^{\mathcal{O}(\tau)}$. Then, by employing the algorithm for FIXED-ORDER BOOK DRAWING with a bounded number of crossings per edge [20], we obtain a fixed-parameter tractable algorithm running in time $\mathcal{O}((3b + 1) \cdot 2^{(3b+2) \cdot 2^{\mathcal{O}(\tau)}} + \tau \cdot |V|)$.

As far as we know, our result is the first fixed-parameter tractable algorithm for the BDBC problem. Together with our previous result for FIXED-ORDER BOOK DRAWING with a bounded number of crossings per edge [20], our result provides a more complete answer to a question posed by Bhore et al. [6]. Our result also puts the considered problem among a frontier research area of graph drawing where few edge crossings are allowed per edge (see, e.g., [1,8,14,16]).

2   Preliminaries

We consider only undirected simple graphs. For a graph $G = (V, E)$, let $n = |V|$, $V(G)$ be the vertex set of G and $E(G)$ be the edge set of G. For two vertices u and v, we use uv to denote the edge between u and v. Given two integers a, b in $\mathbb{N}$, we use $[a, b]$ to denote the set $\{a, a + 1, \ldots, b\}$. Given two sets A and B, we also use $A \backslash B$ to denote the set of all elements that belong to A but not to B. Given a vertex v in graph G, we denote by $d_G(v)$ the degree of v in G and by $N_G(v)$ the set of all neighbors of v in G. Furthermore, given a set S in $V(G)$, we denote by $N_G(S)$ the set $\{v \mid v \in N_G(u) \text{ and } u \in S\}$.

Given a graph $G = (V, E)$, we use $\langle \prec, \sigma, b \rangle$ to denote a k-page book drawing of G with at most b crossings per edge, where $\prec$ is a linear order of V and

$\sigma : E \rightarrow [1, k]$ is a function that maps each edge of E to one of k pages denoted by $[1, k]$. Assume that uv is an edge assigned to page p with $p \in [1, k]$. The edge uv is said to *produce* b *crossings on page* p if uv intersects b other edges that are assigned to page p.

A *vertex cover* C of a graph $G = (V, E)$ is a subset $C \subseteq V$ such that each edge in E has at least one endpoint in C. A vertex $v \in V$ is a *cover vertex* if $v \in C$. The *vertex cover number* of G, denoted by τ, is the size of a minimum vertex cover of G. Given a graph G, a vertex cover C with size τ can be computed in time $\mathcal{O}(2^\tau + \tau \cdot |V|)$ [9]. In the rest of this paper, we will use C to denote a minimum vertex cover of size τ.

Let $W \subseteq C$. A vertex in $V(G)\backslash C$ is of type W if its set of neighbors is equal to W [6]. By this definition, the vertices in $V(G)\backslash C$ are partitioned into at most $2^\tau - 1$ distinct types. We use V_W to denote the set of vertices of type W.

3   A Class of 1-Page Graphs

Let (G, k, b, τ) be a yes-instance of the BDBC problem and let $\langle \prec, \sigma, b \rangle$ be a book drawing of G. Assume that C is a minimum vertex cover of G and that $W \subseteq C$ with $|W| \geq 2$. In this section, we introduce a class of 1-page graphs, by which the number of vertices in V_W can be estimated page by page.

Let $D \subseteq [1, |W|]$ and b be a positive integer. A graph H drawn on one page is called a (U, W, b, D)-*crossing graph* if the following properties hold: ① all vertices in $V(H)$ are placed on a horizontal line (also called a spine) and all edges in $E(H)$ are drawn as semicircles on this page; ② $U \cup W$ is a vertex cover of H and each edge in $E(H)$ produces at most b crossings; and ③ $d_H(v) \in D$ for any $v \in V(H)\backslash(U \cup W)$ (see Fig. 2 for examples).

To estimate the number of vertices in V_W, we distinguish two cases based on whether $|W| > k$ or not. Correspondingly, we employ two specific kinds of (U, W, b, D)-*crossing graphs* as tools.

Case (1): $|W| > k$. Then, each vertex in V_W must have at least 2 incident edges assigned to the same page. Assume that v (for $v \in V_W$) has d (for $d \in [2, |W|]$) incident edges that are assigned to page p. Then the vertex v has degree d within page p. To estimate the number of vertices with the same edge assignments as v, we only need to consider one kind of (U, W, b, D)-crossing graphs which meet the special condition that $U \cup W = W$ (i.e., $U = \emptyset$). Hence, we employ a kind of $(\emptyset, W, b, [2, |W|])$-crossing graphs. Suppose that H_1 is an arbitrary $(\emptyset, W, b, [2, |W|])$-crossing graph. Then, an upper bound for the size of $V(H_1)\backslash W$ can be used to bound the number of vertices in V_W that have at least two edges assigned to a particular page.

Case (2): $|W| \leq k$. Let u be an arbitrary vertex in V_W. If u has at least two incident edges assigned to the same page, then we can deal with it by the same approach used in case (1). Herein, we mainly consider the situation that each page in $\langle \prec, \sigma, b \rangle$ contains at most one edge of u. In this situation, the vertex u can be viewed as a vertex of degree 1 within each page that contains one edge of u. Assume that one edge uc is assigned to page p and produces at least one crossing.

Considering some edge crossed by uc may not be covered only by the vertices in W, we introduce another kind of (U, W, b, D)-crossing graphs as follows. A (U, W, b, D)-crossing graph H is called a $(U, W, b, [1, |U|])$-*strict crossing graph* if H additionally meets the properties: ① every vertex in $N_H(W) \backslash W$ has degree 1; and ② every edge in E_W produces at least one but at most b crossings, where $E_W = \{vw \mid vw \in E(H), v \in N_H(W) \backslash W, \text{ and } w \in W\}$. Suppose that H_2 is an arbitrary $(U, W, b, [1, |U|])$-crossing graph. Then, an upper bound for the size of $N_{H_2}(W) \backslash W$ can be used to bound the number of vertices in V_W that have at most one edge on each page and have at least one edge which produces crossing on a particular page.

Figure 2 depicts a $(\emptyset, W, b, [2, |W|])$-crossing graph and a $(U, W, b, [1, |U|])$-strict crossing graph respectively.

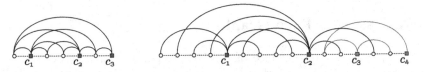

Fig. 2. A $(\emptyset, W, b, [2, |W|])$-crossing graph with $W = \{c_1, c_2, c_3\}$ and $b = 1$ (left) and a $(U, W, b, [1, |U|])$-strict crossing graph with $U = \{c_3, c_4\}$, $W = \{c_1, c_2\}$, and $b = 2$ (right). The edges with blue color in the right figure are covered by the vertices in U. (Color figure online)

In the subsequent sections, we will give a detailed description of how to bound the number of vertices in two kinds of (U, W, b, D)-crossing graphs respectively.

4   An Upper Bound for the Number of Vertices in Any $(\emptyset, W, b, [2, |W|])$-Crossing Graph

Let G be a $(\emptyset, W, b, [2, |W|])$-crossing graph. Observe that G can be decomposed into $\binom{|W|}{2}$ $(\emptyset, \{w_i, w_j\}, b, \{2\})$-crossing graphs, where $w_i \in W$ and $w_j \in W$. Thus, we first analyze the maximum number of vertices in any $(\emptyset, \{w_i, w_j\}, b, \{2\})$-crossing graph.

Let $W' = \{w_i, w_j\}$ be a subset of W. Assume that $H_{W'}$ is a $(\emptyset, W', b, \{2\})$-crossing graph. To facilitate estimating the number of vertices in $H_{W'}$, we consider its variant description, that is, the *round* $(\emptyset, W', b, \{2\})$-crossing graph $R_{W'}$. In $R_{W'}$, the vertices are placed in a circular spine and the edges are drawn on the outer side of the circular spine. Moreover, the number of crossings per edge in $R_{W'}$ is the same as that of the corresponding edge in $H_{W'}$ (see Fig. 3 for an illustration). In the following, we will not distinguish between $H_{W'}$ and $R_{W'}$. For the spine in $R_{W'}$, the portion on the left (resp. right) of the vertical axis connecting w_1 with w_2 is called the left (resp. right) semi-circular spine.

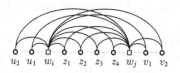 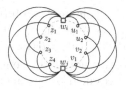

Fig. 3. A $(\emptyset, W', 3, \{2\})$-crossing graph, where $W' = \{w_i, w_j\}$ (left) and the corresponding round $(\emptyset, W', 3, \{2\})$-crossing graph (right).

Given an arbitrary $(\emptyset, W', b, \{2\})$-crossing graph, we can derive an upper bound on the number of vertices of degree 2 in it.

Lemma 1. *Let* $W' = \{w_i, w_j\}$ *and let* b *be an arbitrary non-negative integer. Assume that* $H_{W'}$ *is an arbitrary* $(\emptyset, W', b, \{2\})$-*crossing graph. Then* $|V(H_{W'})\backslash W'| \le 2(b+1)$.

Proof. Suppose by contradiction that there are at least $2(b+1)+1$ vertices of degree 2 in $V(H_{W'})\backslash W'$. By the pigeonhole principle, at least $b+2$ vertices of degree 2 lie on one of the two semi-circular spines. Without loss of generality, assume that $z_1, z_2, \ldots z_{b+2}$ are such vertices and lie on the left semi-circular spine consecutively (see Fig. 3 (right) for an illustration). Then the edge $z_1 w_j$ will cross each edge $w_i z_r$ for $r = 2, 3, \ldots, b+2$, leading to $b+1$ crossings. But this is not possible in a $(\emptyset, W', b, \{2\})$-crossing graph. $\square$

Based on Lemma 1, we can derive a bound on the number of vertices of degree 2 for an arbitrary $(\emptyset, W, b, \{2\})$-crossing graph.

Lemma 2. *Assume that* H_W *is an arbitrary* $(\emptyset, W, b, \{2\})$-*crossing graph. Then* $|V(H_W)\backslash W| \le (b+1) \cdot (|W|^2 - |W|)$.

Proof. Assume that $|W| \ge 2$. Let $W_{ij} = \{w_i, w_j\}$ be a subset of W and let D_{ij} be the set of vertices of degree 2 that connect with w_i and w_j simultaneously. We use $H_{W_{ij}}$ to denote the subgraph of H_W induced by the vertices in $D_{ij} \cup W_{ij}$. Since there are in total $\binom{|W|}{2}$ combinations of two vertices from W, the graph H_W can be decomposed into $\binom{|W|}{2}$ subgraphs such that each vertex of degree 2 and its two incident edges are partitioned into one of the subgraphs. Correspondingly, the vertices in $V(H_W)\backslash W$ can be classified into $\binom{|W|}{2}$ subsets. By the assumption that H_W is a $(\emptyset, W, b, \{2\})$-crossing graph, there is at most b crossings per edge in H_W. Since $H_{W_{ij}}$ is a subgraph of H_W, there is at most b crossings per edge in $H_{W_{ij}}$. Thus, $H_{W_{ij}}$ is a $(\emptyset, W_{ij}, b, \{2\})$-crossing graph. From Lemma 1, there are at most $2(b+1)$ vertices in D_{ij}. Thus, it holds that $|V(H_W)\backslash W| \le \binom{|W|}{2} \cdot 2(b+1)$ $= (b+1) \cdot (|W|^2 - |W|)$. $\square$

By Lemma 2, we further consider the $(\emptyset, W, b, [2, |W|])$-crossing graphs. Let H'_W be an arbitrary $(\emptyset, W, b, [2, |W|])$-crossing graph and let $v \in V(H'_W)\backslash W$ with $d_{H'_W}(v) = t$ (for $t \in [2, |W|]$). Observe that the vertex v can be seen as a compound vertex composed of $\binom{t}{2}$ vertices of degree 2. Thus, the number of vertices in H'_W is no more than that of the corresponding $(\emptyset, W, b, \{2\})$-crossing graph. We can immediately obtain the following conclusion.

Theorem 1. *Assume that H'_W is an arbitrary $(\emptyset, W, b, [2, |W|])$-crossing graph. Then $|V(H'_W)\backslash W| \le (b + 1) \cdot (|W|^2 - |W|)$.*

5 An Upper Bound for the Number of Vertices in Any $(U, W, b, [1, |U|])$-Strict Crossing Graph

Let G be a $(U, W, b, [1, |U|])$-strict crossing graph. We estimate the number of vertices in $N_G(W)\backslash W$ by decomposing G into some $(\emptyset, \{w_i, w_j\}, b, \{1\})$-strict crossing graphs and some $(\{u_i\}, \{w_h\}, b, \{1\})$-strict crossing graphs, where $u_i \in U$ and $\{w_i, w_j, w_h\} \subseteq W$.

Let $W' = \{w_i, w_j\}$ be a subset of W and let $A_{W'}$ be a $(\emptyset, W', b, \{1\})$-strict crossing graph. To ease of estimating the number of vertices in $N_{A_{W'}}(W')\backslash W'$, we also consider its corresponding *round* $(\emptyset, W', b, \{1\})$-strict crossing graph $R_{W'}$ (see Fig. 4 for an illustration). Note that the number of crossings per edge in $R_{W'}$ is the same as that of the corresponding edge in $A_{W'}$. In the following, we will also not distinguish between $A_{W'}$ and $R_{W'}$.

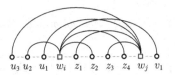
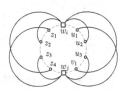

Fig. 4. A $(\emptyset, W', 2, \{1\})$-strict crossing graph, where $W' = \{w_i, w_j\}$ (left) and the corresponding round $(\emptyset, W', 2, \{1\})$-strict crossing graph (right).

For a given integer b, there may be many $(\emptyset, W', b, \{1\})$-strict crossing graphs that are not isomorphic. However, the following statement holds.

Lemma 3. *Let $W' = \{w_i, w_j\}$ and let b be an arbitrary positive integer. Assume that $A_{W'}$ is an arbitrary $(\emptyset, W', b, \{1\})$-strict crossing graph. Then $|V(A_{W'})\backslash W'| \le 4b$.*

Proof. Suppose by contradiction that there are at least $4b + 1$ vertices of degree 1 in $V(A_{W'})\backslash W'$. By the pigeonhole principle, at least $2b + 1$ vertices separate w_i from w_j on one of the two semi-circular spines. Without loss of generality, assume that $z_1, z_2, \ldots, z_{2b+1}$ are such vertices and lie on the left semi-circular spine consecutively (refer to Fig. 4 (right) for an illustration). Since each of the $2b+1$ vertices is adjacent to either w_i or w_j, we further assume that w_i has $b+1$ neighbors among these vertices, denoted by $z_1^*, z_2^*, \ldots, z_{b+1}^*$ respectively. Since each edge in $E(A_{W'})$ is involved in at least one crossing, there must exist one vertex z_q such that z_q precedes z_1^* when walking counterclockwise on the left semi-circular spine, and that $w_i z_1^*$ crosses $z_q w_j$. Thus, $z_q w_j$ will cross each edge $w_i z_t^*$ for $t = 1, 2, \ldots, b+1$, producing $b+1$ crossings. But this is not possible in a $(\emptyset, W', b, \{1\})$-strict crossing graph. $\qquad\square$

Similarly, we can obtain the following statement for any $(\{u\}, \{w\}, b, \{1\})$-strict crossing graph.

Lemma 4. *Let $Y = \{w, u\}$ and let b be an arbitrary positive integer. Assume that B_Y is an arbitrary $(\{u\}, \{w\}, b, \{1\})$-strict crossing graph. Then $|N_{B_Y}(w)| \leq 2b$.*

Based on Lemma 3 and 4, we prove the following conclusion.

Theorem 2. *Assume that G is an arbitrary $(U, W, b, [1, |U|])$-strict crossing graph. Then $|N_G(W) \backslash W| \leq 2b \cdot |W| \cdot (|W| + |U| - 1)$.*

Proof. Let $w \in W, v \in U \cup W$ and $R = \{w, v\}$. Let E_w (resp. E_v) be the set of edges incident to w (resp. v). Let $S = \{e_x \mid e_x \in E_w$ and e_x crosses at least one edge in $E_v\} \cup \{e_y \mid e_y \in E_v$ and e_y crosses at least one edge in $E_w\}$. If $S \neq \emptyset$, we can obtain one subgraph D_R of G induced by the edges in S. Moreover, if $v \in W$, then D_R is a $(\emptyset, \{w, v\}, b, \{1\})$-strict crossing graph. Otherwise, D_R is a $(\{v\}, \{w\}, b, \{1\})$-strict crossing graph.

By the approach above, we can construct at most $\binom{|W|}{2}$ $(\emptyset, \{w, v\}, b, \{1\})$-strict crossing graphs and at most $|W| \cdot |U|$ $(\{v\}, \{w\}, b, \{1\})$-strict crossing graphs from G such that each vertex in $N_G(W) \backslash W$ occurs in at least one constructed subgraph. By Lemma 3, for an arbitrary $(\emptyset, \{w, v\}, b, \{1\})$-strict crossing graph, say H_1, it holds that $|V(H_1) \backslash \{w, v\}| \leq 4b$, and by Lemma 4, for an arbitrary $(\{v\}, \{w\}, b, \{1\})$-strict crossing graph, say H_2, it holds that $|N_{H_2}(w)| \leq 2b$. Therefore, it follows that $|N_G(W) \backslash W| \leq 4b \cdot \binom{|W|}{2} + 2b \cdot |W| \cdot |U| = 2b \cdot |W| \cdot (|W| + |U| - 1)$. $\qquad\square$

6 A Parameterized Algorithm Based on Kernelization

In this section, we present a parameterized algorithm based on kernelization for the BDBC problem. Given an instance (G, k, b, τ), we first locate a minimum vertex cover C of G. Then for each subset W of C, we count the number of vertices in V_W and deal with V_W by some reduction rules (this strategy has been used in [6]). By Theorem 1 and 2, our specific reduction rules for the BDBC problem are described as follows. If $|W| > k$ and $|V_W| > k \cdot (b+1) \cdot (|W|^2 - |W|)$, then (G, k, b, τ) must be a no-instance. If $|W| \leq k$ and $|V_W| > k \cdot (b+1) \cdot (|W|^2 - |W|) + k \cdot 2b \cdot |W| \cdot (|C| - 1) + 1$, then delete all but $k \cdot (b+1) \cdot (|W|^2 - |W|) + k \cdot 2b \cdot |W| \cdot (|C| - 1) + 1$ vertices. The reduced instance is denoted by (G^*, k, b, τ).

Theorem 3. (G, k, b, τ) *is a yes-instance of the BDBC problem if and only if* (G^*, k, b, τ) *is a yes-instance of the BDBC problem. Moreover, the size of* G^* *can be bounded by* $(3b + 1) \cdot 2^{\mathcal{O}(\tau)}$.

Proof. ($\Rightarrow$) Assume that (G, k, b, τ) is a yes-instance. Then, (G^*, k, b, τ) must be a yes-instance because deleting some vertex from a book drawing with at most b crossings per edge keeps the property of being a book drawing with at most b crossings per edge.

($\Leftarrow$) Assume that (G^*, k, b, τ) is a yes-instance. Let $\langle \prec, \sigma, b \rangle$ be a k-page book drawing of G^* with at most b crossings per edge. We distinguish two cases based on whether $|W| > k$ or not.

(1) Suppose that $|W| > k$. Then, for each vertex in V_W, it has at least two edges assigned to the same page. Let p be an arbitrary page in $\langle \prec, \sigma, b \rangle$ and let V_W^p be the subset of vertices in V_W that have at least two edges assigned to p. We first estimate the size of V_W^p. Observe that the subgraph induced by the edges that are incident to the vertices in V_W^p and that are assigned to page p is exactly a $(\emptyset, W, b, [2, |W|])$-crossing graph. By Theorem 1, it holds that $|V_W^p| \leq (b+1) \cdot (|W|^2 - |W|)$. Furthermore, by the fact that each vertex in V_W belongs to at least one subset V_W^p (for $p \in [1, k]$), it follows that $|V_W| \leq k \cdot |V_W^p| \leq k \cdot (b+1) \cdot (|W|^2 - |W|)$.

(2) Suppose that $|W| \leq k$. Let $v \in V_W$ and let E_v be the set of edges that are incident to v. There are three types of assignments for the edges in E_v. Case ①: at least two edges in E_v are assigned to the same page. Case ②: each edge in E_v is assigned to a distinct page and there exists at least one edge that produces at least one but at most b crossings on a particular page. Case ③: each edge in E_v is assigned to a distinct page but no edge is involved in any crossing. Accordingly, the set V_W can be partitioned into three subsets, respectively denoted by B_1, B_2 and B_3, such that for each vertex in B_i, the assignment of its edges can be described by case ⓘ for $i = 1, 2, 3$. By the same analysis as that in (1), the number of vertices in B_1 is at most $k \cdot (b+1) \cdot (|W|^2 - |W|)$. Next, we mainly estimate the size of B_2. Let B_2^p be the subset of vertices in B_2 that have one edge assigned to page p and producing at most b crossings on page p. Let E_1 be the set of edges that are incident to the vertices in B_2^p and are assigned to page p and let E_2 be the set of edges that are assigned to page p and crossed by one edge in E_1. Observe that the subgraph induced by the edges in $E_1 \cup E_2$ is exactly a $(C \backslash W, W, b, [1, |C \backslash W|])$-strict crossing graph. By Theorem 2, the number of vertices in B_2 is at most $k \cdot 2b \cdot |W| \cdot (|C| - 1)$. Thus, if there are at least $k \cdot (b+1) \cdot (|W|^2 - |W|) + k \cdot 2b \cdot |W| \cdot (|C| - 1) + 1$ vertices in V_W, then there must exist at least one vertex that belongs to B_3. Let u be such a vertex. Then, no edge that incident to u is involved in any crossing in the k-page book drawing $\langle \prec, \sigma, b \rangle$. Therefore, the vertex u can be taken as a *"reference"* vertex, by which $\langle \prec, \sigma, b \rangle$ can be safely extended. More precisely, let s be a vertex in V_W reduced by our rule. Now, we can extend $\langle \prec, \sigma, b \rangle$ by inserting s right next to u and assigning each edge sw (for $w \in W$) to

the same page as uw such that sw runs arbitrarily close to uw. Obviously, the extended assignment is exactly a k-page book drawing with at most b crossings per edge for the subgraph induced by $V(G^*) \cup \{s\}$. After inserting all reduced vertices one by one, we obtain a k-page book drawing with at most b crossings per edge for graph G.

Finally, the size of G^* can be estimated as follows. We can assume that $k \leq \tau$ because G^* admits a k-page book embedding [6] in case $k > \tau$, which can be seen as a k-page book drawing with 0 crossing per edge. Since there are at most $2^\tau - 1$ nonempty subsets of C and $|W| \leq \tau$, the size of G^* can be bounded by $2^\tau \times (k \cdot (b+1) \cdot (|W|^2 - |W|) + k \cdot 2b \cdot |W| \cdot (|C|-1)+1)+\tau \leq (3b+1) \cdot 2^\tau \cdot \tau^3 + \tau$
$= (3b+1) \cdot 2^{\mathcal{O}(\tau)}$.   □

Then, we decide whether G^* admits a k-page book drawing with at most b crossings per edge by trying all possible linear orders of $V(G^*)$ and by employing the algorithm for solving FIXED-ORDER BOOK DRAWING with a bounded number of crossings per edge in [20]. We denote by fo-bd$(G, \prec)$ the minimum number of pages in a fixed-order book drawing with at most b crossings per edge.

Lemma 5 ([20]). *There is an algorithm which takes as input a graph $G = (V, E)$ with a fixed vertex order $\prec$, and two integers b, k, decides whether fo-bd$(G, \prec) \leq k$ with running time $(b + 2)^{\mathcal{O}(\tau^3)} \cdot |V|$ where b denotes the maximum number of crossings per edge and τ denotes the vertex cover number of G. If (G, k, b, τ) is a yes-instance, this algorithm can also return a k-page book drawing of G with respect to $\prec$.*

Based on Theorem 3 and Lemma 5, we can obtain the following conclusion.

Theorem 4. *The* BDBC *problem admits an algorithm that, given a graph $G = (V, E)$ and integers k, b, decides whether bd$(G, b) \leq k$ with running time $\mathcal{O}((3b + 1) \cdot 2^{(3b+2) \cdot 2^{\mathcal{O}(\tau)}} + \tau \cdot |V|)$ where τ denotes the vertex cover number of G. If (G, k, b, τ) is a yes-instance, this algorithm can also return a k-page book drawing of G with at most b crossings per edge.*

Proof. Based on Theorem 3, we obtain an equivalent instance with size $(3b+1) \cdot 2^{\mathcal{O}(\tau)}$. The next step is to solve the equivalent instance by guessing all possible linear orders and for each order call the algorithm in Lemma 5.

The running time can be analyzed as follows. The time for computing a vertex cover of size τ can be bounded by $\mathcal{O}(2^\tau + \tau \cdot |V|)$ [9]. Given a vertex cover C of size τ, we can enumerate all subsets in C in time 2^τ. Partitioning all vertices in $V \backslash C$ into 2^τ types can be done in time $\tau \cdot |V|$ and deleting all redundant vertices by our reduction rule can also be done in time $\tau \cdot |V|$. Thus, the running time of the kernelization procedure is bounded by $\mathcal{O}(2^\tau + \tau \cdot |V|)$. By Theorem 3, the size of G^* is $(3b + 1) \cdot 2^{\mathcal{O}(\tau)}$. Since any two vertices of the same type can be interchanged in $\prec$ [6], the number of fixed linear orders on $V(G^*)$ can be bounded by $(2^\tau)^{(3b+1) \cdot 2^{\mathcal{O}(\tau)}} = 2^{(3b+1) \cdot 2^{\mathcal{O}(\tau)}}$. By Lemma 5, the procedure solving FIXED-ORDER BOOK DRAWING can be done in time $(b + 2)^{O(\tau^3)} \cdot |V|$.

Therefore, whether G^* admits a k-page book drawing with at most b crossings per edge can be determined in time $\mathcal{O}(2^{(3b+1)\cdot 2^{\mathcal{O}(\tau)}}\cdot(b+2)^{\mathcal{O}(\tau^3)}\cdot(3b+1)\cdot 2^{\mathcal{O}(\tau)}) = \mathcal{O}((3b+1)\cdot 2^{(3b+2)\cdot 2^{\mathcal{O}(\tau)}})$. If (G^*, k, b, τ) is a yes-instance, then we can obtain a k-page book drawing of G by extending the k-page book drawing of G^* in time $\mathcal{O}(\tau\cdot|V|)$. □

A book drawing with the minimum number of pages and with at most b crossings per edge can be obtained by trying all possible choices for $k \in [1, \tau]$. Herein, by applying a binary search for the number k of pages (this technique has been used in [6,7]), we obtain the following claim.

Corollary 1. *Let $G = (V, E)$ be a graph with n vertices and vertex cover number τ. A book drawing of G with at most b crossings per edge and with minimum number of pages can be computed in $\mathcal{O}((3b+1)\cdot 2^{(3b+2)\cdot 2^{\mathcal{O}(\tau)}} + \tau\log\tau\cdot|V|)$ time.*

7   Conclusion

In this work, we have shown the fixed-parameter tractability result for the problem BOOK DRAWING parameterized by both the maximum number of crossings per edge and the vertex cover number of the input graph. Together with our previous result for FIXED-ORDER BOOK DRAWING with a bounded number of crossings per edge [20], our result provides a more complete answer to a question posed by Bhore et al. in [6].

Some problems are interesting and deserve further research. (1). In evaluating the number of vertices of the same type, we adopt the subgraph decomposition strategy and consider the edge-crossings that are only contained in some subgraphs, which makes the upper functions seem to have a lot of room for improvement. (2). Does the problem we considered in this paper admit a kernel of polynomial size?

Acknowledgements. The authors thank the anonymous referees for their valuable comments and suggestions.

References

1. Angelini, P., Bekos, M.A., Kaufmann, M., Montecchianib, F.: On 3D visibility representations of graphs with few crossings per edge. Theor. Comput. Sci. **784**, 11–20 (2019)
2. Bannister, M.J., Cabello, S., Eppstein, D.: Parameterized complexity of 1-planarity. J. Graph Algorithms Appl. **22**(1), 23–49 (2018)
3. Bannister, M.J., Eppstein, D.: Crossing minimization for 1-page and 2-page drawing of graphs with bounded treewidth. J. Graph Algorithms Appl. **22**(4), 577–606 (2018)
4. Bannister, M.J., Eppstein, D., Simons, J.A.: Fixed parameter tractability of crossing minimization of almost-trees. In: Wismath, S., Wolff, A. (eds.) GD 2013. LNCS, vol. 8242, pp. 340–351. Springer, Cham (2013). https://doi.org/10.1007/978-3-319-03841-4_30

5. Baur, M., Brandes, U.: Crossing reduction in circular layouts. In: Hromkovič, J., Nagl, M., Westfechtel, B. (eds.) WG 2004. LNCS, vol. 3353, pp. 332–343. Springer, Heidelberg (2004). https://doi.org/10.1007/978-3-540-30559-0_28

6. Bhore, S., Ganian, R., Montecchiani, F., Nöllenburg, M.: Parameterized algorithms for book embedding problems. J. Graph Algorithms Appl. **24**(4), 603–620 (2020)

7. Bhore, S., Ganian, R., Montecchiani, F., Nöllenburg, M.: Parameterized algorithms for queue layouts. In: GD 2020. LNCS, vol. 12590, pp. 40–54. Springer, Cham (2020). https://doi.org/10.1007/978-3-030-68766-3_4

8. Binucci, C., Giacomoa, E.D., Hossainb, M.I., Liotta, G.: 1-page and 2-page drawings with bounded number of crossings per edge. Eur. J. Comb. **68**, 24–37 (2018)

9. Chen, J., Kanj, I.A., Xia, G.: Improved upper bounds for vertex cover. Theor. Comput. Sci. **411**(40–42), 3736–3756 (2010)

10. Chung, F., Leighton, F., Rosenberg, A.: Embedding graphs in books: a layout problem with applications to VLSI design. SIAM J. Algebraic Discrete Methods **8**(1), 33–58 (1987)

11. Clote, P., Dobrev, S., Dotu, I., Kranakis, E., Krizanc, D., Urrutia, J.: On the page number of RNA secondary structures with pseudoknots. J. Math. Biol. **65**(6–7), 1337–1357 (2012). https://doi.org/10.1007/s00285-011-0493-6

12. Da Lozzo, G., Eppstein, D., Goodrich, M.T., Gupta, S.: Subexponential-time and FPT algorithms for embedded flat clustered planarity. In: Brandstädt, A., Köhler, E., Meer, K. (eds.) WG 2018. LNCS, vol. 11159, pp. 111–124. Springer, Cham (2018). https://doi.org/10.1007/978-3-030-00256-5_10

13. Di Giacomo, E., Liotta, G., Montecchiani, F.: Sketched representations and orthogonal planarity of bounded treewidth graphs. In: Archambault, D., Tóth, C.D. (eds.) GD 2019. LNCS, vol. 11904, pp. 379–392. Springer, Cham (2019). https://doi.org/10.1007/978-3-030-35802-0_29

14. Di Giacomo, E., Didimo, W., Liotta, G., Montecchiani, F.: Area requirement of graph drawings with few crossing per edge. Comput. Geom. **46**(8), 909–916 (2013)

15. Dynnikov, I.A.: Three-page approach to knot theory. Encoding and local moves. Funct. Anal. Appl. **33**(4), 260–269 (1999). https://doi.org/10.1007/BF02467109

16. Grigoriev, A., Bodlaender, H.L.: Algorithms for graphs embeddable with few crossings per edge. Algorithmica **49**(1), 1–11 (2007). https://doi.org/10.1007/s00453-007-0010-x

17. Hliněný, P., Sankaran, A.: Exact crossing number parameterized by vertex cover. In: Archambault, D., Tóth, C.D. (eds.) GD 2019. LNCS, vol. 11904, pp. 307–319. Springer, Cham (2019). https://doi.org/10.1007/978-3-030-35802-0_24

18. Klawitter, J., Mchedlidze, T., Nöllenburg, M.: Experimental evaluation of book drawing algorithms. In: Frati, F., Ma, K.-L. (eds.) GD 2017. LNCS, vol. 10692, pp. 224–238. Springer, Cham (2018). https://doi.org/10.1007/978-3-319-73915-1_19

19. Liu, Y., Chen, J., Huang, J.: Fixed-order book thickness with respect to the vertex-cover number: new observations and further analysis. In: Chen, J., Feng, Q., Xu, J. (eds.) TAMC 2020. LNCS, vol. 12337, pp. 414–425. Springer, Cham (2020). https://doi.org/10.1007/978-3-030-59267-7_35

20. Liu, Y., Chen, J., Huang, J.: Parameterized algorithms for fixed-order book drawing with bounded number of crossings per edge. In: Wu, W., Zhang, Z. (eds.) COCOA 2020. LNCS, vol. 12577, pp. 562–576. Springer, Cham (2020). https://doi.org/10.1007/978-3-030-64843-5_38

21. Liu, Y., Chen, J., Huang, J., Wang, J.: On parameterized algorithms for fixed-order book thickness with respect to the pathwidth of the vertex ordering. Theor. Comput. Sci. **873**, 16–24 (2021)

Author Index

Bu, Yuehua 400

Cai, Lijian 38
Chang, Hong 190
Chen, Jianer 83
Chen, Ruijie 356
Chen, Shengminjie 200
Chen, Wenping 229
Chen, Xue-gang 418
Chen, Zhibo 111
Chen, Zihan 159
Cui, Lu 289

Ding, Xingjian 134, 229
Dong, Luobing 356
Du, Donglei 190
Du, Hongmin W. 159, 240
Du, Linda 317
Du, Xiangang 356
Duan, Zhenhua 344

Gao, Guichen 71
Gao, Suixiang 123, 303
Gong, Ye 406
Gu, Qian-Ping 406
Guo, Jianxiong 134, 240
Guo, Longkun 265
Guo, Yin 83

Han, Xinxin 71
Hao, Chunlin 47, 265
Harutyunyan, Hovhannes A. 57
Hong, Yi 111
Hsieh, Sun-Yuan 252
Huang, Jingui 438
Huang, Minjie 83
Hung, Ling-Ju 252

Ji, Sai 15

Kao, Shih-Shun 252
Klasing, Ralf 252

Li, Deying 229
Li, Gaidi 15, 212
Li, Jianping 38
Li, Min 179
Li, Quan-Lin 329
Li, Weidong 27, 146
Li, Xiao 111, 317
Li, Xinghua 303
Li, Yajie 27
Li, Yan 395
Li, Yixuan 438
Li, Zhiyuan 57
Liang, Wei 3
Lichen, Junran 38
Liu, Bin 159, 426
Liu, Lei 265
Liu, Qian 179
Liu, Suding 38
Liu, Xiaofei 27
Liu, Xiaoqiao 146
Liu, Yunlong 438
Liu, Zhen 303
Liu, Zhenming 426
Liu, Zhicheng 190
Lu, Lingfa 96
Luo, Chuanwen 111
Luo, Haobin 356

Ma, Jing-Yu 329
Meng, Liang 289
Meng, Xiangguang 134

Ni, Qiufen 240

Ochi, Luiz Satoru 380
Ou, Jinwen 96

Pan, Pengxiang   38

Ren, Chunying   369

Si, Yu   123, 303
Silva, Janio Carlos Nascimento   380
Sohn, Moo Young   418
Souza, Uéverton S.   380
Su, Guowei   356
Sun, Xin   212

Tang, Shaojie   170

Wang, Ailian   289
Wang, Huizhen   277
Wang, Wencheng   38
Wang, Xiyun   111
Wang, Yajie   344
Wang, Yang   71
Wang, Ye   395
Wang, Yongcai   229
Wu, Chenchen   47
Wu, Zijun   369

Xiao, Man   146
Xu, Dachuan   369
Xu, Wenqing   200, 369
Xu, Yicheng   47

Yang, Jinhua   27
Yang, Wenguo   123, 303

Zhang, Dongmei   15
Zhang, Feiteng   426
Zhang, Liqi   96
Zhang, Nan   344
Zhang, Xianzhao   15
Zhang, Xiaojuan   179
Zhang, Xiaoyan   190
Zhang, Xiujuan   134, 229
Zhang, Yapu   200, 212
Zhang, Yong   47, 71
Zhang, Zhao   3
Zhang, Zhenning   200, 212
Zhou, Yang   179
Zhu, Hongguo   400
Zhu, Junlei   400
Zou, Wenjie   265